Roger Buvat

Ontogeny, Cell Differentiation, and Structure of Vascular Plants

With 107 Plates and 283 Figures

Springer-Verlag
Berlin Heidelberg New York
London Paris Tokyo

Professor Dr. ROGER BUVAT
Faculté de Médecine de Marseille
Laboratoire de Génétique
27, Boulevard Jean-Moulin
F-13385 Marseille Cedex 5

ISBN 3-540-19213-1 Springer-Verlag Berlin Heidelberg New York
ISBN 0-387-19213-1 Springer-Verlag New York Berlin Heidelberg

Library of Congress Cataloging-in-Publication Data.
Buvat, Roger. Ontogeny, cell differentiation and structure of vascular plants.
1. Plant cells and tissues. 2. Plants – Ontogeny. 3. Plant cell differentiation. I. Title.
QK725.B85   1989   582'.08   88-24802

Reproduction of figures: Gustav Dreher GmbH, Stuttgart
Typesetting, printing, and bookbinding by Appl, Wemding
2131/3130-543210 – Printed on acid-free paper

Preface

With improved microscope and preparation techniques, studies of histological structures of plant organisms experienced a revival of interest at the end of the 19th century. From that time, histological data have substantially studies of the pioneers in botanical science. From the beginning of the 20th century, the microscope allowed research in cell structure, the general functional unit of living beings. Advances in cytology gradually influenced histology, at first, however, rather timidly. Only the new and spectacular progress in ultrastructural cytology and cytochemistry led to a great increase in modern work on the structures of vascular plants and the related ontogenical and physiological data, thanks to the use of the electron microscope and the contribution of molecular biology.

Not only did new techniques lead to new approaches, but achievements in general biology shifted the orientation of research, linking investigation to the physiological aspects of cell and tissue differentiation. Among these, the demonstration of the general principles of development, and the characterization of molecules common to plants and animals, which control and govern the main basic functions of cells and tissues, have widened the scope of modern research on plant structures.

Present trends in biological research show that it is necessary to know the structures thoroughly, from the ultrastructural cytological scale to the scale of tissue and organ arrangement, even for physiological research on either cells, tissues, or whole organs. The study of growth factors, differentiation, or organogenesis can be mentioned as an example.

Obviously, this handbook cannot develop any question of plant physiology that goes beyond the author's specialization; its aim is to present the structural aspects of the development of cell and tissue constituents, and their ontogenic differentiation. However, the physiological importance of results in cytochemistry or biochemistry is frequently mentioned or discussed.

Research in the field of vascular plant ontogenesis is far from being concluded. Many unsolved problems give rise to hypotheses that in their turn call for new studies, which will either confirm or invalidate them. To realize this, we must trust the techniques often referred to in this book, cytochemistry, cytoenzymology, histoautoradiography, microspectrophotometry, statistics, and others.

References to the literature extend over more than 150 years, the quotation of older results, classical or of historical interest, is not only a homage to the wisdom of the pioneers in plant histology and embryology and their patient labour, but also a reminder of the often artistic quality of their writing.

Later publications related to development, structures, and organizing of vascular plants are so numerous, that to mention all would be impossible. The selection aims at avoiding too large a scattering of examples, by making occasional use of monographs. This method allows the presentation of valuable, often injustly forgotten, data. Figures have been taken over as much as possible from monographs, and are thus restricted to particularly closely studied species. Where possible, illustrations are original: most of them are microphotographs.

This book is not an exhaustive treatise, as numerous features of vascular plant organization and development are omitted, and it is not a manual of anatomy. The author's purpose was to gather the main results and concepts and to place them within the framework of new ideas on the processes of differentiation that intervene during the development and histological organization of plants.

In the introduction, the concluding remarks to each chapter, and the final remarks, in addition to the structural descriptions, the author emphasizes general concepts related either to developmental or general biology, going beyond the plant world. Some of these considerations, the result of reflections in the course of a long career, are not at present classical; they are put forward to be submitted to the test of time, and are an important part of the justification of the book.

This work seeks to bring accurate information to scientists variously specialized in plant biology, to teachers and students in either biology or pharmacy. May it show them that plant development remains a matter of modern research and of reflection, raising unsolved problems and requiring advanced techniques.

Finally, the author hopes that in this handbook, teachers will find not only precise facts and examples, but also general concepts that give these examples and facts a place in the field of modern biology.

Acknowledgements. This handbook has profited from the material help of administrators and colleagues to whom I express my gratitude. Professor GEORGES SERRATRICE, President of the University of Aix-Marseille, offered the means of translating the text into English, in the person of Mrs. C. VIDAL-NAQUET; this translation posed trying problems.

This book would not have been completed without the generous hospitality of Professor MAURICE TOGA, Dean of the Faculty of Medicine of Marseille, and of Professor ANDRÉ STAHL, who took the initiative of receiving me in his laboratory and enabling me to maintain my acitivities after retirement. I wish to assure them of my warm and faithful gratitude.

Thanks to Doctor MARCEL BESSIS, member of the Académie des Sciences of the Institut de France, through whose help I was introduced to the competent services of Springer-Verlag. I address him my cordial gratitude.

I thank the association "Naturalia et Biologia" for help in the preparation of the manuscript.

I thank particularly all colleagues who had the kindness to send me original figures, and allow me to reproduce them. Their names and references are added to the legends of figures and plates.

Also I owe a debt of gratitude to those of my colleagues and students who have examined the manuscript and given me useful criticism, advice and support: Mrs. ARLETTE NOUGARÈDE, Mr. and Mrs. MAX PELLEGRINI, Mrs. ANNE-MARIE CATESSON and YVETTE CZANINSKI.

Marseille, October, 1988   ROGER BUVAT

Contents

XIV

Part III   Conclusions

Introduction

1 Growth and Tissues

The development and the shape and size of plant organisms do not appear as a "finite" phenomenon as they do for animals. In animals, growth ceases long before death and as soon as an organism presents a more or less completed morphology.

For instance, a vertebrate, at the adult stage, keeps the same features and size by means of exclusive processes which maintain and renew its cells, while, each year, a tree develops new branches and grows indefinitely until death. In the animal organism, it seems that no cells are capable of elaborating new organs by means of an organized proliferation, whereas such cells occur in plants, similar by their function to the embryonic cells which determines the primary edification of the organism.

In less developed Thallophytes, all cells have the same characteristics, e.g. the filaments of *Spirogyra* or the undulated lamina of *Ulva*. Particularly, each cell can divide and contributes to thallus enlargement: there is no area specialized in growth. The same is observed in Fungi: when they constitute huge masses by the association of numerous hyphae, this association is not realized by especially organized selected areas which ensure the rising of the bulky organism. In such plants without localized "foci of organogenesis", cells are not particularly varied and there are no *true tissues:* histology, i.e. the study of the various types of cells and their arrangement in tissues, is hardly concerned by them.

Leaving all phylogenic intentions aside, if we try to classify the plant kingdom by looking for the modalities through which the organisms become more and more complex, we note that the diversity mainly affects the external shapes of the organisms rather than the structural characteristics of their cells, as long as there are no growth-devoted cells localized in determined areas.

This phenomenon is remarkably illustrated by a siphonal alga such as *Caulerpa* which often presents complicated as well as elegant shapes, sometimes simulating the organs of advanced plants, whereas their internal, cytological and histological organizations remain very rudimentary, since it is only constituted of a coenocytic "siphon" with no individualized cells, *therefore without tissues.* In the same way, one could quote the defined and complex shapes of the capillitiums of Myxomycetes and, even more interesting, those of the disseminating apparatus of Myxobacteriales: such systems are organized in the first case by a plasmodium, in the second case by similar cells living separately in a common mucilage.

In contrast to these morphological differentiations without any histocytological counterpart are the great majority of plants in which the complexity of morphology is linked to the appearance of *areas devoted to growth;* thus, the differentiation is morphological, anatomical and histocytological. The histogenic and organogenic areas are called *meristems*. Their proliferating activities supply the cellular material necessary for growth.

Relatively homogeneous at the beginning, this material diversifies by means of "cellular differentiation". Groups of similarly originated cells become more particularly concerned with one or several functions useful to the whole organism, while in other groups, all the cells especially prepare themselves for other activities.

Each group of cells, either similarly differentiated or participating in the same special physiological processes, forms a tissue.

Thus, under such conditions, the essentially physiological concept of *tissue* is related in plants to the concept of *meristem*. For this reason, we shall consider that only the plants which have meristems have true tissues, in contrast to Thallophytes without meristems which have only *false tissues* such as the plectenchyma of fungi.

Meristems, and the true tissues, exist in the most differentiated Thallophytes (Fucales, Laminariales) and in Bryophytes, but the histological diversification becomes pronounced only in vascular plants.

As for vascular plants, the above mentioned opposition between the growth of plants and that of animals (except for certain colonial organisms, mainly Coelenterae) implies two possibilities of behaviour of the growth areas. Proliferating, undifferentiated cells supplying the necessary material for the edification of organs and localized in the meristems may persist in areas where organogenesis goes on during the whole development; but similar cells may also rise again in such places by means of *dedifferentiation* of cells belonging to already differentiated tissues.

In vascular plants, both processes exist normally. On the one hand, organogenesis centres, i.e. the storage of actively proliferating cells which constitute the *primary meristems,* are found at the extremity or in the subterminal zone of the axial vegetative organs, stems and roots of all these plants. The extension of the stems and roots depends upon the functioning of these *primary meristems* or *apical meristems*. On the other hand, the branching of these axial organs often requires the presence of more or less mature cells, which dedifferentiate and give rise to new primary meristems.

Thus, one could think that, as in Thallophytes, all the cells of vascular plants or most of them can contribute to the growth. Therefore, the characteristic features of these "superior" plants are found elsewhere.

2 General Characteristics of Vascular Plants

We have just mentioned two organs characteristic of these plants: the stem and the root. The first common feature to vascular plants is in fact morphological: they are sporophytic organisms constituted of stems, which generally bear leaves, and roots. Only two living genera are an exception to this rule: the *Psilotum* and the *Tmesipteris,* which are the most ancient vascular plants and have no roots.

With the exception of such "relics", vascular plants are thus opposed to Thallophytes by their three types of organs: stems, leaves, roots. The sporophytic apparatus of Bryophytes does not show such a diversification; at best, it presents an axis similar to the stem but deprived of leaves and roots. The *thallus* of other plants comprises none of these three organs even in the case of a sporophyte *(Fucus)*.

The second general feature is histological: it is so obvious that it is used to refer to all the plants concerned here. In fact, these plants are the only ones in which the differentiation of conducting tissues is continued until the completion of two main cellular types: the *sieve cells,* characteristic of the phloem, arranged in sieve tubes in advanced forms, and the *tracheids,* characteristic of the xylem. In the species most differentiated with respect to xylem, the tracheids develop into constitutive elements of *vessels.*

These two essential features seem to be the consequence of the same phenomenon, i. e. the conquest of the aerial medium by plants, and represent two aspects of the adaptation to life in the atmosphere. In spite of this fundamental organographic and histological uniformity, really opposed to the great morphological and cytological variety to Thallophytes, the vascular plants can be considered as a prodigiously varied group mainly if we remember the exhuberant flora of past times. The multiplicity of forms, summarized in the table below, results from a general fact itself related to the conquest of the continents; the more and more pronounced development of the sporophytes entailing its diversification.

Standing, as if it was a wager, in the most unfavourable medium for living, the sporophyte of vascular plants survives only by fighting the conditions of the medium that it invades, owing to this creation of tissues which were useless in sea plants. So the aerial medium seems to entail both the *anatomical diversification* and the *histological differentiation* which form a wide gap between vascular plants and all other plants.

3 Succinct Classification of Vascular Plants

We only intend to give the main lines of this classification, details will be given when necessary.

We shall follow Embergers's suggestions and underline the peculiarities of some present-day plants, relics of a fossil flora which was abundant, and we shall separate Cycadales and Ginkgoales from true Phanerogams as well as the extinct plants which had some similarities with these two types. Vascular plants will be divided in three embranchments: Pteridophytes, Prephanerogams (Emberger) and Phanerogams or Spermaphytes.

Classification of Vascular Plants. Each subkingdom embranchment comprises several classes, among which many are exclusively fossil, especially in the first two classes. The table below summarizes this classification. *Underlined* unities are the only ones to include still living species in the present flora.

PTERIDOPHYTES

A. Nematophytes		
B. Psilophytinead	Rhyniales	
	Psilophytales	
	Asteroxylales	
C. *Psilotinead*	*Psilotales*	
D. *Lycopodineae*	Lepidophytales	
	Lepidospermales	
	Lycopodiales	
E. *Articuleae*	Protoarticulales	
	Sphenophyllales	
	Equisetales	
F. Noeggerathiales		
G. *Filicineae*	Phyllophoreae	Iridopteridales
		Stauropteridales
		Cladoxylales
		Zygopteridales
	Transition forms	*Osmundales*
	Aphyllophoreae	*Eusporangiate*
		Inversicatenales
		Marattiales
		Ophioglossales
		Leptosporangiate
		Filicales
		Hydropteridales

PREPHANEROGAMS

A. *Pteridosperms*	Pteridospermales
	Cycadales
B. *Cordaites*	Cordaitales
	Ginkgoales

PHANEROGAMS

A. *Gymnosperms*	Bennettitales	
	Nilssoniales	
	Coniferales	
B. *Preangiosperms*	Gnetales	
(Chlamydosperms)	Piperales	
	Juglandales	
C. *Angiosperms*	Monocotyledons	(Numerous orders in each of
	Dicotyledons	these subclasses)

4 The Two Phases of Sporophyte Development

These plants are therefore the "dominant" forms of the continental flora. Although they include a few species which have "gone back" to aquatic medium, most of them live on the ground and in the air, a medium often unfavourable to life. So it is not surprising to see that all these plants use more or less sophisticated means to avoid problems of the medium to the first stages of their sporophyte development. This development can be divided into two phases: first, a phase of *embryogenesis,* similar to that found in animals, during which the original zygote segments and proliferation builds the first organs of the young sporophyte. Incapable of living alone during this period, the embryonic sporophyte is a parasite of the organism on which it was born; it is said to have an *heterotrophic* life.

Then, when the first organs necessary for an independent and *autotrophic* life appear, the sporophyte can become free from the mother organism. It can develop alone and the activity of its *meristems* ensures its "continued" growth.

We shall first briefly discuss the most general modalities of initial embryogenesis in the various groups of vascular plants, then, dealing with meristems, we shall discuss the continued embryogenesis and cellular differentiation.

PART I

Embryogeny and Post-embryogenic Ontogeny

Segmentation and First Edification of the Sporophyte

1.1 Embryogeny in Atmospheric Medium

1.1.1 The Environment of the Zygote

The zygote of vascular plants always stems from an oosphere enclosed in the body of a prothallus and divides itself therein. Thus, it is never free, never *scattered*. This can be considered as a feature of adjustment; the zygote is at no time faced with the external medium, as for instance the zygote of a *Fucus* which develops in the sea and is directly affected by this medium.

However, although indirectly, the severity of the atmospheric medium may affect the zygote or the young embryo more or less easily depending on the efficiency of the protection ensured by the prothallus.

Several cases have thus to be distinguished, whether this prothallus is free or itself enclosed in the tissues of a pre-existing sporophyte.

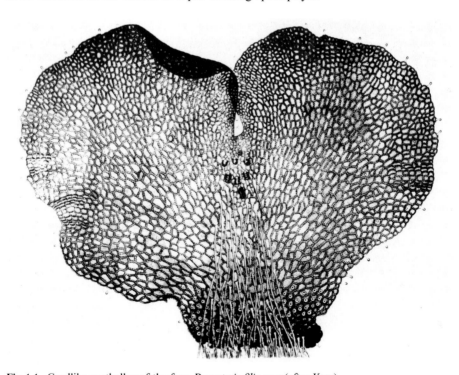

Fig. 1.1. Cordlike prothallus of the fern *Dryopteris filix mas* (after KNY)

1.1.1.1 The Prothallus is Free

In Pteridophytes, tetraspores are scattered and germinate on the ground, producing a prothallus with a free life, most often autotrophic but sometimes symbiotic, with endophyte fungus like in Psilotales, *Lycopodium* and a few primitive Filicales (Ophioglossales) (cf. Fig. 1.24).

These symbiotic prothalli are massive, entirely or partially underground and sometimes have a long life. They are provided with reserve-rich tissues and can endure poor seasons. They can shelter and efficiently feed the young embryos enclosed for a long time.

The autotrophic prothalli of Filicales and Equisetales are of reduced size and form only small green lamina of a few mm^2 which vegetate on the damp ground (Fig. 1.1).

They are fragile organisms sensitive to dryness and excess light; the protection they give to embryogeny is thus much more precarious than in the case of symbiotic prothalli. However, the young sporophyte benefits from the autotrophic activity of the prothallium which provides it with the necessary substances, but the duration of the embryogeny is necessarily shorter. Thus, as long as the external medium is favourable to the prothallus, the embryo finds a nutritive medium relatively constant in the "cushion" where it develops.

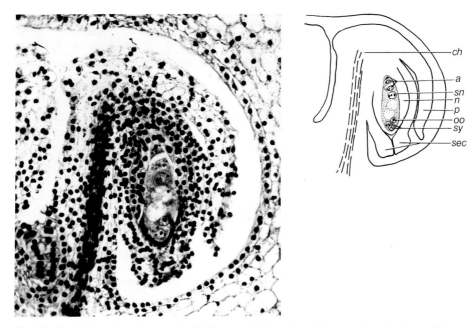

Fig. 1.2. Sagittal section of an ovule of *Lilium candidum*, microphotograph and scheme of interpretation. *ch* Chalaza; *a* antipodal cells; *n* nucellus; *sn* secondary nucleus; *p* primine (outer integument); *oo* egg (oosphere); *sec* secondine (inner integument); *sy* synergids. The enclosed gametophyte (embryo sac) is shown in *grey*

1.1.1.2 The Prothallus is Enclosed

The other vascular plants save their tetraspores and their ♀ prothalli the trouble of scattering out in the atmospheric medium and of a free life. This retention of the spores, then of the ♀ gametophytes, is the rule in the Prephanerogams and Phanerogams, i.e. in ovule plants; some exceptional cases are found in Selaginella (e.g. *Selaginella apus*). The ♀ tetraspore, or macrospore, develops into the sporangium where it stems, which here is the *nucellus* of the ovule, and remains enclosed in it as well as the prothallus it produces. The nucellus, a homolog of a sporangium wall, is itself protected by one or two teguments, thus elaborating a system implemented by means of *the tissues of the anterior sporophyte,* which encloses the prothallus, i.e. the *endosperm* (Prephanerogams and Gymnosperms) or the *embryo sac* (Angiosperms) (Figs. 1.2, 1.3).

What the prothallus becomes after fertilization justifies the use of two different terms.

The Prothallus is Acrescent. In Prephanerogams and Gymnosperms, the prothallus grows and surrounds the embryo it feeds, forming the endosperm. It is itself the medium where the young sporophyte develops from the division of the zygote until reaching autotrophic life (Fig. 1.4).

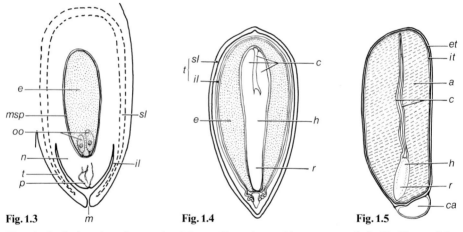

Fig. 1.3   m   Fig. 1.4   Fig. 1.5

Fig. 1.3. Sagittal section of an ovule of *Pinus. il* inner layer of integument; *sl* sclerified layer of the integument; *e* endosperm; *msp* megaspore wall; *n* nucellus; *oo* two eggs; *p* trace of pollen tubes in the nucellus; *t* integument. The enclosed gametophyte (endosperm) is shown in *grey*

Fig. 1.4. Sagittal section of a seed of *Pinus maritima;* overlapping tissues of three generations: sporophyte 1 (integument *t*), gametophyte, acrescent and persistent (endosperm *e,* shown in *grey*); sporophyte 2 (embryo, in the *centre, white*). *il* internal layer of integument; *sl* sclerified layer; *c* cotyledons; *h* hypocotyl; *r* radicle

Fig. 1.5. Sagittal section of a seed of *Ricinus communis:* overlapping tissues of two persistent generations: sporophyte 1 (integuments, external *et* internal *it*), sporophyte 2 (embryo, *white*). The intermediate gametophyte (embryo sac of the ovule) has disappeared but is physiologically replaced by the albumen *a* (in *grey*) issued from the "double fertilization". *c* cotyledons; *ca* caruncle (differentiation of external integument); *h* hypocotyl; *r* radicle

The Prothallus is Resorbed. This is the case with Angiosperms and it is suggestive to compare this behaviour to the exceptional characteristic of fertilization in these beings: in addition to the usual fusion of the gametes, a second antherozoon merges with a non-sexual cell of the prothallus. As a result, a "secondary" or "accessory egg" develops and divides, thus producing a particular nutritive tissue, the *albumen,* which replaces the prothallus and becomes *physiologically* the equivalent of the endosperm of Gymnosperms (Fig. 1.5).

If one considers the extreme reduction of the ♀ prothallus of Angiosperms, a kind of degeneration, the *"double fertilization",* appears as a complementary process, a process for protecting the embryo, particularly improved. In ovule plants, at the beginning, sporophytes are thus sheltered and nourished by the gametophytic body where they are enclosed, the latter being itself protected by the tissues of the anterior sporophyte. Whether ovule or seed, the scattering organ, especially adjusted to the atmospheric medium, is thus composed of tissues supplied by *three overlapping generations.*

1.1.2 The Regularity of the Segmentation

Thus, at the beginning, the embryo develops in a physiological medium relatively constant in spite of the possible effects of the atmospheric medium. This internal medium can be compared to the medium in which the embryos of non-aquatic animals develop. In some way it is not entirely different from the aquatic medium where the embryos of most aquatic plants and relatively primitive aquatic animals develop freely.

Under such relatively regular physiological conditions, the first stages of division sometimes show, not surprisingly, a geometrical regularity often very marked, particularly in Angiosperms. This regularity is a manifestation of the genetic factors and their regulation which can be expressed freely as the medium does not change significantly.

Nevertheless, this regularity allowed the geometrical study of cell filiation, a study which, contrarily, was found to be useless in the analysis of the apical meristem function. In fact, in these meristems, the research on cell filiation, which predominates over all the work done for over a century on the "vegetative apex", resulted only in contradictions and uncertainties.

1.1.3 Seed Plants and Plants Without Seeds

In Phanerogams, the scattering organ, i.e. the seed, presents the noticeable particularity to adapt itself to a slow condition of life and *to dormancy.* The mature seed remains in this condition, and germinates only when the environment is favourable. The enclosed embryo defers its development and when ready, it will become free of the organ where it was prisoner, but sheltered. The development of the Phanerogam sporophyte shows two stages separated by a period of latency: a development strictly *"embryonic"* which is *intraseminal* and a further stage or postseminal stage beginning after seed germination.

This is an undeniable advantage over the other vascular plants in which, once it has started, the development must go on without a possibility of dormancy. It is then the transition, probably progressive, from the heterotrophic life to the autotrophic life which best characterizes both phases of their vegetation.

1.2 Embryogeny of Pteridophytes

The fecundation is aquatic also in Pteridophytes. Ciliated, free antherozoa propel themselves in the pellicle of water which covers the prothallus surface and penetrate into the canal of archegonia where they are attracted by chimiotactism. The union between an antherozoon and an oosphere results in a zygote which typically soon divides.

The conditions of division are very diversified in Pteridophytes, certainly more than in Angiosperms for which, however, a detailed and often fastidious systematization leads embryologists to establish a highly complex classification. In spite of these mainly geometrical data, the histocytology of embryogeny is still poorly known and we will not extend this subject.

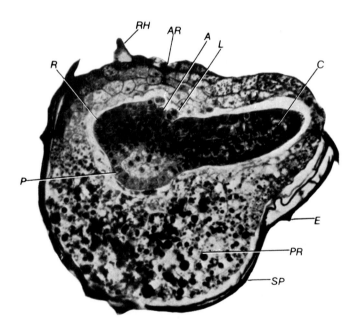

Fig. 1.6. Embryo of *Isoetes lithophila*, sagittal section, *in situ* in the prothallus *(PR)*, which is enclosed in the double coat of the megaspore *(SP)*. The apex *(A)* arises on the side of the *archegonium* canal *(AR)* (exoscopic embryo). *R* first root; *C* cotyledon and *L* its ligule; *P* foot; *RH* rhizoid; *E* external coat of the spore, separated from the internal one at this site (After LA MOTTE 1933)

1.2.1 Early Segmentations of the Zygote

Generally, the authors paid much attention to the orientation of the first division of the zygote relative to the axis of archegonium and to the surface of the prothallus where this archegonium is fixed or embedded. In most archegoniates without ovules, the first division is perpendicular to the axis of archegonium (cf. e.g. Figs. 1.9, 1.15).

When a suspensor, an organ which forces the embryo deep into the body of the prothallus forms, it originates from the cell located on the side of the archegonium canal. Then it is the most internal cell which gives rise to the embryo itself: the embryo is said to be *endoscopic* (cf. Figs. 1.14, 1.15).

When no suspensor forms, both cells unite to produce the embryo, but it is the cell situated on the canal side which produces the first leafy organs and the shoot meristem of the future stem: the embryo is said to be *exoscopic* (Fig. 1.6).

Living *Bryophytes, Psilotales* and *Equisetales* as well as *certain Ophioglossales* have exoscopic embryos. *Lycopodiaceae* (except for *Isoetes*), *the other Ophioglossales* and *Marattiaceae* have endoscopic embryos.

Finally, in the other Filicineae (*Osmunda* und Leptosporangiates) the first cleavage is longitudinal, in a plane including the archegonium axis. In Leptosporangiales, it is the cell turned towards the apical extremity of the prothallus which supplies the apex of the leafy stem, so the embryo is said to be *"acroscopic"* (cf. Figs. 1.31, 1.32, 1.33).

The *Isoetes* embryo develops from a first oblique division, but it is the cell oriented towards the canal which produces the caulinary apex, this embryo is therefore said to be almost exoscopic, contrarily to those of the other Lycopodineae (cf. Figs. 1.6, 1.22).

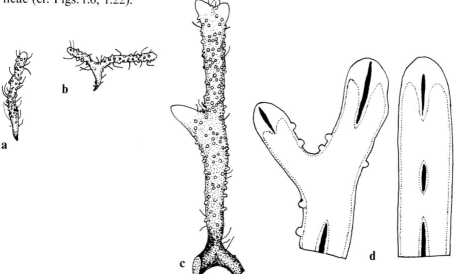

Fig. 1.7 a–d. Prothalli of *Psilotum triquetrum* Swartz. **a** and **b** Entire young prothalli, with rhizoids and antheridia (×4); **c** part of the entirely developed prothallus, covered with antheridia (×4); **d** schemes of two portions of large vascularized prothalli; the tracheid sites are *black; the dotted line* limits the zone of cells containing endotrophic fungi (after HOLLOWAY 1939)

13

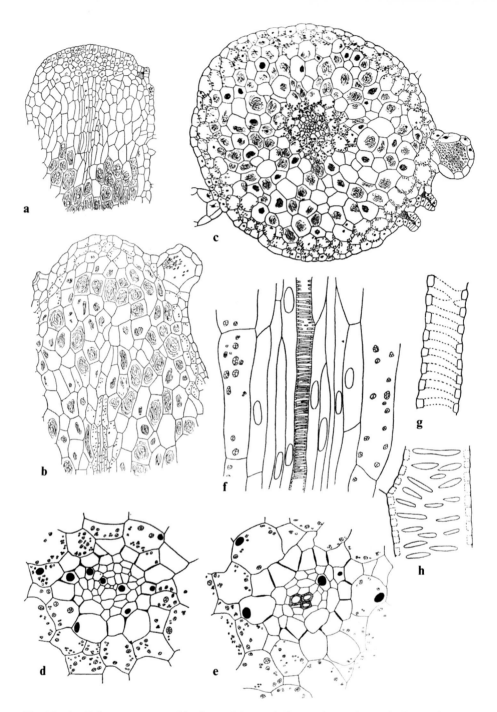

Fig. 1.8 a–h. *Psilotum triquetrum*, histology of the prothallus. **a** Apex of an actively growing prothallus: apical meristem, axial proconducting strand, distal cells invaded by the endophytic fungus. **b** Apex of a resting prothallus, differentiation of the apical and cortical cells, which are loaded with starch, and acropetal advance of the endophytic fungus. **c** Transverse section in an advanced prothallus: peripheral starch zone, fungus-containing cell area and, in the *centre,* a rudi-

In all cases, on beginning, segmentation always results in a group of cells forming a more or less massive and heterotrophic organism. It is from this proembryo that early organs originate, enabling the young sporophyte to reach sooner or later an autotrophic life.

The development which precedes the emergence of the first leaves and roots is more or less progressive and extended, according to the groups.

1.2.2 Embryogeny of Psilotales

The prothallus of *Psilotum* (e.g. Fig. 1.7) and *Tmesipteris* is massive and underground. It has a saprophytic and symbiotic life, its vegetative cells being invaded by an endophytic fungus. Archegonia are distributed throughout and forced into the tissue (cf. Fig. 1.10).

These underground prothallia have a relatively long life before producing embryos and they are the most differentiated gametophytes in vascular plants.

From outside to inside, their structure includes: first, several layers of starch-rich cells (Fig. 1.8 c), then an area of cells invaded by the endophytic fungus (Fig. 1.8 a–c); in the centre again starch cells and an axial cord composed of narrow and long cells simulating a stele of elementary conducting tissues are found (Fig. 1.8 c–e). Besides, HOLLOWAY (1939) showed that the prothalli of *Psilotum triquetrum* include true annular and reticulated tracheids distributed among the axial cells (Fig. 1.8 e–h). Up to now it is the only example known of a *vascularized gametophyte*. This outstanding feature, in relationship with the massive structure and the way of life of the prothallus, suggests that the determinism of xylem differentiation is essentially physiological.

After fertilization, the zygote produces an exoscopic embryo composed only of a "foot" which lobes itself inside the prothallium, and an apical apex where the segmentation appears to be irregular (HOLLOWAY 1918, 1921) (Fig. 1.9). This apex grows in an underground axis which branches out sometimes very early (Fig. 1.11) (HOLLOWAY 1915).

The young sporophyte is then an underground flexous and sharp-pointed, hair-covered, simple rhizome without any leaves or roots and with a diameter of about 1 mm. It grows slowly by means of the terminal meristems, the organization of which appears rather undefined. Sometimes larger and more or less irregular apical cells can be found. The separation from the prothallus occurs on the basis of the foot which is excluded with the gametophyte.

After a period of underground growth, one or two meristems, stronger than the others, give rise to an about 1-cm-high stem, first without leaves. A few scally leaves appear at the end, then this stem dies. Once it has grown sufficiently, the

---

mentary stele, without differentiated tracheids. **d,e** Central parts of advanced gametophytes, transverse sections, the first one in the absence of vascular differentiation, the second one with three tracheids; an endodermis is well differentiated. **f** Longitudinal section of a stele surrounded by starch cells and bearing only one axial tracheid. **g** Part of an annular tracheid. **h** Part of a reticulate tracheid (after HOLLOWAY 1939)

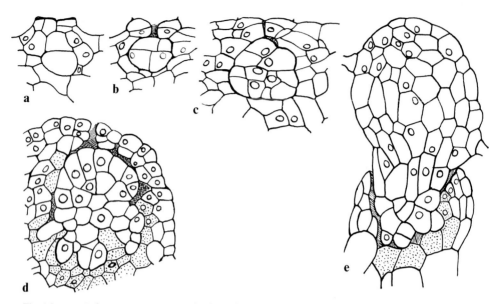

Fig. 1.9 a–e. *Psilotum triquetrum.* Beginning of the development of the sporophyte. **a** Two-celled proembryo, the first septum is perpendicular to the archegonium axis. **b** four-celled stage; **c** beginning of differentiation of the embryo itself (*upper half,* exoscopic) and of the foot *(lower half);* **d** more developed embryo, the foot produces nutritive extensions which sink into the prothallus; **e** embryo yet undifferentiated, rising out of the prothallus (the cells of the prothallus, situated at the contact of the sporophyte, are shown in *grey* in **d** and **e**) (after Holloway 1939)

underground rhizome, which has no roots, produces normal stems where sporangia will form (Fig. 1.11 c).

From the embryologic standpoint, this development is characterized by the early formation of a globular organism where the disposition of the segmentations, if it exists, is not clear (Fig. 1.9 d, e). The embryo proper discontinuously shows a distinct apical cell in the primary apex, then in those resulting from its branching. The figures published by Holloway demonstrate that very early elongated cells indicate a rough axial conducting strand. Surrounding this strand, "cortical" cells will soon be invaded by the symbiotic fungus (Fig. 1.8 a).

1.2.3 Embryogeny of Lycopodineae

It is necessary, first of all, to distinguish the Lycopodineae Aliguleae from those which have ligules at the basis of the leaf.

1.2.3.1 Aliguleae *(Lycopodium-Phylloglossum)*

Aliguleae themselves are varied because of the structure and mode of life of their prothallus, resulting in variations in their embryogeny. We shall deal with three typical examples.

16

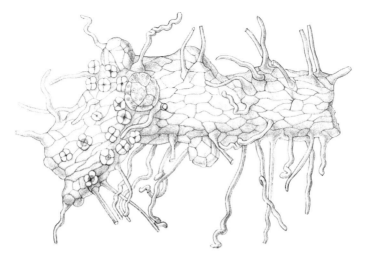

Fig. 1.10. Fragment of prothallus of *Tmesipteris tannensis* (after LAWSON 1917)

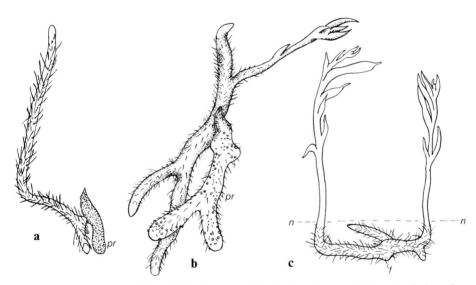

Fig. 1.11 a-c. Young sporophyte of *Tmesipteris*. **a** Needle-shaped young plant, yet entirely underground, with two apices on both sides of the foot; **b** more advanced sporophyte, bearing four apices, one of them producing a leafy aerial stem; **c** rhizome, two apices of which are yet underground, the two others having edified aerial upright stems. *f* foot; *pr* prothallus; *n-n* level of the ground (after HOLLOWAY **b**: 1915, **a** and **c**: 1921)

Type Lycopodium selago. The prothallus may develop underground or at the surface. It is associated with an endophytic fungus and its shape, always quite massive, is quite variable (Fig. 1.12).

The development of the zygote begins similarly in all Lycopodia: the embryo is endoscopic and undergoes a phase of three cell layers: the suspensor cell, turned towards the archegonium canal, and two 4-cell layers (Fig. 1.15 c).

17

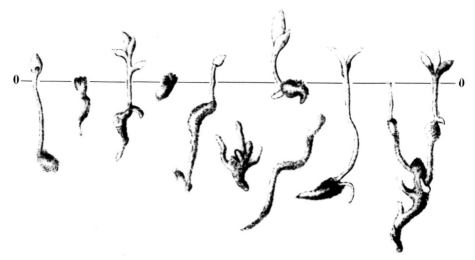

Fig. 1.12. *Lycopodium selago* L., prothalli with or without young plants, showing the situation with respect to the ground level, and particularly the lengthening of the hypocotyl when the prothallus is deeply sunk (after BRUCHMANN 1898, natural size)

In all cases, the median four cells produce the "foot" by which the embryo is attached to the food-supplying prothallus and the upper four cells organize the sporophytic organism itself.

Then there are differences: (1) in the foot development and (2) in the upper layer development *prior to the formation of the organs for an autotrophic life.*

In *L. selago,* the foot is moderately developed (Fig. 1.13); the embryo itself soon becomes dissymmetric, the side growing faster produces an early leaf or "cotyledon", the other side gives rise to an *apical meristem* (Fig. 1.13 d). From the latter, the successive leaves and the stem will soon originate. The cells located under the apex and under the cotyledon proliferate and extend to form a *hypocotyl* rising above the foot (Fig. 1.13 e).

Radicular meristems develop on this hypocotyl, the primary root exactly at the base above the foot, the others higher (Fig. 1.13 e). The length of the hypocotyl depends mainly on how deep the prothallium is. It is clearly meant to push the shoot apex out of the soil.

The regularization of the apical organization seems to be slow and undecided at the first stages of development when the proliferation produces a relatively homogeneous mass of small cells, as shown in the accurate drawings of BRUCHMANN (1898, 1910).

The development of *Lycopodium phlegmaria* studied by TREUB (1886) is very similar to the previous one and demonstrates also the progressive organization of the apex around which the first leaves grow. However, the prothallus is epiphytic like the sporophyte.

Type Lycopodium clavatum. The massive prothallus of this species, in the shape of a flattened cone (Fig. 1.14) is totally underground and sometimes deep in the ground. The hypocotyl does not extend enough to raise the apex up into light, thus autotrophic life is delayed.

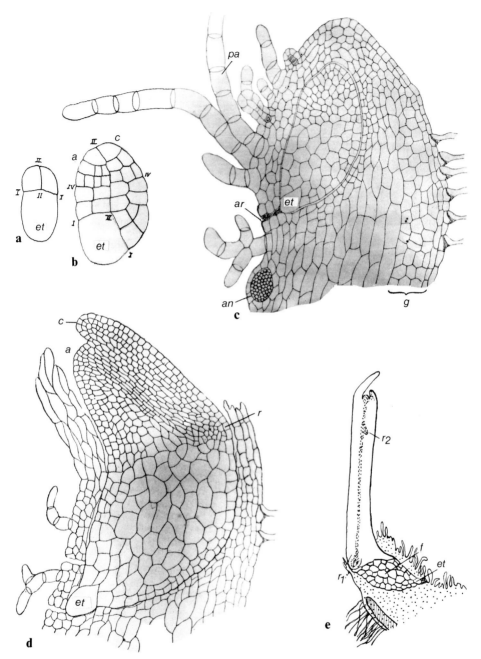

Fig. 1.13a-e. Segmentation of the zygote and development of *Lycopodium selago;* **a** After the two first cleavages (*I. I* and *II. II*); **b** more advanced state; **c** embryo, in situ in the prothallus showing the endoscopic arrangement; **d** more developed embryo rising out of the prothallus tissues; **e** development of the hypocotyl bringing the apex of the young sporophyte into light. *et* suspensor; *ar* archegonium canal; *an* antheridium; *a* apex; *c* cotyledon; *pa* papillae of the prothallus; *g* zone of the cells invaded by the endophytic fungus; *r* first root primordium; r_1, r_2 primordia of the first and second roots; *f* foot; **a-d**: ×about 100; **e**: ×20 (after Bruchmann 1910, 1898)

19

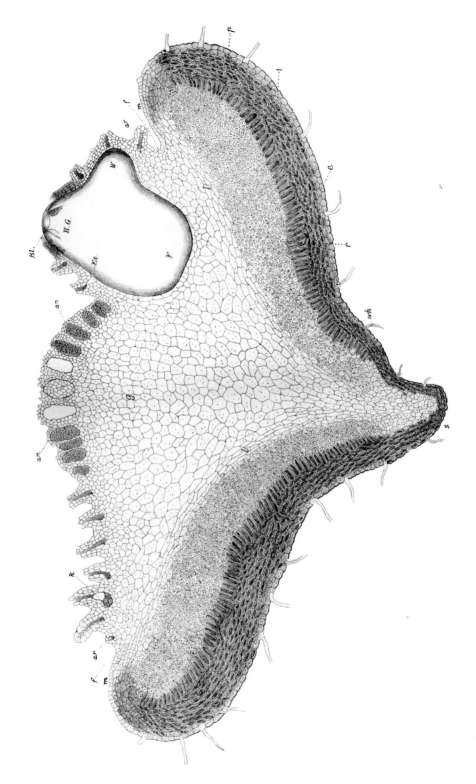

20

Fig. 1.14. Prothallus of *Lycopodium clavatum* L., sagittal section. *e* Superficial zone, playing the part of epidermis; *wh* rhizoid; *t* cortical zone of cells containing endotrophic fungus; *l* flat cells between the peripheral starch zone and the central tissue *g*; *m* bordering meristem; *f* prothallus groove; *an* antheridia; *ar* archegonia; *k* very young embryo; *Et* embryo bearer; *F* foot; *W* first root; *Bl* the first one among the leaves covering the apex; *HG* hypocotyl segment; authors's original lettering; actual diameter 0.5 cm (after BRUCHMANN 1898)

◁───

In fact, the extension of the epicotyl stem replaces that of the hypocotyl but in such cases, the foot develops previously, always from the middle layer of the embryo and much more widely than mentioned above. The foot turns into a pulpy globular mass which goes on growing in the prothallus, mainly on the food-conducting side (Fig. 1.15).

The upper layer of the embryo grows more slowly, it is distended at the surface of the foot and gives rise to three, then five protuberances. The first three are two opposite leaves with an apex inbetween. A third leaf grows forming a cross with the two others, then opposite and towards the base of the embryonic stage the first root rises (Fig. 1.15f). Later the hypocotyl, then the epicotyl axis extend, above the opposite primordia of leaves. This young stem always stays underground for a while and grows, producing small scale-like leaves (Fig. 1.15g). Normal leaves will only develop when the apex reaches the atmospheric medium.

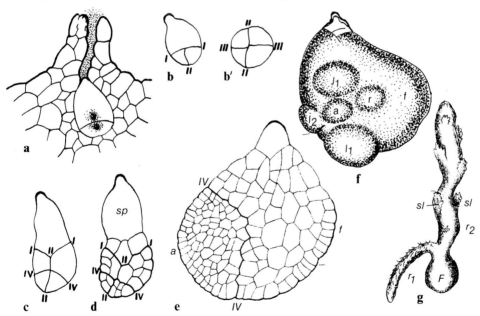

Fig. 1.15a–g. Segmentation of the zygote and development of *Lycopodium clavatum*. **a** First cleavage, perpendicular to the axis of the archegonium; **b, b'** formation of four cells at the lower pole; **c** two-octant state; **d** beginning of foot growth between the suspensor *sp* and the cleavage No. IV; **e** more advanced proembryo showing the hypertrophic extension of the foot *f* with respect to the apex *a*; **f** apical face of the embryo with distended primordia on the foot surface; *l₁* cotyledon leaves; *l₂* first epicotyl leaf; *r* radicle; **g** development of the vegetative axis, yet subterranean, and of the first root *r₁*; *F* foot; *r₂* second root; *sl* underground scaly leaves; **a–e** × 150; **f** × 52; **g** × 10 (after BRUCHMANN 1898)

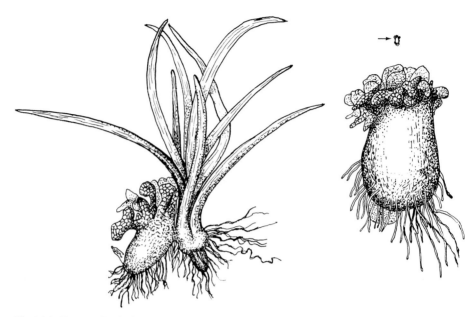

Fig. 1.16. Two prothalli of *Lycopodium cernuum*. The one on the *right* bears a young sporophyte; *above* the *left* one, the same in real size *(arrow)* (after Treub 1884)

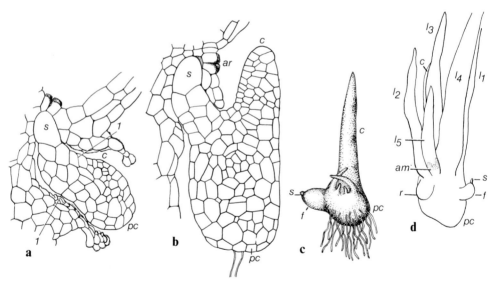

Fig. 1.17 a-d. Development of *Lycopodium cernuum*. **a** Young embryo when it perforates the prothallus; **b** more developed embryo, formation of the protocorm, before the differentiation of an apex; **c** isolated embryo before the appearance of the apex, at the initiation of successive little leaves on the protocorm; **d** scheme of a more advanced young sporophyte, the protocorm of which has formed five leaves after the cotyledon; appearance of an apical meristem and of the first root. *1-1* : first partition of the zygote; *am* apical meristem; *ar* archegonium; *c* cotyledon; l_1-l_5 successive leaves; *f* foot; *pc* protocorm; *r* root; *s* suspensor; **a, b**: ×200; **c, d**: ×24 (after Treub 1890)

Fig. 1.18. Protocorms of *Lycopodium laterale* upon which vegetative axes ♀ and root *r* originate late (after HOLLOWAY 1915)

In the meantime, the first root, then a second one develop deep into the soil (Fig. 1.15 g). *Lycopodium annotinum* develops in the same way.

Type Lycopodium cernuum. The prothallus lives on the surface of the soil, sinking into it by means of a cone-shaped pivot with rhizoids (Fig. 1.16). The aerial part differentiates chlorophyllous lobes, while an endophytic fungus invades the peripheral cells of the cortex of the underground part.

In this example, the development of the median stage of the embryo results in a foot of a few cells, whereas the upper stage, on the contrary, proliferates into a pulpy mass which extends from the prothallus and develops freely on the soil, clinging to it by rhizoids but without forming leaves or roots at that tier. This parenchymatous mass where cells are associated with an endocellular fungus is the *protocorm* (Fig. 1.17, *pc*). Later, a raised cylindric papilla is constituted to become the first leaf or "protophyll" and other leaves appear on the axil of this "cotyledon". A first root grows from the base of the protocorm. The vegetative shoot apex is even more slowly structured, from the residual cells located between the crown formed by the first leaves. It is the same process with the *Lycopodium innundatum*. Other species have poorly defined protocorms sometimes branched, with several apex distributed at random (*L. laterale*, Fig. 1.18).

Finally, the single species of the only other type of living Lypocodiacea, *Phylloglossum drummondi*, presents a symbiotic tubercle analogous to the former protocorms, but which arises and dies every year with the remaining sporophyte. This type of growth is similar to that of orchids.

1.2.3.2 Liguleae

They only include two living types, a numerous one, *Selaginella,* and another less numerous, *Isoetes.*

Both types are very dissimilar and it would have been very difficult to bring them together without the help of paleobotany.

Genus Selaginella. S. denticulata, studied by BRUCHMANN in 1912, is a good example.

Unisexual prothalli remain enclosed within the envelope of the spore, tearing it only on the side where archegonia develop, if it is a ♀ prothallium (cf. Fig. 1.19 a).

After fertilization, the zygote divides with a wall perpendicular to the axis of the archegonium and the cell turned towards the canal proliferates actively, pro-

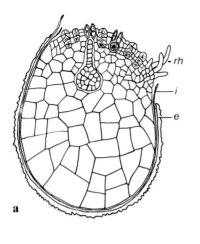

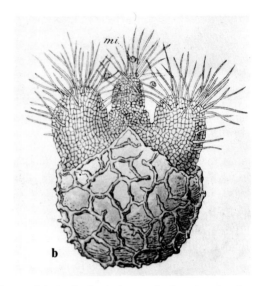

Fig. 1.19a,b. Female prothalli of Selaginellae. **a** *Selaginella denticulata*, sagittal section showing the two integuments of the spore, the thin internal one *i*, and the adorned and cutinized external one *e*. Rhizoids, *rh*, archegonia at various stages of development and a young embryo, forced by the suspensor into the vegetative, large-celled tissues, are seen at the *upper pole*. **b** *S. Galeottei*, external aspect, prothallus emerging from the splits of the spore coat, bearing excrescences with rhizoids and, in the *centre*, a dome bearing the archegonia (hardly visible); *mi* microspores; × about 200 (after BRUCHMANN 1912)

ducing a series of cells which form the suspensor and forces the internal cell into the prothallus. This cell grows and divides to produce the embryo itself which is thus said to be endoscopic. The first mitosis occurs in a sagittal plane (Fig. 1.20b) the following ones in an oblique plane. Very early, at the end of the cellular mass issued from the internal cell, an apex and two opposite primordia of cotyledons appear (Fig. 1.20f), then at the axil of each cotyledon, a ligule *(1)* is differentiated (Fig. 1.20f).

Meanwhile, the base of the embryonic mass increases dissymmetrically, forming a curve which straightens the apical end towards the free part of the prothallium (Fig. 1.20g). The region of maximal growth produces the foot opposite the suspensor; near the base of the suspensor, between these two organs, appears a first root (Fig. 1.20g, *r*).

The elongation of the hypocotyl pushes the apex out of the prothallus, the root arising on its side. It is soon possible, in the hypocotyl, to distinguish a central area with narrow and long cells, an outline of the conducting tissue system, and a cortical area with larger and shorter cells (Fig. 1.20g).

S. denticulata is an average type. In other species, like in *S. spinulosa*, the foot does not develop much and does not bend the embryo (Fig. 1.21). Other variations concern the mode of apical division which seems always to be more rigorous than in the former types; the similarity or difference of cotyledons; the importance of development of the foot; and the location of the first root in relation to the other parts of the embryo and the suspensor.

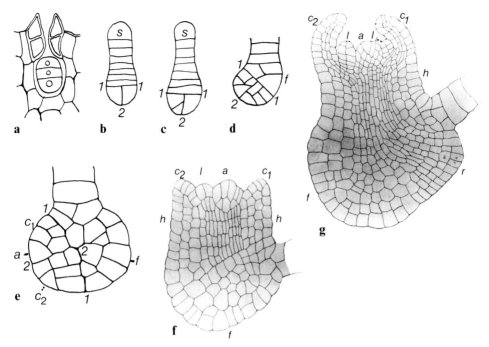

Fig. 1.20 a–g. Development of *Selaginella denticulata*. **a** Three-celled stage, in situ into the archegonium, only the lower cell will give rise to the true embryo; **b** first sagittal partition of the embryo; **c** slightly more advanced state; **d** beginning of curving, due to hypertrophy of the cell which will produce the foot *f*; **e** more developed embryo showing the straightening of the apex *a* and the sites of edification of the two cotyledons c_1 and c_2; **f,g** two successive states of the more and more differentiated embryo. *l* ligule; *h* hypocotyl; *r* primordium of the first root; *s* suspensor; *1-1, 2-2* first two partitions of the true embryo

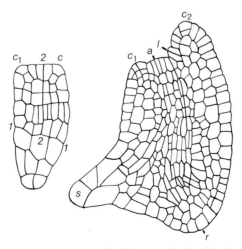

Fig. 1.21. *Selaginella spinulosa*. Two stages of embryogenesis showing the formation of a straight embryo, contrarily to the embryo of *S. denticulata* (abbreviations as in Fig. 1.20) (after BRUCH-MANN 1897)

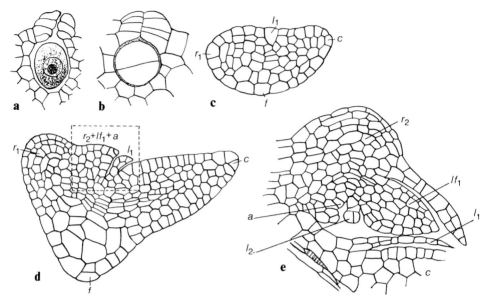

Fig. 1.22 a–e. Development of *Isoetes echinospora*. **a, b** Zygote and first cleavage, slightly oblique; **c** very young embryo, already showing the origin of the cotyledon *c,* the first root r_1, the ligule of the cotyledon l_1 and the foot *f;* **d** more advanced embryo with the site of initiation of the second root r_2, the first leaf lf_1 and the apex *a;* **e** detail of this last area on a more developed embryo; l_2 ligule of the leaf lf_1. The orientation of the figures being approximately constant, one sees that the embryo is exoscopic (after Campbell 1895)

Genus Isoetes. The unisexual prothalli of *Isoetes* are similar to those of *Selaginella* and also remain enclosed in the envelope of spores.

However, the embryogeny is very different. At first, the embryo is exoscopic and there is no suspensor. The first division of the zygote is more or less oblique, producing an oblong mass of cells, the great axis of which is parallel to the surface of the prothallus (Fig. 1.22). At one end a single "cotyledon" arises, at the other end the first root, and in between, on the surface, the first ligule. Between the ligule and the root is a small depression, the cells of this will give rise to a second leaf opposite the "cotyledon", then to a third one, etc. Thus, this apex is rather exiguous and only develops later.

1.2.4 Embryogeny of Equisetales

Unfortunately, among Articulateae, only the genus *Equisetum* with an exoscopic embryo can be found in nature presently.

In *Equisetum arvense,* studied by Sadebeck (1878), the first cleavage, perpendicular to the axis of the archegonium, oblique according to Laroche (1968), produces on the canal side a cell which will produce the stem and on the bottom side a cell which will produce the foot and the first root (Fig. 1.23 a). There is *no suspensor.*

The upper cell divides itself twice sagittally resulting in four quadrants arranged in the shape of a crown, but one of them proliferates quicker than the others so that its productions overlap the top and constitute the apex where a tetrahedral apical cell is soon differentiated. The basal part of the cells originating from this quadrant, together with the three other quadrants, will give rise to the first crown of leaves (Fig. 1.23).

The lower cell also divides sagittally. One of the two daughter cells divides slightly quicker and the foot is organized as a part of a moderately dilated sphere, while the second cell produces a smaller pile of cells from which the first root is organized (Fig. 1.23 d).

The other species show some variants. Thus, the first root may be organized from the upper cell *(E. hiemale)* but these details are not important especially because in the same species, the first segmentations are not arranged regularly enough to fit the requirements of a strict geometrical plane.

The growth of the first stem of the young *Equisetum* is limited to a few internodes only (6, 12). Below the first foliar whorl, a second apex is organized from which a second more developed stem arises and then a third one appears at the

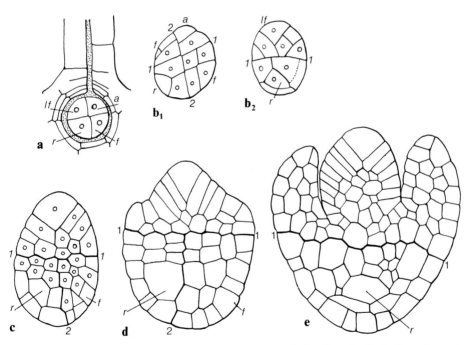

Fig. 1.23 a–e. Development of *Equisetum arvense*. **a** Four-celled embryo in situ in the archegonium; *1-1* first cleavage, perpendicular to the archegonium axis; *a* cell which generates the apex (exoscopic embryo); *lf* cell from which the first leaf verticil will arise; *r* and *f* same for the root and the foot; **b₁, b₂** two orthogonal orientations of the more developed embryo; *2-2* second cleavage, same indications as for **a**; **c** still more advanced embryo, differentiation of the apex, the foot *f* and the root apical cell; **d** beginning of differentiation of the first crown of leaves at the base of the apex; **e** more differentiated embryo showing the characteristic aspect of the genus *Equisetum* and the "buttress" of the first leaf verticil (after Sadebeck 1878)

basis of the latter etc. It is only after these stems that the rhizome, well settled, produces normal sterile and fertile stems each year.

1.2.5 Embryogeny of Ferns

Ferns constitute a heterogeneous group in which ancient species of past eras and advanced living species can be found. The first species belong to Eusporangiates, others to Leptosporangiates.

1.2.5.1 Eusporangiate Ferns

Ophioglossaeae seem to be the most ancient ferns in nature at present. These pteridophytes have massive, heterotrophic underground prothalli living in association with endophytic fungi; like the prothallus of Psilotales they are rough cylindrical masses with irregular lobes and sometimes rhizoids (Figs. 1.24, 1.25).

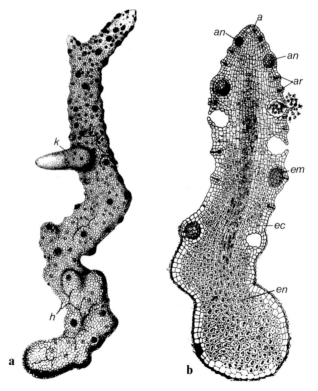

Fig. 1.24a, b. Prothallus of *Ophioglossum vulgatum* L. **a** External aspect; **b** section, approximately sagittal, showing a thin coat *ec* of uncontaminated cells, whereas the whole *centre* is occupied by endophytic fungus invaded cells *en*; *a* apex; *an* antheridia; *ar* archegonia; *em* embryo; *h* fungal hyphae; *k* young embryo, the root of which springs through the prothallus and is the first organ to emerge (after BRUCHMANN 1904)

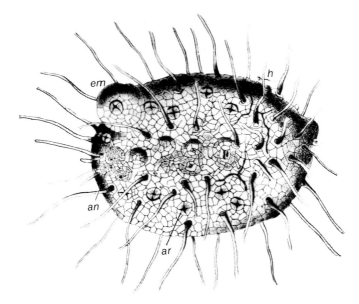

Fig. 1.25. Massive prothallus of *Botrychium lunaria* surrounded by rhizoids. Same abbreviations as in Fig. 1.24 (after BRUCHMANN 1906)

In the *Ophioglossum vulgatum,* studied by BRUCHMANN, the embryo is exoscopic. The internal cell, issued from the first division of the zygote, produces only the foot which is poorly developed and the first root. The later prevails soon and extends from the prothallus first (Fig. 1.26 c). The "cotyledon" is organized from the external cell and remains rudimentary so that it is only the following leaf which will reach light. This occurs only late as the apex and the leafy stem develop only very slowly, sometimes many years after fertilization and the first root organization. Meanwhile, like the prothallus, the young sporophyte lives associated with an endocellular fungus.

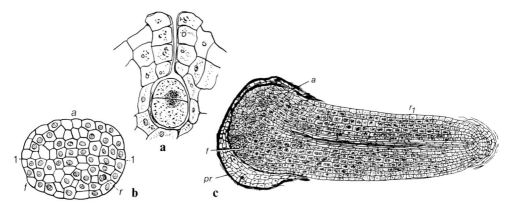

Fig. 1.26 a–c. Development of *Ophioglossum vulgatum* L. **a** First cleavage of the zygote, perpendicular to the archegonium axis; **b** massive and undifferentiated embryo. *1-1* First cleavage; *a, r, f* areas from which the apex, the first root and the foot will be edificated respectively. **c** More advanced embryo showing the early differentiation of the root r_1, when the apex *a* is still undifferentiated; *pr* prothallus (after BRUCHMANN 1904)

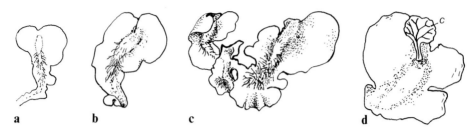

Fig. 1.27 a–d. Prothalli of Marattiales. **a,b,c** Three prothalli of *Danaea elliptica* (lower face) of increasing size; **d** prothallus of *Angiopteris sp.* (upper face) torn by the cotyledon *c* of a young sporophyte, borne into the prothallus (after CAMPBELL 1911)

This exoscopic species has no suspensor. The same fact is also observed in other *Ophioglossum* and in certain *Botrychium* (e.g. *B. lunaria*). Other *Botrychium* (e.g. *B. obliquum*) and *Helminthostachys* have endoscopic embryos and a suspensor. The variety of orientation of the embryo and the correlative absence or presence of a suspensor demonstrate the transitional feature of these still primitive ferns which appear presently as relics of past epochs.

A similar variability can be found in Marattiaceae, but it concerns only the presence or absence of a suspensor, the embryo being always *endoscopic*. The prothallus of these plants is usually horizontal, laminalike but it is larger than the cordlike prothallus of Leptosporangiate ferns (Fig. 1.27 a–c).

Antheridia are enclosed in the prothallus which is thicker than that of Leptosporangiate ferns, the antheridia of which are external (Fig. 1.28). Archegonia open on the internal face, their canal is oriented downwards and after the first division of the zygote, the apex of the future rhizome and the first leaf are organized from the upper cell (internal).

The leaf will break out on the upper face of the prothallus (Fig. 1.27 d). The embryo becomes massive and the organ differentiation is achieved slowly and late, especially for the first root (Fig. 1.29). There is no suspensor in the *Marattia,* but a monocellular suspensor exists in *Danaea* (Fig. 1.29 a) and may be found in some species of *Angiopteris*.

The embryologic feature of *Osmondaceae* are intermediary between the above-mentioned and those of Leptosporangiate. The prothallus is rather of the "cord-like" type but often thick and plump and shows an "axial vein" as seen on the thallus lamina of the liverwort *Pellia epiphylla* (Fig. 1.30). Archegonia are located

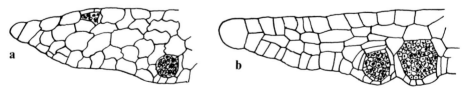

Fig. 1.28 a,b. Transversal section of prothallus fragments of Marattiales, in regions composed of several layers of cells and bearing antheridia. **a** *Danaea jamaicensis;* **b** *Kaulfussia* sp. (after CAMPBELL 1911)

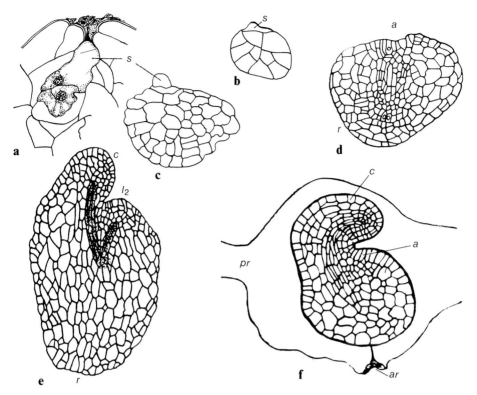

Fig. 1.29 a–f. Embryos of Marattiales. **a–e** *Danaea jamaicensis*. **a** Three-celled embryo, first cleavage perpendicular to the archegonium axis, embryo endoscopic; **b** more developed embryo; **c** advanced, but yet unorganized embryo; **d** onset of organization of the primary axis; **e** more developed embryo showing the initiation of the first organs: cotyledon or first leaf *c,* first root *r* and first epicotyl leaf *l₂*. **f** Advanced embryo of *Marattia douglasii;* the first leaf (or cotyledon) pierces the upper face of the prothallus *pr* into which the embryo is enclosed. *a* Apex; *ar* archegonium; *s* suspensor (after CAMPBELL 1911)

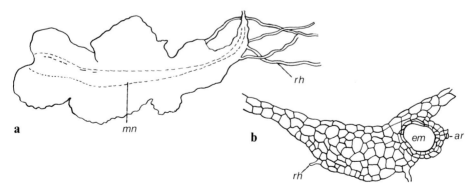

Fig. 1.30 a. Prothallus of *Osmunda claytoniana,* about 2-months-old. *mn* Axial region thickened in the shape of a medial nervure; *rh* rhizoids. **b** Transversal section in an aged prothallus showing the median thickening and the lateral situation of the embryo *em; ar* remnant of the archegonium in which the embryo was born (after CAMPBELL 1892)

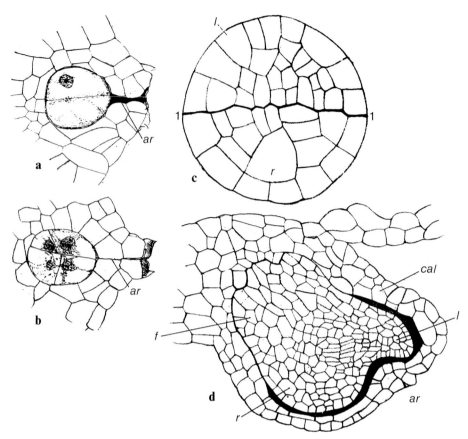

Fig. 1.31 a, b. Beginning of the development of *Osmunda claytoniana,* stage of four and eight nuclei. The two first cleavages are parallel to the archegonium axis. **c, d** More developed embryos of *Osmunda cinnamomea* showing the early onset of organogenesis; *l* initium of the first leaf; *r* buttress of the radicle; *cal* calyptra, kind of envelope formed by the prothallus, acrescent around the embryo; *f* foot (after CAMPBELL 1892)

on the edges, their axis being thus horizontal. The first division of the zygote is also horizontal. The upper cell will give rise to the first leaf and apex, the lower cell to the foot and the first root (Fig. 1.31).

The embryo of Osmondaceae is less massive than the former. The first divisions appear to be more "regular", i.e. the orientation of segmentation is less hazardous and, moreover, the organ differentiation is organized earlier.

1.2.5.2 Leptosporangiate Ferns

Contrarily to the groups mentioned above, the Leptosporangiate ferns have a remarkably uniform embryogeny.

The prothallus usually spreads over the surface of the soil and archegonia are more or less vertical, the canal orientated downwards (Fig. 1.32 a).

32

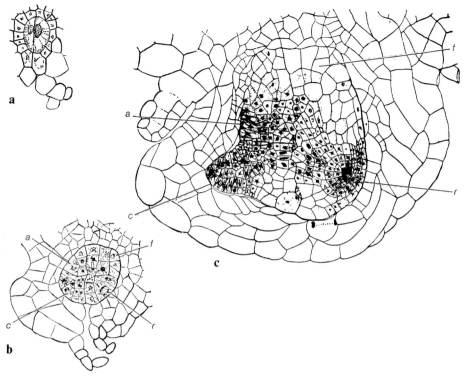

Fig. 1.32 a–c. Development of *Adiantum concinnum* (Filicaceae). **a** First cleavage of the zygote parallel to the archegonium axis; **b** still young embryo, but the localization of the first organs are indicated; **c** more advanced embryo showing the buttresses of the first leaf (or cotyledon *c*) of the first root *r*, and the arrangement of an apex *a*; *f* foot (after ATKINSON 1894)

The plane of the first zygote cleavage includes the archegonium axis, thus being quite vertical. Moreover, it is determined by the apex of the prothallus. In fact, the first division separates one cell orientated towards the apex and one cell oriented towards the prothallus basis.

The first cell will generate the first leaf and apex of the future rhizome. The embryo is said to be *acroscopic*. The other cell will generate the foot and the first root.

Each of the two cells divides into four quadrants and the following divisions soon reveal the single apical cells of the axis, the first leaf and the first root (Fig. 1.32 b, c). There is no distinct apical cell for the foot.

The first leaf and root are soon released from the prothallium, while the apex develops later. These first organs tear the lower face of the prothallus and soon become functional. The young sporophyte quickly reaches *autotrophic life*, while the prothallus degenerates.

Such a forwardness, together with the division regularity differentiate Leptosporangiates from Eusporangiates. These two features are similarly found in aquatic Filicales and heterosporangiate ferns *(Marsilea, Pilularia)* (Figs. 1.33, 1.34).

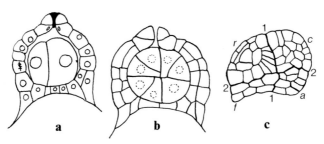

Fig. 1.33 a–c. Development of the embryo of *Marsilea vestita*. **a** Two-celled embryo, in situ in the archegonium; **b** sagittal section of a more advanced embryo; **c** beginning of differentiation of the first organs on a rather exiguous embryo. *a* apex; *c* cotyledon initium; *f* foot; *r* initium of the first root (after CAMPBELL 1895)

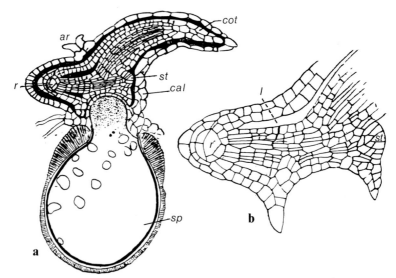

Fig. 1.34 a, b. *Pilularia globulifera.* **a** Young sporophyte, still fixed to the prothallus, which surrounds it by a thin layer of cells, the calyptra, *cal;* the canal of the initial archegonium *ar* is still visible; *cot* cotyledon, *r* root; *st* stem apex; *sp* megaspore, which has become the enclosed prothallus. **b** Detail of the primordial axis of the young plant; *r* apical cell of the root; *st* apical cell of the stem; *l* lacunae

From a *cytological* standpoint, most studies deal more with the maturation of oosphere and fertilization than with embryogeny. TOURTE's studies (1975 a, b) revealed original phenomena such as nucleolar emissions in the cytoplasm where they degenerate, limited processes of autophagy and formation of mitochondria complexes possibly generated by the oosphere at rest. These complexes originate in the flattening of mitochondria, in the shape of flexible discs piling up as they become cuplike. (Plate 1.1 a–f).

Plastids are poorly structured but store starch in the mature oosphere. Thus, at the time of fertilization, the oosphere presents several features of differentiation, sometimes unusual, which allow its distinction from the meristem cells.

A small number of publications deal with the evolution of cellular organelles, from the zygote to the arrangement of meristems. The pioneer research carried out by EMBERGER (1921) and BONNET (1955-1957) with light microscopy yield some data on the evolution of chondriosomes and plastids at the beginning of segmentation.

In Filicales, the zygote has a granulous, homogeneous chondriome without plastids clearly differentiated from mitochondria in light microscopy. Small vacuoles have a concentrated, often chromophil, content. They grow from the first mitoses while certain granules increase and turn into plastids quite different from chondriosomes; they produce chlorophyll in the aerial parts, starch in the underground organs (EMBERGER 1921). These data should be completed by the cytological evolution of the end of embryogeny, when apical meristems become functional. These meristems enclose smaller and more meristematic cells than those resulting from the first zygote divisions.

In Hydropteridaceae, BONNET showed very subtile changes in the differentiation of plastids in the cells which, from the zygote, develop the meristems quickly. The organogen cells located under the apical cell are more "meristematic" than the zygote and their mother cells.

The autoradiographic study of *Marsilea vestita* (Hydropteridaceae) studied by KULIGOWSKI-ANDRES (1975a, b) and KULIGOWSKI-ANDRES and TOURTE (1978) displayed original features concerning the fertilization and embryonic segmentation. The marking, either of ♂ chromatin or ♀ chromatin, showed that these chromatins mingle only progressively during the segmentation and that certain areas of the embryo react preferably to one or the other. Thus, the early individualized cauline and root apical cells, with a relatively long cellular cycle, catch mainly ♂ chromatin; the foliar initials and foot cells mainly ♀ chromatin.

The same technics applied to mitochondria and spermatic plastids suggest that ♂ mitochondria, after penetration into the zygote, may participate in the formation of the chondriome. The results remain unclear for the plastids.

1.2.6 Conclusions

This review is especially brief because it only gives a summary of the most diversified group of vascular plants. In such a wide and heterogeneous group, phylogenists have contributed several general ideas from the modes of embryogeny. We do not intend to deal with phylogeny. We would like to make a few remarks only.

First, the most primitive forms are probably characterized by the development of slowly growing massive embryos in which the organization, i.e. the formation of the first organs, occurs late, e.g. the undifferentiated heterotrophic and symbiotic rhizome of *Psilotum;* "protocorms" of *Lycopodium cernuum;* the nearly acephalous *Ophioglossum* embryo, etc.

In contrast, the evolved forms have exiguous embryos because organs are differentiated earlier. This is the case in Filicales and Hydropteridales. Hence, it is easier to observe the geometrical, stereotyped arrangement of divisions, while segmentation appears to be at random in lower forms.

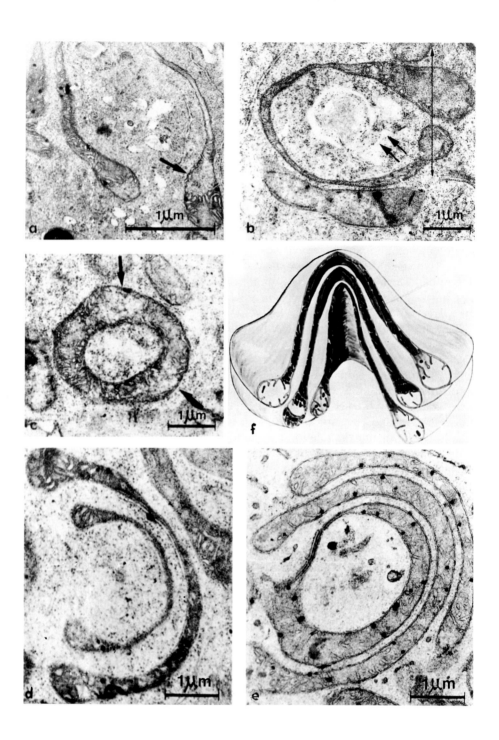

36

The second remark concerns the modes of growth specialization during this embryogeny. In fact, the constitution of massive embryos without previous organization of determined organs, implies a non-localized, slow but continued proliferation in which all cells are similarly "embryonic".

In contrast, the arising of well-defined organs is the result of orientation and specialization in proliferation and growth.

Still for some time, all cells are really embryonic because all of them divide and grow but, according to their location in the embryo, they soon all become different.

As soon as the first organs are sufficiently developed, a part of the cellular material from which they originate and which has not been used for their initiation, has a reduced activity in growth; they constitute selected areas which become then specialized in the function of *organized proliferation*.

Although the authors did not insist on this matter, their description and features clearly show that as soon as *it adjusts itself*, the proliferation first gives rise to a *leaf* and a *root* or to several of these organs.

Only then, sooner or later, but progressively, at the basis of the first leaves, an area develops and specializes in growth. Its activity will give rise either to the further part of the stem and the further leaves or to the root.

The apical meristems arise from this progressive localization of growth. Thus, they appear as the last shelters of proliferating cells. They continue providing the cellular material required for indefinite ontogeny, but their progressive arrangement is a *result* of embryogeny and is not at all necessary in the development of the first organs.

The production of so-called initials, permanent cells with a specialized and determining activity in organogenesis, is unlikely compatible with this mode of origination of meristems.

Nevertheless, one can consider that the young sporophyte includes, on the one hand, a primordial body composed of cells and organs produced by embryogeny *sensu stricto* and, on the other hand, *further* organs produced by the apical meristems. The latter organs are derived from a particular embryogeny which determines the indefinite mode of growth.

◁ ──

Plate 1.1a–e. Evolution of mitochondria during the oosphere maturation of *Pteridium aquilinum.*

a Flattening of mitochondria, in the shape of discs, peripherally thickened in the shape of a pad *(arrow);* × 24000;

b Deformation of mitochondria, which acquire a cupule aspect and enclose part of the cytoplasm *(arrows);* × 13000;

c Transversal section at the level of the pad, according to the *double arrow* of **b,** showing the light spaces, which are the sites of mitochondrial DNA *(arrows);*

c × 13000;

d,e "mitochondrial complexes", due to the grouping of cupule-shaped mitochondria; × 16000;

f three-dimensional reconstruction of a "mitochondrial complex".

Fixation: glutaraldehyde-O$_5$O$_4$; contrast: uranyle acetate-lead citrate (after TOURTE 1975b)

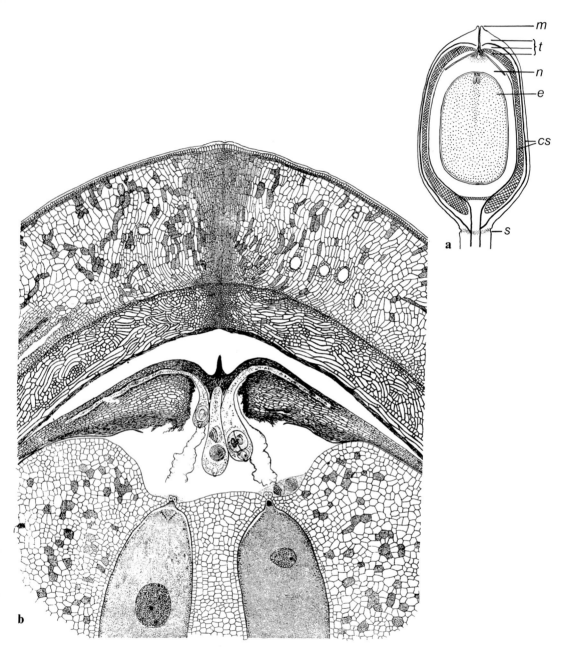

Fig. 1.35a, b. *Dioon edule.* **a** Axial section of an ovule, in approximately true size, after pollination, but 3 months before fertilization (the upper part of the nucellus is still intact); *m* micropyle; *t* integument formed of three layers, the inner one becomes sclerified and contains medial and internal conducting strands *cs; n* nucellus; *s* area of separating cells; *e* endosperm, containing two archegonia. **b** Enlarged upper part of an ovule at the time of fertilization: the upper endosperm is left bare by a lysis of the top of the nucellus. The pollen tubes pour their contents in the cup over the archegonia. Each one encloses an enormous egg, the upper half and the nucleus of which are visible; ×about 20 (after CHAMBERLAIN 1906, 1910)

1.3 Embryogeny of Prephanerogams

Presently in nature, these plants are mainly characterized by the large size of their ovules (Fig. 1.35a) and by their so-called zoidogame mode of fertilization.

The pollen germinates in the pollen chamber, forming a tube: an extremity of the pollen tube which may become branched invades the upper layers of the nucellus, while the other is suspended in the pollen chamber (Fig. 1.35b). Here, ciliated antherozoa elaborated into the pollen tube are released with the liquid of the tube in which they move.

The whole formation, antherozoa with the liquid medium, drains off and collects above the endosperm which, at this time, is exposed by a lysis of the upper part of the nucellus and of the megaspore membrane. An antherozoon penetrates an archegonium and fertilizes the oosphere (Fig. 1.35b). The zygote is a huge ellipsoidal cell (about 3500×1000 µm in *Cycas*) rich in nutritive reserves. In *Ginkgo*, this cell includes numerous lipido-protein inclusions (FAVRE-DUCHARTRE, 1956).

1.3.1 Embryogeny of Cycadales

The zygote nucleus begins a series of divisions without segmentation of the cytoplasm (Fig. 1.36a). These successive divisions are simultaneous and occur eight to ten times (256 to 1024 nuclei).

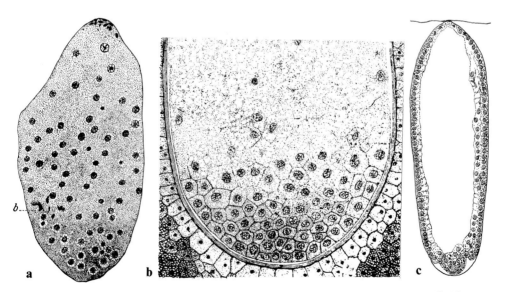

Fig. 1.36a–c. Three aspects of the beginning of the embryogeny of Cycadales. **a** *Zamia floridana;* phase of free multiplication of nuclei; *b* remnants of the spermatozoon blepharoplast (after COULTER and CHAMBERLAIN 1903). **b** *Dioon edule;* accumulation of nuclei and cytoplasmic segmentation at the pole opposite the micropyle (after CHAMBERLAIN 1910). **c** *Cycas circinalis;* peripheral distribution of the cytoplasm and nuclei, around a central vacuole, and appearance of walls (after TREUB 1884)

39

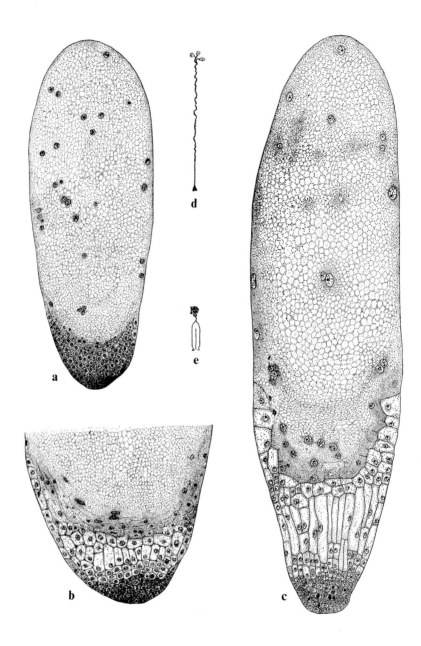

Fig. 1.37 a–e. *Zamia floridana;* development of the proembryo from an isolated cell at the apical pole (opposite the micropyle). **a** Constitution of a homogeneous mass of small cells; **b** beginning of elongation of the suspensor cells; **c** differentiation of nutritive cells (in contact with the anhyst remnants of the zygote), of the suspensor (elongated cells) and of the true embryo (homogeneous mass of small cells becoming meristematic); **d** young embryo and its suspensor, in real size; **e** more advanced embryo, with the suspensor, now coiled into the residual space of the zygote (after COULTER and CHAMBERLAIN 1903)

The development of a large central vacuole shifts the cytoplasm and the nuclei towards the periphery (maybe the most central nuclei degenerate). The cellular walls then appear simultaneously; the whole process resembles a kind of goatskin bottle limited by one or two layers of cells, except for the basis where the "wall" is thicker (Fig. 1.36 b, c). This basis, oppositely to the micropyle, will produce the embryo while the other part degenerates. It is the proembryo. A globular body is initiated with three different parts: (1) the cells located at the edge of the central lacuna of the proembryo form an absorption area, this area being in contact with the nutritive material (Fig. 1.37 a, b); (2) the central part will sometimes extend excessively (up to 70 mm) to form a filamentous and twisted suspensor (Fig. 1.37 c, d); (3) the apical part proliferates and produces the embryo proper (Figs. 1.38, 1.39). Several archegonia of the same ovule can be fertilized. The consequence is the formation of as many suspensors which entangle and push their embryo into the mass of the endosperm. Only one of them subsists and develops. The suspensor coils along, filling the cavity left vacant by achegonium destruction.

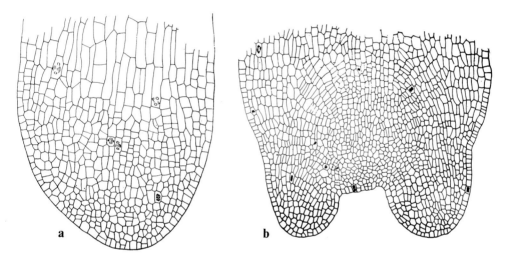

a   **b**

Fig. 1.38 a, b. *Dioon edule;* more advanced embryos. **a** Apical body still undifferentiated, but already composed of a great number of cells; **b** onset of organogenesis, the raising up of cotyledons yields a residual apical material which is particularly abundant (after CHAMBERLAIN 1910)

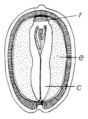

Fig. 1.39. *Dioon edule;* sagittal section of the "seed"; sclerified layers of the integument are *hachured;* embryo is *white; c* cotyledons; *e* endosperm; *r* radicle, covered by the coleorhize (after CHAMBERLAIN 1910)

41

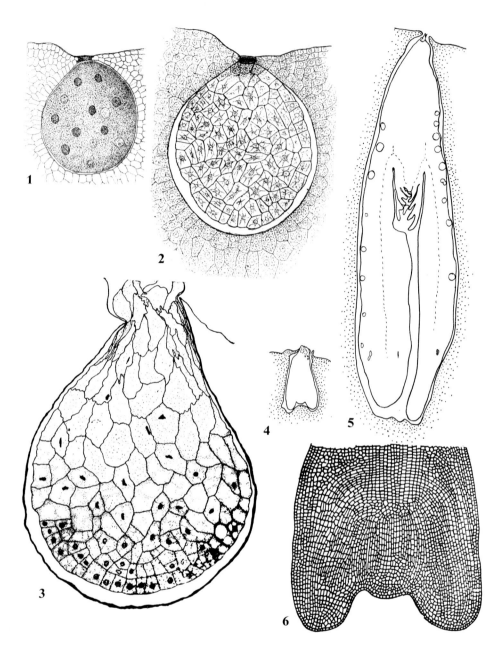

Fig. 1.40. *Ginkgo biloba,* embryogeny. **1** Phase of free divisions of the nuclei; **2** segmentation of the cytoplasm, partitioning the totality of the zygote content; **3** onset of differentiation of the two poles (short multicellular suspensor, small apical cells becoming meristematic, numerous and all similar); **4,5** aspects of the embryo at the beginning and at the end of the plantlet edification, short radicle in the upper region (micropylar), two cotyledons and a gemmule with buttresses of numerous leaves; **6** histological aspect of the apical pole, at the stage of **4**, slow differentiation from very numerous small cells (**1,2** after SRASBURGER 1872; **3** after LYON 1904; **4-6** after COULTER and CHAMBERLAIN 1932)

The embryo proper increases first into a nearly undetermined body (Fig. 1.38). Then, at the basis, it organizes a kind of very particular cover called coleorhize and generally, at the apex, two cotyledons (Figs. 1.38, 1.39).

The first root appears later during germination and goes through the coleorhize which itself has arisen from the sclerified shell of the tegument (Fig. 1.39).

1.3.2 Embryogeny of Ginkgoales

It presents some similarities with that of Cycadales but because of a difference, this development was long considered as exceptional among all Gymnosperms: there is no long and thin suspensor and the middle part extends only moderately (Fig. 1.40/3). Then the whole body of the zygote divides into uninucleated cells (Fig. 1.40/1,3).

Here again, it is only the apical part of the proembryo which produces the embryo proper. First, it forms a cylindrical body of small cells where growth and proliferation are organized little by little (Fig. 1.40). At the extremities of this cylinder, the proliferation starts to localize, arranging two sites of growth, one for the stem and the other for the root. Then the primordia of both cotyledons (often three) rise (Fig. 1.40/6).

1.3.3 Conclusions

The most characteristic feature of the embryogeny of these plants is the hazardous proliferation of the zygote, first scattering numerous free nuclei into its cytoplasm, then producing a massive embryo from its internal pole only, in which the geometrical analysis is impossible. Contrarily to that which occurred in Filicales, for instance, the primordia of the organs are immediately constituted of numerous cells and it is impossible to recognize differentiated initials.

The delay of organization, compared with proliferation, suggests here, better than elsewhere, that morphogenesis is epigenetic rather than predetermined.

The *cytological study* of this development seems very interesting a priori. The zygote as well as the oosphere are huge cells, as different as possible from the meristematic cells which originate from them. In fact, more than 500 to 1000 cells are produced in the initial zygote volume by segmentation. The protoplasmic modifications from the zygote to the embryonic cells then to the primary meristematic cells are worth a cytological and ontological study.

Only the cytological process of the mature oosphere of *Ginkgo* and therefore of the zygote during fertilization was studied with light microscopy by FAVRE-DU-CHARTRE and with electron microscopy by CAMEFORT (1965). FAVRE-DUCHARTRE (1956) considered the dense inclusions of the cytoplasm of this oosphere as protein and phospholipid-rich vacuoles, i.e. as paraplasmic storage material, analogous to the *vitellus* of animal ovocytes. CAMEFORT's works (1965) showed that these "vitelline bodies" are cytoplasm elements enclosed in an envelope issued from the endoplasmic reticulum and composed of a double membrane (Plates 1.2, 1.3). Thus, this is neither paraplasm nor true vitellus.

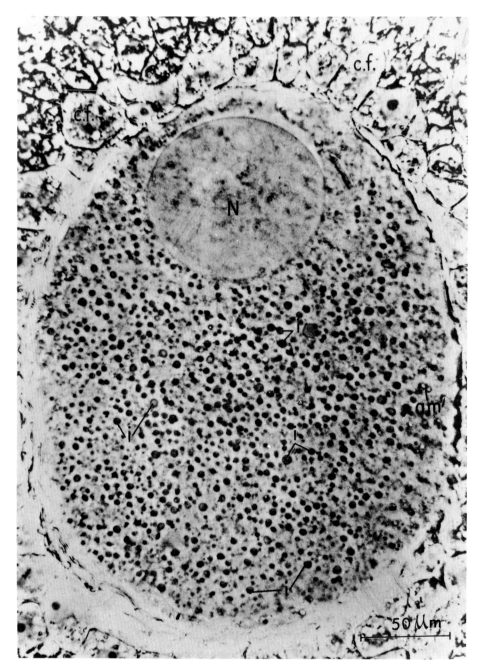

Plate 1.2. Young oosphere of *Ginkgo*

Longitudinal semi-thin section, observed with phase contrast microscope. *N* nucleus; *am* amyloplast; *i* dense inclusions; ×470 (after CAMEFORT 1965)

Plate 1.3. Legend see p. 46

44

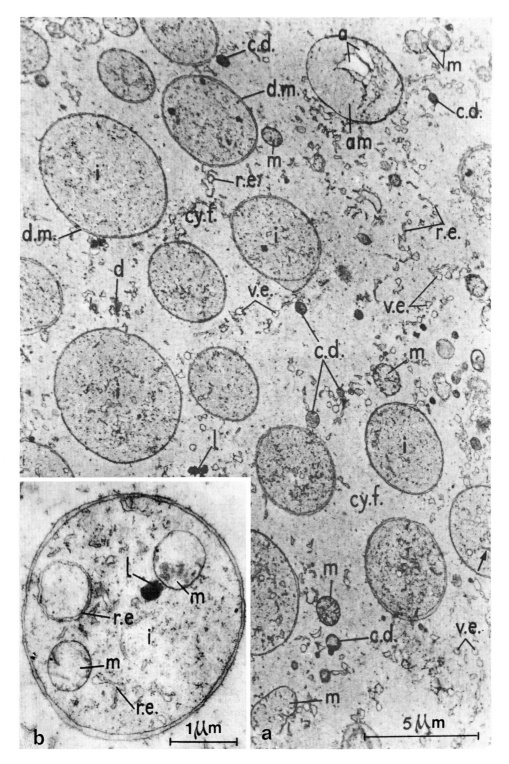

45

Plate 1.3. Oosphere of *Ginkgo,* deep cytoplasm.

a General view and **b** detail of an inclusion *(inset);* dense inclusions *i,* enclosed by a double membrane *dm,* clearly distinct from the amyloplast *am,* and containing cytomembrane structures, probably endoplasmic reticulum, consequently cytoplasmic in nature; *m* mitochondria; *d* dictyosomes; *ve* endoplasmic vesicles; *re* endoplasmic reticulum; *cd* dense bodies; *l* lipids; *cyf* basic cytoplasm; × 6350; (after CAMEFORT 1965; author's original lettering)

Plate 1.4a,b. Plastid evolution in the *Ginkgo* oosphere.

a Completion of a crista *ld:* double lamella, proceeding from the inner membrane, and dividing the plastidal stroma into two compartments. At the upper part of the figure, the outer membrane has torn apart and this zone is now only limited by the inner membrane (*ms* single membrane); *sf* fundamental substance (stroma); *cyf* cytoplasm; *dm* plastid double membrane; *a* starch; × 15000. **b** Separation of the two parts of a plastid, after the breaking of the outer membrane; the *arrow* indicates a persistent site of this membrane, maintaining the contact between the two parts of the plastid; × 19000.

(After CAMEFORT 1965; author's original lettering)

46

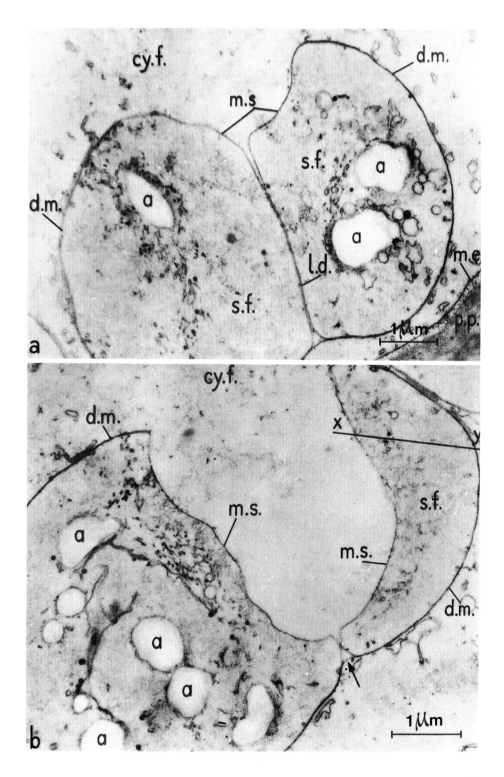

47

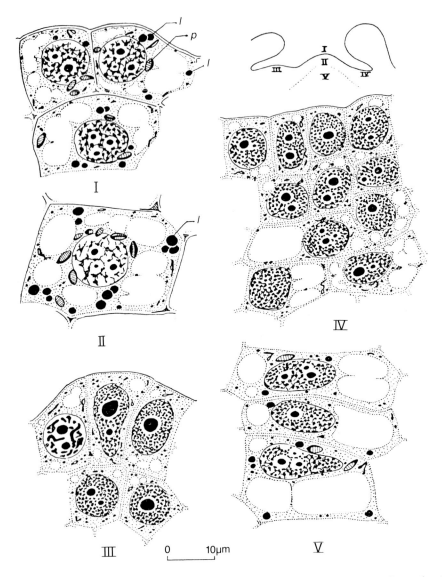

Fig. 1.41. *Ginkgo biloba*, light microscope cytology of the regions of the apical meristem showing the completion of dedifferentiation processes performed during the end of embryology. A gradient of meristematization is obvious from *zone I* to *IV* where the initiating ring initiates the leaves. Fixation: Meves-hematoxylin (after CAMEFORT 1956)

Moreover, CAMEFORT (1965) described a very peculiar evolution of the amyliferous leucoplasts of this oosphere. These bodies, often surrounded by a more or less continuous layer of endoplasmic reticulum, divide with a quite particular process. Cristae issued from the internal membrane of the plastid share the stroma in two parts (Plate 1.4a), then ruptures of the external membrane opposite the cristae tear the plastid apart in two areas (Plate 1.4b).

The *Ginkgo* oosphere, which becomes the zygote after fertilization, is thus a huge cell presenting pronounced, particular features of differentiation of its cytoplasm and plastids. These features are obviously contrary to the meristematic cell, the characteristics of which originate at the end of embryogeny when organogenesis occurs. The cytological modification of embryonic cells into meristematic cells infers therefore a dedifferentiation process (Fig. 1.41).

1.4 Embryogeny in Gymnosperms

In a strict sense, Gymnosperms are presently reduced to Coniferales. The details of embryogeny are quite varied in this very heterogeneous group but it is still possible to distinguish a preliminary phase in which the proembryo forms and a further phase: the arising of the embryo proper.

1.4.1 The Formation of the Proembryo

The *Pinus laricio* (Pinaceae) can be taken as an example.

Here again, the oosphere is a large elongated cell (about 600×200 μm) with a huge nucleus also, of about 100 μm in diameter, within a nuclear and very sinuous membrane, including a great number of poorly chromophil nucleoli (up to 100). But strongly chromophil hematoxylin-positive bodies similar to the vitellus bodies of animal ovocytes are scattered in the cytoplasm. These bodies were considered as intravacuolar lipidoprotein storage material. They are of two types: (1) numerous small spheric or elongated inclusions about 4 to 5 μm in diameter; (2) large inclusions about 40 to 50 μm which, in electron microscopy, appear isolated in the cytoplasm like vacuolar inclusions. They were described as "proteid vacuoles" (Plate 1.5).

CAMEFORT's works (1959, 1960, 1962) in electron microscopy showed that these inclusions were not vacuoles. The small ones, composed of cytoplasm elements, can keep the usual components but become more and more dense after being partially surrounded by a double membrane issued from the fusion of cytoplasmic vesicles belonging to the smooth endoplasmic reticulum (Plate 1.6 *ip*). The large ones result from a particular and excessive development of plastids which are distorted and encircle the cytoplasm elements partially, while the content loses lamellae and stroma density (Plates 1.6, 1.7).

Thus, the so-called vitellus is in fact constituted by specific and original differentiations of the cytoplasm and plastids of the oosphere. This is again a particularly differentiated cell.

49

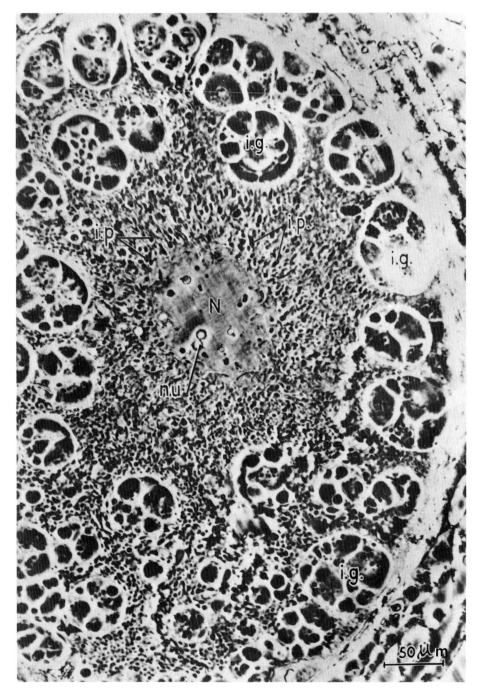

Plate 1.5. *Pinus laricio,* mature oosphere

Semi-thin section observed with a phase contrast microscope. *ig* large inclusions; *ip* small inclusions; *N* nucleus; *nu* one of the numerous nucleoli; ×310 (after CAMEFORT 1962; author's original lettering)

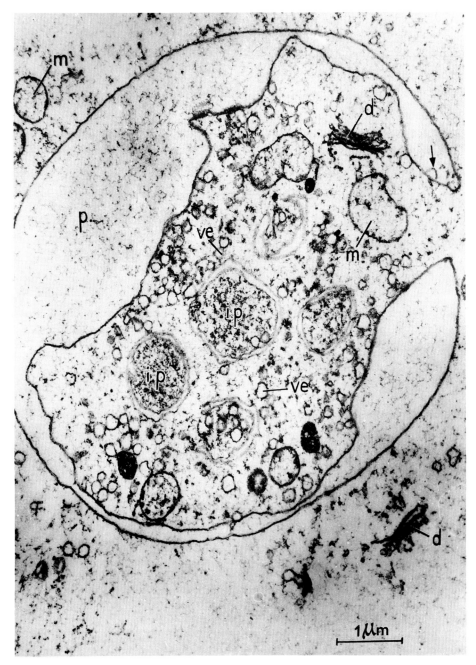

Plate 1.6.

Pinus laricio, oosphere invaginated plastid *p* surrounding a cytoplasmic area in relationship with the fundamental cytoplasm, and developing into "large inclusions"; *d* dictyosome; *ip* "small inclusion"; *m* mitochondrion; *ve* vesicles, some of which merge and form the double envelope of "small inclusions". The *arrow* points out vestigial lamellae; × 17 500 (electron microscopy); (after CAMEFORT 1962; author's original lettering)

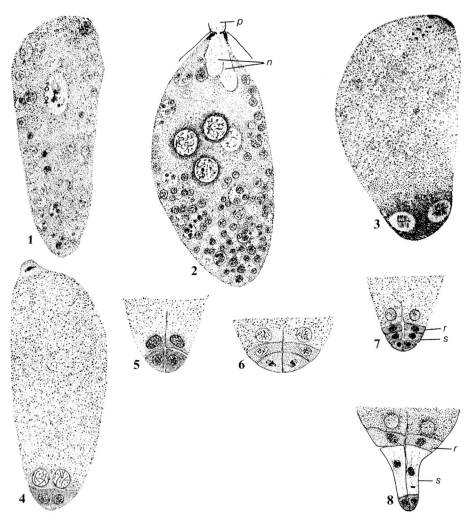

Fig. 1.42. *Pinus laricio,* development of the proembryo. **1** Fertilization of the oosphere at the time of fusion of the male and female nuclei; **2** stage with four nuclei, free in the cytoplasm of the zygote; *n* gaps resulting from the penetration of pollen tube contents *p;* **3** simultaneous mitoses of the four nuclei (only two are visible) after their migration to the apical pole; **4,5** eight nuclei stage; **6** formation of the last proembryonic layers; **7** achieved proembryo; *s* suspensor layer; *r* "rosette" layer; **8** onset of the suspensor elongation (after COULTER and CHAMBERLAIN 1932)

Plate 1.7

Structure of a "large inclusion", actually constituted of a plastid *p,* deeply distorted by several invaginations or branchings which encompass cytoplamic elements containing some cytoplasmic organelles and "small inclusions" *ip.* The contents of these enclaves, which have been compared to vitellus, are therefore composed of cytoplasm. The plastid stroma has become homogeneous and transparent, hence unrecognizable. The *double arrows* point out isthmuses by which the enclosed cytoplasm communicates with the fundamental cytoplasm of the oosphere. *l* Vestigial plastidal lamella. The *single arrows* indicate secondary invaginations of the plastidal body; *m* mitochodrion; *d* dictyosome; ×5000 (electron microscopy); (after CAMEFORT 1962)

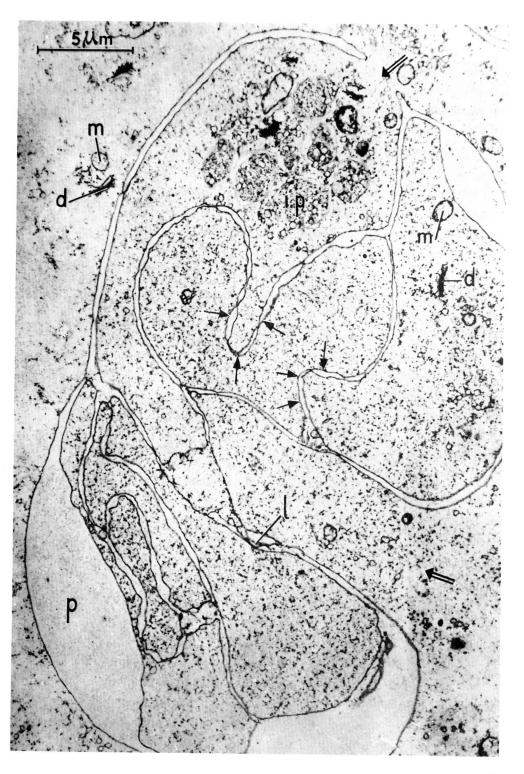

53

After fertilization, which is siphonogamic, a metaphase process occurs in the neighbouring area where the spermatozoon penetrates the nucleus. This process fills only a small part of the nucleus and the mitosis goes on while the nuclear membrane disappears. Two daughter nuclei form from a small part of nucleoplasm and a greater part is left out around them. Each nucleus divides again resulting in four nuclei, free in the former nucleoplasm substance of the oosphere (Fig. 1.42/2).

This set of four nuclei and the former nucleoplasm which remains goes down to the basis of the zygote, to the opposite pole of the micropyle and the four daughter nuclei fit in the same transversal plane (Fig. 1.42/3). They divide simultaneously producing two planes of four nuclei each. Then vertical and horizontal walls separate four lower cells (apical) leaving above four other cells opened on the side of the nutritive material of the zygote (Fig. 1.42/4, 5).

Each plane divides again by simultaneous division of the four cells resulting in four layers of four cells, the upper four cells being opened towards the zygote body (Fig. 1.42/6, 7).

The accurate cyto-enzymatic and ultrastructural cytological study on this onset of embryogeny has revealed noticeable features. It would be interesting to resume the observation of several examples of embryogeny by means of new technics of cell biology. H. CAMEFORT together with L. CHESNOY and M. J. THOMAS demonstrated that mainly in *Pinus laricio* (CAMEFORT 1966b) the cytoplasm of the *Pinus* oosphere degenerates after fertilization and that the nucleoplasm, left aside during the formation of the first four nuclei of the proembryo and which surrounds them during their migration towards the basis of the zygote, changes into the essential part of the "proembryonic" neocytoplasm (CAMEFORT 1958, 1966a).

With several examples, mainly in Cupressaceae, CHESNOY (1969, 1977b) demonstrated a surprising feature: mitochondria and plastids which enter the proembryo cells are probably derived exclusively from the *male gamete* (Plates 1.8, 1.9), the corresponding bodies of the oosphere degenerate with the cytoplasm. This patrocline inheritance is most interesting in the field of plastids, the mutation of which has phenotypic consequences (ornamental varieties of cultivated Cupressaceae).

The embryo proper of *Pinus laricio* originates from the lower stage (apical); the four cells of the stage just above extend to form the suspensor (Fig. 1.42/8); the cells of the third stage constitute a zone of trophic cells, the "rosette" in contact with the residual nutritive material. The open upper cells disappear after some time.

The details of the proembryo formation vary according to families and even to genera. The description of *Pinus* can be applied to Abietaceae.

Generally, in *Taxodiaceae,* the proembryo presents only three layers of cells instead of four and no rosette, this is also the case in Cupressaceae, but the four embryonic cells are sometimes arranged in a tetrahedron and not in a plane, which modifies their behaviour as the embryo proper develops.

The *Sequoia sempervirens* is an exceptional case: the first division of the zygote nucleus is followed by the formation of a wall; there is no phase of free division of the nuclei.

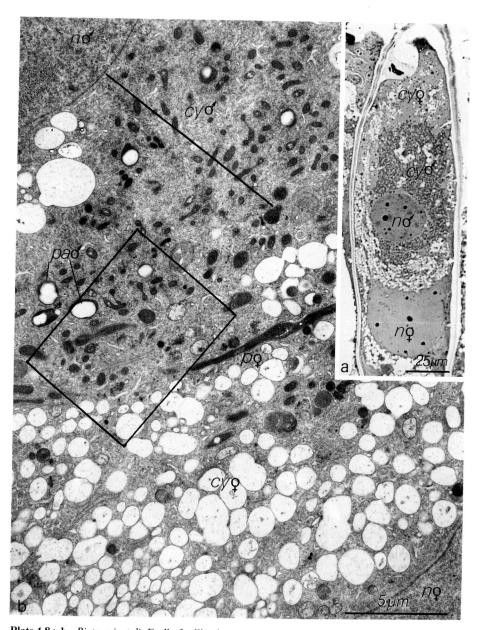

Plate 1.8 a, b. *Biota orientalis* Endl., fertilization.

a The relatively dense male gamete (♂) including the nucleus (*n* ♂) and the cytoplasm around it (*cy* ♂) has penetrated into the oosphere, the lighter cytoplasm of which (*cy* ♀) surrounds the male gamete; × 440; semi-thin section, hematoxylin stained, electron microscopy.

b Detail of *a,* area situated between the two nuclei *n* ♂ and *n* ♀. Differences between male and female cytoplasms, *cy* ♂ and *cy* ♀, and of the male plastids *pa* ♀, (amyliferous plastids), and the female plastids *p* ♀ (without starch). The thickness of the male cytoplasm is indicated by the *bar;* × 5500; electron microscopy; fixation: glutaraldehyde-O$_s$O$_4$; contrast; KMnO$_4$

(After CHESNOY 1977 a)

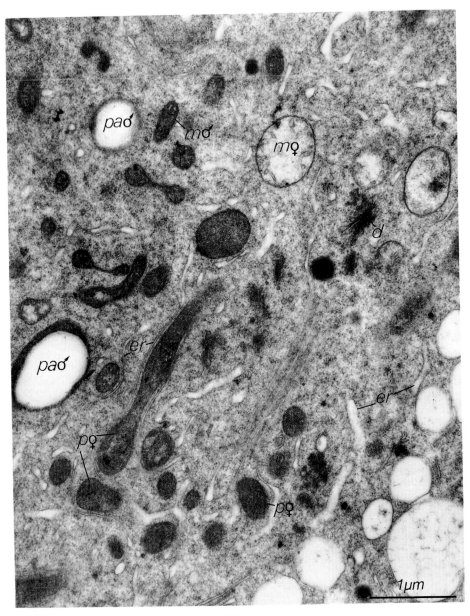

Plate 1.9. Enlargement of the framed area of Plate 1.8 b.

The neighbouring male and female organelles are recognizable due to their different aspects; *pa* ♂ and *p* ♀, *m* ♀ and *m* ♀ have not mingled; *er* endoplasmic reticulum; *d* dictyosome; ×23 200; (after CHESNOY 1977 a)

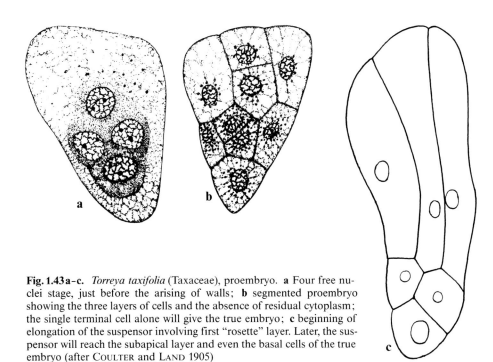

Fig. 1.43 a–c. *Torreya taxifolia* (Taxaceae), proembryo. **a** Four free nuclei stage, just before the arising of walls; **b** segmented proembryo
showing the three layers of cells and the absence of residual cytoplasm;
the single terminal cell alone will give the true embryo; **c** beginning of
elongation of the suspensor involving first "rosette" layer. Later, the suspensor will reach the subapical layer and even the basal cells of the true
embryo (after COULTER and LAND 1905)

In Taxaceae, depending on genera, walls appear after the formation of 4 *(Torreya)*, 8 or 16 nuclei *(Taxus)*. Generally, these walls divide the whole zygote volume, smaller in this case than in the former cases (Fig. 1.43). In such case there are
no "open cells" left; moreover, the stage arrangement of the proembryo cells is
less clear, mainly on the embryonic apical side. Correlatively, the extension which
forms the suspensor reaches all the cellular stages of the proembryo: there is no
"rosette" left and the basal cells of the embryo proper participate in the formation
of the suspensor (Fig. 1.43).

1.4.2 The Formation of the Suspensor and the Embryo

The case of the *Pinus laricio* will serve again as an example (Fig. 1.44).

We left the proembryo with four apical cells, each one covered by a pile of
three other cells. Those in contact with the apical cells extend excessively without
dividing, producing four suspensors which drive the end cells into the endosperm
tissue. Thus, each zygote gives rise to four embryos, but only one will stay in the
mature seed.

The division of the end cell of each element produces a series of small cells
(Fig. 1.44a, b) then by oblique divisions, a massive embryo forms, supplied with a
first root and a three to ten cotyledon crown originate around the apex.

At the beginning, cells of the embryo basis extend, lengthening the suspensor
(Fig. 1.44b). These "embryonic tubes" remain unicellular, like the suspensors, re-

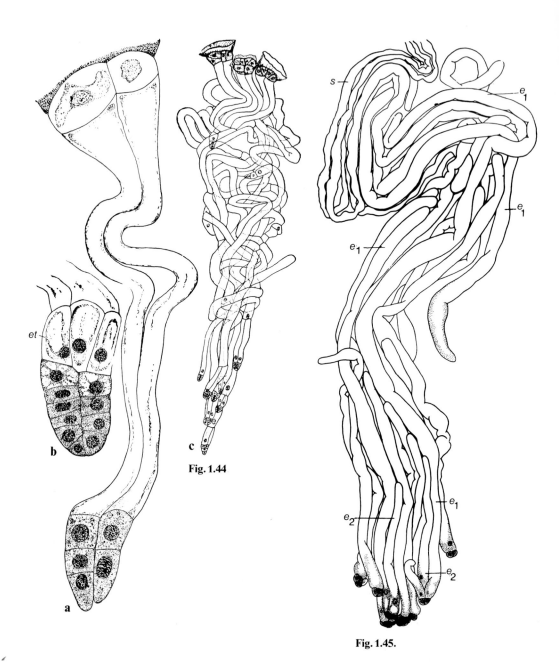

Fig. 1.44. Pinus laricio, development of the embryo. **a** Two very young embryos carried to the end of long suspensors; **b** slightly more advanced embryo; the base has formed three embryonic tubes *et;* **c** intermingling of suspensors pertaining to 12 embryos, sprung from three zygotes born in the same ovule (after COULTER and CHAMBERLAIN 1932)

Fig. 1.45. Sequoia gigantea, extremities of suspensors *s* in an embryonic complex, and embryonic tube bundles, e_1, e_2 successively formed from terminal proembryonic cells (after BUCHHOLZ 1939a,b)

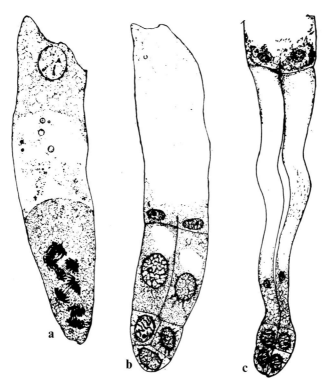

Fig. 1.46a–c. *Thuja occidentalis,* beginning of embryogenesis. **a** Last mitoses giving rise to the eight free nuclei; **b** segmentation of the zygote into three embryonic layers. The cells of the layer opposite the true embryo remain "open cells", the four cells of the embryonic layer are disposed in the form of a tetrahedron, but not in a plane; **c** suspensor lengthening, at the extremity of which only one embryo is formed (after LAND 1902)

gardless of their length. They can become very long and grow in many successive series, as in the *Sequoia gigantea,* studied by BUCHHOLZ (1939a, b) (Fig. 1.45). In extending, the suspensors coil all together into the cavity left by the archegonium (Fig. 1.45).

The embryo formation also varies extensively according to families and genera of Coniferales. For instance, in *Taxodium* (Taxodiaceae) the number of suspensor cells may be different from that of the end tier. The suspensors can either deliver a variable number of embryos or keep the apical cell cohesion, then only one embryo appears.

In certain Cupressaceae like *Thuja* with end cells arranged in a tetrahedron, there is only one embryo formed (Fig. 1.46). It is generally the same in Taxaceae.

We have already mentioned that in some Taxaceae like in *Thuja,* the suspensor includes cells from various stages of the proembryo.

We would also like to mention a particular feature which shows that the geometrical aspect of segmentation has no actual phylogenic significance: in an Araucariacea, *Araucaria brasiliana* (at the exception of the others), and in *Taxaceae* of the *Cephalotaxus* genus, it is the second tier of cells of the proembryo which gives

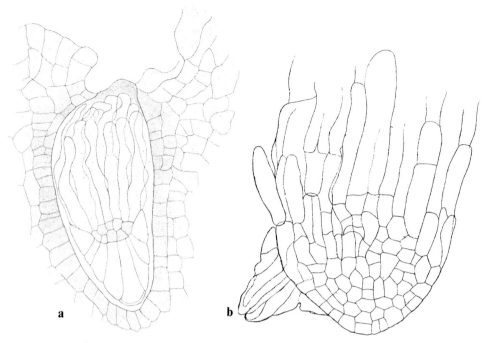

Fig. 1.47a, b. *Araucaria brasiliana.* **a** Proembryo, composed of three layers of cells, and filling the space of the ancient zygote. The middle layer only *(grey)* will produce the embryo; the upper layer becomes a suspensor, the terminal layer edifies a kind of cap. **b** Formation of the true embryo, composed of an apical material of small dedifferentiated cells; the "cap" is pushed aside by the suspensor (after STRASBURGER 1879)

rise to the embryo. The apical tier is only a provisional type of cap. The opposite tier, usually producing the rosette, initiates the suspensor (Fig. 1.47). All these variations are less interesting than the chronological process of proliferation related to organogenesis.

The number of cotyledons growing on the plantlet can also vary; in the same species of *Pinus* or *Abies* more than 12 cotyledons can be numbered, while on certain species this number is both reduced and constant. Two cotyledons are regularly found in various Taxodiaceae (e.g. *Sequoia sempervirens*), more in others (three to six in *Sequoia gigantea*).

Most *Thuja* have two cotyledons, whereas there are two to five of them in various species of *Cupressus* and *Juniperus*.

1.4.3 Conclusions

The variations in the details of embryology of Coniferales show once again that these characteristics have no general interest.

Conversely, it should be remarked that the embryogeny of Coniferales is globally less undetermined at the onset than that of Prephanerogams. The free multiplication phase of the nuclei is more reduced, even inexistent in some species *(Sequoia sempervirens)*.

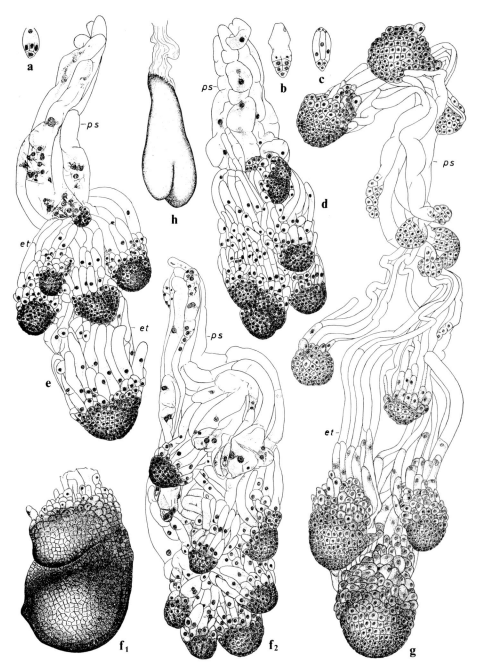

Fig. 1.48 a–h. *Torreya californica.* **a–c** Formation of the proembryo and onset of lengthening of prosuspensors (× about 55); **d–g** complete embryonic systems. The "prosuspensor" *ps*, produced by the elongation of proembryonic cells situated on the micropyle side forms, by means of a kind of budding, small-sized adventitious supernumerary embryos. The greatest embryos are borne by secondary suspensors, surrounded by embryonic tubes *et* (× about 55). **h** Embryo dissected from a mature seed (note the two cotyledons and the remnant of the suspensor) (after BUCHHOLZ 1940)

Moreover, embryos are less massive, suggesting that the proliferation begins a slightly earlier to produce the first organs of the young sporophyte with less cellular material. However, the development remains undefined. A characteristic of this embryogeny is the frequent polyembryony. At first, several archegonia of the same ovule can be fertilized and produce several proembryos, but in addition several embryos (often four) arise from the same proembryo. Thus, in the endosperm, an acute competition occurs between numerous embryos of which one only remains. This feature, a frequent waste in nature, is of interest in showing that the proembryo cells of the "embryonic stage" are still equivalent and omnipotent. In *Taxodium*, for instance, the apical material can remain as such or be divided in a variable number of areas, each of them producing an embryo, although only one remains. Such an indetermination as well as the undefined tendency to polyembryony is again demonstrated by a fact quoted several times by embryologists. In Taxaceae, where a single embryo usually forms, secondary embryos can arise through a kind of budding, either from the body of the embryo itself *(Cephalotaxus)* or from a suspensor cell *(Podocarpus, Torreya)* (Fig. 1.48).

When the whole zygote volume divides, the problem of the distribution of the large cytoplasmic material remains to be studied. In all cases, the primordial cell, the zygote, is a huge cell rich in substances which, even parts of the cytoplasm, provide the embryonic cells with nutritive material. It produces embryonic cells all different from itself, then meristematic, commonly paraplasm-poor cells.

The above mentioned ultrastructural studies of CAMEFORT and co-workers revealed an original decanting process achieved through the modification of nucleoplasm into neocytoplasm and the waste represented by the only trophic use of the living cytoplasmic material of the oosphere.

Later, CAMEFORT and colleagues reported further cytological positive data. For instance, the works of THOMAS (1973, 1978) on *Pinus* elucidated the process of formation of a new cell wall specific to the proembryo, first elaborated against the oosphere wall from which it separates thereafter. In *Cryptomeria japonica* (CAMEFORT 1970), the proembryonic wall is immediately separated from that of the oosphere.

From the first divisions of the proembryo, the process of dedifferentiation increases in end cells which separate, producing one embryo each, while the primary and secondary suspensor cells, excessively elongated, differentiate. These cells seem to have not only a mechanic function but also seem to participate in the trophic supplies meant for the embryo, taken from the remaining oosphere and in the endosperm (cf. THOMAS 1980). After experiments with an in vitro culture of *Pinus* embryos, THOMAS (1972a, b) noticed that the embryonic division was implemented by the maintenance of the cohesion of the proembryo with the suspensor cells.

Plate 1.10 A, B. *Pinus halepensis,* embryos.

A Young embryo at the onset of cotyledon initiation; the axial apical zone *az* becomes less stained, particularly the nuclei, than the initiums of cotyledons *ic* and hypocotyl buttress *h*. The nuclei of the future radicular centre *rc* are already less stainable; *ec* embryonic cap; *s* suspensor.

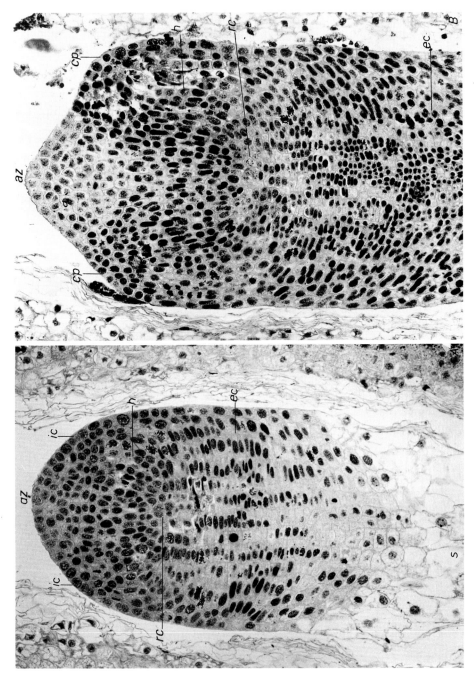

B More advanced embryo; the axial zone *az* becomes more distinct from the cotyledon primordia *cp* and from the hypocotyl *h,* because of its lesser staining. The radicular centre *rc* also becomes more distinct from the hypocotyl *h* and from the embryonic cap *ec* that surrounds it, due to its slight chromophilia.

Fixation; Navaschin-hematoxylin

Thus, like in other vascular plants, the onset of embryogeny in Gymnosperms presents a preliminary dedifferentiation process. Therefore, the cells of globular embryos are uniformly meristematic.

Later, as the primary body of the plantlet grows, only the organogen areas, remain meristematic: cotyledons, hypocotyl, basis of the embryonic cap. Between these areas, i.e. at the cauline end of the embryonic axis and at that of the hypocotyl, a few residual cells left aside differentiate again moderately as shown by pyronine staining of RNA (BRACHET 1940), which fades away in their cytoplasm, as well as hematoxylin staining (Plate 1.10 A, B) which also fades away in the nuclei.

These few residual cells resume their ontogenic activity later, organizing themselves into primary, cauline and root meristems which give rise to the future organs of the plantlet, then of the plant after germination ensuring the continued embryogeny, a characteristic of plants.

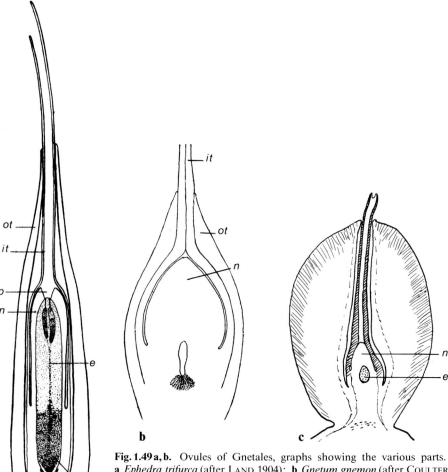

Fig. 1.49 a, b. Ovules of Gnetales, graphs showing the various parts. **a** *Ephedra trifurca* (after LAND 1904); **b** *Gnetum gnemon* (after COULTER 1908); **c** *Welwitschia mirabilis* (after STRASBURGER 1872). *e* Female prothallus (endosperm); *n* nucellus; *p* pollen chamber; *r* accumulation region of endosperm reserves; *ot* outer integument; *it* inner integument

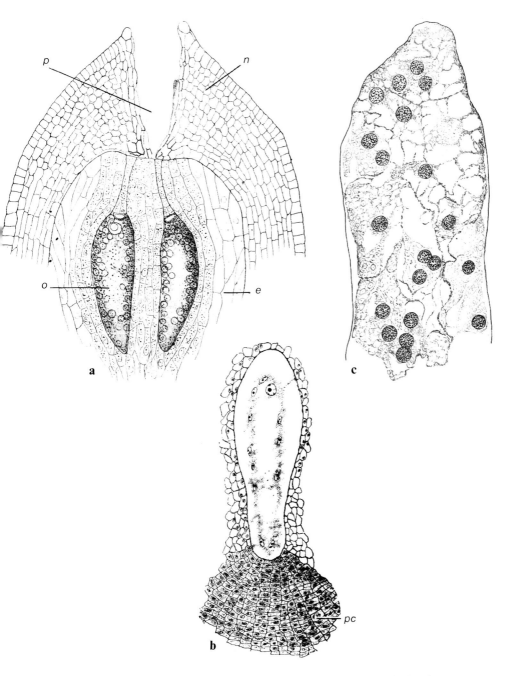

Fig. 1.50 a-c. Ovules of Gnetales, structure of the female prothallus. **a** *Ephedra altissima,* upper part of the ovule; *e* cellular endosperm with thin cell walls, two archegonia with bulky oosphere *o; p* pollen chamber; *n* nucellus (after STRASBURGER 1872). **b** *Gnetum gnemon,* coenocytic "embryo sac", no archegonia; *pc* pavement cells at the bottom of nucellus (after COULTER 1908). **c** *Welwitschia mirabilis,* upper part of the endosperm, partitioned into multinucleated elements; no archegonia (after PEARSON 1909)

1.5 Embryogeny of Preangiosperms

1.5.1 Fertilization in Gnetales

We shall deal with a few data relative to Gnetales to show the particular features of embryogeny of these plants in which Gymnosperm and Angiosperm characteristics are mixed. Because of the particularities of these plants, it is necessary to recall the processes of their fertilization.

The mature ovule encloses an "endosperm" or ♀ gametophyte, the structure of which varies according to genera. On maturity, in *Ephedra,* the endosperm is composed of a great number of cells separated by thin but real walls and encloses two or three archegonia, the long and irregular canal of which is followed by a voluminous and elongated oosphere (Fig. 1.49 a, 1.50 a).

In *Welwitschia* (Fig. 1.49 c, 1.50 c) the endosperm is divided by thin walls enclosing plurinucleated sets. The units located towards the micropyle have nuclei which can be ferlilized, and they extend into the nucellus, with nuclei penetrating the extension. There are no archegonia (Fig. 1.51).

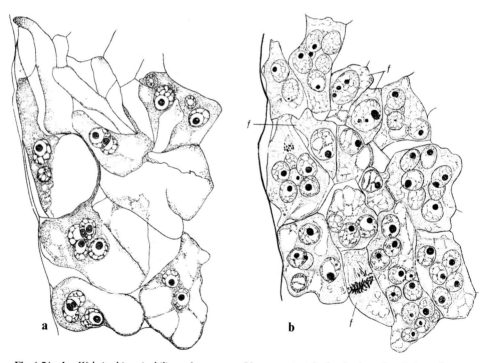

Fig. 1.51 a, b. *Welwitschia mirabilis,* endosperm. **a** Upper part at the beginning of emission of prothallus tubes. These tubes will penetrate into the nucellus and meet pollen tubes; **b** lower part of the same endosperm. Into the multinucleate compartments, nuclear fusions give rise to diploid cells *f* that resemble the albumen cells of Angiosperms (after PEARSON 1909)

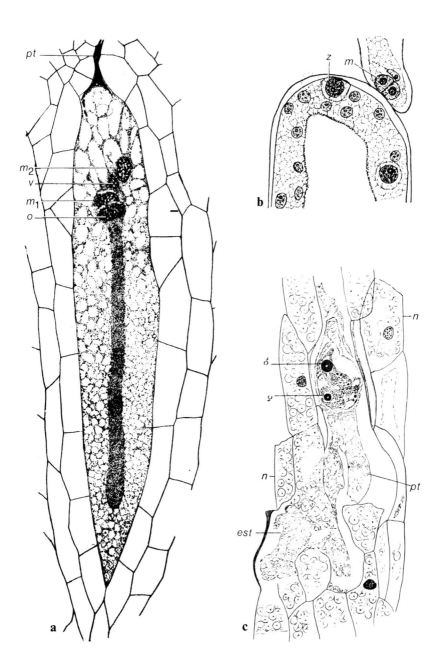

Fig. 1.52 a–c. Fertilization in the Gnetales. **a** *Ephedra trifurca;* *pt* remnants of the pollen tube in the archegonium canal; *o* oosphere nucleus and m^1 male nucleus, in the way of fusion; *v* ventral nucleus of the pollen tube; m^2 second male nucleus, similarly released in the archegonium (after LAND 1907). **b** *Gnetum gnemon;* top of the embryo sac, close to which the pollen tube comes; *m* male nuclei; *z* zygote (after THOMPSON 1916). **c** *Welwitschia mirabilis;* meeting of a pollen tube *pt* and a prothallus tube *est* in the nucellus *n* and fusion of a pollen nucleus (♂) and a prothallus nucleus (♀) (after PEARSON 1909)

The endosperm of *Gnetum* (Fig. 1.49 b, 1.50 b) is a "sac": the centre is formed by a large vacuole and the periphery by a layer of cytoplasm where numerous free nuclei are distributed. Here again there are no archegonia, but the coenocytic structure can be compared to the embryo sac of Angiosperms. However, *each nucleus* may be fertilized.

The pollen grain from Gnetales has usually a greater number of cells than that of Gymnosperms. At the end of the pollen tube, two ♂ nuclei form, next to non-reproductive nuclei.

In *Ephedra,* this tube goes into the canal of an archegonium and releases the whole content of its extremity in the oosphere. One ♂ nucleus only merges with the oosphere: the fertilization results in a single zygote per archegonium, but all the archegonia (two to three) of the same ovule can be fertilized (Fig. 1.52 a).

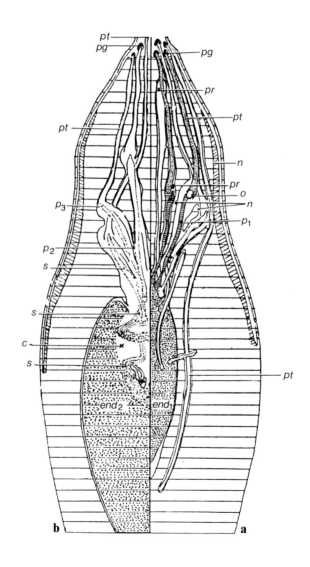

In *Gnetum* (Fig. 1.52b) pollen tubes, often several on the same ovule, release two ♂ nuclei each in the embryo sac. Other non-reproductive nuclei along with extranuclear material may also be introduced in the macroprothallus. Each ♂ nucleus can give rise to a zygote, so each pollen tube can generate two fertilizations, but both zygotes are *equivalent a priori:* this "double fertilization" is therefore different from that which characterizes Angiosperms.

In *Welwitschia* (Fig. 1.52c) fertilization occurs by anastomosis between pollen tubes and tubes extended into the nucellus by the ♀ prothallus. Besides, a "double fertilization" may occur from each pollen tube resulting in two zygotes *a priori equivalent,* so it is again different from the double fertilization of Angiosperms (Fig. 1.53).

1.5.2 Embryogeny of *Ephedra* (Fig. 1.54)

The three living genera of Gnetales present so many embryonic differences that they should be considered separately.

In *Ephedra,* the details of embryogeny were thoroughly described by STRASBURGER already in 1872. The zygote nucleus first divides without cell walls with up to eight nuclei. Two to eight independent cells arise then in the zygote volume. Later, each will function alone (Fig. 1.54). Each extends into a suspensor which penetrates into the endosperm. One mitosis separates one embryo cell at the extremity of the suspensor (Fig. 1.54/2). The embryo rises from this last cell, as it occurs at the extremity of the *Pinus* suspensor cells (Fig. 1.55).

The nuclei of the cells surrounding the former oosphere, or tapetum cells, usually undergo a mitotic or amitotic division, then the membranes disappear leaving the content of these cells to mix with the cytoplasm of the zygote.

The ♂ nucleus left aside grows to form a large cell on the upper part of the former archegonium. When the cell degenerates, its membrane fades and the nucleus divides (maybe after the fusion of the tapetum nuclei) producing a multiple and irregular set of spindles. Then in this apical area a body of small cells appears which have been compared with the albumen of Angiosperms (Fig. 1.56).

While the suspensor cell extends exceedingly (up to 3000 µm) the distal cell thus forced into endosperm form the embryo. It first divides transversally, then oblique cleavages give rise to an ovoid proembryo (Fig. 1.55d). The basal cells of this proembryo also grow in length, producing a multicellular secondary suspensor. The first root initiates at the end and into the body of the secondary suspensor. Several embryos can develop in the same ovule, but only one survives.

◁——————————————————————————————

Fig. 1.53a, b. *Welwitschia mirabilis;* schematic drawing of two half-ovules. **a** At the moment of fertilization: junction of pollen tubes *pt* and prothallus tubes *p'*; **b** beginning of embryogenesis; *pg* pollen grains; *n* nucellus (horizontal lines); *end'*, *end²* endosperm, before and after fertilization; *c* cavity resulting from fertilization at the upper part of the endosperm; *p'*, *p²* etc. canaliculi formed by the prothallus tubes *pr (striped);* *s* suspensors, ending with embryos and their embryonic tubes; *o* zygote; *pt* pollen tubes *(dotted)* (after PEARSON 1909)

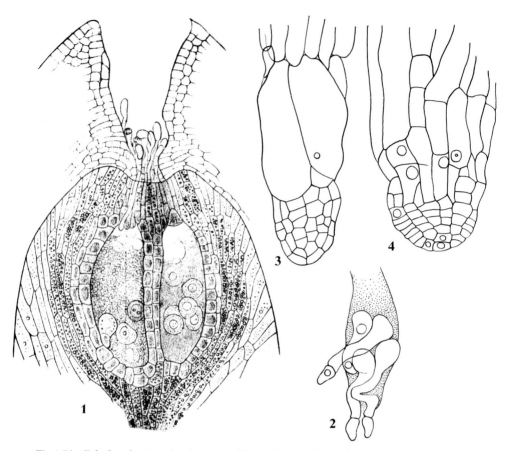

Fig. 1.54. *Ephedra altissima;* development of the embryo. **1** Formation of proembryonic cells following the phase of nuclear-free divisions (six to eight free nuclei); **2** elongation of proembryos and separation of the suspensor from the embryonic cell, which is pushed out of the zygote volume into the endosperm; **3,4** two successive aspects of the completion of the apical meristematic body. This apical body will become organogenic and will build the true embryo (after STRASBURGER 1879)

▷

Fig. 1.55a–d. *Ephedra trifurca;* development of the embryo. **a,b** Separation of the suspensor *s* from the true embryonic cell *e; extension of the suspensor;* **c** end of the first mitosis of the embryonic cell; **d** formation of the still unorganized embryonic body at the suspensor extremity (after LAND 1907)

Fig. 1.56a,b. *Ephedra trifurca;* pseudo-albumen, upper part of the zygote after formation of the free proembryonic cells *(p).* **a** Enlargment of the second male nucleus m^2 the chromatin of which spreads in small groups and mingles with the remnants of degenerating endosperm cells (tapetum cells); **b** formation of spindles and constitution of anarchic cells, with a nutritive function, comparable to the albumen of Angiosperms (after LAND 1907)

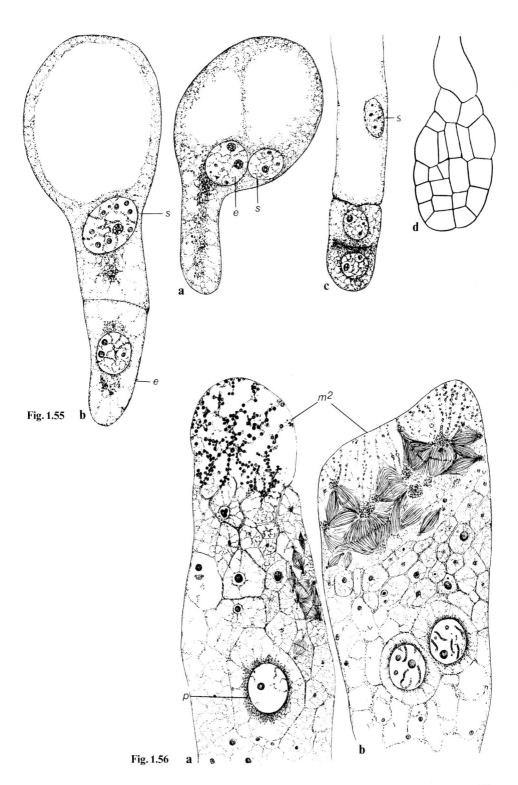

Fig. 1.55

Fig. 1.56

71

1.5.3 Embryogeny of *Welwitschia* (Fig. 1.57)

Fertilization occurs in the tube issued from the ♀ prothallus. There is a nuclear division of the zygote but the wall separating both daughter nuclei forms later. Each zygote acts more or less as one of the *Ephedra* proembryonic cells. It extends exceedingly while the distal cell gives rise to the embryo (Fig. 1.57/1). It produces a more or less conic embryo the basal cells of which extend into "embryonic tubes" around the suspensor extremity (Fig. 1.57). Embryonic tubes have the same origin as the *Ephedra* secondary suspensor.

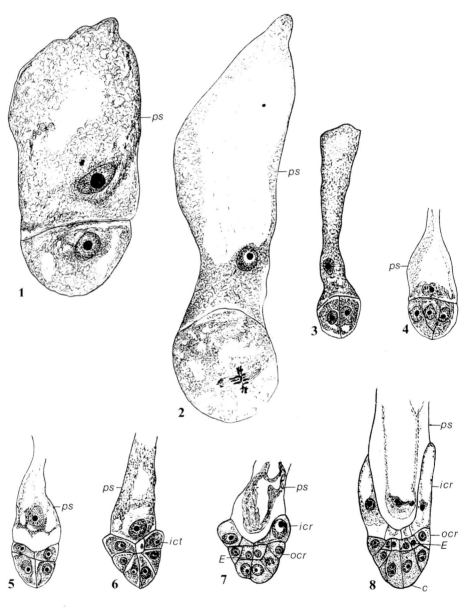

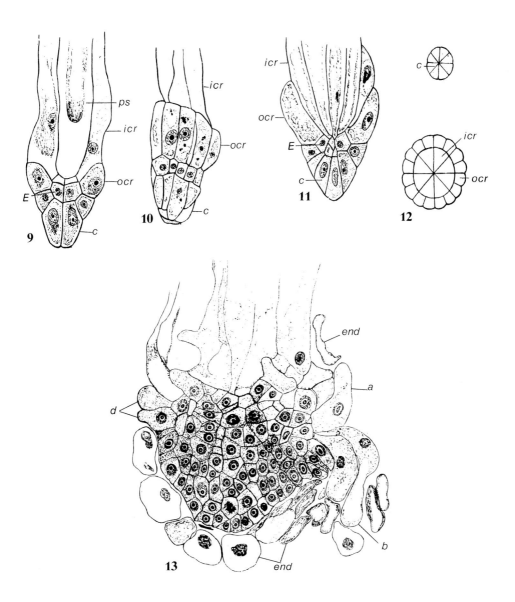

Fig. 1.57. Development of *Welwitschia mirabilis*. **1** First cleavage of the zygote; *ps* primary suspensor; **2** lengthening of the primary suspensor and mitosis of the proembryonic cell; **3** two-celled proembryo; **4-6** edification of the cone-shaped proembryo composed of undifferentiated cells; **7-9** development of inner embryonic tubes (inner cortical ring *icr*) surrounding the primary suspensor extremity by elongation of the basal proembryo cells; **10, 11** development of outer embryonic tubes (outer cortical ring *ocr*) by elongation of new basal proembryo cells. During this time, the extremity of the proembryo builds a kind of cap *c,* the destiny of which is not well known, but it seems that the true embryo *E* is only arising from underlying cells; **12** transversal sections into the cap and at the extremity of the secondary suspensor; **13** formation of the undifferentiated meristematic body from which the embryo will be born; *a,b,d* suspensor cells; *end* endosperm cells (after PEARSON 1909)

The embryo differentiates a radicle covered by a voluminous coleorhize, a rather short hypocotyl and two cotyledons (Fig. 1.58). Two protuberances appear near the apical meristem and later, during germination, a kind of scutellum develops laterally at the hypocotyl basis and penetrates the endosperm in the course of lysis (MARTENS and WATERKEYN (1964).

The development of reproductive organs and the embryogeny of *Welwitschia mirabilis* were thoroughly studied in largely illustrated publications by MARTENS and WATERKEYN (mainly 1964, 1974, 1975).

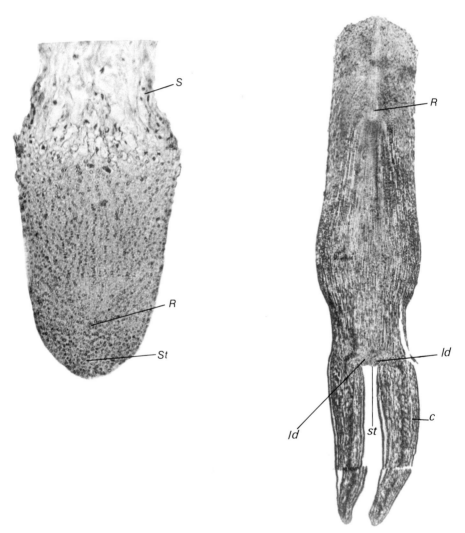

Fig. 1.58. *Welwitschia mirabilis;* end of edification of the embryo; an abundant cellular material is used for the organogenesis; *R* radicle; *st* apex where two lateral domes *ld* were born; *c* cotyledon (after PEARSON 1910)

74

1.5.4 Embryogeny of *Gnetum*

The zygote develops differently depending on the species (HAINING). Most often the zygote extends into spindles or primary suspensors coiling down in the embryo sac (Fig. 1.59/1-4). Meanwhile the nucleus divides and daughter nuclei go into the primary suspensors.

At the end of each branch, a transversal wall separates a cell which proliferates and forms a secondary suspensor on which several embryos can rise (Fig. 1.59/5).

In *Gnetum gnemon,* studied by BOWER (1882) and COULTER (1908) and in *Gnetum funiculare* (HAINING 1920) the zygote extensions penetrate the endosperm separately. Moreover, in *Gnetum gnemon* the terminal cells of branches produce embryos without secondary suspensors. In all cases, proembryonic distal cells are particularly rich in paraplasmic material, mainly starch. In some cases, a new phase of free nuclear divisions indicates the onset of proliferation.

In proliferating, these cells produce an unorganized body of small cells. Those located at the basis differentiate later into "embryonic tubes", extending as in *Welwitschia,* while the others accentuate their meristematic characteristics (Fig. 1.59/6) and later organize a radicle, a long hypocotyl and two tiny cotyledons. Situated between the radicle and hypocotyl, a cylindrical "foot" acts as a sucker, like the *Welwitschia* spatulate overgrowth (Fig. 1.59/7).

1.5.5 Conclusions

The embryogeny of Gnetales presents variations and peculiarities as interesting as the other characteristics of these plants. Despite the differences to the embryogeny of Coniferales, in the three genera, details separate them from Angiosperms. The most notable is the free division of nuclei when the zygote begins to develop. In *Welwitschia* only two free nuclei are constituted and soon separated by a wall; in *Ephedra* there are often eight nuclei which do not always develop into proembryonic cells; in *Gnetum* there are two phases of free nuclear divisions: one from the zygote as it extends into a suspensor at the end of which the proembryonic cell separates, and the other when this cell begins to divide, the young embryo first being coenocytic.

These features again express the lack of predetermination at the onset of development. This is also true of polyembryony *(Ephedra, Gnetum).* Later, an undifferentiated embryo appears with no organized segmentations. Organogenesis occurs only slowly.

The strange morphological aspect of development of these plants held the attention of embryologists more than the cytological evolution. However, the authors' illustrations are very interesting for inciting the study of cytological evolution.

Gnetales were classified by EMBERGER with primitive Angiosperms (Piperales, Juglandales) into a particular group: the Preangiosperms or Chlamydosperms. This classification is justified by morphological features related to female reproductive systems.

From the embryologic standpoint, the development of the Piperales and Juglandales is more similar to that of the other Angiosperms than to that of Gne-

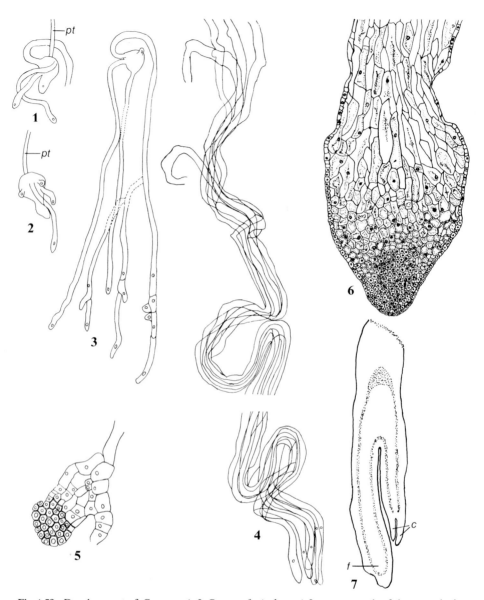

Fig. 1.59. Development of *Gnetum*. **1-3** *Gnetum funiculare:* **1,2** zygotes, each of them producing several suspensors; *pt* pollen tube; **3** reconstitution of the embryonic system proceeding from a zygote, and comprising several branched suspensors. These suspensors force themselves separately into the endosperm or find their way at its side. Embryos will rise from cells which are isolated by a cell wall at the end of ramifications. **4** *Gnetum sp.29;* bundle of suspensors intermingling in the volume of the ancient zygote (actual length: 23 mm). **5** *Gnetum sp.59;* formation of a still unorganized body of small cells at the extremity of a secondary suspensor. **6** *Gnetum sp.15;* embryo more advanced than that shown in *5,* but still unorganized; accentuation of the meristematic aspect of the apex which is composed of numerous cells. **7** *Gnetum sp.15;* section of a nearly completed embryo showing, against the true embryo axis, a foot, which will assume the nutritive function of a scutellum at the time of germination; *c* the two cotyledons; *f* foot (after HAINING 1920)

tales. The phase of free nuclear divisions disappears, but, however, the embryogeny of certain species such as the *Juglans regia* presents individual variations which obviously express a lack of rigorous predetermination. The first segmentation of the zygote in Piperales is longitudinal, a characteristic which can be found in some other Angiosperms only. There is no formation of a suspensor and here again the segmentation does not show any organization disclosing a possible predetermination.

1.6 Embryogeny of Angiosperms

In addition to the double fertilization mentioned above, Angiosperms are characterized by the ♀ prothallium reduction to the few cells of the embryonic sac (often eight), and by a decrease of the cellular material necessary to the production of the first organs of the embryo. In this respect, the precocity of organogenesis and the small size of embryos allow the comparison with modern ferns.

This reduction can explain the more or less regular geometrical aspects of segmentation, allowing a precise but complicated systematization. We shall deal with an example of Dicotyledon and one of Monocotyledon only, mentioning the most interesting variations in the field of Angiosperms.

In the comparison of the various species, it should be noted that the details of segmentation do not prevent a relative uniformity in the sequence of divisions, *at the onset of development.*

On the contrary, after some time, the mechanism of embryogeny shows differences according to the class, the order, the familiy and the species considered.

The development is approximatively uniform as long as the embryo keeps an axial symmetry, whereas the most significative features appear with the passage to bilateral symmetry.

According to SOUÈGES, we shall divide our study into two parts:

1. The *proembryo* development with axial symmetry;
2. The *embryo* development per se, considered when bilateral symmetry appears.

1.6.1 The Proembryo

The onset of segmentation and formation of the proembryo are more or less uniform in the field of Angiosperms. We shall consider one type among the Dicotyledons and indicate a few examples of variation.

1.6.1.1 Edification of the Proembryo of *Myosurus minimus* (SOUÈGES 1911)

The egg, undivided for a long time, lengthens in the embryo sac and becomes cylindric (Fig. 1.60/1). Then it is divided transversally giving a basal cell *bc,* and a smaller apical cell *ac* (Fig. 1.60/2). Another (second) transversal division shares the

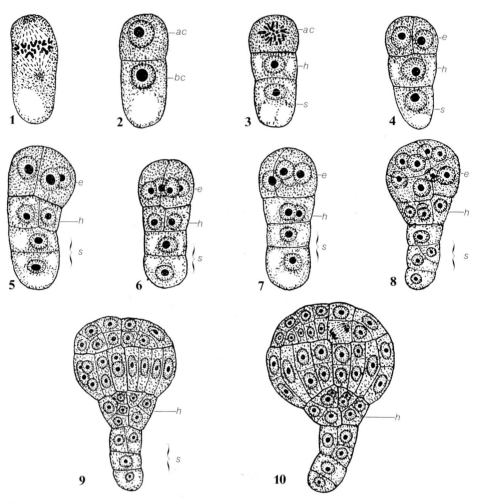

Fig. 1.60. *Myosurus minimus* L., development of the proembryo. **1,2** First cleavage of the zygote (transversal); *ac* apical cell; *bc* basal cell; **3** three-celled embryo resulting from the transversal division of *bc,* giving *h* and *s* (suspensor); *ac* is in a process of longitudinal division; **4-7** formation of the four cells of tier *e,* which will produce the main part of the embryo; the tier *h* follows with a constant delay; **8** proembryo showing the suspensor growth through transversal mitoses, and the three tiers that will edificate the embryo; two tiers *e* (four upper octants and four lower octants) and the tier *h;* **9,10** completion of the proembryo, change from axial symmetry to bilateral symmetry (after SOUÈGES 1911)

bc cell in a median cell *h,* origin of the part called "proembryo hypophysis" and in another basal cell *s,* origin of the "suspensor" (Fig. 1.60/3).

The embryo per se originates from the apical cell *ac* of this three-celled embryo. This cell is divided soon by a longitudinal wall, resulting in two cells which in turn divide longitudinally again, producing four embryonic quadrants (Fig. 1.60/4, 5). In the meantime, the median cell *h* also divides parallel to the embryo axis, while the cell *s* divides transversally (Fig. 1.60/5, 6).

Then, while each quadrant divides transversally into two *octants,* the cells derived from *h* divide again longitudinally, producing a four-celled stage analogous to the four embryonic quadrants of the former tier. Both lower cells continue dividing in a line of eight to ten cells to form the suspensor (Fig. 1.60/7–9).

This suspensor disappears later, digested by the albumen cells next to the micropyle. Thus, the embryo grows from the three remaining four-celled stages:

1. Four upper octants;
2. Four lower octants;
3. Four hypophysis cells.

For some time, it is still possible to follow the schedule and orientation of divisions in the different stages, then this less and less determined arrangement escapes observation (Fig. 1.60/9, 10).

A subspheric, then ovoid, embryo is constituted in which three histological areas can apparently be distinguished: the surface tier, three or four approximately regular tiers and a cord of central cells, hence, the usual description of three areas: epidermis, cortex, central cylinder.

Prior to the appearance of the first organs of the plantlet, the embryo flattens; transversal sections are elliptic (Fig. 1.66); this corresponds to the passage from axial symmetry to bilateral symmetry.

1.6.1.2 Variants and Their Signification

Almost in all cases, the segmentation results in a more or less globular proembryo (sometimes very extended) presenting an axial symmetry and in a suspensor composed of a variable number of cells with respect to species, even to individuals.

Compared to the *Myosurus type* described above, variants affect mainly the following points:

1. Orientation of the zygote first division is generally transversal as in *Myosurus,* but in some cases it is almost longitudinal: Piperaceae (quoted with Preangiosperms), Balanophoraceae (no suspensor) and also *Scabiosa* (rudimentary suspensor) (Fig. 1.61).

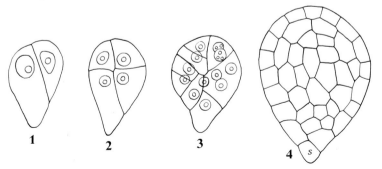

Fig. 1.61. Segmentation of the *Scabiosa succisa* zygote and formation of the embryo. **1** First cleavage, nearly longitudinal; **2,3,4** successive stages showing the residual formation of a rudimentary suspensor *s* (after SOUÈGES 1937)

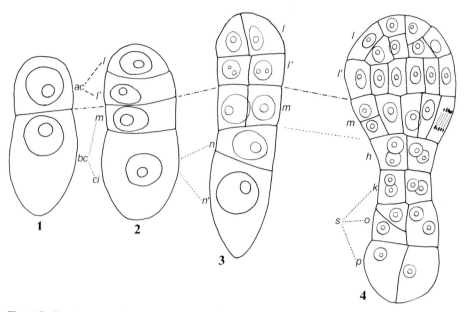

Fig. 1.62. Development of the proembryo of *Sagina procumbens*. Transversal division of *ac* and of the two resulting cells; inertia of *bc*, which becomes a giant suspensor cell; *1, 1', m, n, o, p* embryonic stories completed at the end of the formation of the proembryo (after Souèges 1924a)

2. With the second cleavage, the apical cell *ac* can divide longitudinally *(Myosurus)* or *transversally*, as the zygote (Caryophyllaceae such as *Sagina, Medicago, Sagittaria* etc.) (Fig. 1.62).
3. The basal cell *bc* issued from the zygote can either play an active part in the edification of the proembryo itself *(Chenopodium)* (Fig. 1.63), provide it only with a few cells *(Myosurus)*, produce the suspensor only *(Sagina)* or even grow into a giant cell, at the extremity of the suspensor *(Sagittaria)* (Fig. 1.64).

These considerations enabled embryologists to separate six main types among Angiosperms. This classification shows that the modes of segmentation have only a limited systematic value. In fact, the species of a same family, if not of the same genus, can belong to different types with regards to their embryogeny. For instance, two main types, *Onagraceae* and *Caryophylleae* are found in Papilionaceae in which the *Trifolium* genus alone seems to present several types. Some species have no suspensor *(Hedysarum coronarium)* or have a rudimenatry one *(Soja, some Trifolium)*. In others, the suspensor shows a great variety in size and shape (Fig. 1.65).

Without considering all theses modifications, each one of the six main types has been divided again into a number of variations. We do not intend to deal with these details. The examination of some figures (Figs. 1.60–1.65) show that these variations have no influence on the following fact, i.e. the edification of a proembryo with axial symmetry.

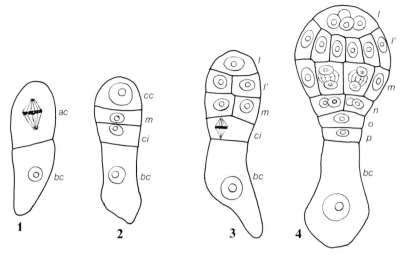

Fig. 1.63. Embryogeny of *Chenopodium Bonus Henricus*. Participation of cells derived from *bc* in the edification of the embryo; *ac* supplies only the *l* and *l'* tiers; *m, h, k, o, p* embryonic tiers proceeding from *bc*; *k, o* and *p* constitute the suspensor *s* (after Souèges 1920)

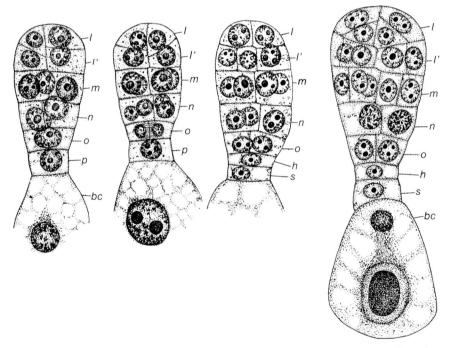

Fig. 1.64. *Sagittaria sagittaefolia*, development of the proembryo showing the formation of successive tiers derived from the apical cell, whereas the cell *bc* no longer divides and produces a giant cell at the suspensor extremity. This suspensor will be completed by the cell *s*, issued from *p* (after Souèges 1931)

81

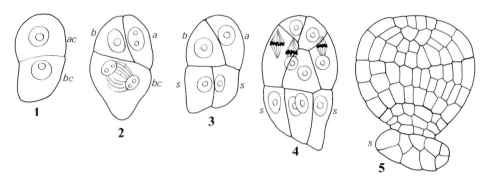

Fig. 1.65. *Trifolium minus,* some stages of embryogeny showing the edification of a short and massive suspensor *s,* generated by the cell *bc; a,b* two of the four blastomeres derived from *ac* after longitudinal mitoses (after SOUÈGES 1929)

1.6.2 The Embryo

The arrangement of the primary organs of the plantlet is indicated by the passage to bilateral symmetry. From this condition, the development becomes different, even in the main features, according to the species. It seems justified to separate the case of Monocotyledons from that of Dicotyledons.

We shall consider again the embryogeny of *Myosurus* as an example of this last group.

1.6.2.1 Dicotyledons: Edification of the Embryo of *Myosurus minimus*

We have already mentioned that the *Myosurus* proembryo flattens so that its transversal section becomes elliptic. The orientation of this flattening is variable and hazardous, in relation to the plane of symmetry of the ovule. The cotyledons rise at the extremities of the great axis of the elliptic section by periclinal divisions of subepidermal cells of the upper octant (Figs. 1.66 and 1.67/1–3); the epidermal cells adjust the growth by anticlinal divisions.

The remainder of the lower octant produces the hypocotyl axis and the upper part of the root, up to the extremity of the central cylinder, while the hypophysis quadrant gives rise to the embryonic extremity of the root and to the cap (Fig. 1.67/4–6). The suspensor disappears into the albumen at the end of this phase of development.

Between both cotyledons and therefore in *prolongation of the axis of symmetry of the proembryo* a narrow groove remains where the *vegetative point* of the shoot plantlet will later rise. This shoot apex is still reduced to a few cells, some of them part of two regular layers which will become the tunical layers (proepidermis + subepidermal layer) and the others part of a small body which ends the central cord of the hypocotyl axis (Fig. 1.67/5, shown in grey). This phase coincides with the maturation of the seed. The intraseminal embryo is completed and on both extremities includes a few cells which will ensure a continued growth after

82

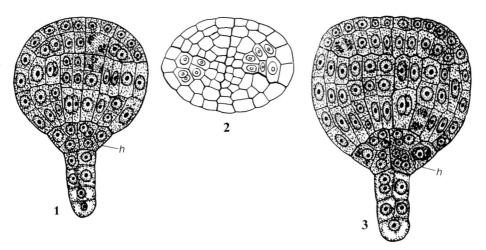

Fig. 1.66. *Myosurus minimus*, transition from axial to bilateral symmetry. **1** Completed proembryo; **2** transversal section showing the flattening of the tiers derived from proembryonic octants; **3** embryo at the onset of initiation of the two cotyledons (apical enlargement and subepidermal mitoses) (after SOUÈGES 1911)

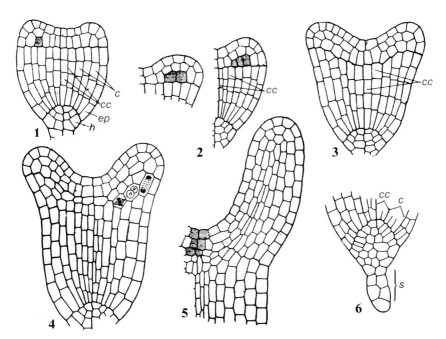

Fig. 1.67. *Myosurus minimus*, edification of the true embryo. **1,2** Birth of cotyledons and onset of differentiation of the embryo primordial axis; *c* cortex; *cc* central cylinder; *ep* epidermis; *h* hypophysis; **3,4** vertical growth of cotyledons; **5** onset of differentiation of cotyledon and hypocotyl conducting tissues; the residual axial apical cells, from which the apical meristem of the leafy stem will rise, are *grey;* **6** radicular extremity of the embryo showing the fate of the hypophysis. The primordial apical cells of the radicle are situated just after the central cylinder *cc* extremity of the primordial axis of the embryo; *s* suspensor, which will degenerate. The primordial axis differentiation, producing the hypocotyl and the radicle, will occur later (after SOUÈGES 1911)

germination; in their most rudimentary condition, these cells constitute the apical meristems. The organization of these meristems is only completed during vegetative growth after germination.

1.6.2.2 Monocotyledons: Edification of the Embryo of *Sagittaria sagittaefolia*

In *Sagittaria*, the basal cell *cb*, derived from the first division of the zygote, does not divide further and does not participate in the embryo development. It becomes a giant cell forming the suspensor basis. The whole proembryo and the upper part of the suspensor are derived from the apical cell *ac* (Fig. 1.64). The onset of division of this single cell is similar to the segmentation of the whole proembryo of *Myosurus* (Fig. 1.60). At the end of its development, the proembryo has six layers of cells (excluding the suspensor):

1. Two upper octants (l–l') which give rise to the *cotyledon;*
2. Two middle stages m and n, points of origin of the hypocotyl area and the *vegetative point;*
3. One stage *o* producing the cortex extremity only and the main part of the radicle cap;
4. Finally, the lower stage including a single cell *h* for a long time, produces the four terminal cells of the cap exclusively (Fig. 1.64).

It should be noted that the aspect of the cotyledon is similar to the aspect of the upper part of the *Myosurus proembryo*. This extremity itself, which is axial, wholly develops to form the cotyledon exclusively. This one flattens and curves, surrounding the gemmule (Fig. 1.68). Moreover, in contrast to Dicotyledons, the vegetative apex is found laterally at the level of stage *m* in *Sagittaria*, i.e. *below the cotyledon basis*. A transverse depression is constituted by a modification of division of a few cells of stage *m*. At the onset two contiguous cells of stage *m* are differentiated by a larger size (Fig. 1.68/1). Two transverse successive mitoses occur in each of these two cells forming two four-celled lines (Fig. 1.68/2), while the other cells divide transversally, then longitudinally. At this point the transverse division stops so that the depression appears. In this groove, the proliferation of subepidermal cells initiates a protuberance (Fig. 1.68/3) point of origin of the first and second leaves, and then the following leaf initiums are individualized by grooves sharing the protuberance in two parts: the initium and the remainder of the vegetative dome. The latter grows on the same process according to the usual activity of a vegetative apex as will be explained below.

The edification of the root, a controversial matter, has been thoroughly studied. Despite the well-acknowledged difference between Monocotyledons and Dicotyledons, the first ones having a cap apparently independent of the cortex and hair layer, the arrangement of radicular meristems presents no fundamental *embryologic* differences between both groups.

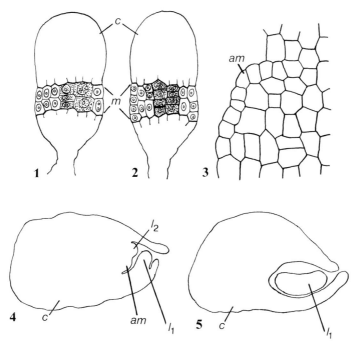

Fig. 1.68. *Sagittaria sagittaefolia*, development of the true embryo. **1** Lateral differentiation of privileged cells into the *m* tier; *c* cotyledon; **2** selectively transversal divisions of the privileged cells, producing the lateral depression and the vegetative cone, or apical meristem, visible on **3** (am); **4** transversal section in an embryo showing the gutter-shaped gap of the cotyledon *c*, and the apical meristem *am*, already surrounded by two leaves successively formed, l_1 and l_2; **5** section of the same embryo, passing above the apical meristem, but through the extremity of the leaf primordium l_1 (after Souèges 1931)

1.6.2.3 Significance of the "Cotyledon" of Monocotyledons

Contrarily to the seminal root, the cotyledon edification of Monocotyledons seems to be quite different from that of both cotyledons of Dicotyledons.

In Dicotyledons, cotyledons are lateral extensions initiated from a part of the cells derived from both octants of the apical area. Cotyledons, similarly to leaves, rise *at the vegetative apex:* in the upper part of the proembryo, horizontal growth goes on before the arising of these lateral organs, likewise the horizontal growth of leaf bases before the vertical development of leaf primordia (see p. 155).

The development of a "cotyledon" in Monocotyledons is very different:

1. It does not arise as a lateral organ, but as the upper part of the *axis itself* of the proembryo. In fact, any leafy organ is essentially a *lateral* organ.
2. There is no process of horizontal enlargement like in Dicotyledons, indicating the passage to bilateral symmetry.

In Monocotyledons, the passage from the axial symmetry of the proembryo to the bilateral symmetry of the embryo follows a different process, through the *lateral*

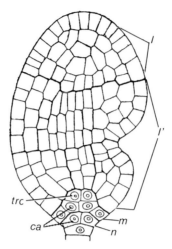

Fig. 1.69. *Luzula forsteri* D. C., embryo, just at the beginning of the lateral formation of the meristem. The top of the embryo will only produce the "cotyledon", which has a limited growth; *trc* top of the radicle cortex; *ca* cap; *l,l',m,n* embryonic tiers proceeding from the proembryo (after SOUÈGES 1933)

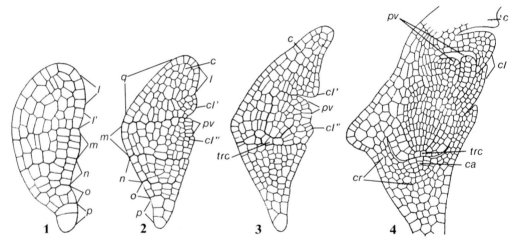

Fig. 1.70. *Poa annua* embryo; lateral edification of the meristem *pv,* the upper part exclusively becoming the "cotyledon" *c; l,l',m,n,o,p,q* embryonic tiers proceeding from the proembryo; *cl* coleoptile, beginning with two lips *cl'* and *cl''*; *trc* top of the radicle cortex; *ca* cap; *cr* coleorhize (after SOUÈGES 1924b)

differentiation of a *vegetative apex* (Figs. 1.69, 1.70). This lateral localization seems to be usual, although the embryonic stage involved is not always the same (e.g. *m* in *Sagittaria* and *Poa,* Figs. 1.68 and 1.70; *l'* in *Luzula,* Fig. 1.69).

This apex produces leaves directly, failing to initiate cotyledons. It appears like an *adventitious meristem* as it occurs sometimes on the hypocotyl axis of certain plants, even Dicotyledons *(Linum).*

It seems that in Monocotyledons, the primitive axis, instead of continuing to increase by means of *an apical meristem,* has only a limited extension in the ab-

sence of such a meristem, producing only the organ called "cotyledon" by an er-
roneous analogy with one of the cotyledons of Dicotyledons. Thus, this false cotyledon could be an *aborted axis*.

Therefore, the stem of the plantlet is *adventitious* from the beginning.

The main root, derived from the radicle, disappears also, but later than the
stem in all Monocotyledons. the vegetative system of these particular plants appears to be *exclusively composed of adventitious organs*. This analysis reported by
the author in the 1960s was recently demonstrated by JACQUES-FÉLIX in 1982.

1.6.3 Some Cytological Data on the Embryogeny of Angiosperms

1.6.3.1 The Zygote

With reference to the small number of cases measured by cytophotometry (WOO-
DARD 1956), the nuclei of both gametes are found to be in phase G_1 of the nuclear
cycle at the time of fertilization. It is only after karyogamy that the zygote nucleus
enters phase S (synthesis phase). According to VALLADE and CORNU (1973) this
phase lasts about 16 h in *Petunia*. Then, the nucleus proceeds to phase G_2 *just before the first segmentation*. In *Petunia,* phase G_2 lasts about 32 h and the nucleus
undergoes a *single period S* which initiates a normal diploid cycle. In contrast, in
other species, there can be a succession of periods S prior to the first segmentation. The zygote nucleus reaches a polyploid condition. This is the case with barley in which the zygote may become 8- or 16-ploid (MERICLE and MERICLE 1970).
A regulation occurs in segmentation and the embryo cells are brought back into a
diploid condition.

In Angiosperms, the zygote, as the pre-existing oosphere, is a cell of reduced
size, relatively poor in paraplasm. With respect to oosphere, fertilization causes a
reactivation of cytoplasmic organelles. JENSEN's accurate ultrastructural studies
(1968) showed many aspects of this fact in the cotton plant but there are variations
in some other species. On the whole, the number of mitochondria increases as well
as the number of mitochondrial cristae; plastids may store starch.

At least in some species (cotton tree) ribosomes arrange themselves at the beginning into very long polysomes, supposedly associated with messenger RNA
pre-existing in the nucleus, then in shorter polysomes supposedly derived from the
nucleus reactivation (JENSEN 1968). The activity of the endoplasmic reticulum is
commonly low and variable but seems to be more associated to ribosomes.

According to the species and in the course of time, the vacuolar apparatus is
subjected to subtle volumetric variations, but usually larger vacuoles gather towards the micropylar pole of the zygote, shifting the nucleus and most of the cytoplasm away towards the chazala pole. Except for some species (*Hordeum,*
NORSTOG 1972) in which this distribution is less noticeable or delayed, vacuoles
constitute the most obvious structures to show the zygote polarity. Therefore, the
first cleavage shares the cytoplasm unevenly between the two first blastomeres and
it is the cell located on the chazala side, richer in ribosomes, mitochondria and
plastids which initiates the main part of the embryo, if not in totality.

Thus, because of several characteristics, i.e. abundance of ribosomes, mitochondria and plastids, the zygote of Angiosperms is almost in a meristematic condition. Because of the characteristics of the vacuolar apparatus, as well as the starch storage, it is still far from the cytological structure of the meristematic organogen cells which will be described later.

As the zygote may become transiently polyploid it cannot be compared with a primary meristematic cell. However, this condition is transient and becomes regular as from the first divisions.

1.6.3.2 The Suspensor

The situation is different with the cell (s) which initiates the suspensor and may reach and keep a high degree of ploidy: 1024n in *Tropaeolum majus,* 4096n in *Phaseolus coccineus* and in several *Lathyrus* (NAGL 1974).

Endopolyploidy of the suspensor cells is accompanied by a marked volumetric growth (Plate 1.12, 3). The endoplasmic reticulum of these cells enlarges during their development, as well as the number and the activity of dictyosomes which probably contribute to the pronounced increase of cell walls. The cell walls may build internal protrusions which increase the plasmalemma surface and make the suspensor cells similar to the *transfer cells* of conducting tissues (SCHULZ and JENSEN 1969). In correlation with endopolyploidy, the suspensor cells provide an overabundance of ribosomal RNA. This RNA may be carried into the embryo, the pyroninophily of which increases at the same time (AVANZI et al. 1970).

The suspensor cells synthesize also proteins and lipids and probably produce hormones and enzymes. In particular, they strongly incorporate tryptophane, a precursor of indole-acetic acid, among other substances.

Due to these cytophysiological aspects, two functions have been acknowledged in suspensor cells: they convey nutrients to the embryo and they secrete hormones and enzymes. Therefore, the suspensor can be compared with the nutritive cells of insect ovocytes and the trophoblast of mammalians (NAGL 1973; BRADY 1973)

Plate 1.11. Evolution of the RNA content of cells during the development of the Angiosperm proembryo.

1 *Myosurus minimus,* zygote, moderately rich in RNA.

2-5 Proembryos of *Alyssum maritimum Lamk;* **2** unicellular proembryo, RNA-rich, but containing paraplasmic inclusions; **3,4** two stages of development of the spheric proembryo *e,* with all cells uniformly basophil; *al* albumen, equally RNA-rich, in contrast to the suspensor *s;* **5** proembryo, just before passing to bilateral symmetry; very beginning of clearing of apical pole cells *ap* and the basal pole *bp;* × 1070.

BRACHET's technique, methyl-green-pyronine; (after RONDET 1961)

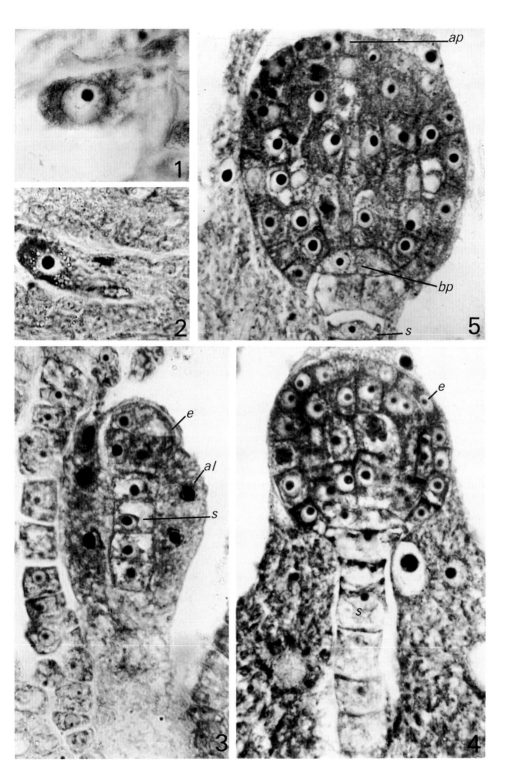

89

1.6.3.3 The Embryo

The proembryonic phase, until the occurrence of bilateral symmetry, is cytologically characterized by the accentuation of characteristics proper to the primary meristematic cells. On the chalaza side, the first division of the zygote soon separates a smaller cell with reduced vacuoles in which the ribosomes concentrate selectively. The reaction of Brachet which stains RNA by means of pyronine is particularly positive in this "apical cell" (Plate 1.11, 1, 2).

Later, the segmentation initiates a globular proembryo in which all the cells are apparently similar, RNA-rich with few vacuoles (Plate 1.11/3, 4). The nucleus is central and includes large nucleoli; mitochondria are numerous and globular plastids are found as undifferentiated proplastids.

Thus, the onset of embryogenesis achieves the dedifferentiated state which is maintained, after the occurrence of bilateral symmetry, in the embryonic areas where organogenesis continues actively: cotyledons, hypocotyl, embryonic cap (Plate 1.12/1, 2).

But the organogenesis does not activate all the embryonic cells in the same way and those left aside soon become less pyroninophil; this indicates the beginning of differentiation. Then, these residual cells, in relative inactivity, can be distinguished at both ends of the embryo axis where they will later initiate the material of the cauline vegetative apex and the root meristem (Plate 1.11/5 and Plate 1.12/1, 2) (cf. RONDET 1958, 1961, 1962; NORREEL 1973). It is only after the edification of the strictly embryonic organs, forming the "primary body" of the young sporophyte, that the residual cells reactivate resulting in the apical meristems of the stem und root. In numerous species, this activity begins in the seed before the phase of latent life, in others, only after germination (RONDET 1965).

In all cases, the cellular activity of the young seedling is soon suspended by the dehydration process which accompanies the seed maturation.

Embryonic organogenesis slows down along with the storage of seminal material. This material is mainly stored in specialized tissues such as albumen or cotyledons, according to species, but the other organs of the young seedling contribute, to a less extent, to this storage, mainly proteins, which change vacuoles into aleurone grains, and lipids which form numerous enclaves into the cytoplasm.

--- ▷

Plate 1.12.

1,2 Evolution of the RNA content of cells during the development of the true embryo of *Myosurus minimus*. **1** Initiation of cotyledons *c,* increasing of the RNA content of the initial cells, whereas the residual cells *az,* which have not been used for this initiation, but which will later produce the apical meristem, contain less RNA, as well as the initials *ri* which edify the root meristem. × 1070. **2** Embryo in the way of completion; accentuation of the basophilia of the hypocotyl axis, now active, and of proliferating regions of cotyledons. In contrast, RNA impoverishment of cells which will be activated only later, and will organize the stem apical meristem and the root meristem. *c* Cotyledons; *ec* embryonic cap, which will later disappear; *ri* residual cells which will initiate the radicle; *hy* hypocotyl; *az* apical zone; *s* remnants of the suspensor; × 370; (after RONDET 1962)

3 Spherical proembryo *e* of lentil *(Lens culinaris),* uniformly basophil, provided with a bulky suspensor cell *s;* × 400; BRACHET's technique, methyl – green – pyronine, 1941; (after RONDET 1958)

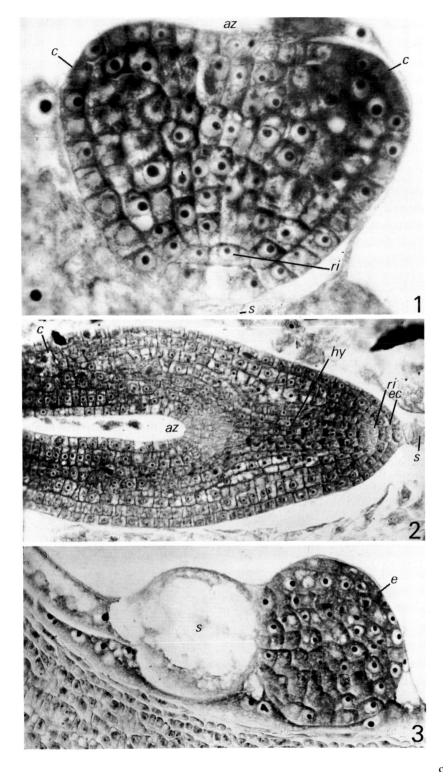

The transporting procedure of the protein storage from the endoplasmic reticulum (e.r.) where they are synthesized, to the vacuoles which become the aleurone grains, has not been clearly demonstrated, even with electron microscopy.

However, in cotyledons of *Phaseolus vulgaris*, TOURMEL-PREVOSTI (1980) observed a pronounced development of e.r. in parallel layers which become osmophilic and from which small vesicles are released to be carried away towards the "protein bodies", different from the aleurone grains, according to the author, since they do not proceed from pre-existing vacuoles.

But this distinction is not clear, the protein bodies seem to be similar to aleurone grains, according to ÖPIK (1968), who described their ontogenesis by fragmentations of vacuoles.

Most authors failed to observe any ultrastructural continuity between the e.r. and the vacuoles; thus, one can suppose that the vesicles released by the e.r. could be collected by dictyosomes, being able to carry them into vacuoles. Moreover, the protein bodies may already be the result of storage proceeding from dictyosomes (TOURMEL-PREVOSTI 1980), and the relationship between Golgi apparatus formations (provacuolar) and vacuoles, (MARTY 1973, BUVAT and ROBERT 1979) could suggest a transfer which has to be demonstrated.

Despite the absence of anatomical confluences between the e.r. and the vacuoles, it is curious to note close associations between both structures when the barley embryo enters a period of latent life, the e.r.-forming sheaths surrounding the future aleurone vacuoles (Plate 1.13a). These features persist in mature and dehydrated seed, into which the regression of the reticular membrane is, however, perceptible (Plate 1.13b) and can also be seen at the onset of germinative hydration (BUVAT and ROBERT 1982b).

Thus, in the course of maturation, the embryonic or meristematic cells of the young seedling become rich in paraplasmic material; they are subjected to a rough differentiation, an inverse evolution from that of segmentation.

▷

Plate 1.13a,b. Mature karyopsis of barley; dry embryo, root meristem cells.

a Persistency of ribosomes, plasmalemma, tonoplasts *t*, mitochondrial and plastidal membranes, but the endoplasmic reticulum and the dictyosomes are not visible. *a* Aleurone grains, partially extracted; *gl* lipid globule; *m* mitochondrion; *p* proplastid, containing plastoglobules; × 26000; anhydrous fixation, according to THOMSON (1979); contrast according to REYNOLDS (1963).
b Same material, ordinary fixation (glutaraldehyde – OsO_4). Cytoplasmic area containing mitochondria *m* and proplastids *p*, some of which enclose cytoplasmic inclusions *cyi*; *lgl* large lipid globule, not associated with the peripheral cytoplasm, as those of **a**; × 26000; contrast according to REYNOLDS (1963)

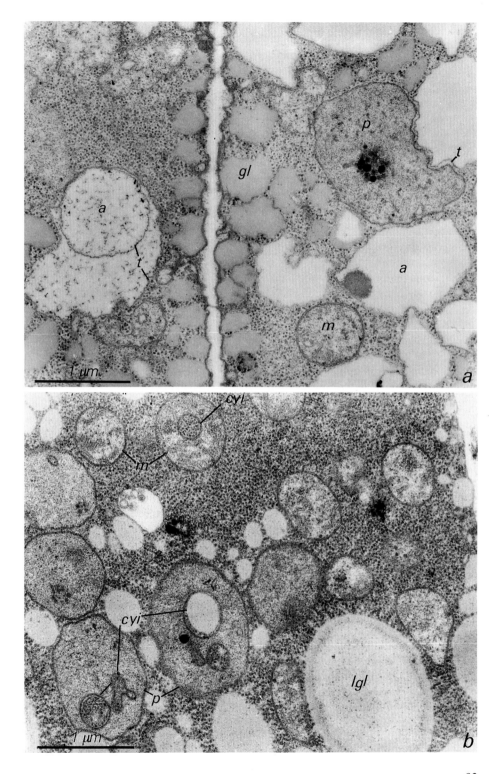

1.6.3.4 The Plantlet in the Mature Seed

The dehydration of maturation which stops the young seedling development, impairs the metabolic activity of its cells almost completely. In addition to the genesis of aleurone grains and lipid storage, modifications of cytoplasmic organelles and the nucleus occur.

These modifications have been studied in a few species only and further research is necessary to assess general features.

In cytoplasm, ribosomes remain as isolated particles, non-associated in polysomes and are certainly inactive (Plates 1.13, 1.14). The tonoplast, which encircles the aleurone grains, does not show any notable modification and despite a few differences due perhaps to the method of preparation; results are the same with the plasmalemma.

It seems logical for these structures to be spared. Tonoplasts must escape from the hydrolytic processes increasing in aleurone vacuoles during germination, plasmalemma must ensure the selective permeability for the survival of the cells as they resume their activity as well as necessary exchanges (Plate 1.14a)

In contrast, dynamic structures such as the endoplasmic reticulum and mainly the dictyosomes disorganize when they cease to function (BUVAT and ROBERT 1982a, b) (Plate 1.15).

Proplastids undergo distortions which commonly lead to a kind of ingestion of cytoplasmic drops (Plate 1.14b). The same thing occurs sometimes with mitochondria and, in both types of organelles, the peripheral membranes and cristae become fuzzy. The stroma of mitochondria has an unusual heterogeneous consistency (BUVAT and ROBERT 1982b).

The outline of the nucleus becomes more or less curved and sinuous; chromatin is more strongly condensated than in hydrated and active cells. Nucleoli are diversely changed; sometimes they present the same fibrillar and granulous parts, but the ground nucleolar substance is denser and hides the usual heterogeneity; sometimes granules seem to disintegrate and nucleoli become homogeneous, slightly granulofibrillar and very osmophilic. In the nucleoplasm, for example in barley, *perichromatin granules* are constantly observed. These granules are known to be the sites of maturation for the messenger RNAs (not published) (Plate 1.15b).

▷

Plate 1.14a, b. Root meristem of barley embryo.

a Embryo at the beginning of seed maturation, the dehydration has not yet started, but the vacuoles are already transformed into aleurone grains *a,* with a concentrated content, and begin to link with endoplasmic reticulum sheets *er; m* mitochondrion; × 54000.

b Dehydrated embryo in mature state. Persistency of ribosomes, with a density increased by drying, *er* endoplasmic reticulum remnants, remaining associated with aleurone grains *a;* × 26000.

Fixation: glutaraldehyde – OsO_4; contrast according to REYNOLDS (1963)

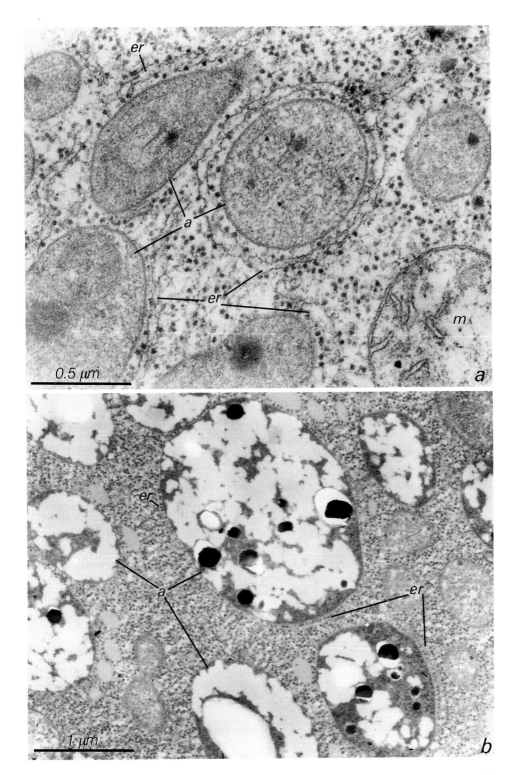

This fact is interesting when compared to the early recurrence of protein synthesis, at the onset of germination, suggesting during the prolonged period of inactivity of the young seedling, the presence of messenger RNAs (or premessengers), ready to migrate directly towards the cytoplasm.

Finally, in the course of dehydration of maturation, when mitoses cease little by little, the nuclei of meristematic cells are progressively stopped in phase G_1 of their cycle (BRUNORI and D'AMATO 1967). This is understandable since it was demonstrated that this phase preserves the proliferation potentialities (REMBUR 1970; NOUGARÈDE and REMBUR 1977; REMBUR and NOUGARÈDE 1977).

Cytology and cytochemistry of the embryogeny of Phanerogams were studied in two outstanding publications of THOMAS (1975, 1980). It is advisable to refer to these works for further information on the results obtained before 1980.

1.7 General Considerations on the Embryogeny of Vascular Plants

1.7.1 The Geometry of Segmentation and its Results

We have only briefly reported the results obtained from many works, most often very detailed, in which the researchers attempted to explain the mechanisms of segmentation and embryogeny and to define some rules. They often attempted to recognize a regular and therefore foreseeable sequence of partitions of the zygote in order to determine whether the strange determinism of the genesis of specific forms is achieved by means of geometrical orders. We have already mentioned that this geometrical regularity can be found in species in which the embryo remains small and organizes early enough with a reduced cellular material such as in Filicales, Hydropteridales and Angiosperms. There are generally no stereotyped segmentations in plants in which the embryo becomes massive and organizes late.

Another matter of study for embryologists was the classification of the types of embryogeny and to compare it with the systematics of vascular plants.

The few attempts that we have made or perhaps only quoted show that these classifications, although very complex and subtle in Angiosperms, only revealed a rather deceiving feature: the spatial mechanism, i.e. the geometry of embryogeny does not provide valuable fundamental data for a better classification.

However, if one distinguishes the essential features from the secondary aspects, it is possible to define types of embryogeny which characterize *the main groups* of vascular plants.

Plate 1.15a, b. Root meristem of dehydrated barley embryos in mature state.

a Cytoplasmic area without ribosomes showing remnants of vesicles *gv*, interpreted as being Golgi material, disorganized at the time when the seed passes to dormancy; × 54000.
b Part of dry seed nucleus, in latent life state; persistency of perichromatin grains *(pchrg)* ordinarily present in the nuclei of active cells. *nu* nucleolus, nearly exclusively in the form of *"pars granulosa"*; *chr* chromatin; × 26000.

Fixation: glutaraldehyde – O_sO_4; contrast according to REYNOLDS (1963)

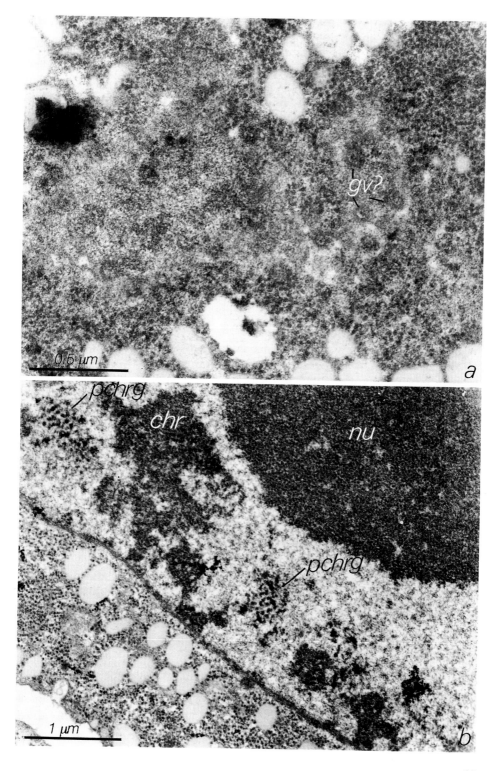

97

Thus, with an example of each type, we were able to distinguish the embryogeny of Filicales, Coniferales, Dicotyledons, Monocotyledons, etc.

Besides these examples, further knowledge is necessary to discover still more general facts allowing further advancement in the field of the fundamental processes of development.

1.7.2 Predestination or Epigenesis

In plants with early organogenesis, the very first divisions generally separate the zygote into very different cells, the function of which in normal embryogeny is often strictly determined, e.g. the apical cell and the basal cell of the bicellular embryo of Angiosperms. The advances in the experimental embryogeny of vascular plants do not allow embryologists to tell whether this determination is immutable or whether it only proceeds from normal relations between these cells and the surrounding tissue. However, it seems that the distribution of the *cytoplasmic substances* influences the specific modalities of the first cleavage.

In contrast, many natural cases show that the cells resulting from the first mitoses of the zygote are more or less equivalent. This is the case of embryos with late organogenesis, such as those of Gymnosperms in which the equivalence is even more manifested by the polyembryony.

The determination is very progressive and is perhaps never irreversible. The function of a cell seems to depend only on its location, itself coming from a haphazardly proliferating and growing: then, the embryo organizes through an epigenetic process. This fundamental lack of predetermination is well exemplified by Taxaceae in which suspensors sometimes give rise to extra embryos.

1.7.3 Segmentation and Organogenesis

The histological study of the embryos of vascular plants proved that, depending on the particular case, the proliferation producing almost similar cells *without initiating differentiated organs* is more or less maintained.

Thus, organogenesis starts early in *Myosurus:* the small embryo, still composed of a few cells, initiates a young plantlet with a hypocotyl axis, two cotyledons and a radicle. In contrast, over 1000 equivalent nuclei are derived from the zygote nucleus of *Cycas* or *Dioon* and scatter at the periphery of this huge cell. Then the proembryo produces numerous small cells still mostly similar at the extremity of an oversized suspensor and the organogenesis begins only thereafter.

All the intermediate types can fit between these two cases, and more examples of both can be found. Histology alone cannot explain these differences but two facts can be underlined:

1. As a whole, voluminous zygotes, overloaded with storage material undergo an extended phase of proliferation before organogenesis occurs.
2. Organogenesis lasts longer when segmentation is slower. This is illustrated by the comparison between primitive Filicineae (Ophioglossales) and advanced ferns (Filicales leptosporangiate).

1.7.4 Formation of Organogenic Cells, Cytological Results of Embryogenesis

Female gametes and the zygotes derived from them are often more or less voluminous cells with storage material in the cytoplasm, with sometimes differentiated plastids and in many cases presenting a pronounced vacuolar system.

These cells are rich in "paraplasm" and in substances to be used as such, and the *ratio* between the volume of the nucleus and the volume of the cell is *small*.

When organogenesis occurs, after a period of segmentation, the cells which initiate the first organs are very small (compared to the zygote), poor in paraplasm and have a relatively voluminous central nucleus. These characteristics, conventionally expressed by a high nucleocytoplasmic ratio, are those of the so-called primary meristematic cells which, after embryogeny, give rise to the organs of the adult plant.

Thus, the processes of embryogeny allow the passage of the zygote, a cell rich in storage material, capable of proliferating but not of initiating organs, from its cytological condition to the cytological condition of primary meristematic cells which, in addition to their capability of proliferating, can initiate organs.

1.7.5 Dedifferentiation at the Origin of Ontogenesis

This aspect of embryogeny seems general but further accurate and extensive studies are required in order to understand the evolution of the cellular components of embryos, from the zygote to the post-embryonic apical development. In 1965, JENSEN published an accurate ultrastructural cytological study concerning the oosphere and the "central cell" of the cotton plant. The author underlined the relatively *differentiated* characteristics of the oosphere: large vacuoles, abundant reticulum, rather low ribosomal density. After fertilization (JENSEN 1963), the zygote divisions, unequal with regards to the cytoplasm, eventually result in embryonic cells without vacuoles, which have more and more meristematic features.

Present data lead to the conclusion that the onset of development of sporophytes consists in a *progressive dedifferentiation,* prior to post-embryonic organogenesis, until the edification of cells is actually capable of initiating the typical organs of adult plants: stems, leaves and roots.

In another chapter, we will define the processes of dedifferentiation and show their importance in the continued ontogenesis of vascular plants.

It is already remarkable that such processes trigger the onset of this ontogenesis. However, a distinction must be made: it is usual to compare the meristematic cells of the apices with "embryonic" cells, which remain as such and in the same places during the whole development. The above considerations lead to more subtility. The zygote and the true "embryonic" cells, at the onset of segmentation, *do not present the characteristics of meristematic cells;* these characteristics only appear at the end of embryogeny.

In seed plants in which the development of the young sporophyte is stopped for a variable period of time, the meristematic cells of the plantlet, when they are settled before the passage to slow life, "differentiate" again, i.e. they take a small part in accumulating storage material which affects the whole seed. During germination, after rehydration, another "dedifferentiation" of meristematic cells occurs (BUVAT 1952, p.223).

1.7.6 The Embryonic Differentiation

This expression is not paradoxical; while the organogenic cells are elaborated in the areas where the first apical meristems of the plantlet will arise, *other cells which do not participate in the formation of these meristems* form the remaining part of the primitive body of the embryo, i.e. an axis including the hypocotyl, the basis of the radicle and the cotyledons. They are the only tissues of the sporophyte *which are not derived from the meristems*. They can produce anatomical structures completely different from those produced later by the functioning of the vegetative meristems. For example, the primordial body of *Pteris aquilina* includes a protostelic conducting system, while the subsequent apical activity will form a dictyostele; these differences will be analyzed in another chapter.

Moreover, the cotyledons, which expand during the germination of epigeal seeds, have most often shapes different from those of epicotyl leaves produced by the vegetative meristem.

The differentiation of the cells directly placed by embryogenesis is variably marked according to species. It is sometimes hardly significant.

However, these cells constitute *only a transitory relationship* between the organs derived from the apical meristems. When the meristems organize the main part of the sporophyte, these cells will stop their activity and will disappear or will be affected again by ontogenesis owing to the impulse of the meristematic activities, discussed in the next chapter.

References

Atkinson GF (1894) The biology of ferns. Macmillan, London

Avanzi S, Cionini PG, d'Amato F (1970) Cytochemical and autoradiographic analyses on the embryo-suspensor cells of *Phaseolus coccineus*. Caryologia 23: 605–638

Bonnet ALM (1955a) Contribution à l'étude des Hydroptéridées. I. Recherches sur *Pilularia globulifera* L. et *P.minuta* Durr. Cellule 57: 131–239

Bonnet ALM (1955b) Contribution à l'étude des Hydroptéridées. II. Recherches sur *Salvinia auriculata* Aubl. Ann Sci Bot 11 ème sér 16: 529–600

Bonnet ALM (1957) Contribution à l'étude des Hydroptéridées. III. Recherches sur *Azolla filiculoides* Lamk. Rev Cytol Biol Vég 18: 1–88

Bower FO (1882) The germination and embryogeny of *Gnetum gnemon*. Q J Microsc Sci 21: 328

Brachet J (1940) La détection histochimique des acides pentosenucléiques. C R Soc Biol 133: 88–90

Brady T (1973) Cytological studies on the suspensor polytene chromosomes of *Phaseolus*: DNA contents and synthesis and the ribosomal cistrons. Caryologia 25 supp. From ovule to seed; Ultrastructural and biochemical aspects, pp 233–259

Bruchmann H (1897) *Selaginella spinulosa*. Gotha

Bruchmann H (1898) Ueber die Prothallien und die Keimpflanze mehrerer europaischer Lycopodien. Gotha

Bruchmann H (1904) Ueber das Prothallium und die Keimpflanze von *Ophioglossum vulgatum* L. Bot Zeit 62: 227–247

Bruchmann H (1906) Ueber das Prothallium und Sporenpflanze von *Botrychium lunaria* S.W. Flora 96: 203-230

Bruchmann H (1910) Die Keimung der Sporen und die Entwicklung der Prothallien von *Lycopodium clavatum* L., *L. annotinum* L. und L. *selago* L. Flora 101: 220-267

Bruchmann H (1912) Zur Embryologie der Selaginellaceen. Flora 104: 180-224

Brunori A, d'Amato F (1967) The DNA content of nuclei in the embryo of dry seeds of *Pinus pinea* and *Lactuca sativa*. Caryologia 20: 153-161

Buchholz JT (1939a) The morphology and embryogeny of *Sequoia sempervirens*. Am J Bot 26: 93-101

Buchholz JT (1939b) The embryogeny of *Sequoia sempervirens* with a comparison of the Sequoias. Am J Bot 26: 248-257

Buchholz JT (1940) The embryogeny of *Torreya*, with a note on *Austrotaxus*. Bull Torrey Bot Club 67: 731-754

Buvat R (1952) Structure, évolution et fonctionnement du méristème apical de quelques Dicotylédones. Ann Sci Nat Bot 11 ème série: 199-300

Buvat R, Robert G (1979) Vacuole formation in the actively growing root meristem of barley *(Hordeum sativum)*. Am J Bot 66: 1219-1237

Buvat R, Robert G (1982a) Recherches cytologiques sur la déshydratation de maturation des ébauches de racines de l'embryon de l'Orge (*Hordeum vulgare* L) I. Evolution du réticulum endoplasmique et de l'appareil vacuolaire. Ann Sci Nat Bot Paris 13 ème sér 4: 1-14

Buvat R, Robert G (1982b) Id. II. Rapports entre les vacuoles et le réticulum endoplasmique vestigial, déformations des mitochondries et des proplastes, désorganisation de l'appareil de Golgi. Ann Sci Nat Bot Paris 13 ème sér 4: 73-90

Camefort H (1956) Etude de la structure du point végétatif et des variations phyllotaxiques chez quelques Gymnospermes. Ann Sci Nat Bot 11e sér 17: 1-185

Camefort H (1958) Rôle du suc nucléaire et des nucléoles dans la formation du cytoplasme du proembryon chez le *Pinus laricio* (var. *austriaca*). C R Acad Sci Paris 246: 2014-2017

Camefort H (1959) Sur la nature cytoplasmique des inclusions dites "vitellines" de l'oosphère du *Pinus laricio* (var. *austriaca*): étude en microscopie électronique. C R Acad Sci Paris 248: 1568-1570

Camefort H (1960) Evolution de l'organisation du cytoplasme dans la cellule centrale et l'oosphère du *Pinus laricio* (var. *austriaca*) C R Acad Sci Paris 250: 3707-3709

Camefort H (1962) L'organisation du cytoplasme dans l'oosphère et la cellule centrale du *Pinus laricio* Poir. (var. *austriaca*). Ann Sci Nat Bot Biol Vég 12 ème sér 3: 265-291

Camefort H (1965) L'organisation du protoplasme dans le gamète femelle, ou oosphère du *Ginkgo biloba* L. J Microsc 4: 531-546

Camefort H (1966a) Etude en microscopie électronique du néocytoplasme des proembryons coenocytiques du *Pinus laricio* Poir. var. *austriaca* (P. *nigra* Arn.) dont les noyaux ont émigré à la base de l'oosphère. C R Acad Sci 263: 1371-1374

Camefort H (1966b) Etude en microscopie électronique de la dégénérescence du cytoplasme maternel dans les oosphères embryonnées du *Pinus laricio* Poir. var. *austriaca* Arn. C R Acad Sci 263: 1443-1446

Camefort H (1970) Sur l'origine et l'activité des structures de type lomasome observées dans les embryons du *Cryptomeria japonica* D. DON. (Taxodiacées). 7 ème Congrès international de microscopie électronique Grenoble. III. Biologie, pp 449-450

Campbell DH (1892) On the prothallium and embryo of *Osmunda claytoniata* L. and *O. cinnamomea* L. Ann Bot 6: 49-94

Campbell DH (1895) The structure and development of the mosses and ferns. Macmillan, London

Campbell DH (1911) The Eusporangiate. Carnegie Inst Washington 140: 1-229

Chamberlain CJ (1906) The ovule and female gametophyte of Dioon. Bot Gaz 42: 321-358

Chamberlain CJ (1910) Fertilization and embryogeny of *Dioon edule*. Bot Gaz 50: 415-429

Chesnoy L (1969) Sur la participation du gamète mâle à la constitution du cytoplasme de l'embryon chez le *Biota orientalis* Endl. Rev Cytol Biol Vég 32: 273-294

Chesnoy L (1977a) Etude cytologique des gamètes, de la fécondation et de la proembryogenèse chez le *Biota orientalis* Endl. III. - Fécondation et proembryogenèse; transmission du cytoplasme du gamète mâle au proembryon. Rev Cytol Biol Vég 40: 293-396

Chesnoy L (1977b) L'origine du cytoplasme de l'embryon chez les Cupressacées: rôle prédondérant du cytoplasme du gamète mâle. Comptes rendus du 102 ème Congrès national des Sociétés savantes – Sciences, fasc I, pp 291–299, Limoges

Coulter JM (1908) The embryo sac and embryo of *Gnetum gnemon*. Bot Gaz 46: 43–49

Coulter JM, Chamberlain CJ (1903) The embryogeny of *Zamia*. Bot Gaz 35: 184–194

Coulter JM, Chamberlain CJ (1932) Morphology of Gymnosperms, 5th ed. The University of Chicago Press

Coulter JM, Land WJG (1905) Gametophytes and embryo of *Torreya taxifolia*. Bot Gaz 39: 161–178

Emberger L (1921) Recherches sur l'origine et l'évolution des plastides chez les Ptéridophytes. Contribution à l'étude de la cellule végétale. Arch Morphol Gén Exp 1: 1–186

Favre-Duchartre M (1956) Contribution à l'étude de la reproduction chez le *Ginkgo biloba*. Rev Cytol Biol Vég 17: 1–213

Haining HI (1920) Development of embryo of *Gnetum*. Bot Gaz 70: 436–445

Holloway JE (1915) Studies of the New Zealand species of *Lycopodium*. Part I. Trans New Zeal Inst 48: 253–303

Holloway JE (1918) The prothallus and young plant of *Tmesipteris*. Trans New Zeal Inst 50: 1–44

Holloway JE (1921) Further studies on the prothallus, embryo and young sporophyte of *Tmesipteris*. Trans New Zeal Inst 53: 386–422

Holloway JE (1939) The gametophyte, embryo and young rhizome of *Psilotum triquetrum* Swartz. Ann Bot N S 3: 313–336

Jacques-Felix H (1982) Les Monocotylédones n'ont pas de cotylédon. Bull Mus Natl Hist Nat Paris 4 ème sér 4 Section B Adamsonia 3: 40

Jensen WA (1963) Cell development during plant embryogenesis. Meristems and differentiation. Brookhaven Symposia in Biology 16: 179–202

Jensen WA (1965) The ultrastructure and composition of the egg and central cell of cotton. Am J Bot 52: 781–797

Jensen Wa (1968) Cotton embryogenesis: the zygote. Planta (Berl) 79: 346–366

Kuligowski-Andrès J (1975a, b) Contribution à l'étude d'une Fougère, le *Marsilea vestita*, du stade embryon au stade du sporophyte adulte. *a:* Mise en place des territoires organogènes. Ann Sci Nat Bot 12 ème sér 16: 151–216. *b:* L'embryon organisé. Ibid, pp 249–307

Kuligowski-Andrès J, Tourte Y (1978) Première étude, par autoradiographie, de la fécondation chez une Ptéridophyte. Biol Cell 31: 101–108

La Motte C (1933) Morphology of the megagametophyte and the embryo sporophyte of *Isoetes lithophila*. Am J Bot 20: 217–233

Land WJG (1902) A morphological study of *Thuja*. Bot Gaz 34: 249–259

Land WJG (1904) Spermatogenesis and oogenesis in *Ephedra trifurca*. Bot Gaz 38: 1–18

Land WJG (1907) Fertilization and embryogeny in *Ephedra trifurca*. Bot Gaz 44: 273–292

Laroche J (1968) Contribution à l'étude de l'*Equisetum arvense* L. II. – Etude embryologique. Caractères morphologiques, histologiques et anatomiques de la première pousse transitoire. Rev Cytol Biol Vég 31: 155–216

Lawson AA (1917) The prothallus of *Tmesipteris tannensis*. Trans R Soc Edinb 51 (3): 785–794

Lawson AA (1917), (1921) The gametophyte generation of the Psilotaceae. Trans R Soc Edinb 52: 93–113

Lyon HL (1904) The embryogeny of *Ginkgo*. Minn Bot Stud 3: 275–290

Martens P, Waterkeyn L (1964) Recherches sur *Welwitschia mirabilis*. – IV: germination et plantules; structure, fonctionnement et productions du méristème caulinaire apical. Cellule 65: 5–68

Martens P, Waterkeyn L (1974) Recherches sur *Welwitschia mirabilis*. V: Evolution ovulaire et embryogenèse. Cellule 70: 163–258

Martens P, Waterkeyn L (1975) Recherches sur *Welwitschia mirabilis*. VII. Histologie et histogenèse de la fleur femelle. Cellule 71: 102–144

Marty F (1973) Mise en évidence d'un appareil provacuolaire et de son rôle dans l'autophagie cellulaire et l'origine des vacuoles. C R Acad Sci Paris 276 D: 1549–1552

Mericle LW, Mericle RP (1970) Nuclear DNA complement in young proembryos of barley. Mutat Res 10: 515–518

Nagl W (1973) The angiosperm suspensor and the mammalian trophoblast: organs with similar cell structure and function. Soc Bot Fr Mém Coll Morphol: Naissance de la forme chez l'embryon, pp 289–302

Nagl W (1974) The phaseolus suspensor and its polytene chromosomes. Z Pflanzenphysiol 73: 1–45

Norreel B (1973) Etude comparative de la répartition des acides ribonucléiques au cours de l'embryogenèse zygotique et de l'embryogenèse androgénétique chez le *Nicotiana tabacum* L. C R Acad Sci Paris 275: 1219–1222

Norstog K (1972) Early development of the barley embryo: fine structure. Am J Bot 59: 123–132

Nougarède A, Rembur J (1977) Determination of cell cycle and DNA synthesis duration in the shoot apex of *Chrysanthemum segetum* L. by double labelling autoradiographic techniques. Z Pflanzenphysiol 85: 283–295

Nougarède A, Rembur J (1978) Variations of the cell cycle phases in the shoot apex of *Chrysanthemum segetum* L. Z Pflanzenphysiol 90: 379–389

Öpik K (1968) Development of cotyledon cell structure in ripening *Phaseolus vulgaris* seeds. J Exp Bot 19: 64–76

Pearson HHW (1909) Further observations on *Welwitschia*. Philos Trans R Soc Lond Biol Sci 200: 331–402

Pearson HHW (1910) On the embryo of *Welwitschia*, Ann Bot 24: 759–766

Rembur J (1970) Etude autoradiographique et cytophotométrique de la synthèse du DNA au cours des premières heures de germination chez le *Xanthium pennsylvanicum* Wallr. (Ambrosiacées). C R Acad Sci Fr 271: 908–911

Rembur J, Nougarède A (1977) Duration of cell cycles in the shoot apex of *Chrysanthemum segetum* L. Z Pflanzenphysiol 81: 173–179

Reynolds ES (1963) The use of lead citrate at high pH as an electron opaque stain in electron microscopy. J Cell Biol 17: 208–212

Rondet P (1958) Répartition et signification des acides ribonucléiques au cours de l'embryogenèse chez *Lens culinaris* L. C R Acad Sci Fr 246: 2396–2399

Rondet P (1961) Répartition et signification des acides ribonucléiques au cours de l'embryogenèse chez *Myosurus minimus* L. C R Acad Sci Fr 253: 1725–1727

Rondet P (1962) L'organogenèse au cours de l'embryogenèse chez *Alyssum maritimum* Lamk. C R Acad Sci Fr 255: 2278–2280

Rondet P (1965) Mise en place des méristèmes chez les Angiospermes au cours de l'embryogenèse. Bull Soc Fr Physiol Vég 11: 175–186

Sadebeck R (1878) Die Entwicklung des Keimes der Schachtelhalme. Jahr Wiss Bot 11: 575–602

Schulz SP, Jensen WA (1969) *Capsella embryogenesis:* the suspensor and the basal cell. Protoplasma 67: 139–163

Souèges R (1911) Recherches sur l'embryogénie des Renonculacées. Bull Soc Bot Fr 58: 542–549, 629–636, 718–725

Souèges R (1920) Embryogénie des Chénopodiacées. Développement de l'embryon chez le *Chenopodium Bonus Henricus* L. Bull Soc Bot Fr 67: 233–257

Souèges R (1924a) Embryogénie des Caryophyllacées. Développement de l'embryon chez le *Sagina procumbens* L. Bull Soc Bot Fr 71: 590–614

Souèges R (1924b) Embryogénie des Graminées. Développement de l'embryon chez le Poa annua L. C R Acad Sci Paris 178: 860–862

Souèges R (1929) Recherches sur l'embryogénie des Légumineuses (*Trifolium minus* Rehl.) Bull Soc Bot Fr 76: 338–346

Souèges R (1931) L'embryon de *Sagittaria sagittaefolia* L. Le cône végétatif de la tige et l'extrémité radiculaire chez les Monocotylédones. Ann Sci Nat Bot 10e sér 13: 353–402

Souèges R (1933) Recherches sur l'embryogénie des Joncacées. Bull Soc Bot Fr 80: 51–69

Souèges R (1937) Développement de l'embryon de *Scabiosa succisa* L. C R Acad Sci Paris 204: 292–294

Strasburger E (1872) Die Coniferen und die Gnetaceen. Ambr Abel Leipzig, Bermühler, Berlin

Strasburger E (1879) Die Angiospermen und die Gymnospermen. Fisher, *Jena*

Thomas MJ (1972a) Comportement des embryons de 3 espèces de Pins (*Pinus mugo* Turra, P. *silvestris* L. et P. *nigra* Arn.) isolés au moment de leur clivage et cultivés *in vitro,* en présence de cultures-nourrices. C R Acad Sci Fr 274: 2655–2658

Thomas MJ (1972b) Etude comparée des développements *in situ* et *in vitro* des embryons de Pins: influence des cellules - nourrices sur les embryons isolés durant les premiers stades de leur développement. Soc Bot Fr Mém 1973 Coll Morphol: Naissance de la forme chez l'embryon, pp 147–178

Thomas MJ (1973) Etude cytologique de l'embryogenèse du *Pinus silvestris* L. - I. Du proembryon cellulaire au clivage embryonnaire. Rev Cytol Biol Vég 36: 165-252

Thomas MJ (1975) Apports de la cytologie et de la Cytochimie à l'étude du développement embryonnaire chez les Phanérogames. L'Année biol 14: 294-352

Thomas MJ (1978) Etude cytologique de l'embryogenèse du *Pinus silvestris* L. II. Du clivage embryonnaire à la graine mûre. Rev Cytol Biol Vég Bot 1: 313-403

Thomas MJ (1980) Contributions récentes de la cytologie et de la Cytochimie à l'étude de l'embryogenèse des Spermaphytes. Bull Soc Bot Fr, Actual Bot 127: 5-18

Thompson WP (1916) The morphology and affinities of *Gnetum*. Am J Bot 3: 135-184

Thomson WW (1979) Ultrastructure of dry seed tissue after a non aqueous primary fixation. New Phytol 82: 207-212

Tourmel-Prevosti AM (1980) Embryogenèse du *Phaseolus vulgaris* L. var. *Contender*. Thèse Fac Sci Genève (N° 1940)

Tourte Y (1975a) Etude ultrastructurale de l'oogenèse chez une Ptéridophyte, le *Pteridium aquilinum* L. Kuln I. Evolution des structures nucléaires. J Microsc Biol Cell 22: 87-108

Tourte Y (1975b) Etude infrastructurale de l'oogenèse chez une Ptéridophyte. II. Evolution des mitochondries et des plastes. J Microsc Biol Cell 23: 301-316

Treub M (1884) Etudes sur les Lycopodiacées. Ann Jard Bot Buitenzorg 4: 107-138

Treub M (1886) Etudes sur les Lycopodiacées. II. le prothalle de *Lycopodium phlegmaria* L. Ann Jard Bot Buitenzorg 5: 87-139

Treub M (1890) Etudes sur les Lycopodiacées. L'embryon et la plantule du *Lycopodium cernuum* L. VII. Les tubercules radicaux du *Lycopodium cernuum* L. Ann Jard Bot Buitenzorg 8: 1-37

Vallade J, Cornu A (1973) Etude cytophotométrique des stades zygotiques chez le *Petunia*. C R Acad Sci Fr 216: 2793-2796

Woodard JW (1956) DNA in gametogenesis and embryogeny in *Tradescantia*. J Biophys Biochem 2: 765-776

CHAPTER 2

Meristems and the Indefinite Ontogenesis of Plants

The distribution in the sporophytic organism as well as the histological and cyto-logical structures allow the distinction of two main groups of meristems: *the apical meristems* or primary meristems situated at the extremities of the axes, and the *secondary meristems* or *cambiums* which elaborate layers of proliferating cells enclosed within the tissues they produce on each side. We will discuss these groups separately.

2.1 Apical Meristems

All vascular plants have apical meristems at the extremity of their axial members.

2.1.1 Ontogenetic Function of Apical Meristems

The functioning of the apical meristems allows the lengthening of axial organs, shoots and roots and the initiation of lateral members of bilateral symmetry such as leaves.

They are composed of a cluster of cells which can divide as long as growth continues, i.e. theoretically during the whole life of the plant.

Thus, in apical meristems, a permanent, continued embryogenesis takes place which relates the zygote segmentation and the intraseminal edification of the first organs of the embryo.

In Phanerogams, the sporophyte develops in two stages, separated by a latency period of the seed. The apical meristems ensure then *the post-seminal ontogenesis* which extends the intraseminal development by which the plantlet develops inside the seed.

In Pteridophytes, we know that the zygote segmentation results in a young sporophyte with a buttress of shoot and with a first root. Both organs have an apical meristem similar to that of the seed plantlets, but their functioning can be permanent if circumstances are favourable. Thus, the embryonic proliferation progresses along in the meristems and the ontogenesis goes on without any latency phase.

The apical meristem of the shoot and that of the root in the same plant have different structures which shall be discussed separately (cauline and root meristems).

2.1.2 Histological Structure of Shoot Apical Meristems

The role of the apical meristems has led to the consideration that it is as important to know their structure as it is to know the modalities of embryogenesis at the beginning of zygote segmentation. In fact, the histology of these meristems has been largely studied and discussed. We shall only give a short chronological account of the past studies.

2.1.2.1 Successive Concepts

The "Apical Cell", Single Initial. The first accurate studies on the structure of the vegetative apices were those of NAEGELI (1845) and HOFMEISTER (1851) in vascular cryptogams. They showed that there is a large apical pyramidal cell on top of the fern axes, well differentiated from the surrounding ones. The shape and location of the neighbouring cells led to the concept that this cell initiates all the tissues of the organ in partitioning parallel to its various faces (Fig. 2.1).

From these results these authors and a number of others tried to recognize the single apical initial cell in the apices of other vascular plants.

These studies had a phylogenetic aim. At this time, due to the paleobotanic advances, Gymnosperms were thought to be derived from Pteridophytes and Angiosperms from Gymnosperms and in the intimate processes of organogenesis, survival of past times were looked for.

Thus, many authors attempted to generalize the structure acknowledged in ferns, in other plants and mainly in Gymnosperms. While discussions were going on, another concept was elaborated from the Angiosperm study.

The Histogens. In 1858–59, for the first time, CASPARY suggested the existence of several superposed initials. But two important reports of HANSTEIN, fundamental because of the number of examples (48 genera) and the accuracy of interpretation were published in 1868 and 1870.

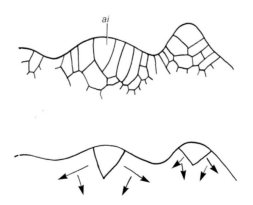

Fig. 2.1. Vegetative apex of an *Adiantum* rhizome, histological aspect and past interpretation, according to HOFMEISTER's concepts (1851). *ai* Apical initial

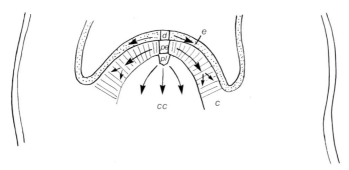

Fig. 2.2. Vegetative apex of Angiosperm, interpreted according to HANSTEIN's "histogen" theory (1868). *d Dermatogen* initial, from which the epidermis *e* originates; *pe periblem* initial, from which the cortex *c* originates; *pl plerome* initial, from which the central cylinder *cc* originates

In the course of embryogenesis, HANSTEIN reported the appearance of three superposed cells at the apex, in the organ axis, each cell producing a distinct meristematic layer. These three layers, *plerome, periblem* and *dermatogen* (Fig. 2.2) constitute the three *histogens* from which all the organ tissues derive. They respectively initiate the central cylinder, the cortex and the epidermis.

But HANSTEIN's contemporaries were not long in arguing that, mainly in the shoot, the separation between the cortex and the central cylinder was difficult to establish. They considered the presence of two independent internal histogens, plerome and periblem, as doubtful. In 1890, DOULIOT distinguished Angiosperms with three initials from Angiosperms with two initials, the second initial being common to the cortex and to the central cylinder.

Although this concept of apical structure of the shoots was considered doubtful shortly after its publication, it has become conventional and is still largely dealt with in books. However, almost all anatomists have chosen a more satisfactory interpretation.

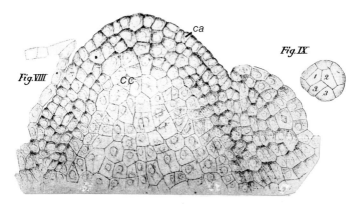

Fig. 2.3. Vegetative apex of *Taxus,* longitudinal section *(Fig. VIII),* showing two cytologically distinct zones: the central core *cc* and the hood (cap) *ca.* The absence of notably differentiated top cells is visible on the scalp *(Fig. IX).* Reproduction of a drawing after KOCH (1891)

SCHMIDT's *Interpretation*. Already in 1891, KOCH who was among the first to deal with the cellular contents and not only with cell walls, described the vegetative points of Gymnosperms and *Ephedra,* showing that they were composed of two *cytologically* well-defined areas: a *central body* with relatively large cells, including many vacuoles, and a hood of small cells with a denser protoplasm (Fig. 2.3). These two areas cannot be compared with the histogens of HANSTEIN, because KOCH indicated that the first ones produced only the medulla and the second all the *remaining parts* i.e. the cortex as well as the conducting tissues for instance.

However, KOCH admitted that the whole set was derived from an apical cell, sometimes divided into four; but the shapes and mode of segmentation of this cell are not homologous of the apical cell described by HOFMEISTER. KOCH's interpretation was very close to that which was accepted almost unanimously but formulated only 33 years later by SCHMIDT (1924). This author distinguished two *cytologically* and *histologically* well-distinguished areas:

1. A mass of isodiametric cells, i.e. of about the same size in all directions, arranged at random, with relatively large vacuoles, dividing in various planes. This mass constitutes the *corpus* (Fig. 2.4);
2. The *corpus* is covered by one or several layers of smaller cells with reduced vacuoles, which form the *tunica*. In the *tunica,* partitions are always perpendicular to the surface of the apex, i.e. anticline, except in places initiating leaves. The arrangement of the *tunica* in regular layers depends upon the orientation of mitoses. The author did not give the cells of each area any precise filiation, i.e. determined and strict. Therefore, these two zones are in no way comparable to two histogens according to HANSTEIN. Moreover, although SCHMIDT did not discuss the concept of an axial apical initiation, he denied the concept of a determined number of initials. Beside, the number of tunical layers, almost constant for a selected species varies from one species to the other, e.g. *Avena sativa,* one; *Syringa vulgaris,* two; *Rubus idaeus,* three; *Hippuris vulgaris;* four to six.

Contrarily to HANSTEIN's scheme, SCHMIDT's interpretation was moderate enough to deal with quite varied structures so that it was accepted by several histologists.

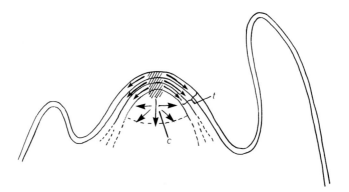

Fig. 2.4. Vegetative apex of Angiosperm, interpreted according to SCHMIDT (1924). *t* Tunica; *c* corpus; the two zones are restored by the activity of axial apical cells *(hachured)*

Nevertheless, some plants showed apical structures difficult to determine with regards to the tunica and corpus, such as Cycadeae. In these plants FOSTER and colleagues considered the apices as variably arranged zones, cytologically and histologically recognizable but interdependent and poorly delimited (see p. 117).

In attempting to determine the apex zonation precisely, other authors had to resume research on cellular filiation, but their criteria were not always of value. This fact resulted in the excessive multiplication of the structural types of cauline meristems (POPHAM 1951). New concepts issued from phyllotaxic research carried out by L. PLANTEFOL, simultaneously contradicted this tendency to diversify, marked by methods of past times.

The Initiating Ring (Anneau Initial) and the Critic of the Apical Initiation. We do not intend to give a detailed study of L. PLANTEFOL's phyllotaxic concepts (1947 and 1948). Presently, we only need to know that this author denounced the unreal characteristic of the phyllotaxic theory which had been acknowledged for more than a century, i.e. the suggestion that leaves originate at the vegetative apex in such loci that their points of insertion are situated on a single helix, "the genetic spiral".

PLANTEFOL demonstrated that, generally, leaves were initiated on the stem according to *several* helices and not to a single one. Between two leaves, the conventional theory defined a supposedly constant central angle, called "divergence". PLANTEFOL showed that on *each foliar helix,* successive leaves arise *side by side,* as if an induction transmitted from one to the other settled a *meristematic continuity* between the genesis of the successive leaves of the same helix. Therefore, the author substituted the concept of *contiguity* to that of *divergence*.

Thus, in the vegetative point, several organogenesis centres, as many as the number of helices on the stem, may have a more or less autonomous functioning. This multiplicity of generative centres does not fit the concept of *apical initiation,* usually admitted by anatomists.

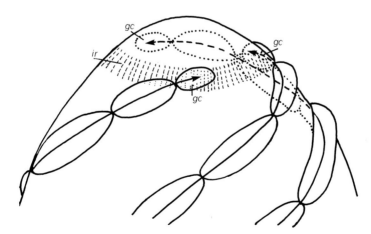

Fig. 2.5. Vegetative apex interpreted in accordance with the concept of *foliar helices* and contiguity. Three foliar helices are represented, each one proceeding from a generating centre *gc,* moving in the "initiating ring" (anneau initial) *ir;* (after PLANTEFOL 1947)

When he resumed his studies on apex functioning, PLANTEFOL denied the apical initiation to which he substituted a subapical and peripheral initiating zone giving the "anneau initial" or "initiating ring" (Fig. 2.5).

The illustrations of numerous former works, reviewed with these new concepts in mind, confirm the existence of a rough ring of particularly meristematic cells from which the leaf primordia are derived.

These new concepts, recently compared with the histological and cytological researches going on, led to a more satisfactory and more general interpretation of the structures and of the role of cauline vegetative apices of Dicotyledons. Moreover, the subsequent works of CAMEFORT (1956a) extended these data to Gymnosperms. Since then, numerous works have dealt with the cytological and kinetic study of the cauline meristems of the vascular plants belonging to all the large systematic groups.

The present state of knowledge lacks some data but it allows a quite satisfactory review of the apical structures of shoots.

2.1.2.2 Shoot Apical Meristems of Pteridophytes

Advanced Pteridophytes. The existence of the tetrahedral or pyramidal apical cell, described by NAEGELI (1868) and HOFMEISTER (1851), has been confirmed many times over one century in the Pteridophytes that are usually considered as the most advanced, e.g. leptosporangiate ferns and Equisetineae.

On both sides of this particular cell are flattened cells interpreted as the production of the apical cell. According to this interpretation, this cell could undergo unequal segmentations resulting in successive segments, piled up, which then partition again in variable ways.

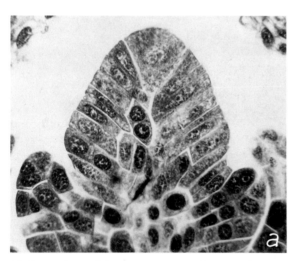
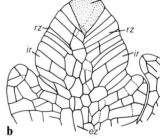

Fig. 2.6. a Apex of *Equisetum arvense;* **b** interpretation scheme. *az* Apical zone, usually reduced to a single tetrahedral cell; *rz* regeneration zone of the initial ring *ir; ez* elongation zone of medullary cells. The oblique cleavages hardly visible on the microphotograph are shown by *dotted lines*

110

The Apical Meristem of Equisetum. Histologically, due to its size, its shape and that of the apex itself, it is in the *Equisetum* that the apical cell is best demonstrated, the most distinct from the neighbouring ones (Fig. 2.6).

The study of the cellular division frequency in the various areas of this apex reveals that the apical cell divides only rarely in adult plants. The highest proliferating activity is found on the sides of the apical cone at the points of leaf initiation. This lateral area is composed of a series of flattened cells which are the most meristematic of the apex and have the most important histogenetic role: they are the homolog of the *initiating ring* which will be described in Dicotyledons more easily. Pericline divisions of these lateral cells separate central smalls cells which will produce the medullary tissue.

Considering the results of mitosis frequency, the quantitative measurements of the DNA content of the *Equisetum* apical cells carried out by means of cytophotometry led D'AMATO and AVANZI (1968) to detect that numerous apical cells are polyploid and accumulate paraplasmic material such as starch.

The authors concluded that the apical cell is active at the onset of the edification of the meristem to which it gives a shape but it has almost no further activity, neither itself nor the close neighbouring cells, in initiating the leaf stem of adult plants. But in the apical functioning, the apical cell has a possibly endocrine role or plays a part in the apical organization (WARDLAW 1955).

The Apical Meristem of Filicales Leptosporangiate. The apex of advanced ferns is usually much more flattened that that of *Equisetum* (Fig. 2.7) but it can be briefly interpreted in the same terms. The studies of N. MICHAUX-FERRIÈRE (1968, 1971) provided accurate information on the apical activity of *Pteris cretica*. The young vegetative points, still in formation, have a tetrahedral apical cell, which segments with a frequency two times less than in lateral cells at the places where leaves are initiated. Its regular activity determines the histological aspect of the meristem (Fig. 2.7) which will be maintained by a further activity. But when the plant becomes adult and grows old, the segmentation of the apical cell occurs more and more rarely and the nuclear divisions result in a progressive endopolyploidy. The apical cell, loaded with storage material, loses its role of initial (MICHAUX 1970a, b, 1971). The initiation of leaves begins in zone 2 (Fig. 2.7b), the most proliferating. It starts with an oblique mitosis resulting in a leaf apical cell (Fig. 2.7).

Cytologically, a dedifferentiation gradient is visible from the apical cell to the cells of zone 2 which initiate the leaf primordia: these cells form the equivalent of the *"initiating ring"* which will be described in Angiosperms. A similar gradient exists between zone 1 and zone 3 from which the axial tissues of the rhizome are derived.

When a leaf primordium rises over zone 2, encroaching on it more or less, this zone regenerates through anticline mitoses of the cells spared by the leaf arising. This maintenance mechanism of the vegetative point, independent from the activity of apical cells, will be found again in spermaphytes.

Primitive Pteridophytes. Among the other Pteridophytes considered as rather primitive according to the anatomical structure of their vegetative system (sporiferous apparatus excluded), a particular apical cell is not always found. Such is the case

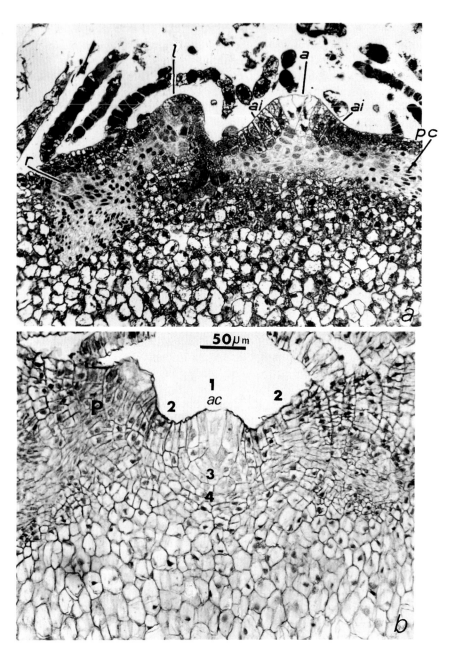

Fig. 2.7 a, b. Vegetative apex of *Pteris cretica* rhizome (leptosporangiate fern), lying at the extremity of an axis, rapidly enlarged by the growth of the massive ground parenchyma. **a** Apical cells, with large vacuoles; *a* central tetahedral cell; *ai* flank cells with tanniferous small vacuoles, the proliferation of which restores the "anneau initial" (initiating ring); *l* leaf buttress, already with a procambium; *r* initiation of an adventitious root, born from the procambium of the basis of leaf buttress (gemmar root). Below the apex, clear cells (not tanniferous) which proliferate periclinally contribute to the longitudinal growth by the production of ground parenchyma cells. In this parenchyma, and only laterally, proconducting cells *pc* occur. The leaf initiation takes place above

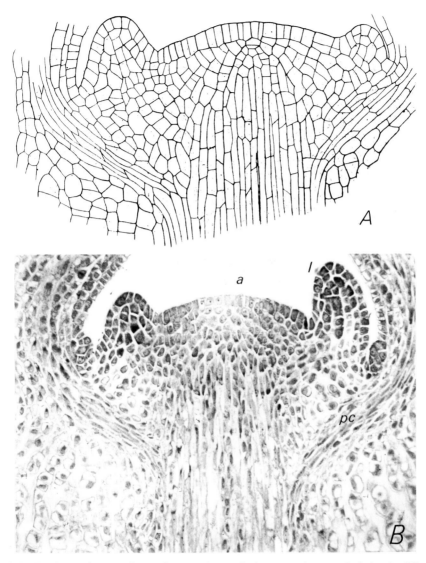

Fig. 2.8 A, B. Apex of *Lycopodium selago*. **A** Anatomical aspect; absence of obviously differentiated apical cells; axial origin of the stele; the leaves (microphylls) and their proconducting strand are born *independently* of the stele; (after STRASBURGER 1872). **B** Apex treated according to BRACHET's technique, showing the RNA enrichment of lateral leaf primordia, as well as subterminal initials of the stele. Abbreviations as in Fig. 2.7 a; (after NOUGAREDE and LOISEAU 1963)

these cells (REGAUD's technique). **b** Same material, apical zonation according to MICHAUX (1970b), as it appears with BRACHET's techniques for RNA detection; *ac* apical cells (zone *1*); zone *2* lateral cells, showing a dedifferentiation gradient; zone *3* the cells are less differentiated than the overlying apical cells; zone *4* "shell zone" at the origin of medullary parenchyma

113

of many Lycopodinae. Among the Psilotales, presently limited to two genera *(Psi-lotum* and *Tmesipteris),* the apical cell is not always very different from the neighbouring ones. Moreover, there is no quite distinct apical cell in *Eusporangiate Filicinae,* such as Osmondales and Ophioglossales. Thus in *Botrychium,* the apical cell, not very different from the neighbouring ones, seems to be in relationship only with the superficial tissues of the leafy stem, as the internal tissues constitute an apparently independent system.

In many cases, the apex shows several similar cells, the functioning of which is not established, the cell wall arrangement, being a wrong criterion, e.g. *Lycopodium selago* (Fig.2.8). Insufficient research has been devoted to the functional study of these meristems.

The basophilic variations in the areas of the vegetative point of *Selaginella caulescens* (Fig.2.9a) suggest that the top cells, in which no particular initium can be singled out, are less active than the subapical cells which are much more meristematic and constitute the initiating sites of the leaves and stele (in such plants, the

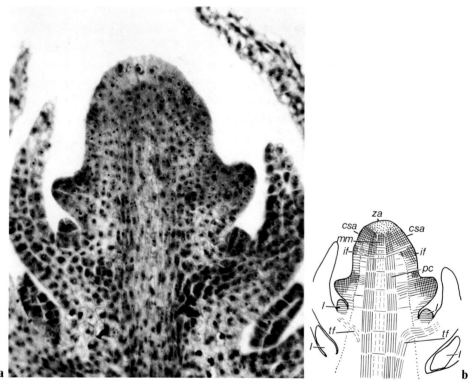

Fig. 2.9 a, b. *Selaginella caulescens.* **a** Section of the apex, axial at the upper part; some apical cells are larger and more vacuolated, the most central one is triangular in section. The diverse areas are all the more meristematic as they are darker (Fixation: Navaschin-hematoxylin). **b** Schema of functional interpretation; *za* more or less passive apical zone; *csa* crown of subapical cells, initiating leaves, cortex and, independently, conducting tissues; *mm* medullary meristem, initiating the pith; *if* leaf initiums; *pc* proconducting cylinder; *l* ligules; *tf* leaf traces (original lettering). The initiating sites are indicated by *crossed lines;* (after BUVAT 1955)

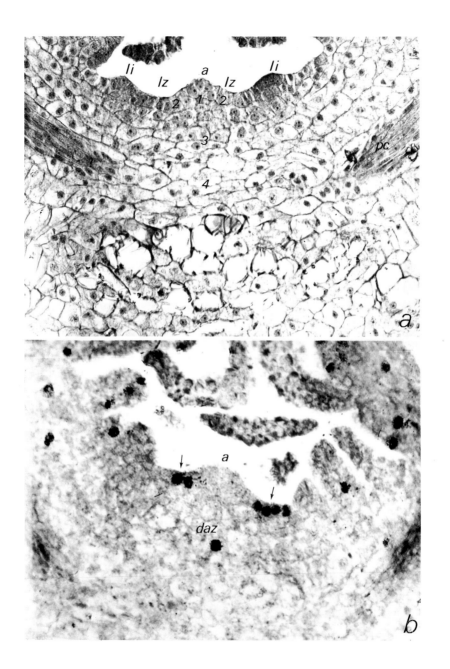

Fig. 2.10a, b. *Isoetes setacea* Lam. **a** Vegetative apex treated according to Brachet's technique. The zonation is similar to that of *Pteris cretica* (Fig. 2.7). *a* Axial cells; *lz* lateral zone; *li* leaf initium; *pc* procambium. The dedifferentiation gradient is visible between *a* and *lz*. Correlatively, the cell cycles are shorter and shorter; (after Michaux 1970b). **b** Vegetative apex after ^{3}H-thymidine incorporation over 3 h; fixation after 36 h. The labelling is prevalent in the lateral zone *(arrows)* and poor in the deep apical zone *(daz);* (after Michaux 1967)

latter initiates separately from the leaves and the leaf traces then join to it by apposition) (Fig. 2.9 b).

MICHAUX's more accurate investigations (1966, 1967) on the vegetative point of the *Isoetes setacea,* using histoautoradiography, showed that the top cells of zone 1 (Fig. 2.10 b) divide only rarely in the adult plant; zone 3 is the site of leaf initiation and its regeneration fundamentally results from pericline mitoses of zones 3 and 2 residual cells. Zone 4, moderately active, generates the central parenchyma (Fig. 2.10 a).

Let us recall that the existence of a *clearly distinct* apical cell *is rather a characteristic of highly advanced plant than of a primitive one.*

2.1.2.3 Apical Meristems of Prephanerogams (Prespermaphytes)

The vegetative point of various *Cycadales* and *Ginkgo* was studied by FOSTER (1938, 1941 a, b). It concerns large meristems with depressed domes, in which the author described a more complex zonation than in true Phanerogams. In fact, the diameter of the vegetative point may be about 2 or 3 mm in some of these plants. Each histological area distinguished by FOSTER is composed of numerous cells. No particular initials can be singled out.

Figures 2.11 b and d illustrate the author's interpretations concerning *Gingko biloba* and *Dioon* edule. Areas 1 (zia) are considered as *initiating areas,* therefore likely axial apical cells. Larger underlying cells derived from zone 2 are called *central mother cells (cmc).* FOSTER described them as thickened cell-walled where mitoses are rare but which play an initiating part. Zone 3 initiates peripheral tissues (zp); zone 4 is the meristem of central tissues *(mc)* or *"rib meristem".* In zones 3 and 4 the cells themselves should derive from the initial cells of zones 1 and 2.

The works of CAMEFORT (1951, 1956 a, b), a colleague of PLANTEFOL, confirm the relative inertia of the so-called axial or central initials of *Gingko* and CAMEFORT's interpretation of the apical structure of this plant is illustrated in Fig. 2.11 c, inferred from Fig. 2.11 a.

It shows that the larger part of the stem tissues and all those of leaves are produced by the activity of an initiating ring (roughly FOSTER's zone 3). In the centre, under the more or less passive area, is the medullary meristem (FOSTER's zone 4) which initiates the medulla. Moreover, this zone is very developed in Prephanerogams, which is considered a primitive feature (with regards to Phanerogams). The organogenic initiating ring is constantly maintained by flank anticline mitoses

---

Fig. 2.11 a–d. Apical meristems of *Prephanerogams.* **a–c** *Ginkgo biloba.* **a** Photomicrograph of a longitudinal axial section (after CAMEFORT 1956); **b** interpretation of the same apex according to FOSTER's concept (1938); *zai* zone of apical "initials"; *cmc* central mother cells; *zp* zone of peripheral tissue; *mc* central meristem; **c** interpretation according to CAMEFORT (1951); *za* apical zone, practically inactive; *ai* "anneau initial" (initiating ring) topped by a maintenance zone, functioning through flank mitoses, indicated by *double arrows; mm* medullary meristem. **d** *Dioon edule,* FOSTER's interpretation (1941 a). *1* Zone of apical initials; *2* central mother cells; *3* zone of peripheral tissue; *4* central meristem

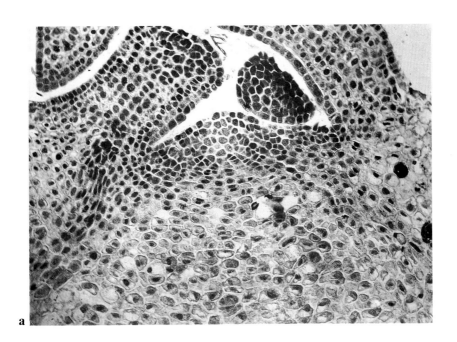

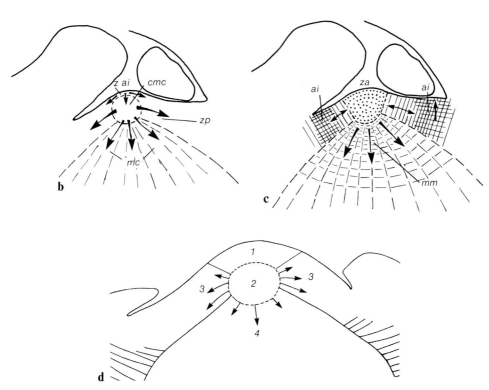

(partitions perpendicular to the apex surface) at its upper limit, i.e. between the zone of leaf initiation and the passive or slightly active central zone. To date, the significance of the latter has not been elucidated.

2.1.2.4 Apical Meristems of Gymnosperms

The vegetative point of Gymnosperms soon attracted the attention of biologists, however, its structure was much debated. When the "Tunica-Corpus" concept was explicated by SCHMIDT, a number of authors tried to interprete the apical structure of Gymnosperms in such terms. Although the two areas described by KOCH (cf. Fig. 2.3) cannot be denied, this interpretation caused difficulties, mainly raised by the fact that pericline mitoses occur here and there in the superficial layer, apart from the genesis of leaves. This fact is not consistent with the definition of the tunica given by SCHMIDT and accounts for KORODY's concept (1937) that the Gymnosperm apex could be a bare corpus.

Too much importance was given to illusory geometrical considerations and to undetermined cellular filiations. With reference to the works of KOCH, KORODY, CROSS and through his personal research, FOSTER (1941b) established a scheme of the apex zonation in Gymnosperms (Fig. 2.12) close to that of Prephanerogams, including a body of central mother cells, themselves derived from the axial apical cells (e.g. *Abies venusta*, Fig. 2.12).

As in *Ginkgo*, CAMEFORT (1950) observed a relative inertia in the apical cells and the presence of a phyllogen initiating ring. The scheme established by this author (Fig. 2.13) for *Picea excelsa* is easily comparable to that concerning the *Ginkgo*. The accurate cytological studies carried out by CAMEFORT (1950, 1951, 1956b) on various Gymnosperms *(Pinus, Picea, Cupressus, Taxus, Sequoia)* as well as on *Ginkgo*, yielded particularly clear results: Gymnosperms gave the best demonstrations of the autonomous functioning of the initiating ring. In fact, during the period of activity of the vegetative point, CAMEFORT provided evidence of a gradient of dedifferentiation growing from top to bottom on the apex flanks. The proliferating activity of the upper cells of the ring along with a continued dedifferentiation of the cells produced downwards, constantly without any help from the vege-

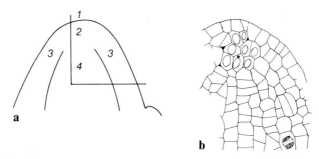

Fig. 2.12 a, b. *Abies venusta*, apical meristem, interpreted according to FOSTER (1941b). **a** Interpretation scheme; *1* apical "initials"; *2* central mother cells; *3* zone of peripheral tissue; *4* rib meristem. **b** Detail of the part of **a** limited by the right angle

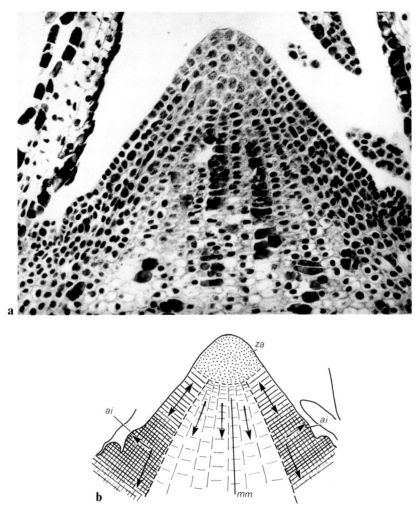

Fig. 2.13a, b. *Picea excelsa.* **a** Apical meristem; **b** interpretation scheme, according to H. CAME-
FORT (1950). *za* Apical zone, practically inactive; *ai* "anneau initial" (initiating ring), where
the leaves are initiated and which is regenerated by anticlinal flank mitoses *(double arrows);*
mm medullary meristem.

   The sites of initiation are represented by *crossed lines.* The cell density, increasing from the top
towards the lower part of the flanks, corresponds to the accentuation of the meristematic state

tative cone top, results in the production of organogenic primary meristematic
cells initiating the leaves and their basement.

   Dedifferentiation is mainly characterized by the enlargement of nucleoli, along
with the reduction of plastids, the pulverization of chondriosomes and the decrease of tannins. Moreover, these phenomena are consistent with the considerable enrichment of the cells in ribonucleic acids, checked by the technique of BRA-
CHET using pyronine staining (CAMEFORT 1954).

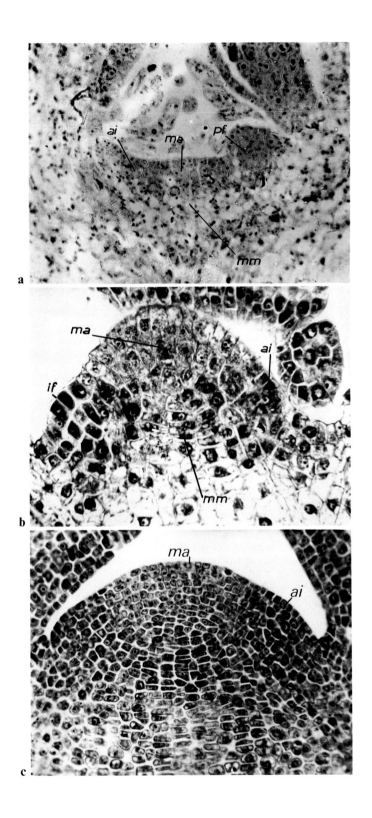

Fig. 2.14 a-c. Apical meristems of Dicotyledons, histological aspect. **a** *Myosurus minimus:* flat apex, raised on the edges by the "anneau initial" (initiating ring) *ai; ma* "meristème d'attente" (waiting meristem); *mm* medullary meristem; *pf* leaf primordium. **b** *Chrysanthemum segetum,* convex apex, the anneau initial *ai* is situated at the base and encompasses the medullary meristem *mm; if* leaf initium; (after LANCE 1957). **c** *Cheiranthus cheiri,* apical meristem of an aged stem about to become a reproductive meristem. The meristème d'attente (waiting meristem) *ma* is much enlarged, the tunical layers, two in number at first, have become numerous, but the anneau initial and the medullary meristem remain clearly visible

Inside the cone-shaped ring, an active medullary meristem exclusively produces an abundant medulla.

The relative inactivity of most apical cells lessens the importance of the pericline mitoses which may affect these cells. In the maintenance zone of the initiating ring, those mitoses slightly alter the seried aspect of the "tunica". It is normal to observe pericline segmentation slightly downwards in the initiating ring itself: they will give rise to the leaves.

2.1.2.5 Apical Meristems of Angiosperms

Although most work on the Angiosperm apex was performed on Dicotyledons, it is possible, with reference to recent works, to determine the structures and functioning of the vegetative point of Monocotyledons.

Typically, the vegetative point of Dicotyledons presents functional areas comparable to those of the Gymnosperm apex, but is generally smaller in size (Fig. 2.14).

The top of the axis may be dome-shaped *(Lupinus, Vicia, Chrysanthemum),* often very flat *(Cheiranthus)* or it forms a plane surface (*Myosurus,* Fig. 2.14).

But the shape and structure of the apex can vary during the development of the plant, after germination. The organization of the vegetative point is sometimes completed only slowly, at the beginning of post-seminal ontogenesis. Below, we shall deal only with vegetative meristems which have completed their zonation, except for special quotations.

Two zones defined by SCHMIDT are usually distinguished: *tunica* and *corpus,* most often the tunica includes two layers, but sometimes more (*Rubus idaeus:* three, *Heracleum spondylium:* six to eight).

However, in both zones, several areas with different functions must be typically distinguished. The more axial cells, in the tunica as well as in the corpus, are more or less quiescent during the phase of vegetative growth. In this upper area mitoses are more or less frequent according to the species and, most often, the zone, where proliferation is by far the most active, is located around this area (Fig. 2.15).

This zone forms a subterminal ring at the level of which the leaves initiate during the organogenesis of the leafy shoot. It constitutes the initiating ring (PLANTE-FOL 1947, Fig. 2.17).

On the border of the three areas: apical-axial, initiating ring, medullary meristem (cf. Figs. 2.16 and 2.17) "growth harmonization zones" establish arranged transition zones (Fig. 2.15, stippled).

121

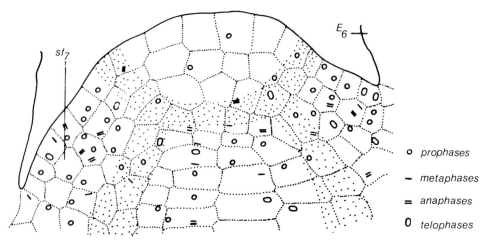

o   prophases

—   metaphases

=   anaphases

0   telophases

Fig. 2.15. *Chrysanthemum segetum;* superposition of reports of mitoses from axial longitudinal sections of ten apical meristems, on the initiation of the seventh leaf initium *Sf 7* and the regeneration of the initiating ring at the sixth leaf primordium axil *E 6*. The transition zones between the three regions (apical zone, initiating ring and medullary meristem) or "growth harmonization zones" are *stippled;* (after LANCE 1957)

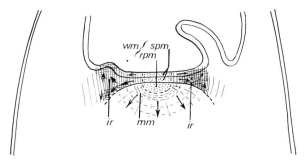

Fig. 2.16. Interpretation of the flat vegetative apex of *Myosurus minimus. wm* Waiting meristem ("meristème d'attente"), composed of the tunical sporogen promeristem *spm* and the receptacle promeristem composed of the upper part of the corpus, *rpm; ir* initiating ring (anneau initial); *mm* medullary meristem

English speaking authors discussed the concept of initiating ring, putting forward the existence of sporadic mitoses in the top area. In fact, such mitoses are only frequent in particular cases and mainly in aquatic Angiosperms in which the stele reduction inaugurates the recurrence of a condition similar to that of Selaginellaceae (Fig. 2.18).

In such cases, the works of LANCE-NOUGAREDE and LOISEAU (1960) assessed that there was no initiating ring since its existence is linked to leaf gaps, therefore to the stele structure as will be shown later (Fig. 2.18).

In addition, the initiating ring is a structure of the phase of vegetative growth of leafy shoots. It disappears when the reproductive phase begins. Thus, it is clearly determined in species in which these two phases are well separated but it is

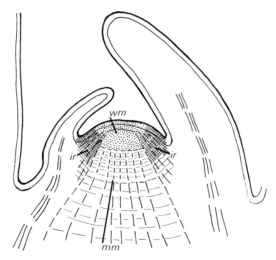

Fig. 2.17. Interpretation of the slightly convex apical meristem of a still young stem of *Cheiranthus cheiri* (same abbreviations as in Figure 2.16)

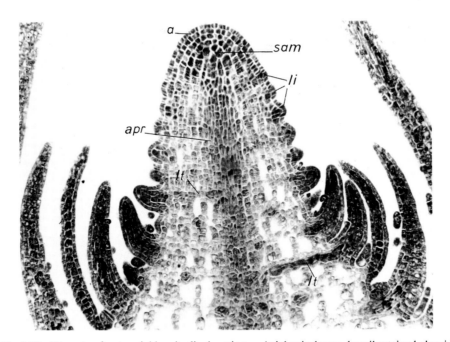

Fig. 2.18. *Hippuris vulgaris*, axial longitudinal section. *a* Axial apical zone, heavily stained, devoid of initiating ring; *sam* subapical meristem, from which the axial provascular strand *apr* is derived; *li* subepidermal leaf initiums; *lt* leaf traces linked to the provascular strand without leaf gap. BRACHET's technique (1941) was used for RNA detection; (after LANCE-NOUGAREDE and LOISEAU 1960)

123

sometimes less characterized in plants in which the passage from the vegetative condition to the reproductive one occurs either very early or very progressively (*Tropaeolum,* numerous Cruciferaceae).

Besides, the cauline meristem of preferring or strict photoperiodic plants, cultivated for a continued duration under a photoperiod which is not favourable to flowering, develops into a particular state referred to as the "intermediate state" (LANCE 1957, p 375; NOUGAREDE et al. 1965). The corpus grows and is covered by a tunica with more or less zonation, while the meristem basis keeps on initiating leaves. A few favourable photoperiods, one only in some species, are necessary to set off the passage to the reproductive state. Such a particular structure with reduced zonation concerning meristems slowly activated in view of preflowering, often led to wrong interpretations, mitoses being more or less uniformly distributed.

Thus, the upper zone of the meristem may be more or less passive according to the species, however, during the vegetative activity, it seems to be activated only by impulsions induced by areas of maximal proliferating activity. These impulsions are necessary to restore the activity in this zone.

Finally, this upper zone will be considerably activated as the apex becomes a reproductive meristem. In spite of its previous low activity, this zone is prepared for this modification, and was given the name of "waiting meristem", an expression which has been contested by English speaking authors.

The initiating ring is mainly composed of the lateral cells of the tunica, but lateral cells derived from the corpus might be casually included. The limit between the tunica and the corpus becomes in fact very fuzzy at the level of the initial ring, due to the activity of this whole zone where the enlargement of other areas must necessarily become harmonized (LANCE 1957, cf. p 190).

Under the "waiting meristem", cells occur with a proliferating activity which is oriented: partitions, perpendicular to the axis, produce the longitudinal cellular lines which constitute the medulla. This axial meristem, surrounded by the initiating ring and the leaf segments derived from it, constitutes the *medullary meristem.* The most apical part of this meristem is the basis of the corpus (Figs. 2.16, 2.17).

These various areas are usually differentiated by cytological characteristics, mainly by the characteristics of the cytoplasm, the vacuoles and chondriosomes; these characteristics reflect the histogenic nature of the cells. The distinctions mentioned above are mainly functional and physiological. Moreover, such areas cannot be precisely delimited though some authors attempted to do this by determining the productions of the various apical initials. On the contrary, it seems that the functioning of these meristems does not follow any geometrical law ruling the sequence of segmentations in a stereotyped way. The apical proliferation is thus rather ruled by a trophic and humoral determinism and is not divided by any immutably defined limits.

Thus, the inactive apical area may be enlarged or reduced according to the development of the plant or the time of the day. Above all, this area may show a more obvious activity in the species which have a reduced apical meristem with rapid growth (e.g. *Pisum sativum,* NOUGAREDE and RONDET 1973) or when the arising of leaf primordia breaks the initiating ring more or less completely (e.g. *Impatiens biflora,* LOISEAU 1959, *Aquilegia vulgaris,* ROUGIER 1955). All these facts do not affect the essence per se of the apical functioning (LANCE 1975).

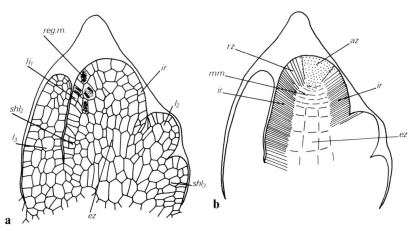

Fig. 2.19 a, b. *Triticum vulgare*, apical meristem. **a** Histology; *ir* initiating ring; *reg.m.* regeneration mitoses of the initiating ring, which will be encroached by the development of the last leaf initium *li₁*; *l₂* leaf primordium, just preceding *li₁*; *shl₂* initiation of the *l₂* sheath base; *l₃* and *shl₃* preceding leaf primordium and its sheath base; *ez* medullary elongation zone before the differentiation of nodes and internodes. **b** Interpretation schema; *az* practically passive apical zone; *rz* zone of the initiating ring *(ir)* regeneration; *mm* medullary meristem, reduced to some cells and *ez* its elongation zone

Finally, there is nothing comparable to the histogens of HANSTEIN, or the independent "layers" other than the generative superficial layer of epidermis in the vegetative apical meristem of the Dicotyledon leafy stem.

The available data reveal a similar apical functioning in Monocotyledons. However, a few differences must be mentioned, perhaps not in all cases, due to the fact that the leaves which encircle the stem widely encroach the apex in rising, which occurs rarely in Dicotyledons (Ombellifers). As a result, the apical meristem is often very small in these plants (*Triticum*, Fig. 2.19). Its structure is also simplified. The waiting meristem may be reduced to a few cells only and may not be permanent. Thus, in *Luzula pedemontana* an activity of mitoses progressing from the axil of primordium up to the top of the small apical dome restores the initiating ring (cf. sects. 2.1.5 and 2.1.6) after each leaf erection (Fig. 2.20 a, stage II). This activity causes oscillations of the initiating ring and the momentarily inactive zone (Fig. 2.20 b) (CATESSON 1953). It should be noted that the reduced apex sometimes looks like that of *Equisetum*, which is also very small (cf. Figs. 2.6 and 2.19).

This comparison suggests that the single apical cell of *Equisetum* can result from the evolutionary reduction of the vegetative point rather than the primitive completion of a peculiar apical initial. Thus, it would be a characteristic of evolution, very pronounced in certain Monocotyledons but at most in *Equisetum*. It should be recalled that the stem anatomy of these two groups of vascular plants also has very advanced and similar characteristics (e.g. the peripheral vessels).

In Graminaceae, the *tunica* is reduced to a single superficial layer which initiates a majority or the totality of the leaves by means of superficial pericline segmentations (Fig. 2.19). The apparent autonomy of the epidermis, which appeared as the only "histogen" in Dicotyledons, does not exist in these plants.

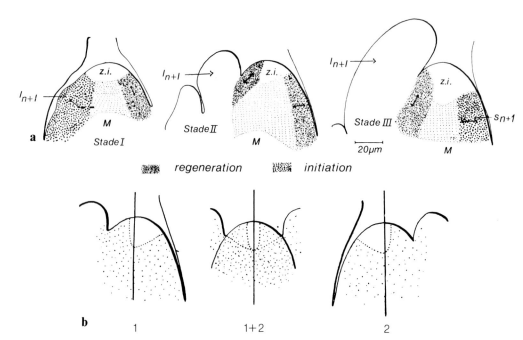

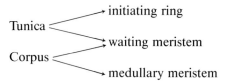

Fig. 2.20 a, b. *Luzula pedemontana.* **a** Apex zonation at three successive stages of leaf development *(ln + 1)* according to the mitosis frequency; *zi* inactive zone; *M* medullary meristem: *sn + 1* initiation site of the sheath base of *ln + 1*. At stage *II*, the regeneration mitoses of the initiating ring involve the top of the apex. **b** Oscillation of the initiating ring, due to the upwards progression of regeneration mitoses at the axil of each new leaf primordium. The inactive region is left *white*; (after CATESSON 1953)

Thus, as a whole, the cauline vegetative points of the vascular plants have a group of poorly active apical cells, less meristematic than the lateral cells. The latter form an initiating ring in which the leaves and the main tissues of the stem are initiated as we shall see further on. The centre of the stem, represented by the medullary parenchyma, is derived from an axial and subterminal medullary meristem (Figs. 2.16, 2.17).

Considering the average case of Dicotyledons, the various functional areas can be illustrated as follows:

Tunica —→ initiating ring
     —→ waiting meristem
Corpus —→ medullary meristem

Generally, the cauline meristems are all the more reduced as the plants are more advanced. The reduction affects mainly the surface of the top, above the initiating ring, which is sometimes reduced to a single cell *(Equisetum)* and divides then more frequently, at least in young plants. The previous interpretations were often confirmed by statistical histoautoradiographic studies and measurements of cellular cycles. SAINT CÔME (1966) was among the first who studied the incorporation

126

of tritiated thymidine in the different areas of the vegetative point of *Coleus blumei* and at the time of the passage to a reproductive condition.

The author showed that after a germinative phase, during which the apex is constituted by evenly distributed mitoses, the apical zonation establishes and a parallel is created between the cellular basophilia and the proliferating activity manifested by the number of nuclei marked after 6 h of incorporation in the different apex areas (Plate 2.1/1–4).

Although axial cells may be absorbed here and there in the lateral zone but mainly contribute to enlarge the top meristem, the statistical study shows that the low percentage of axial mitoses is insufficient *"to ensure the generation of cells necessary to the foliar initiation and to allow at the same time the clear increase of the volume of the axial zone"*. The marking is most pronounced during the regeneration phase (see p. 150, 151 and following, and Figs. 2.32, 2.33) starting at the axil of the last primordia, in the cells which have not been used and going on along the meristem flanks up to the top (Plate 2.1/4). It is at this phase that a few nuclei may be labelled in the axial zone; but for the main part, the top area which is almost passive is not involved in the apical regeneration (Plate 2.1/1).

These results confirm or are confirmed by a number of other autohistoradiographic studies (e.g. BERNIER and BRONCHART 1963; BERNIER 1964, with *Sinapis alba*; NOUGAREDE 1965 with *Teucrium scorodonia*; NOUGAREDE 1967 with *Chrysanthemum segetum*; TAILLANDIER 1969 with *Anagallis arvensis*; BESNARD-WIBAUT 1970, 1977 with *Arabidopsis thaliana*, etc.).

The measurement of the cell cycle duration in the three zones of the vegetative point, i.e. apical zone, lateral zone or initiating ring, and medullary meristem, provided further data to show the poor activity of the apical zone. In *Chrysanthemum segetum* in which the proliferating activity gives evidence of a zonation, regardless of the method used (colchicine, double-labelling ^{3}H or ^{14}C $-$ ^{3}H) NOUGAREDE and REMBUR (1977, 1978) and REMBUR and NOUGAREDE (1977) showed that the cell

Table 2.1. Duration of cellular cycles in the three zones of the cauline meristems of *Chrysanthemum segetum* and *Pisum sativum*

Material	Technique	Total duration of cycle (h)		
		Axial zone	Lateral zone	Medullary meristem
Chrysanthemum segetum[a]	Colchicine 0,5%	189	48	70
Chrysanthemum segetum[b]	Double labelling Thymidine ^{14}C + Thymidine ^{3}H	126	52	7
Chrysanthemum segetum	Double Labelling Thymidine ^{3}H	139	54	70
Pisum sativum[c]				
Vegetative point	Continued labelling	49	31	43
Inhibited axiallary	Continued labelling	127	65	55
Reactivated axillary	Continued labelling	40	33	41

[a] According to REMBUR and NOUGAREDE (1977).
[b] According to NOUGAREDE and REMBUR (1977).
[c] According to NOUGAREDE and RONDET (1976).

cycle is about three times longer in the axial zone than in the lateral zone and about two times longer in the medullary meristem than in the lateral zone (Table 2.1). In a species with a particularly rapid plastochronic rhythm, e.g. *Pisum sativum*, the cell cycles differ also in the same way according to the zones (NOUGA- REDE and RONDET 1976, Table 2.1).

The differences in cycle duration between the zones are more precise and mainly due to the duration of phase G1, a phase of presynthesis which controls the proliferation. When a meristem is temporarily inhibited, either by climatic con- ditions (*Isoetes*, MICHAUX 1972), apical dominance (*Pisum sativum*, NOUGAREDE and RONDET 1975), seasonal dormancy (*Fraxinus excelsior*, COTTIGNIES 1974, 1979; NOUGAREDE et al. 1973) or otherwise (bulbils of *Bryophyllum daigremontianum*, BROSSARD 1973), cells are blocked in phase G1 of the cycle. This phase saves the potentialities of the meristematic condition and allows the recurrence of prolifera- tion when the conditions are favourable. The blocking in phase G1 is therefore re- versible (NOUGAREDE and REMBUR 1985).

In contrast, phase G2 activates the differentiation and tends to impair the pro- liferating activity during normal development, sometimes definitely, for instance when such a phase favours the formation of endopolyploid cells (MICHAUX 1970a). In spite of the variations in cycle duration between the zones, the duration of mitoses is constant (Table 2.2).

Thus, in *Chrysanthemum* the relative variations of the mitotic index,

$$\frac{number\ of\ mitoses}{number\ of\ cells\ of\ selected\ area} \times 100,$$

and of the ratios of cellular division $\left(\frac{1}{T} \times 100;\ T = total\ duration\ of\ cycle\right)$ are

parallel. Therefore, in this example as in other species (NOUGAREDE and REMBUR 1985), the mitotic index is a reliable indication of the proliferation rate.

With regards to phase S, the synthesis phase, although the duration is longer in cells which divide less frequently, this duration does not increase proportionally to the extension of the cycle. Phase S represents 7.5% of the cycles of axial cells, 15.4 and 14% of the shorter cycles of the lateral zone and medullary meristem. This can explain the rarity of labellings obtained in the axial zone, after incorporation of tritiated thymidine. The use of both autoradiographic methods and microspectro-

---

Plate 2.1. Vegetative apical meristems of *Coleus blumei* (Labiateae). Histo autoradiographs ob- tained after 6-h incorporation of ^{3}H-thymidine.

1 Minimal area, beginning of regeneration at the axil of the leaf pair No.9, l_9; the labelled nuclei are localized on the meristem flanks, to the exclusion of the apical zone *az* nuclei.
2 End of the regeneration period, thus maximal area recovering; some nuclei are labelled in the medullary meristem and in the young medulla, a prelude to the apex rising; no labelling in the ax- ial zone *az*.
3 Maximal area, initiation of leaf primordia No.11 (p_{11}), where the labelled nuclei are localized.
4 Beginning of rising of the primordia No.11, accompanied by syntheses which occur at their ax- ils *(arrows)*, indicating the precocity of the regeneration by flank mitoses. No labelling in the axial zone *az*. (After SAINT COME 1966)

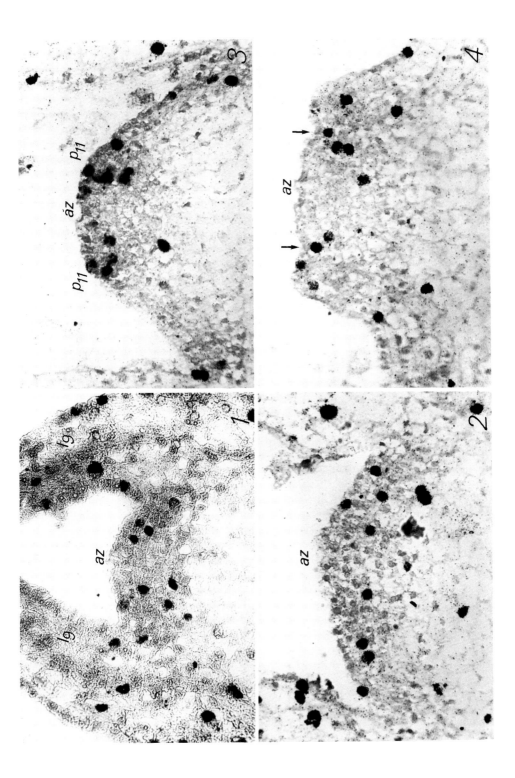

129

Table 2.2. Phase durations of the cellular cycle in the different zones of the cauline meristem of *Chrysanthemum segetum* (h) (NOUGARÈDE and REMBUR 1978)

	Total duration of cycle	Duration of the various phases			
		G1	S	G2	M
Axial zone	135	112.4	10.1	9.3	3.2
Lateral zone	51.4	30.9	7.9	9.4	3.2
Medullary meristem	72.5	51.7	8.9	8.6	3.3

photometry of the DNA in the cellular cycle of the various zones of the *Chrysanthemum* vegetative point show that the sum of phase S + G2 + M has an almost constant duration (Table 2.2).

Thus, the usual characteristics of the meristematic condition, i.e. cytoplasmic basophilia, size of nucleoli, exiguity of vacuoles, structures of proplastids and mitochondria, accurately reflect the importance of the metablic and proliferating activities of the apical cells.

2.1.3 Histological Structure of Root Meristems

The root meristems have been less studied than the cauline meristems. This is only due to the structural uniformity of the roots compared to the diversity of aerial organs. In fact, although the morphology of young roots, their origin and their primary structure present some features which characterized the large groups of vascular plants, the changes of the meristem are not very significant.

The structure of the root meristem was first studied in comparison with that of the aerial vegetative apex, then, the latter alone was considered and during many years there were more advances in the study of the vegetative point than in the study of the root apex.

Since 1950, further work has yielded more data concerning the initiation in roots and about unknown zonation in the root meristems, at least, in Angiosperms.

2.1.3.1 Pteridophytes

The authors (NAEGELI, HOFMEISTER, etc.) who acknowledged the existence of a particular apical cell in the cauline meristems of Pteridophytes found a similar cell most often tetrahedral on the root apex (Fig. 2.21) (NAEGELI and LEITGEB 1868).

The existence of a particular cell was proved in many cases and obviously demonstrated in Pteridophytes which have a single apical cell at the top of their stems (Filicineae, Equisetineae). In the most primitive forms considered as such presently in nature (Lycopodineae), no tetrahedral cell can be distinguished from the neighbouring cells. As regards the apical cell, the parallelism is striking between the cauline meristem and root meristem. The same plant has a distinct apical cell in both meristems or in neither. When no tetrahedral cell is found at the extremity of the root, under the cap, one can find a group of small cells with no determined arrangement (e.g.: *Isoetes*).

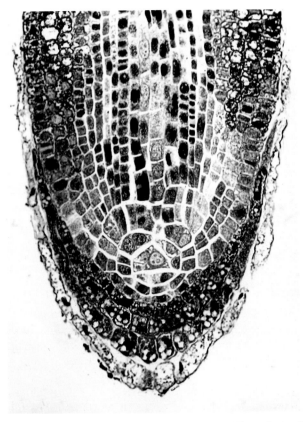

Fig. 2.21. *Equisetum arvense;* root meristem, histological aspect. The apical cell is triangular in section and the root cap is independent of the cortex and the root hair layer, as in Monocotyledons (see Fig. 2.24)

According to current opinion, the tetrahedral apical cell, in dividing unequally by partitions parallel to its various faces, would be the initial of the cap and the different areas of the root body (Fig. 2.21).

The systematic study of the mitosis frequency in the root meristem of *Equisetum arvense* shows that the apical cell divides in some cases but does not show a continued proliferation, even in young roots (BUVAT and ROGER-LIARD 1954).

By means of ^{3}H-histoautoradiographic labelling and cytophotometric techniques, D'AMATO and AVANZI (1967) reported (1) that the particular tetrahedral apical cell of the root meristems of several Filicineae leptosporangiate *(Marsilea strigosa, Blechnum brasiliense, B. gibbum, Polypodium aureum* and *Ceratopteris thalictroides)* can divide regularly during meristem edification, (2) that its activity decreases while the mitosis frequency is reduced by the extension of phase G2 and (3) that nuclei tend to become endopolyploid, thus rare apical mitoses could then produce chimeric apices. Besides, CLOWES (1965 a, b) suggested that the meristem organization could be maintained indefinitely without any segmentation of the tet-

131

rahedral apical cell which created this organization. The organogenic activity is transferred to the cells surrounding the apical one. In roots with no particular apical cell, like those of *Isoetes,* the group of small cells at the top which replaces the apical cell probably shows the same behaviour. However, it would be interesting to verify this fact by studying the mitosis frequency in this area. The importance of this proof will be shown in the study of the roots of Phanerogams.

2.1.3.2 Phanerogams

In Phanerogams, after a description of the apical cell in the root meristem of various Pteridophytes, NAEGELI and LEITGEB (1868) attempted to find this single initial in the root extremity of Phanerogams. They concluded that this initial cell could exist momentarily at the top of young roots but that only an irregular meristem could be found at the top of the adult root.

In the same year, a paper on this matter was published, and then in 1870, HANSTEIN's report already mentioned for cauline meristems. Studying the development of the rootlet in the embryo, the author contested the existence, even temporary, of a single apical cell and replaced it by three systems of independent initial cells, three *histogens* like in the stem, i.e. the plerome, the periblem and the dermatogen, producing respectively (1) the central cylinder, (2) the cortex, and (3) the cap and the external layer (hairy layer, often erroneously named "epidermis") (Fig. 2.22). REINKE, a student of HANSTEIN found the same structures in adult roots (1871).

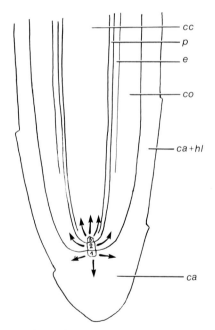

Fig. 2.22. Dicotyledon root meristem, interpreted according to HANSTEIN (1868). **1** Dermatogen initial (cap + hair layer *ca + hl)*; **2** periblem initial (cortex *co*); **3** plerome initial (central cylinder *cc*); *e* endodermis; *p* pericycle

132

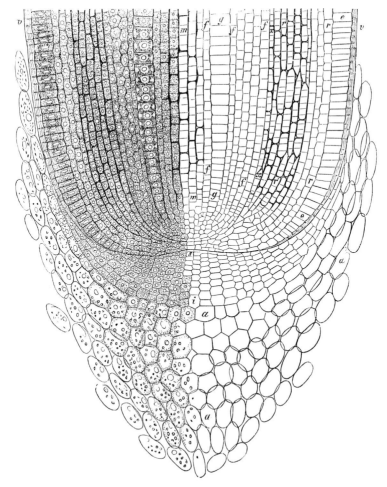

Fig. 2.23. Root meristem of *Zea mays*, reproduction of the classical drawing of SACHS (1868, 1874)

This concept of several independent *histogens* prevailed for almost a century but as soon as HANSTEIN's and REINKE's findings were published, differences appeared regarding the number and the role of the various groups of initial cells.

Already in 1868, SACHS published a drawing of the root meristem of corn, obviously showing that the cap and the hairy layer do not have the same initial cells (Fig. 2.23) (cf. SACHS 1874).

Soon, the works of STRASBURGER (1872; Conifers, Gnetaceae, Cycadeae), RUSSOW (1872; comparisons between Pteridopytes and Phanerogams), PRANTL (1874) and above all DE JANCZEWSKI (1874); HOLLE (1876); TREUB (1875) and ERIKSSON (1878), among many others, manifested many disagreements with HANSTEIN's figure, depending upon the species considered. These works led to the determination of an increasing number of root meristems types. FLAHAULT (1878) remarked that

133

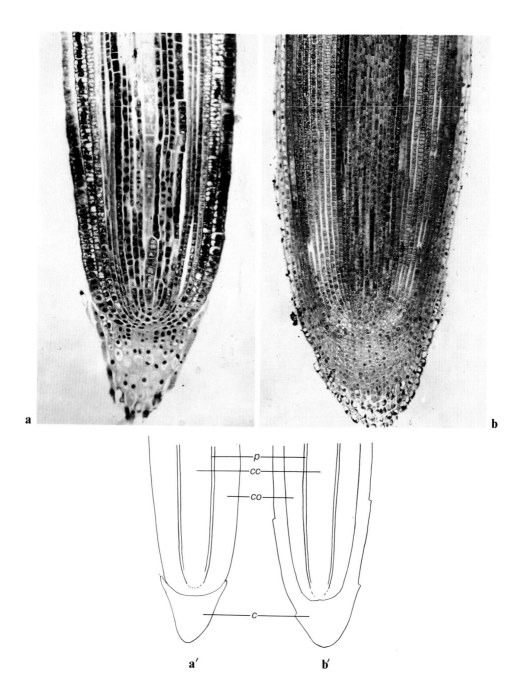

Fig. 2.24. Root meristem of a Monocotyledon (*Triticum vulgare,* **a, a′**) and a Dicotyledon (*Lycopersicum esculentum,* **b, b′**). In the first example, the root cap *c* does not extend along the flanks, so that the hair layer proceeds from the cortex meristem *co.* In the second example, in contrast, the root cap extends on the flanks, and its most inner layer will produce the hair layer, after desquamation of the outer layers (climacorhize root); *p* pericycle; *cc* central cylinder

by combining TREUB's results on Monocotyledons and ERIKSSON's on Dicotyledons, seven different types of root meristem structures in Phanerogams can be distinguished. As in the types of cauline meristems singled out by POPHAM, from this time it was acknowledged that the structure of a young radicle, or a rising radicle, could be of a different type than that of the adult organ.

The existence of numerous typical and intermediate cases has contributed much confusion in the study of root meristem structures. FLAHAULT (1878) elucidated this problem by selecting the most general facts, among this diversity, and distinguishing the essential features from the secondary ones.

We intend to reproduce only the conclusions of the important FLAHAULT's report:

"Above all it is the activity of the cap and the role of the cortex initial cells relatively to the formation of epidermis and of the cap which give the main distinctive features.
In Monocotyledons, the *cap* regenerates independently from the cortex and the epidermis. Most often, it *seems to be connected with the epidermis in a very young state of development* and to derive from a tangential cleavage of the epidermis, but from this moment it remains absolutely independent and regenerates by the activity of its own internal layer" (Fig. 2.24a, a').
The epidermis is generally produced by the cortex initial cells; in some cases it appears to be separated from it as from a very young state; once produced *it never gives rise to the cap.*
In Dicotyledons, the epidermis is almost always completely independent from the cortex: *the cap is always initiated by the cortex or the epidermis of the root; it keeps on regenerating at the expense of the tangential divisions of the cortex or epidermis layers.* When further divisions occur in the cap layers, they are always relatively rare" (Fig. 2.24b, b').
"*The cap derives most often from the epidermis,* but in some cases, the cortex contributes in its formation, sometimes the cortex produces it completely."

FLAHAULT's results have become classical and have been published in all manuals. Thus, the root meristem is usually considered to be composed of three layers, issued from three groups of initial cells.

In Dicotyledons the three groups are the initial cells (1) of the central cylinder, (2) of the cortex, (3) of the hair layer and the cap (Fig. 2.22).

In Monocotyledons the three groups are slightly different: (1) initial cells of the central cylinder, (2) of the cortex and the hair layer, (3) of the cap (Fig. 2.23 a, a).

However, since the time of FLAHAULT, contemporary authors as well as FLAHAULT himself noticed that the distinction between the three layers was not always easy at the extremity of the root. The limits separating the three anatomical areas may become more or less continued towards the meristem top. In many cases, just below the cap, the meristem top presents a body of small cells arranged *with no apparent order, and where no independent groups of initial cells can be singled out* (ALLEN 1947; REEVE 1948; YARBROUGH 1949).

Although the central cylinder, the cortex and the hair layer are usually much more delimited at the extremity of the roots than at the top of the stems (Fig. 2.25 b), the lack of limits in the body of subterminal small cells raised questions with regards to the existence of such three layers in the area of the so-called initial cells (Fig. 2.25 a).

For example, VON GUTTENBERG and colleagues attempted to discover the cellular filiation of the meristem and, in a geometrical consideration, concluded that a cell "Z" in some Dicotyledons, a cell which would be at the origin of all root tis-

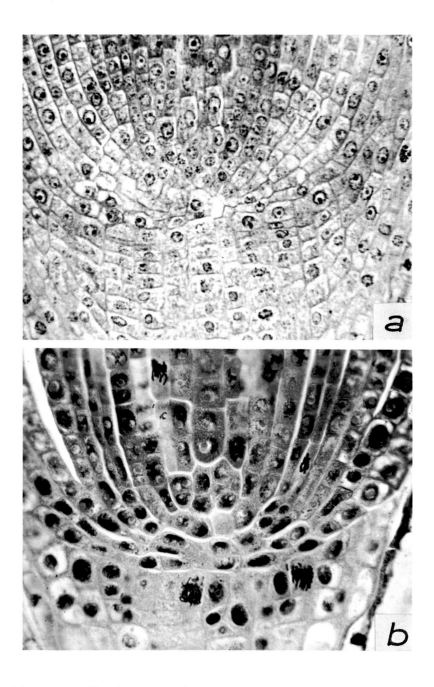

Fig. 2.25 a, b. Extremities of two root meristems.

a Dicotyledon: *Lycopersicum esculentum;* **b** Monocotyledon: *Triticum vulgare.* The first one shows a cell resembling the Z cell of Fig. 2.26, and suggests a common origin for the plerome and the periblem, if not the root cap. The second one suggests a total independence of the three sheets up to the top. This diversity probably hides the general physiological mechanisms of the functioning of root meristems

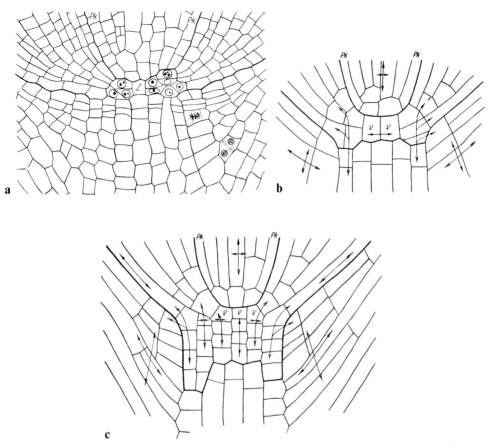

Fig. 2.26 a–c. *Helianthus annuus,* tip of root meristem and its functional interpretation, according to VON GUTTENBERG et al. (1955). **a** Disposition of the cells, showing the central cell *Z* (Zentralzelle) from which the pleriblem and the dermatogen would proceed. **b,c** Onset and development of the entity periblem – root cap that could be produced by "binding cells" *V* (Verbindungszellen), themselves proceeding from the *Z* cell. The plerome could have its own initials. *Pk* pericycle

sues (VON GUTTENBERG et al., 1955) exists (Fig. 2.26 a). However, things are more confused in other Dicotyledons and in Monocotyledons (Fig. 2.26 b, c).

More indirect methods such as the study of the filiation of X-ray irradiated apical cells with chromosomal abnormalities led to different interpretations. BRUMFIELD (1943) assessed the existence of a single layer of three or four initiating cells of the whole root.

These results were not confirmed and have now been rejected due to more advanced research on the real functioning of the subapical areas of the root meristems.

As regards the organogenic activity of a set of meristematic cells, the most direct way to evaluate its functioning consisted in the assessment of the mitosis frequency in the various areas of the meristem.

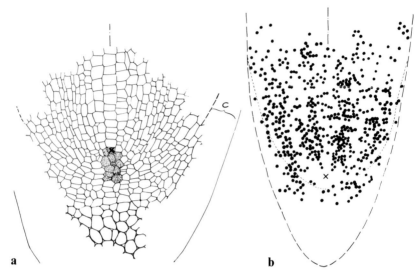

Fig. 2.27 a, b. Statistical distribution of mitoses in the root meristem of *Allium cepa*. **a** Histological aspect, the axial apical cells are *dotted*. **b** Superposition of mitosis reports from 20 axial sections showing the mitoses localization around the apical cells, themselves inactive, and at the root cap base. *c* Lateral root cap extension

Two methods were used for this purpose. The simplest was to superimpose a sufficient number of longitudinal median sections of roots in active growth and to draw the sites of mitoses on the same tracing paper.

The second method, which was more accurate, was to measure the duration of the cellular cycle in the meristem cells by means of ^{3}H-thymidine incorporation of short duration ("pulse labelling").

The superimposition of 20 longitudinal median sections of roots of *Allium cepa* (Fig. 2.27) showed that, on the one hand, the mitoses are distributed in the subterminal area of the meristem and, on the other hand, at the base of the cap. Between the two areas, a limited space subsists, where divisions are rare or absent. Therefore, it seems that the cells considered as "apical initial cells" by past authors, are in fact almost quiescent (BUVAT and GENEVES 1951; BUVAT and LIARD (1951).

Moreover, using techniques of histoautoradiography and microspectrophotometry, CLOWES (1954, 1961 a, b, 1963) found a "quiescent centre" in the functional roots at the site of the so-called apical initial cells. The duration of cellular cycles confirms this relative inertia (Table 2.3).

In some way, the root meristem covers the quiescent centre and takes an inverted *cup-shaped aspect,* for what concerns the main body of the root. It is completed by the cells which initiate the cap alone or together with the hair layer and it covers the apical extremity of the quiescent centre (Fig. 2.28). CLOWES elucidated that this structure characterizes the completed roots in active growth. The top cells contribute actively in the primary edification of these roots, i.e. the meristem in formation does not present any quiescent centre.

138

Table 2.3. Duration of the cellular cycle and its phases (h) in four areas of the root meristem of *Zea mays*. The values in parentheses represent the total value T obtained by accumulation of metaphases, the other values are the results obtained with the "pulse-labelling" techniques (CLOWES 1965 a)[a]

	Initial cells of the cap	Quiescent centre	Stele just above the quiescent centre	Stele 200 μm from the quiescent centre
T	14	(174)	22	23
G1	−1	(151)	2	4
S	8	9	11	9
G2	5	11	7	6
M	2	(3)	2	4
Mitotic index	15.7%	1.9%	10.8%	16.7%

[a] This table shows the longer duration of G1 in the quiescent centre and the low value of the mitotic index with respect to the initiating sites.

Moreover, the cytophotometric studies reveal that the whole meristem has diploid cellular cycles and that the quiescent centre cells are blocked in the G1 phase of the cycle. Presently, this phase is known to maintain the possible recurrence of proliferation, the meristematic feactures, and avoids the risk of endopolyploidy. In case of traumatism, the quiescent centre can be reactivated and generates another complete meristem.

The genuine initiating cells of the root histogenesis are therefore situated: (1) on the sides, (2) above and (3) below a body usually composed of a few cells, apparently passive.

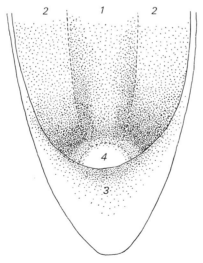

Fig. 2.28. Zonation schema of the root meristem. *1* Zone of central cylinder initiation; *2* zone of the cortex initiation; *3* zone of the root cap (and eventually hair layer) initiation; *4* quiescent centre

The proliferating zones 1 and 2 generate respectively, the cortex and the central cylinder; zone 3 produces the cap (Fig. 2.28).

It should be noted that this meristem does not include an *initiating ring* as that of the stem, and this fact is related to the *absence of exogenous lateral organs* with a bilateral symmetry, which would be the homologs of leaves, thus the body of the root can be considered as being the counterpart of the only cauline part of the stem, reduced to the medulla in advanced plants. Of course, this is only a hypothesis.

2.1.4 Cytological Structure of Apical Meristems

Two types of meristems are usually distinguished: the primary and the secondary meristems which differ according to their location, their ontogenic role and also in the cytological structure of their cells. Apical meristems are primary meristems, but the cytological structure is not the same in all apical cells. In fact, the acknowledged structure of primary meristematic cells is typical only in the *organogenic areas showing active proliferation*. These two conditions are required. Therefore, typically "primary meristematic cells" can be found, on the one hand, in the *initiating ring* of the cauline meristem as well as in the initiums and the foliar primordiums it produces and, on the other hand, in the subterminal zone of the root meristem in the phase of active proliferation.

In light microscopy, the cytological feactures of the "primary meristematic" condition are found to be generally less marked in the central apical area of the vegetative apex and strongly opposed to the cytological features of the underlying "medullary meristem" cells. These different structures can be briefly summarized as follow.

2.1.4.1 Primary Meristematic Cells (Fig. 2.29 and Plate 2.2/1–4)

They correspond to small cells (about 5 to 15 µm) almost of the same dimensions in all directions (so-called isodiametric cells) including a dense cytoplasm and a *central* bulky and voluminous nucleus, compared to the cell volume. The pectocellulosic walls are very thin and there are practically no intercellular spaces between cells (Fig. 2.29 a).

Plate 2.2.

A Apical meristem of *Cheiranthus cheiri* (developed stem); transversal section, locating areas *1* to *4*. *1* Apical cells of the top of the *tunica,* relatively less meristematic than those of regions *2* and *3*; large vacuoles, small nucleoli (about 1.4 µm); not very active cells. *2* Cells of the initiating ring: tiny vacuoles, very short mitochondria, larger nucleoli (about 2 µm); active cells. *3* Cells of the last initium in formation; pronounced primary meristematic state; very small vacuoles, tiny mitochondria, particularly voluminous nucleoli (about 2.6 µm). *4* Axial cells of the *corpus,* practically inactive; they resemble the axial *tunica* cells (*1*) but the vacuoles are thinner, the nucleoli as in *2,* and the plastid differentiation begins.

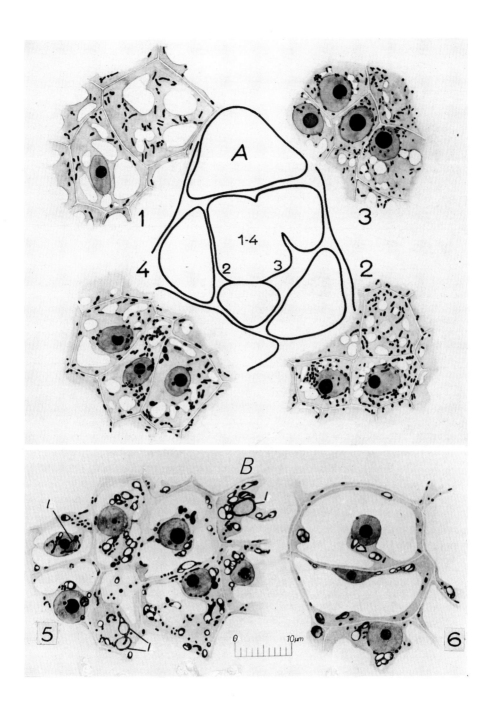

B Medullary meristem of *Lupinus albus* L. *5* Young plant, medullary meristem near the corpus; largely vacuolized cells, amyliferous small plastids; *l* loop-shaped chondriosomes. *6* Cells more distant from the corpus (600 μm from the top), larger and more vacuolized than in *5,* but similar chondriosomes and plastids, typical cells of the medullary meristem.

Fixation: Regaud; staining: hematoxylin (BUVAT 1952)

141

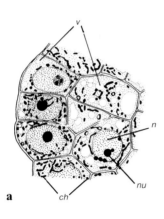

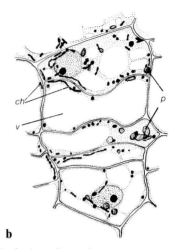

Fig. 2.29. a Cytology of primary meristematic cells; leaf primordium of *Myosurus minimus;* voluminous nuclei *n,* with large nucleoli *nu;* short and small chondriosomes (mitochondria) *ch;* small and numerous vacuoles *v.* **b** Medullary meristem of *Myosurus minimus;* flattened prismatic cells, resulting from the transversal orientation of cleavages, smaller nucleoli, larger vacuoles *v,* sometimes only one per cell; chondriome (*sensu* GUILLIERMOND) differentiated in chondriosomes *ch* and small plastids *p*

The nucleoplasmic ratio $\dfrac{(\text{nucleus volume})}{\text{cell volume}}$ is high. The nucleus includes one or several nucleoli larger than those found in differentiated cells compared to the nucleus volume. The large amount of RNA-rich nucleolus substance seems to be a reliable sign of the growth and proliferation activities of these cells. In fact, the cytochemical method of J. BRACHET reveals that the cytoplasm of the meristematic cells is more pyronine-positive (RNA staining) than that of the neigbouring cells (CAMEFORT 1954; LANCE 1954, 1957) (Fig. 2.30).

The cytoplasm encompasses only very small vacuoles, in great number, with a concentrated component. The chondriome is mainly composed of mitochondria (granules) and usually short rodlike chondriosomes resulting from the association of a small number of granular elements. In most cases, no plastids can be distinguished, or the plastids are difficult to differentiate from the chondriosomes and they are very small (Fig. 2.29 a).

Usually the cells of the axial apical areas have larger vacuoles (Plate 2.2/1, 4) a clearer cytoplasm, poorly stained by pyronine (Fig. 2.30) and more or less differentiated plastids within.

More sophisticated techniques, using microspectrophotometry with specific staining of RNA (NOUGAREDE and REMBUR 1977) and morphometric methods lead to two distinct essential concepts: the RNA concentration and the amount of RNA by cell. The most meristematic cells of the vegetative apex of *Chrysanthemum* (in the initiating ring) have the higher cytoplasmic RNA concentration, but the RNA amount depends upon the cytoplasmic volume of the cell. Thus, the greatest RNA amount by cell appears in the second axial tunical layer where the

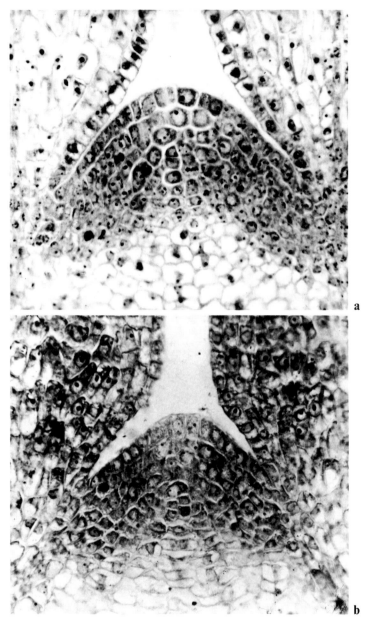

Fig. 2.30 a, b. *Scabiosa ukranica*, apical meristems of stems which are almost in the same stage of development. **a** Prepared according to an ordinary histological technique (Fixation: Navaschin; staining: hematoxylin). **b** Treated according to BRACHET's technique for RNA detection (1940). The staining is all the more accentuated as the cell is RNA-rich. This figure accurately reproduces the histological image of the various apical regions; (after LANCE 1957)

cytoplasm volume is the highest and the lowest in the medullary meristem where this volume is the lowest. These data are confirmed by ribosome counting with electron microscopy (NOUGAREDE and REMBUR 1984).

One of the main characteristics of the primary meristematic cells is their poorness in paraplasmic materials (amorphous materials produced by living matter). This living matter fills the cell almost completely (very small vacuoles, small storage, very thin cell walls).

2.1.4.2 Rib (Medullary) Meristem Cells

These cells are larger than the primary meristematic cells (15–30 μm) and have a prismatic shape, often very flat, arranged in piles parallel to the axis of the stem (Fig. 2.29 b; Plate 2.2/5, 6).

They grow longitudinally and sometimes the cell wall becomes thickened, then they segment transversally and the daughter cells increase again until further segmentation always with the same orientation.

They are cytologically opposed to the typical primary meristematic cells by the characteristics of their cytoplasm and chondriome. The cytoplasm encompasses a large vacuole, or a few large vacuoles, and becomes mainly parietal: it forms only a thin pellicle, with a thickness often inferior to 1 μm, applied on the cell wall and from which trabeculae extend, crossing the vacuole sometimes with a suspended central body including the nucleus (Plate 2.2/5). More often, the nucleus is applied on the horizontal cell wall generated by the same mitosis which gave rise to the cell (Plate 2.2/6).

This cytoplasm with large vacuoles includes short and small chondriosomes and generally amyliferous, more or less chlorophyllous plastids not wholly differentiated (Plate 2.2/5, 6; Fig. 2.29 b).

These various features justify the terms used by English speaking authors (FOSTER 1938, 1941 a, b; MAJUMDAR 1942; PHILIPSON 1949, etc.) to designate such a meristem: *rib meristem, file meristem* or even better *vacuolated meristem.*

Because of the constant orientation of segmentation and the vacuolization of the cytoplasm, the rib meristem can be compared to secondary meristems. We shall return to this matter later.

The rib meristem, characterized by the above mentioned cytological features, is well visible only in the stem apex. In root meristems, the proliferating cells have primary meristematic features. The very progressive development of the vacuoles and the segmentation modalities let medullar cells appear later, in particular in-

---▷

Plate 2.3. *Hordeum vulgare,* root meristem.

Typical primary meristematic cell. Great density of ribosomes (very basophilic cells); nucleus *n* in the centre of the cell, with voluminous nucleolus *nu,* and little condensed chromatin *chr;* plastids *p* undifferentiated (proplastids); mitochondria *m* containing poorly developed cristae; dictyosomes *d* emitting small clear vesicles; young vacuoles *v* containing inclusions, most of them osmiophilic; beginning of sequestration double membranes *au* at the origin of autophagic vacuoles; rough endoplasmic reticulum *er* inconspicuous; *ne* nuclear envelope; *pl* plasmodesmata; × 14 500

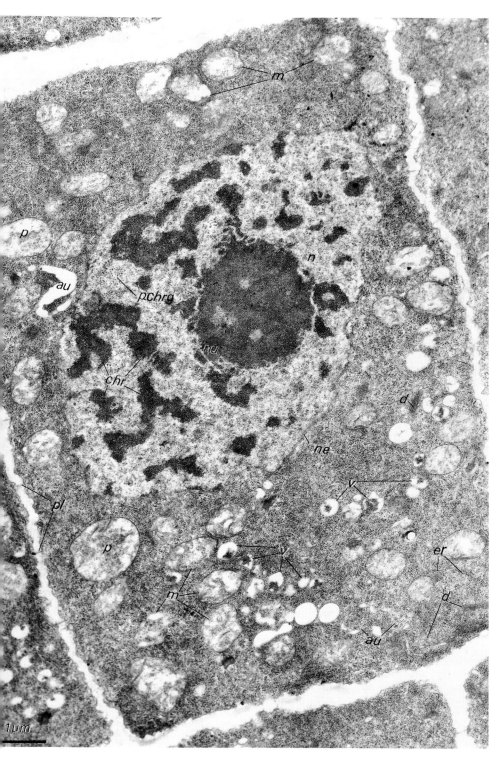

1 μm

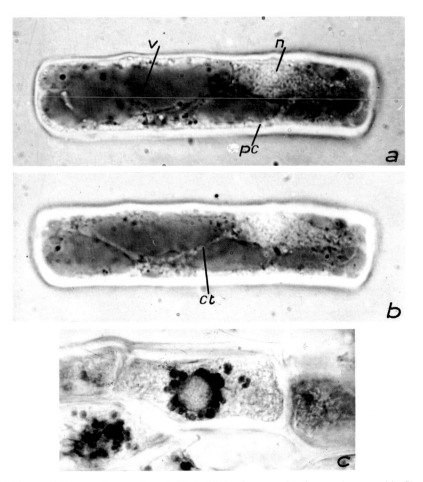

Fig. 2.31 a–c. *Triticum* root cap cells. **a, b** Vital staining by neutral red; *n* nucleus; *pc* thin film of peripheral cytoplasm; *v* large vacuole, crossed by cytoplasmic threads *ct*. **c** Cells treated by iodine – iodide reagent (lugol): numerous starch grains, most of them around the nucleus

▷

Plate 2.4 A, B. *Raphanus sativus*, root meristem. Two aspects of the activites of the Golgi apparatus.

A Dictyosome *d* emitting dark vesicles *dv*, small in diameter, and larger clear vesicles *cv*. The membrane of the latter presents the same tripartite aspect as shown in the plasmalemma (well visible with greater magnification). One of the clear vesicles merges into the plasmalemma *(arrow)*. The work of VIAN and ROLAND et al. (see p. 208) suggests that these vesicles contain cell wall polysaccharide precursors. *mvb* Multivesicular body interpreted as the result of fusions of Golgi vesicles of both kinds, and typically rich in hydrolases; *mt* microtubules; × 50 000. **B** Several dictyosomes *d*, emitting clear and dark Golgi vesicles, *cv* and *dv*, next to a sequestration figure of cytoplasm, where one can see that the double isolation membranes proceed from the fusion of vesicles considered as Golgi vesicles. The sequestrated cytoplasma seems to be still spared, it contains a well-recognizable mitochondrion *m*, but the enclave *p* is no longer determinable; × 41 000

Fixation: glutaraldehyde – O_sO_4; contrast: KMnO$_4$

146

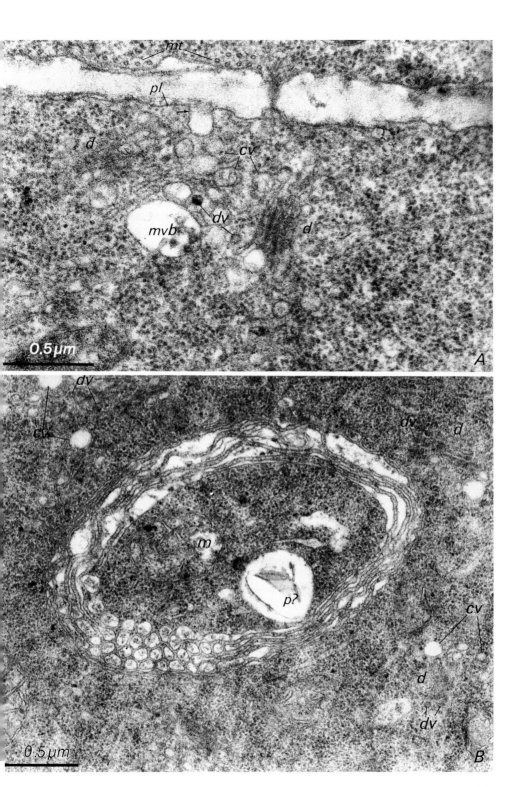

cluding amyliferous leucoplasts which are more developed than in the neighbouring cells.

In contrast, the cap cells differentiate very early by means of the enlargement of vacuoles and the formation of big amyliferous leucoplasts (Fig. 2.31). It is well known that particularly in the central cells of the cap (the columella) the amyloplasts play a role in the geotropic perception of the root. They are called "statoliths" (cf. PERBAL 1974).

Electron microscopy yielded further data on the ultrastructure of primary meristematic cells. First, the positivity of the cytoplasm staining in light microscopy either hematoxylin- (Fig. 2.30a) or pyronine-stained (Fig. 2.30b), indicating its basophilia, comes from the ribosomes which have a maximal density in the most active organogenic cells of the *initiating ring,* the leaf primordia, or the growing, cuplike root meristem (Plate 2.3). Such a ribosomal density diminishes during differentiation as shown in the quantitative and statistical studies of NOUGAREDE and PILET (1965). Mitochondrial cristae are moderately or poorly developed and the leucoplasts, slightly larger than mitochondria, include a few thylacoids only, sometimes forming reduced grana (in the vegetative stem apex), some plastoglobules and sometimes phytoferritin inclusions. They are typically deprived of starch (cf. Plate 3.2A, B).

The Golgi apparatus is well defined and reveals its activity in issuing small osmophilic vesicles and some larger transparent ones. The Golgi activity mainly but not exclusively affects the vacuole genesis (Plate 2.4B) (cf. MARTY 1973, 1978; BUVAT 1977; BUVAT and ROBERT 1979). These vacuoles of reduced size may sometimes be absent on the whole extent of highly meristematic cells.

The moderately developed endoplasmic reticulum is distributed in the whole cytoplasm issuing diverticules which cross the plasmodesmata in the shape of a narrow axial tubule (desmotubule) (see Plate 3.4).

The ultrastructural images allowed the evaluation of up to 20000, the number of plasmodesmata which pass through the thin pectocellulosic cell wall of the meristematic cells. Thus, the cytoplasms are largely connected between these cells.

Independently of the frequent karyokineses, the interphasic nucleus shows signs of metabolic activities: nucleoli are relatively voluminous and include the classical fibrillar and granular zones; neighbouring the chromatin, "perichromatin granules" can be observed, considered as the site of maturation of the messenger RNAs (Plate 1.15B).

The plasmalemma is usually applied on the wall except where pinocytosis or exocytosis processes are taking place, more particularly at the sites of Golgi activity, which brings material to the parietal structure. Numerous microtubules are often visible near its cytoplasmic face (Plate 2.5).

Plate 2.5 A, B. *Raphanus sativus,* root meristem.

Microtubules, in the neighbourhood of the plasmalemma, are seen in transversal section in **A** *(mtt* and *arrows)* and in one of the cells of **B**; and in longitudinal section in the other cell *(mtl)* where one of them is very long; *mvb* multivesicular body; *d* dictyosome; *i* plasmalemma invagination; *p* proplastid; *er* rough endoplasmic reticulum; *av* autophagic vacuole.

×63000; Fixation: glutaraldehyde – O_sO_4; contrast: $KMnO_4$

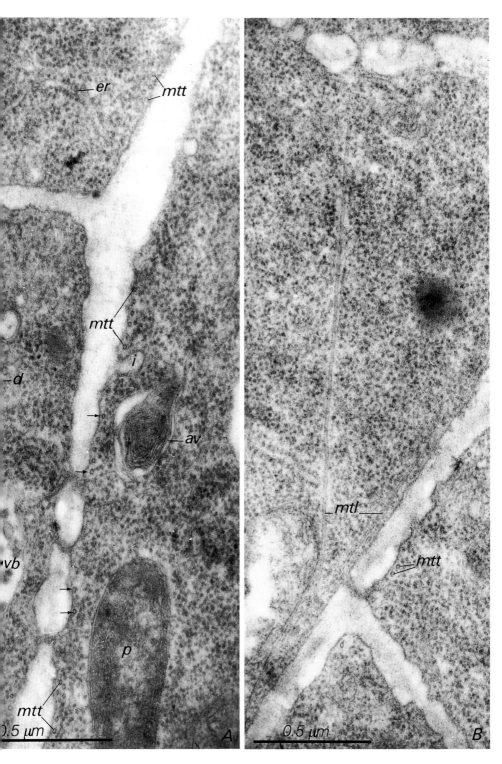

149

2.1.5 Functioning of Shoot Apical Meristems

The functioning of the vegetative apices in the initiation of the primary structure of the leafy stem has been extensively studied. The results obtained by SCHMIDT, LOUIS and PLANTEFOL, are particularly interesting. It should be recalled that PLANTEFOL formulated the concept of *initiating ring* and the histogenetic passivity of the axial apical cell.

Let us consider this "initiating ring" in the case of a stem with alternate leaves (Fig. 2.32). When a leaf (as a primordium P) has just risen above it, the initiating ring is encroached by the leaf production and consequently the apical surface is reduced.

Prior to the initiation of the following leaf encroaching the initiating ring again in another point, the apical surface must regenerate to be in the same condition as before.

This is achieved partly by the enlarging of the side where the next leaf will originate (Fig. 2.32, arrows *1*), i.e. by a "horizontal growth" of the apex. Moreover, at the axil of the newly built primordium, the initiating ring regenerates by anticlinal mitoses (meaning that the cell walls are generated perpendicularly to the dome surface) (Fig. 2.32, double arrows *3*) which occur on the flanks of the apical dome.

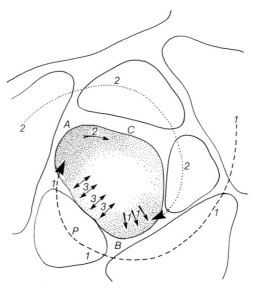

Fig. 2.32. Schema of the activity of the "anneau initial" (initiating ring) in the *Cheiranthus cheiri* apical meristem. **1-1** and **2-2** Materialization of both foliar helices, the first one ending, at this moment, in the leaf initium *A*, the second one in the initium *B*. The leaf initiation is preceded by the horizontal growth of the anneau initial, materialized by the *arrows 1*. At the axils of the previous primordium, the anneau initial is regenerated by flank anticlinal mitoses. This regeneration is shown only at the primordium *P* axil, by *arrows 3*, but it will also take place facing the last primordium of the helix 2.

Finally, before rising into a primordium, the leaf initium *A* induces a wave of dedifferentiation in the ring at the *C* site, where the regenerations has just finished and where the next initiation will occur on the helix *1 (arrow 2)*. A similar induction will occur between *B* and *P*

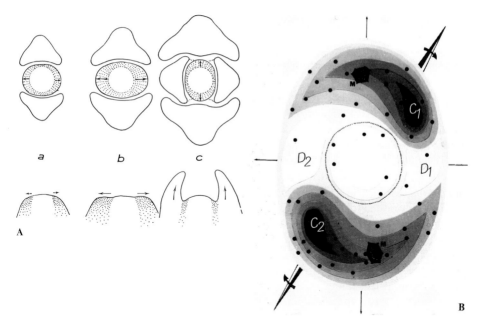

Fig. 2.33. A Schema of the plastochron activity of a Dicotyledon apical meristem with decussate opposite leaves. **a** Minimal area, immediately after the rising of the two last primordia; **b** passage to the maximal area by horizontal growth, i.e. the leaf initiation; **c** return to the minimal area, following the rising of leaf initiums, i.e. the vertical growth. In addition, the initiating ring *(stippled)* starts to regenerate at the axil of the last primordia.
B Distribution of proliferation activities of the initiating ring cells of a plant with decussate opposite leaves, *Fraxinus excelsior,* obtained by statistical analysis of the mitosis proximity and frequency. This distribution materializes the rotation of the centres of maximal activities, *C1* and *C2*. Note that the centres are not in orthogonal position with respect to the meristem axes, but at the stage of maximal area, in front of the large axis where the leaf initiation mitoses *M* occur. *D1* and *D2* are zones in relative and temporary rest towards which the generating centres are moving, *C1* towards *D1, C2* towards *D2,* this movement being due to a meristematic state propagation. The *dots* indicate histoautoradiographic labelled mitoses; (after COTTIGNIES 1984)

Meanwhile, the extension of the foliar bases and the rib meristem raises this dome above the insertion point of the last primordium.

Thus, the apex encroached by a rising leaf is reduced to a *minimal surface*. By means of a *horizontal growth* and a regenerating phase performed by anticlinal mitoses at the axil of the last primordium, the apex recovers its previous surface, or *maximal surface,* produces the next leaf, finding itself again in the condition of minimal surface, and so on (LOUIS 1935).

The process is the same whether one leaf (alternate phyllotaxis), two leaves (opposite leaves) (Fig. 2.33) or several (verticillated leaves) spring up; the organogenetic functioning of the apex is periodical. Referring to an expression of ASKENASY (1880), SCHMIDT gave the name of *plastochron* to the functioning period which enables the apex to recover the same condition (Fig. 2.33 a).

The plastochronic functioning is a main characteristic of the vegetative meristem of the leafy stem.

During each plastochron, the initiating ring generates one, two or several foliar primordiums according to whether the stem has an alternate, opposite or verticillated phyllotaxy. The existence of phyllotaxic rules, i.e. of determined distributions of leaves on each stem, implies that, in the initiating ring, the initiating sites of the successive leaves are conditioned by the organogenesis processes.

For example, consider the relatively simplest case: the birth of one leaf at one time. On the transversal section of the apex (Fig. 2.32) the materialization of two foliar helices is illustrated. The last completed *initium* i.e. "the foliar base" from which the *primordium* will rise, is situated at the extremity of helix 1.

The following initium hardly appears at the extremity of helix 2 (beginning of horizontal growth). The next initium will appear on helix 1 adjacent to *initium A,* in point *C* of Fig. 2.32.

The accurate cytological study allows the detection of an organogenesis induction between an *initium A* and the point *C* where the following will form. In fact, each leaf initiation is usually marked by the *accentuation of the meristematic features* of cells in the initiating ring. This is the case in *Cheiranthus cheiri*. It is then possible to detect a wave of meristematization running from initium A to point C as shown on arrow 2, and confirming PLANTEFOL's concept of "contiguity".

In such a case, the direction of this induction is found to be clearly determined. This can be explained by the fact that, on the other side, the initiating ring encroached by the primordium outcome *P,* has not yet regenerated and therefore is not receptive. In fact, anticlinal regenerating mitoses on the apex flank occur in this place, mitoses apparently induced by primordium *P* (arrows 3). With the generalization of these observations, it is necessary to consider the cases in which the phyllotaxy is decussate opposite. In such a case the histocytological symmetry of the meristem (Fig. 2.33 a) is not sufficient to determine the direction of the foliar helices.

Recent studies on *Fraxinus* (COTTIGNIES 1984) led to the recognition of this direction using the method of the "closest neighbour". The orientation of the leaf helices and the real progression of the leaf-generating centres (PLANTEFOL 1947) are rendered visible and can be determined. Following the phase of maximal area, and along with the leaf initiation, a mitosis wave moves around in an acropetal process, from the site of leaf initiation (Fig. 2.33 b).

2.1.6 Organization of the Primary Structure of the Leafy Stem

2.1.6.1 Initiation of Leaves

We have already explained that on the flank of the apical meristem, during the passage from a minimal surface to a maximal surface, an *horizontal* growth process occurs which is characteristic of the vegetative activity of such a meristem. This growth results from two types of mitoses both of which take place mainly in the *tunica layers*.

On the other hand, *anticlinal* mitoses (walls perpendicular to the surface) enlarge the apex surface, soon followed by *periclinal* mitoses generally occurring in the second layer ("subepidermal") which can involve deeper layers and harmonize the growth of this part of the apex. Thus, the horizontal growth and the thickening of the involved area of the initiating ring are simultaneous (Fig. 2.34a, b). The set of meristematic cells thus produced forms the *"leaf base"* described by Louis (1935). It also represents a leaf initium on the onset (Fig. 2.34b, *li;* Fig. 2.35/1.2), initium which will be completed by an emergence rising above the ring (Fig. 2.34b; Fig. 2.35/2, 3).

Periclinal and anticlinal mitoses can frequently occur in the external layer (Gymnosperms, Graminaceae), but most often the *tunica* has two or several layers and the external one generates only the *epidermis* and is only the site of anticlinal divisions. Then the leaf initiation is soon revealed by subepidermal periclinal mitoses (Fig. 2.35/1, 2, *h*). The subepidermal layer frequently shows a feature of the initiating layer.

2.1.6.2 Formation of Leaf Buttresses

Independently of the shape and structures of the adult leaf, the leaf initium, laterally formed on the initiating ring, grows as a small crista which progressively increases during rising. This emergence is, at first, exclusively composed of clearly primary meristematic cells (cf p. 140) (Figs. 2.34b; 2.14a, *pf*).

Later, the cell differentiation involves three areas: (1) the ventral face; (2) the dorsal face where cells obtain both large vacuoles and well-differentiated chloroplasts ("parenchymatization phenomenon"; Figs. 2.34a, c, *dp, vp;* 2.35/4); (3) moreover, in a median strand and sometimes lateral strands, cell cleavages parallel to the organ axis initiate proconducting cell bundles (Figs. 2.34a, c, *pr;* 2.35/3, 4).

Therefore, the typical primary meristematic cells stand only at the apex, the base and on the outline of the crista of the *leaf buttress* (Fig. 2.35/4).

The further functioning of the marginal meristem and the basal meristem entails the differentiation of the various parts: first the lamina, then the petiole and the sheath. When the leaf is stipulated, stipules rise very early from the initiating ring in small emergences on both sides of the leaf buttress.

2.1.6.3 Evolution of the Leaf Base

The leaf base appears on the onset of the leaf initiation, during the "horizontal growth" which takes place before the initiation of the leaf primordium. If we consider that the primary leaf development includes three stages, i.e. initium, primordium and leaf buttress, we can assess that the initium is still mainly composed of the *leaf base*. A few cells of this initium will generate the appendix part of the leaf system during the phase of "vertical growth".

During the development, the remaining part of the leaf base becomes part of the stem but in fact it corresponds to the basal part of the leaf system mentioned as *"leaf segment"*.

Very early narrow and elongated cells appears in the leaf base and constitute the *procambial strands* then the first conducting tissues of the leaf segment (Fig. 2.34a–c, *pr*).

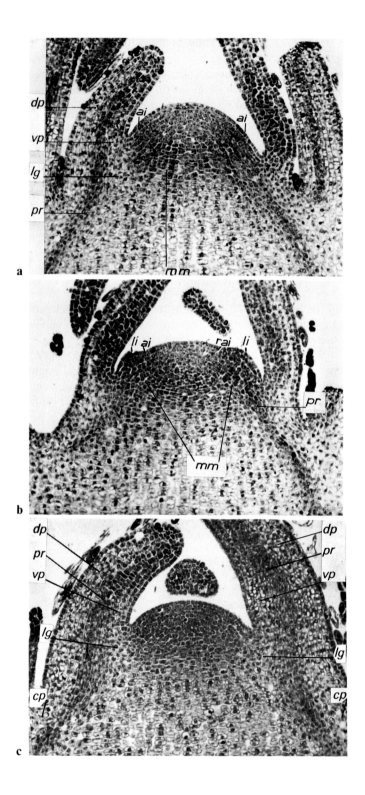

154

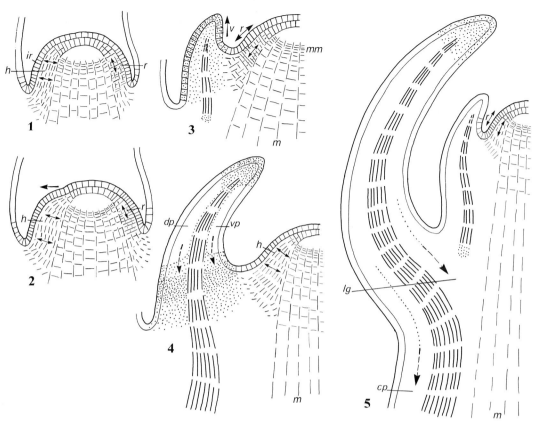

Fig. 2.35. Diagrams of the processes of leaf genesis at the apical meristem. **1,2** Horizontal growth *h*, producing the leaf initium at the expense of the initiating ring *ir*. **3** Vertical growth *v*, raising up the leaf primordium, which encroaches the initiating ring. The latter regenerates itself through flank anticlinal mitoses *r*, before being able to initiate another leaf. **4,5** Evolution of the primordium into leaf buttress, by dorsal *(dp)* then ventral parenchymatization *(vp)*, spreading towards the leaf basement, producing the cortical parenchyma *cp* and the leaf gap *lg* respectively; *mm* medullary meristem; *m* medullary parenchyma (medulla)

◁ ——

Fig. 2.34 a–c. *Cheiranthus cheiri,* aspects of leaf genesis on the apical meristem flanks. **a** Section of the apex through relative resting areas; *ai* anneau initial (initiating ring); *mm* medullary meristem. On the *left* primordium, *pr* procambial strand; *dp* dorsal parenchymatization; *vp* ventral parenchymatization; *lg* leaf gap. **b** Section passing by two areas in the leaf initiation period; the sagittal section of the *right* initium *li* is already provided with its procambial strand *pr; rai* regeneration zone of the anneau initial. **c** Lateral section, tangent to the initiating ring of the same apex as in **a.** On the *left,* a young leaf buttress shows a marked ventral parenchymatization compared to the leaf section of **a.** This parenchymatization is even more pronounced in the *right* older buttress. *cp* Stem cortical parenchyma, continuation of the dorsal leaf parenchyma in the leaf segment

155

Most anatomists (ESAU 1942; LAWALREE 1948; PHILIPSON 1949) consider that, in Dicotyledons and in conifers at least, the differentiation of leaf segment cells is induced by the underlying areas where conducting strands are connected with the already formed conducting tissues, and spreads towards the apex to the point where a leaf is going to rise. Due to this mode of formation, several authors suggested that the extension of the procambium differentiation would play a part in the phyllotaxic organization of the stem in determining the initiating sites of the leaves.

The earliness of such processes makes it difficult to determine whether the first periclinal mitoses initiating the leaf system occur before or whether they follow the procambial mitoses. However, it seems that frequently in Angiosperms, the first periclinal mitoses of leaf initiation occur before the neighbouring procambium, but this one is strongly constituted in a single plastochron. Its further development seems to be rather acropetal while, at the onset, its formation seems to be simultaneous in the whole base and does not show any orientation of extension (BERSILLON 1951).

In any case, when the leaf buttress differentiates in rising above the apex, the procambial strands are connected with those of the base; the tissues with large vacuoles of both faces (ventral and dorsal) of the buttress, resulting from the parenchymatization process, progress towards the base. On the internal side, the parenchymatous tissues form the "leaf gap" which establishes a parenchymatous connection with the medulla and on the external side, with the so-called cortical tissues of the stem (Fig. 2.35/4, 5).

According to species, leaf buttresses individualize one or several procambial strands which develop into *conducting bundles,* the extension of which constitute the *leaf traces* in the stem. Each conducting bundle is then related to a leaf gap and isolated by the parenchymatization of the interbundle cells.

The differentiation of proconducting tissues is not only achieved longitudinally. The initial strands are thickened by longitudinal mitoses of the first proconducting cells as well as by the addition of other cells issued from more or less differentiated and "parenchymatized" cells surrounding the strands.

In numerous Dicotyledons and Gymnosperms, procambial strands are individualized farther off the top, in a ring or a cylinder of isodiametric meristematic cells underlying the initiating ring, the *prodesmogen* (LOUIS 1935). This prodesmogenic ring appears as a remainder of the initiating ring (*Restmeristem,* KAPLAN 1937) or simply results from the delayed differentiation of the leaf base cells. The proconducting cells differ from those which become parenchymatized by their elongation through longitudinal mitosis without any diameter increase and by the RNA enrichment of the cells.

Further on, these initial proconducting strands may extend transversally and form a continued procambial cylinder except in front of the leaf gaps.

Using light microscopy, CATESSON (1964) carried out the cytological study of the proconducting strands in *Acer pseudoplatanus* and elucidated its meristematic characterics: a small amount of paraplasm, RNA richness, multiple but variable vacuoles, undifferentiated plastids, except for slight seasonal variations.

In the vegetative apex of the sycamore, during its progression into the leaf base the proconducting tissue is produced by *dedifferentiation* of the more or less developed cells of the leaf bases (CATESSON 1964).

2.1.6.4 The Two Ontogenic Composing Parts of the Stem

The structure of the vegetative apex (cf. p.155) and the development of the leaf bases result in two sets of tissues which compose the stem and differ in origin:

1. A properly cauline part is derived from the *"rib meristem"*, the medulla, a parenchyma into which, secondarily, other tissues can differentiate but which remains poorly developed in advanced plants.
2. *The remaining part of the stem,* i.e. mainly the cortical tissues, the conducting tissues with a small amount of parenchyma within them, constitute a set which is derived from the initiating ring. *This set is not autonomous since it is produced simultaneously with leaves.*

This notion, according to which the main part of the stem in Angiosperms, Gymnosperms and also Pteridophytes with "large leaves" is linked to the leaf organization, is in agreement with the paleobotanic data reported by EMBERGER (1968).

According to this author, the first vascular plants which appeared on the continents in paleozoic times were composed of axes, branched by dichotomy. Their conducting apparatus was a protostele, the vascular part of which formed a solid cylinder with a central pole e.g. the *Rhynia;* Fig.2.36/1). The centrifuge migration

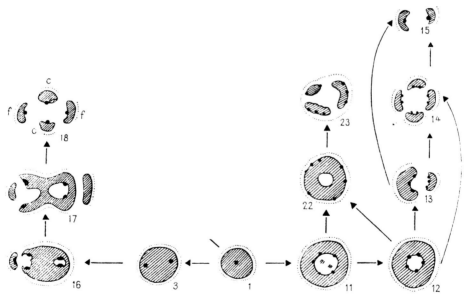

Fig.2.36. Schemata illustrating the phylogenetic relations between the stem anatomical structures. **1** Protostele of *Rhynia,* central vascular pole; **3** *Psilotum* rhizome, centrifugal migration of vascular poles; **11** protostele of *Hymenophyllaceae, appearance of a central medulla;* **22** *Lepidodendron* siphonostele, exarch vascular poles; **23** dislocation of the exarch siphonostele (transition form to the root anatomy); **12** endarch siphonostele; **13-15** fragmentation of the endarch siphonostele, resulting in the stem structure; **16-18** emission of fronds, entailing the formation, then the opening of the loop, and the leaf gaps. *c* Cauline strands, not very distinct from leaf strands *f;* (after EMBERGER 1968, Fig.447, *pro parte*)

of the vascular poles associated with the appearance of a central medulla results in a siphonostele in which the vascular tissue forms a "hollow" cylinder (e.g. *Lepidodendron;* Fig. 2.36/22).

In "megaphyll" plants (ferns, spermaphytes), the origin of the leaves would be settled in the axes which progressively become bilaterally symmetrical, a symmetry inaugurated by the last ramifications, extending down to their base and accentuated by a palmation phenomenon resulting in the formation of the leaf lamina (Figs. 2.37, 2.38).

The basipetal progression of bilateral symmetry spreads into the main leaf axis, the stem, and determines the stele opening, then causing the formation of holes or "leaf gaps" related to the emergence of leaves (Fig. 2.36/13–18.23). The siphonostele is thus perforated or dislocated and becomes a dictyostele (Fig. 2.36/18).

When the leaf-conducting system goes further down the stem, the "perforated" cylinder of the dictyostele is itself dislocated, and the stele stays open. Then there is no real cauline material left in the conducting system of the stem which is derived only from the association of the conducting tissues of the "leaf segments".

The medulla only, initiated by the medullarization of the primary stele, represents the true cauline remainder.

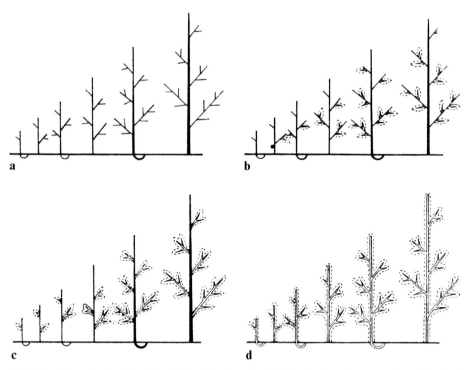

Fig. 2.37. Schemata showing the origin of the fern frond (**d**) by basipetal progression of the leaf structure (**a,b,c**). The "foliarization" begins on the last ramifications (**b**); on **c**, the "foliar" invasion does not yet reach the rachis (*"phyllophore"*), which is only reached in **d**. The cauline parts are represented by *full lines,* the foliarized parts by *double lines;* (after EMBERGER 1968, Fig. 443)

158

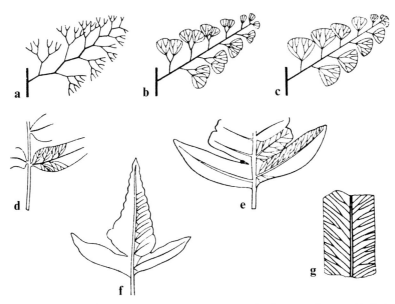

Fig. 2.38 a–g. Origin of leaflets, quills, and sometimes fronds of present ferns. **a** Entirely cauline state (phyllophored); **b,c** foliarization of the last ramifications, producing the pinnules; **d,e,f** palmation joining the pinnules into quills and the quills together; **g** *Scolopendrium* frond; (after EMBERGER 1968, Fig. 444)

The leaves and their bases arise as *lateral buttresses* with bilateral symmetry, the dichotomous ramification, strictly axial apical, disappears and it is easy to understand that the development of the apical meristem results in the formation of an initiating ring which generates the whole foliar organization: lamina, petiole (and its subsidiaries) and the leaf segment, the latter being part of the stem.

Thus, the paleontologic development of the foliated stem of "megaphyll" plants leads to the same interpretation as the ontogenic analysis.

It should also be noted that the histogenic study brings us far from the classical concept of the primary stem structure. The two parts generally singled out, i. e. the *cortex* and the *central cylinder,* to not correspond to the two units of the apical ontogenesis: rib meristem and initiating ring. Therefore, one should consider that the classically defined structures, i. e. cortex and central cylinder, which result from the physiological differentiation of the organ, are *histophysiological functional units* of the young stem, but do not correspond to the two sets elaborated by apical ontogenesis, i. e. medulla and leaf segments.

In fact, the classical concept was influenced by past authors who wished to assimilate the anatomy of the stem to that of the root. But, there is a fundamental difference between the two organs which was not duly appreciated by these authors: the root does not generate any organ with bilateral symmetry, in contrast to the stem and, to be even more rigorous, the expressions cortex and, above all, central cylinder can only concern the root and not the stem.

Besides, the differentiation of a true endodermis with its "Casparian strips" involves only the root (except for the aquatic stems) and is apparently linked to the physiologically aquatic mode of life of this organ. The typical endodermis is "par excellence" the cellular tissue which surrounds a genuine central cylinder in the root.

2.1.6.5 The Organization of Leaf Segments

In Dicotyledons and Gymnosperms, leaf segments grow in length during the phase of "intercalary growth" of the stem internodes, when this stem is developed. Leaf segments are typically grouped in a cylinderlike zone around the medulla. The leaf segment extends downwards to the insertion sites of the lower leaves which are approximately on the same generating line of the stem. In some way, segments are superimposed (Fig. 2.39).

In Monocotyledons, the organization is different: segments thicken in their upper part, they become conic during the elongation of internodes and are encased one in the other (Fig. 2.40).

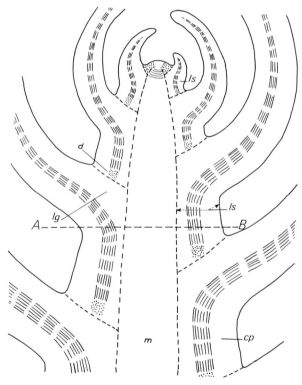

Fig. 2.39. Primary structure of the stem; arrangement of leaf segments in Dicotyledons. The leaf segments *ls* are arranged into a cylinder surrounding the medullary parenchyma *m. d* Shifting of the conducting strand before it becomes laterally connected with the strands of older leaf segments. This deviation spares the leaf gap *lg; cp* cortical parenchyma, in continuity with the dorsal parenchyma of the young leaf

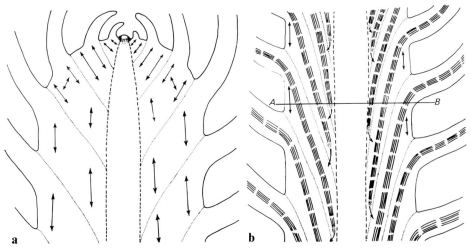

Fig. 2.40. Primary structure of the stem; disposition of leaf segments in Monocotyledons. A very early thickening of the leaf basements, generally very "amplexicaul", corresponds, already in the meristem, to the final thickness of the adult stem (**a**). Subsequently, the elongation changes the leaf basements into cones encased one in the other, resulting in several circles of bundles, the most recent is the most internal (**b**)

This difference may appear to be related to the very "embracing" arrangement of the leaf bases of Monocotyledons but, in fact, the organization is the same in species with a narrow leaf base and, moreover, Dicotyledons with embracing leaves have the typical arrangement of their group, as shown in the case of *Heracleum* (MAJUMDAR 1942).

The differences seem to be rather related to the processes of *thickening* of the stems of these two groups. The thickening of Monocotyledons stems is mainly achieved through periclinal divisions of the the leaf base cells, divisions which take place very close to the apex, entailing the enlargement of the bases into flared cones (Fig. 2.40a).

The periclinal proliferation takes place partially before the formation of procambial strands. The latter rise then in the tissue derived from the periclinal proliferation, whereas the segments grow in length (Fig. 2.40b).

In Dicotyledons and Gymnosperms, the early periclinal proliferation is more limited and the procambial strands arrange themselves in a cylindrical sheath where they are connected. The further thickening will be initiated by a cambium which, rising between the first vessels and the first sieve tubes, is enclosed in a single cylinder. At first the cambium functions in *the advanced parts of the stem* which thickens *at its base*. In Monocotyledons, the early periclinal growth thickens the stem *from the top*.

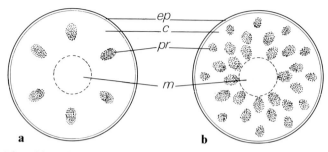

Fig. 2.41 a, b. Disposition of the elements of the stem primary structure. **a** Dicotyledons; **b** Monocotyledons; transversal sections according to the line *AB* of Figs. 2.39 and 2.40 b; *c* Cortex; *ep* epidermis; *pr* procambial strands; *m* medullary parenchyma (medulla)

2.1.6.6 Components of the "Primary Structure"

From inside to outside, a transversal section in a very young area of the stem will show (Fig. 2.41):

1. The medulla: parenchyma with large cells derived from the medullary meristem;
2. Tissues issued from the initiating ring or leaf segments located on a single circle in the case of a Dicotyledon (Fig. 2.41 a) or a Gymnosperm, on several circles in the case of a Monocotyledons, giving sections of encased conic segments (Fig. 2.41 b).

Each leaf segment includes proconducting strands. They are more or less in the process of differentiation and are thus disposed on one circle in Dicotyledons and Gymnosperms, on several in Monocotyledons.

In Dicotyledons and Gymnosperms, the procambial strands may thicken very early to form a continued cylinder, except at the leaf axil, the place of the "leaf gap". If they remain isolated, the cells which separate them develop into parenchyma (primary medullary rays).

Outside the procambial strands, the dorsal parenchymatization of leaves, extended to the stem, results in parenchymatous tissues and supporting tissues generally designated as *cortex*.

Finally, the young stem is limited by a layer in continuity with the external layer of leaves and apical buttresses, which constitute the epidermis.

The whole set being arranged, the in situ cellular differentiation will generate the anatomical features of the primary structure: differentiation of vessels and sieve tubes in procambial strands, periphloemian differentiations (pericycle and its fibres, possibly endodermis) and superficial differentiations (collenchyma, epidermal formations).

2.1.7 Establishment of the Primary Structure of the Root

The development of the root is simpler than that of the stem as the meristem activity and the first phases of differentiation are generally not subjected to the effects of the initiation of lateral organs.

The study of structure differentiation in the root is less advanced than that of the stem. At the extremity of the root two independent areas are clearly singled out: the cap, which extends or does not extend into the hair layer, and a unit of internal isodiametric cells. Around this small unit, the proliferation is high and

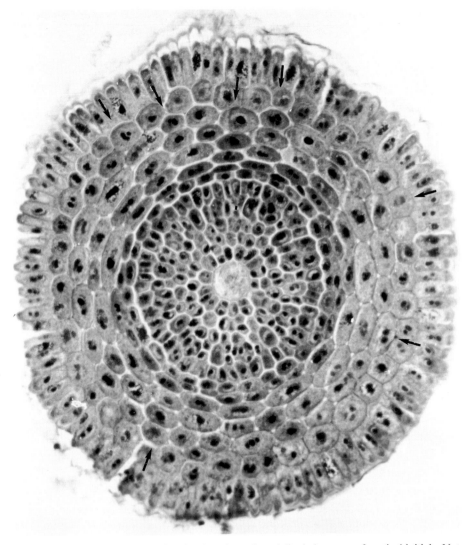

Fig. 2.42. Terminal thickening of barley root through periclinal cleavages of cortical initials. Note the lines of cells marked with *arrows*

163

cells are quickly arranged so that, up to the very top, the main areas of the root can be distinguished: central cylinder and pericycle, endodermis and cortex.

Cortex and central cylinder are soon delimited by repeated periclinal mitoses in the internal layers of the cortex, showing cells arranged in radial files on transversal sections (Fig. 2.42).

Conversely, the central cylinder has no well-delimited proconducting strands as those of the stem. Under the pericycle, cells grow and slowly become vacuolated while alternately, others show very few vacuoles for a longer period and longitudinal divisions so that their section tends to decrease.

The first cells will develop into tracheids or vessels, the others will build the primary phloem. The remaining cells of the central cylinder change into parenchyma slower than in the stem, producing the medulla and the primary medullary rays. The pericycle remains meristematic for a longer period.

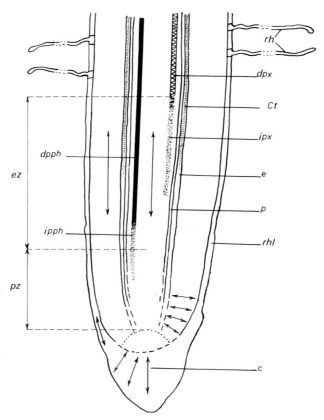

Fig. 2.43. Root histogenesis *(Lycopersicum esculentum)*. *rh* Root hairs; *rhl* root hair layer, here in continuity with the root cap *c; e* endodermis with Casparian strips *Ct; p* pericycle; *dpph* differentiated protophloem; *ipph* immature protophloem; *dpx* differentiated protoxylem; *ipx* immature protoxylem; *ez* intensive elongation zone; *pz* intensive proliferation zone. *Double arrows* indicate the predominating growth orientation

In contrast, endodermal cells increase and differentiate quicker: the vacuoles enlarge and the dividing cell walls are loaded with characteristic subero-lignified inclusions *(Casparian strips)*. The differentiation of sieve tubes occur very early, whereas that of vessels, endodermis and root hairs begins only beyond the zone of the most rapid elongation. The thickening periclinal mitoses of the cortex end at this very level (Fig. 2.43).

2.2 Secondary Meristems

2.2.1 Origin and Function of Secondary Meristems

The apical activity results in a set of young cells which almost differentiate "in situ" and give the primary structure.

However, in most Dicotyledons and in all Gymnosperms, the differentiation of the material produced by the apical activity spares a few deeply located cells which, so to say, remain half way. These cells, part of the procambium, may apparently remain quiescent for some time, and then proliferate again initiating a further growth of the organ called *"secondary growth"*.

In addition to these few cells stopped in their differentiation, other cells which have become more or less parenchymatous, regress, divide again and mingle to the first mentioned cells: this occurs in numerous continued cambial zones; their setting depends upon dedifferentiation processes which will be studied later. Moreover, such processes generate the other *secondary meristems*, the subero-phellodermal layers, or *phellogen*, of generally less deep origin.

These data allow the acknowledgement of two main types of secondary meristems or cambiums: the vascular generating zones or vascular cambium and the subero-phellodermal layers. A less frequent type should be added which ensures the thickening growth of some Monocotyledons *(Dracaena, Cordyline)*. The common feature is the production of radial cellular lines by means of tangential longitudinal segmentations of their cells, an organization generally and erroneously considered characteristic of secondary tissues.

The secondary growth processes concern mainly the increase in thickness, i.e. the transversal growth, of axial organs. Such an orientation of growth gives the "secondary meristems" quite similar and characteristic aspects.

2.2.2 The "Prodesmogen" and the Vascular Cambium

Very soon during leaf histogenesis, some cells of the base are distinguished from the neighbouring ones due to their delayed differentiation; they remain smaller and their vacuoles do not increase as much as those of the neighbouring cells. In many species, these nearly apical cells appear as a "reserved meristem" directly derived from the initiating ring. KAPLAN (1937); BERSILLON (1951); GREGOIRE (1935a, b, 1938) and LOUIS (1935) singled out these cells under the term of *prodesmogen*.

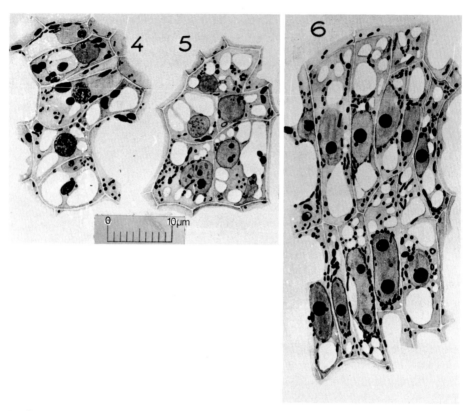

Fig. 2.44. *Myosurus minimus,* vegetative apical meristem and passage to the prefloral state. **4** Beginning of formation of a proconducting strand below a leaf initium. The cells, which had started to differentiate, have big chloroplasts; **5** slightly more advanced stage showing the dedifferentiation of plastids; **6** fragment of procambial conducting strand at the base of a prefloral apex, derived from cells similar to those of **4**; dedifferentiated plastids, divided vacuoles, large nucleoli. The repeated longitudinal divisions result in tiers of procambial cells.

Fixation: Regaud; staining: hematoxylin; (after BUVAT 1952)

   The *procambial* cells appear within the prodesmogen, in such places where the first conducting elements of the leaf and leaf segment are constituted.

   They grow in length and mainly show repeated longitudinal mitoses generating storied groups of long and narrow cells (Fig. 2.44).

   These cells are still very meristematic (large nucleoli, small vacuoles, poorly developed plastids) and compose the *procambium* or proconducting tissue. Initially appearing in isolated bundles or "procambial strands", the procambial tissue extends laterally sooner or later in Dicotyledons and Gymnosperms and constitute the procambial cylinder.

   But meanwhile, the prodesmogen cells situated between the procambial strands may become more or less parenchymatous according to the species. The general existence of the prodesmogen has been discussed in the cases in which the parenchymatization of the cells occurs very soon.

166

Table 2.4. Distribution of the various types of procambial mitoses in growing internodes of a sycamore branch, on the initiation of the seventh pair of leaves (in CATESSON 1964)

Internode range (from the base of the branch)	Length at the moment of cutting (cm)	Average length reached by adult internode	Mitoses[a]		
			Periclinal	Transversal	Anticlinal
3rd	0.8	14.3	35%	50%	15%
2sd	0.8	8	31%	57%	12%
1st	0.9	2.3	82%	15%	3%

[a] Periclinal mitoses are prevalent when the internode is closer to the final development, contrarily to anticlinal mitoses of diameter increase and to transversal mitoses which harmonize together with the elongation of the stem and produce both long and short initials.

Thus, according to the case, the further extension of the procambium into a continued generating zone requires more or less accentuated dedifferentiation processes which are not very well established from a cytological standpoint.

However, part of the procambium cells develop into the vascular cells (tracheids, vessels and related parenchyma) and another part into the phloem cells (sieve tubes and related cells). This in situ differentiation establishes the primary conducting tissues. Between these two types of tissues, procambium cells subsist *where differentiation has stopped,* whether they continue or temporarily cease to proliferate. They have somehow only undergone a *first phase* of differentiation. Generally, particularly during stem ontogenesis, the cellular reserved material, which remains between the first xylem elements (protoxylem) and the first phloem elements (protophloem), increases by means of longitudinal divisions more and more periclinal. This fact results in the organization of more or less regular radial lines of cells which thicken the conducting bundle issued from the procambium. In this cellular material, the cells close to the first vessels and the first sieve tubes differentiate in situ into metaxylem and metaphloem. Therefore, LARSON (1976) used the term metacambium for this material generated by the procambium, in which periclinal mitoses become prevalent (Table 2.4), whereas the elongation of the organ continues and finishes (Fig. 2.45 A).

The metacambium cells grow in length by means of intrusive extension of their extremities which become more and more oblique, mainly in species with a "non-storied cambium" in which this intrusive growth goes on even after the phase of elongation of internodes. In contrast, this process is almost inexistent in species with "a storied cambium" such as *Robinia pseudoacacia* (Fig. 2.51 c). Some of these fusiform cells segment transversally into short cells and the metacambium becomes progressively a completed cambium (Scheme 2.1).

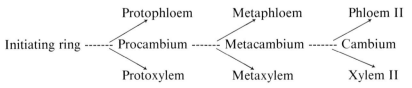

Scheme 2.1. Continued development of procambium into metacambium and vascular cambium and origin of their secondary elements (after LARSON 1976, in CATESSON 1984)

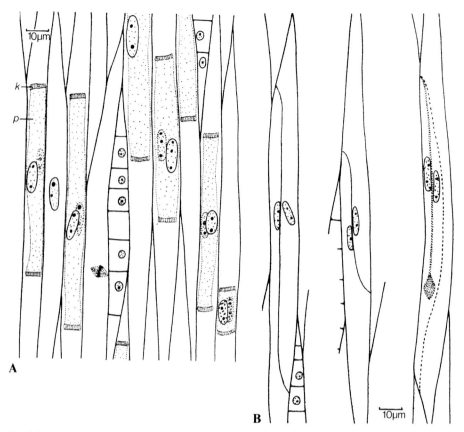

Fig. 2.45 A, B. Proliferative activities in the cambium of *Acer pseudoplatanus*. Two *tangential* sections. **A** Simultaneous periclinal mitoses in numerous fusiform initials; front view of the phragmoplasts *p* (fixation in May). **B** Anticlinal cleavages, nearly vertical, of fusiform cells, allowing the cambium growth in diameter, and the regeneration of the amount of long cells; the phragmoplasts are seen in profile. *k* Kinoplasmosomes.

Fixation: formaldehyde, acetic acid, alcohol, staining: methyl green, pyronine; (after A. M. CATESSON 1964)

Thus, when the elongation of the stem is completed, the reserved cells between the metaxylem and the metaphloem include associations of short and long cells. This set represents the *intrafascicular cambium* made of long initials (or fusiform) and *short initials;* the latter will generate the so-called *horizontal* systems or rays of the phloem and xylem (see p. 348 and Fig. 2.51).

A similar evolution casually spreads to the interfascicular zones where parenchymatous cells resume a proliferating activity which change them into cambial cells; they produce the *interfascicular cambium* (Gymnosperms, numerous Dicotyledons). This does not always include long initials, particularly if the interfascicular zone is narrow, but when long initials are present, they are derived either from the intrusion of fusiform cells located at the periphery of the intrafascicular cambium or from the modification of interfascicular parenchymatous cells achieved by intruding elongation. Moreover, the establishment of the vascular cambium constitutes a kind of *differentiation* of the cells more or less directly issued from the apical meristems and from the procambium. The differentiation into "secondary meristems" gives the cambium a real *histological* identity (CATESSON 1974).

The study of mitotic activity enabled CATESSON (1964) to demonstrate that the cambium is a zone and not a simple generating layer, like the phellogen, (suberophellodermal layer). The increased RNA richness and the poor differentiation of plastids confirm this fact. However, with regards to Gymnosperms, there is still controversy.

The cambium zone increases in diameter by means of anticlinal divisions almost vertically and, moreover, in sufficient number to make up for the loss of fusiform initials due to their modification into short initials, mainly in summer (BANNAN 1950, 1951, 1953, 1957; CATESSON 1964) (Figs. 2.45 B, 2.51 c, d, *arrows*).

Therefore, a zone of meristematic tissues is thus constituted, derived from the "prodesmogen" or, partially, from the "procambium", the vascular generating zone: its activity will exclusively produce new conducting and complex tissues such as the secondary xylem, i.e. the wood.

2.2.3 The Phellogen (Cork Cambium)

This layer is also called the subero-phellodermal layer and shows both histogenetic features common to those of the vascular generating zone and different histogenetic and cytological features. At the periphery it generates the *secondary suber* or *secondary cork* and, towards the interior, the *phelloderm* which is a secondary parenchyma. All these tissues constitute the *periderm*.

The mode of formation and functioning by means of periclinal cleavages is a common feature. Conversely, the origin of the phellogen varies according to species and for the same plant, according to the age of the organ in which it is initiated. In fact, contrarily to the vascular generating zone which is permanent, the phellogen can only function for some time and, in perennating organs, it is then exfoliated and periodically replaced by a more internal, newly formed layer. We shall therefore consider separately the origin of the first subero-phellodermal layer of an organ and the origin of the following ones.

2.2.3.1 The First Phellogen Layer

This layer appears on the stem or the young root still presenting the anatomical features of primary structure. For simplification, we shall deal only with the case of the stem. This organ is surrounded by an epidermis, under which a few so-called cortical layers are found, composed of parenchymatous cells and sometimes collenchyma.

Most often, the appearance of the "phellogen" is marked by periclinal mitoses which take place in one of these cortical layers. In many species *(Prunus, Evonymus),* this layer is the subepidermal layer (Fig. 2.46), but in some cases also, though rarely, the phellogen originates from the epidermis itself (*Pirus,* Fig. 2.47). In other

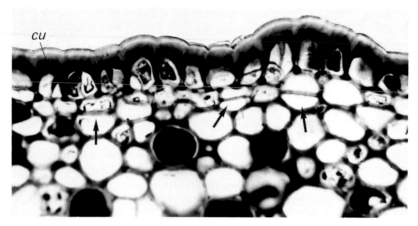

Fig. 2.46. *Evonymus europaeus,* stem. First subepidermal divisions, giving birth to the first phellogen layer *(arrows). cu* Cuticle; the epidermis is strongly cutinized; staining: Sudan black B

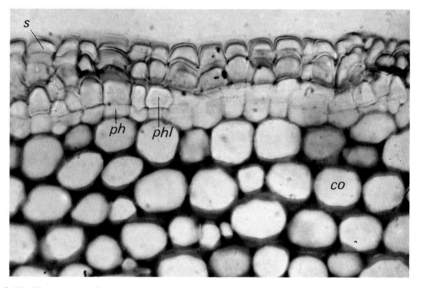

Fig. 2.47. Young stem of pear tree; epidermal origin of phellogen formations. *phl* Phellogen layer; *co* collenchyma; *ph* phelloderm; *s* secondary cork

Fig. 2.48 A, B. *Vitis* stem of 1-year-old, transversal sections. The first phellogen layer, no longer functional at the moment of the fixation (in spring following the stem growth), arises from the primary phloem, underneath the periphloem fibre bundles. So, the first periderm includes a suberized zone *s* and all the more external tissues, the necrosis of which was caused by the suberized zone. These tissues are more or less crushed and will be exfoliated. *ep* Epidermis; *c* cortex; *f* periphloem fibres; *aphl* ancient primary phloem; *fphl* functional phloem of late season; *ca* vascular cambium (A); × ≈ 100 (A); × ≈ 200 (B)

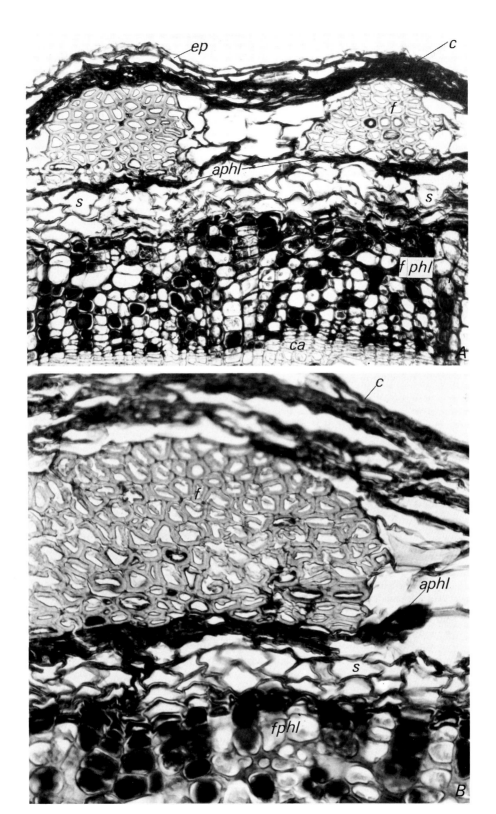

cases, the origin is cortical *(Robinia, Pinus)*. Again in other species, the first layer itself is more internal and originates directly in the phloem. This is the case with *Vitis* (Fig. 2.48). Generally, it concerns cells which have widely differentiated before the recurrence of mitoses: they have the special characteristics of *epidermal, parenchymatous* and even *collenchymatous* cells.

The following new development of these cells requires further studies. According to the present state of knowledge, we can assess that no preliminary cytological preparation involves the cells which undergo the first mitoses: simultaneously with the recurrence of mitoses, the generating cells progressively lose the attributes of their first differentiation (starch, tannins, etc.) (LIER 1955). Here again, the cambium formation depends upon a dedifferentiation process.

The first layer may appear sooner or later according to the species and to the medium conditions. In most cases it is constituted during the first year of growth of the organ. When it is constituted early, it appears simultaneously on the whole outline (still young cells) and, firstly, in selected points. When it is constituted late, the layer spreads slowly around the organ, dedifferentiation processes being apparently more pronounced.

2.2.3.2 Renewal of the Phellogen

When the thickening of the organ lasts a long time, the first phellogen layer and its productions are distended and finally exfoliated. Before they disappear, a more internal layer is constituted. It has been noted that generally the species in which the first epidermis appears in the most external part of the cortex, generate discontinuous following periderms at various levels of the cortical and phloem tissues, i.e. outside the vascular cambium.

In contrast, species with a very internal first periderm generate their following periderms in continued concentric layers in the phloem of successive years *(Vitis)*.

The cytological evolution of the cells which change into phellogen meristematic cells requires further studies. Up to now, only the anatomical and histological aspects of the matter have been studied, e.g. from the first observations of SANIO (1860) to the subsequent works of ESAU (1948).

The formation of the phellogen enters into the framework of essential circumstances in which dedifferentiation processes occur in the normal development of numerous plants (Gymnosperms, Dicotyledons).

In addition, peridermal neoformations appear following injuries scarred over by these periderms or following various parasitic aggressions. Like the periderm, which appears in the normal development of plants, healing periderms are derived from more or less differentiated cells which produce a generating layer of phellogen type by means of dedifferentiation.

On the side of the injury or the parasitic aggression, this layer produces cellular lines which suberize, forming a protective coat which, more or less completely, isolates the underlying tissues from the influences which caused the damage. Thus, healing periderms are defensive reactions of plants. We shall return to this matter with the discussion of protective tissues.

172

2.2.4 The Cambium of Monocotyledons

In some arborescent or herbaceous stems of Monocotyledons, the thickening growth is ensured by generating cells of secondary meristematic type (CHEADLE 1937).

We know (p. 160) that the stem of Monocotyledons is thickend from the vegetative top through the periclinal cleavage of the foliar base cells. They constitute a primary meristem of growth in thickness. In the absence of further data, it seems that the "cambium" of Monocotyledons with secondary formations is derived directly from the cells generated by this first periclinal activity. Although this matter also requires further research, there seems to be no clear distinction, no functional *discontinuity* between the thickening primary meristem and the "cambium".

This cambium is distinguishable from the others by its productions: in some cases it supplies some cells of secondary parenchyma towards the exterior, but mainly, towards the interior, it produces a basic tissue into which conducting strands differentiate, the remaining cells developing into parenchyma of medullary type or into sclerenchyma (Fig. 2.49).

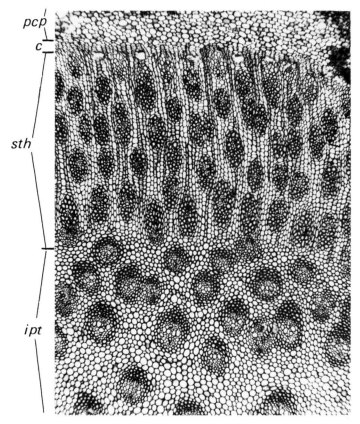

Fig. 2.49. *Dracaena,* transversal section of the stem. *pcp* Primary cortical parenchyma; *c* cambium; *sth* secondary thickening, enclosing conducting strands into a secondary parenchyma; *ipt* internal primary tissues with conducting strands sunk into the ground parenchyma

173

It is this *sclerenchyma* and not "wood" which gives Monocotyledons their feature of "lignous plants".

2.2.5 Cytological Features of Secondary Meristems

The cytology of secondary meristems is not well determined, particularly on the electron microscopy scale. The reason seems mainly due to the technique. It is difficult to obtain good samples of cambial cells which are partially distorted by the neighbouring xylem and phloem tissues with the techniques of ultrastructural cytology. The following data deal only with a small number of species and few generalizations are possible.

We have already described (p. 140, 142) two types of meristematic cells which are generally well distinguishable in the stem apex. The secondary meristematic cells represent a third tpye.

Whereas apical cells are more or less isodiametric, cambial cells are mainly elongated in the same direction as the organ in which they are enclosed. This particular shape is mainly pronounced in the cells of the vascular generating zone. It is markedly demonstrated in cambial cells of trees: with a width inferior to 50 μm, they can extend from 1000 to 4000 μm and more in *Pinus strobus* (BAILEY 1920). When they are very elongated, cambial cells are particularly fusiform.

More precisely, mainly in trees, most vascular generating zones include two kinds of histogen cells. The above mentioned elongated cells are the most numerous and common cells; they constitute the "long initials". But the cambium encloses also shorter cells, sometimes isodiametric in tangential sections, the extremities of which are less oblique or clearly transversal. These "short initials" produce the cells of the so-called horizontal systems of the secondary phloem and xylem; they build the "phloem rays" and the "xylem rays". We shall deal with these cambiums again is the discussion on the secondary conducting tissues.

The phellogen cells or the generating cells of secondary tissues, when existing in Monocotyledons, are generally less elongated.

The cytology of vascular cambial cells associates general features to marked seasonal variations, particularly in trees. The nucleus is ellipsoidal and is found against one face of the cell wall; it has more nucleolar substance than the nuclei of derivative cells. The cytoplasm is always RNA richer than in the latter: besides, the RNA amount increases during the period of proliferating activity (CATESSON 1964).

Fig. 2.50 A–D. *Robinia pseudoacacia,* tangential sections into the cambium; seasonal cytological variations. **A** Fixation April 12, just before the recurrence of activity; mitochondria and vacuoles divided into small elements, radial cell walls thickened between the pits. **B** Fixation May 23, active cambium, widely vacuolated cells, thinner radial cell walls. **C** Fixation June 25, typical secondary meristematic cells, cytoplasm essentially peripheral, tendency of mitochondria to join in chondriocontes. **D** Fixation October 4, return to the inactive condition, mitochondria and vacuoles divided as in **A**, thickening of radial cell walls between the pits. In all cases, plastids remain undifferentiated, more or less indistinct from mitochondria, and do not produce starch

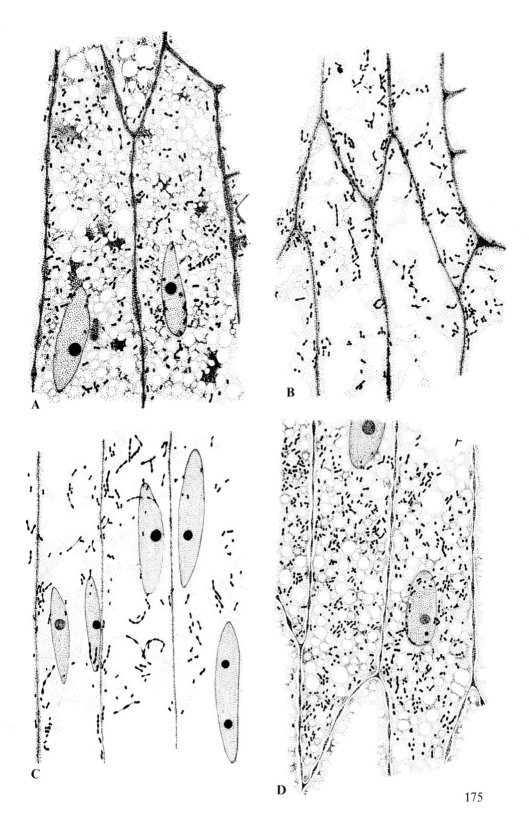

A

B

C

D

175

Another general feature of paraplasmic nature is the poorness in starch. Thus, in the cambium of *Acer pseudoplatanus* (CATESSON 1964) there is no starch from March to June; small amounts are found in the radial initials between July and February–March; it is rare in all seasons in the fusiform initials.

Lipids are only abundant in winter, beginning to increase in July. As for the soluble storage material, it accumulates mainly in vacuoles at the end of summer and disappears quickly in March. Concurrently, the cytoplasmic density increases in winter.

These variations have a large influence on the vacuolar apparatus (Fig. 2.50), when in activity, cambial cells have a large and single vacuole. In September, this vacuole breaks up into numerous inclusions with myelinic characteristics, then these myelinic enclaves turn into small globular vacuoles. From the beginning of spring, a reverse evolution occurs and, after a transition to myelinic forms, the cells have a large single vacuole again (BAILEY 1930).

CATESSON's works (1964) showed that the breaking up of the vacuole in fall occurs concurrently with the stop of cyclosis, whereas the spring evolution corresponds to the recurrence of cytoplasmic movements. The cyclosis reactivation may be experimentally obtained by warming cambial cells up in winter and putting them simultaneously in a hypotonic medium.

The observations concerning the seasonal variations of mitochondria are too restricted to allow a generalization. In the *Robinia* cambium (Fig. 2.50) mitochondria are punctiform or very short in winter; they tend to become associated in filaments (chondriocontes) during spring hydration and are reduced to short elements in fall. In the sycamore (CATESSON 1964, 1974) mitochondria are globular in winter and during the summer phase of activity; they are longer, sometimes "beaded" during spring hydration and when the cambial activity stops at the end of summer. However, the infrastructure of mitochondria is constant (CATESSON 1974).

The study concerning the amount of water and the osmotic pressure of the cambial cells of *Acer pseudoplatanus* (CATESSON 1964) establishes also a parallel between their variations and the morphology of the vacuome and the chondriome.

From January to March–April the osmotic pressure changes from 36 to 12 atm. The cyclosis appears only if the pressure does not exceed 24 atm. As for the amount of water, it is at most in spring, it decreases in September and October, increases slightly in December and is minimal again in January. The association of mitochondria into filamentous chondriocontes is probably caused by the spring hydration of cells when the "mineral" sap springs up.

The high osmotic pressure of cambial cells in winter is probably one of the resistance mechanisms to cold; this resistance mechanism is particularly necessary for these cells which constitute the meristematic material essential to the renewal of activity of long-living plants.

Generally, plastids are not or poorly differentiated (Plate 2.6 A, B). However, as the arrangement of numerous cambial cells depends on a dedifferentiation, the plastids of original cells may preserve their structure for some time. The proliferating activity of the cambial cells is not constant and sometimes it is low. In poorly active cells, plastids can differentiate. The influence of the medium, mainly light, is sometimes favourable to differentiation, particularly in the relatively superficial cells of the phelloderm layers.

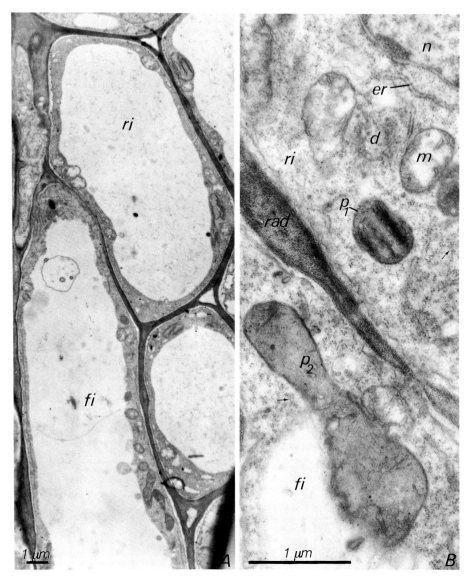

Plate 2.6 A, B. *Acer pseudoplatanus,* cambial cells, tangential sections.

Large and unique vacuole in each cell, cytoplasm forming a thin film against the cell wall. *fi* Fusiform initials; *ri* radial initials; *d* dictyosome; *m* mitochondrion; *n* nucleus; *p₁* lamellated plastid in the radial initial; *p₂* undifferentiated proplastid in the fusiform initial; *rad* radial cell wall; *er* endoplasmic reticulum; numerous polysomes, indicated by *arrows*.

× 5750 (**A**) and 28000 (**B**). Fixation: glutaraldehyde – O$_s$O$_4$; contrast: lead citrate; (after CATESSON 1984)

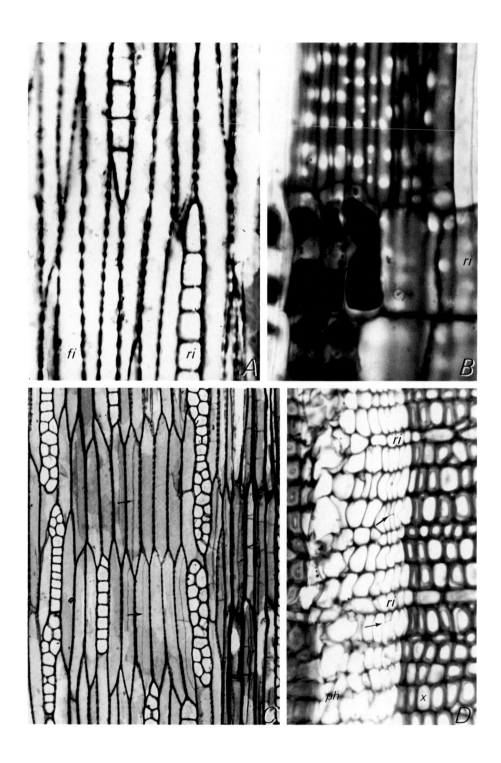

In aerial organs, plastids of cambial cells often enclose chlorophyll, but the enclosed thylacoids are poorly developed and the granar structure does not appear. As for starch, variations according to species are important and no generalization is possible. Plastids of short initials tend to be more amyliferous than those of long initials. Besides, few species have been studied but this fact was confirmed by the observations of SRIVASTAVA (1966) on *Fraxinus americana*.

In this species, the plastids of cambial cells, in addition to the usual components (lamellae and vesicles, plastoglobules and starch grains) encompass a highly osmophilic intralamellar inclusion which could represent a storage material aimed to build granar thylacoids in phloem cells where they are found. Such inclusions exist in various types of cells (CATESSON 1970; D. MARTY 1973) but have not apparently been found in the cambium of other species.

In general, the cell walls of cambial cells are thin and homogenous in herbaceous plants, whereas they show already marked primary differentiations in the vascular cambium of trees. The radial faces of the walls are often thickened but include thin circular areas which are the primary pits (Fig. 2.51). The thickness variations which give the cambium a pearled aspect in tangential section (Fig. 2.51) are more pronounced in winter when the generating zone is at rest than when it is activated (Fig. 2.50).

ROLAND and CATESSON's ultrastructural cytochemical studies (1981) showed that the tangential cambial walls, very rich in fibrillar material, are much more rigid than the radial walls which are not very fibrillar but rich in hemicelluloses. The plasticity and stretching of the latter is thus favourable to the radial growth of the cambium derivatives (CATESSON and ROLAND 1981).

These general features are those of most cambial cells but they can show important fluctuations in the same cambium, depending on the cells involved, the season, and they vary also depending on species. For example, although the characteristics of the vacuolar system are particularly variable, the existence of a large single vacuole is quite general in summer, whereas, in the winter dormancy, the vacuolar system divides in the cambium of most trees (BAILEY 1930; CATESSON 1964) (Fig. 2.50).

The longitudinal division of these cambial cells, which can be about 100 times longer than wide, yield notable figures. BAILEY (1919) described the prolonged functioning of the phragmoplast (Fig. 2.52) which, after the reconstitution of daughter nuclei, ensures the cytodieresis.

◁——

Fig. 2.51 A–D. *Castanea vulgaris* cambium (**A, B, D**). **A** Tangential section; *fi* fusiform initials (long initials); *ri* radial initials (short initials), uniseriate, radial cell wall pits are seen in profile, hence their "beaded aspect". **B** Radial section showing the transversal alignment of the radial cell wall pits of fusiform initials. **D** Transversal section: three to five layers of cambial cells constitute the generating cambial zone. Anticlinal vertical cleavages ensure the maintenance and the diametral growth of the cambium *(arrow)*. *x* Xylem; *ph* phloem. **C** *Robinia pseudoacacia* cambium, fixed in November. Typical storied cambium, fusiform initials relatively short, multiseriate radial initials, radial pits seen in profile (tangential section). The *arrows* indicate anticlinal cleavages ensuring the cambium diametral growth and maintenance

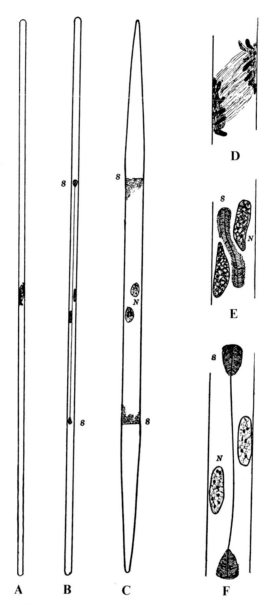

Fig. 2.52 A–F. Tangential longitudinal division of a *Pinus strobus* cambial fusiform initial. **A** Radial view, showing the oblique karyokinetic spindle, at anaphase; **B** id after reconstitution of daughter nuclei and the beginning of phragmoplast extension; **C** same cells as **B**, in tangential view; **D** enlargement of the karyokinetic figure of **A**; **E** beginning of phragmoplast formation and extension; **F** further advanced progression of the phragmoplast through the vacuole. *N* nuclei; *s* fibres of the phragmoplastic pad (kinoplasmosome); (after BAILEY 1919, 1954)

From the site, approximately median to, where karyokinesis takes place, the phragmoplast pad spreads to both lateral edges of the cells on both sides of the nuclear area, then to both extremities (Fig. 2.52 B, C, F). This progression, which results in the "cell plate", is all the more remarkable as it crosses the large single vacuole usually found in such cells.

It is easy to understand that the prolonged progression of the phragmoplast pad (the kinoplasmosome of BAILEY) increases the duration of these mitoses, evaluated to 24 h by WILSON (1964): only 5 for the karyokinesis and 19 for the progression of phragmoplast. This fact shows that the outline of cambial cells does not impair their capacity to proliferate.

Moreover, in long-living organs with a high growth in diameter, several histologists provided evidence of the marked plasticity of the cambium. This plasticity ensures not only the transversal growth but also the continued renewal of cambial initials which maintain this tissue in a young state (HUBER 1952). This kind of fluidity of the cambium induced the thorough research of the Polish team conducted by HEIJNOWICZ since 1971 (HEIJNOWICZ 1971, 1973, 1974 and references in CATESSON 1984). The anticlinal divisions of the cambium are not exactly vertical so that the daughter cells may be slightly oblique, either to the right or to the left. If this phenomenon is repeated in a whole area (a "morphogenetic area") then changes in orientation, in tangential view, the cambium shows ondulations which vary both in amplitude and duration and that can be demonstrated in the wood "grain" derived from these initials. Authors attribute these oscillations to "morphogenetic waves" which might be due to auxins. "Auxinic waves" would spread from the top to the bottom of trunks, starting from the apical meristems and would control the cambial activity (WODZICKI 1980).

Moreover, it is probable that auxins and cytokinins induce the cambial activity in increasing the RNA snytheses, whereas the abscissic acid impairs such syntheses and may play a part in the setting of dormancy (WAREING and PHILLIPS 1970).

2.2.6 Cytology and Meristematic Potentialities

The content of cambial cells allows the comparison with the similarly largely vacuolated cells of medullary meristems. Once more, as in the latter, the cambiums are composed of cells with orientated proliferation, but they mainly segment by means of *tangential longitudinal* walls, whereas medullary meristem cells undergo *transversal* cleavages; this is the main difference between the two types.

Conversely, these two types differ from the "primary meristematic" type by the vacuole features: small and concentrated in this last type, and by the concomitant characteristics of the cytoplasm. We shall come back to this subject in the next chapter.

From a histogenetic standpoint, one can already notice an essential difference between meristems with small vacuoles and meristems with large vacuoles: the first ones or "primary meristems" give rise to *organs:* young leaves and stems, roots, i.e. they are *organogen;* the second ones only *add tissues* to already outlined or wholly completed organs, i.e. they are only *histogen,* and their derivatives are not arranged in completed *organs.*

Thus, the medullary meristem only generates the medulla which completes the stem organization and the cambiums initiate tissues, possibly complex but which fit with already functional organs.

2.2.7 Distribution of Secondary Meristems in Vascular Plants

If all vascular plants develop from the primary meristems, all of them do not acquire secondary meristems during their ontogenesis. In the present state of nature, vascular cambiums do not exist any more except in Prephanerogams, Gymnosperms and Dicotyledons. Living Pteridophytes do not produce any, except in two genera in which such cambiums are very discrete *(Isoetes, Botrychium)*.

The phellogen layers are, on the whole, found in those plants with vascular generating zones.

As for the particular cambium of Monocotyledons, it is sometimes encountered in some Liliaceae *(Aloe, Yucca, Cordyline, Dracaena)* and in some Amaryllidaceae *(Agave)*.

The histological features of these cambiums, as well as the structures of their derivates, have been widely used in the systematic and phylogenetic study of vascular plants. For instance, this is the case for the length of "fusiform initials" of the vascular cambium. We shall briefly mention them with respect to conducting tissues.

References

Allen GS (1947) Embryogeny and the development of the apical meristems of *Pseudotsuga*. III. Development of the apical meristem. Am J Bot 34: 204–211

Askenasy E (1880) Über eine neue Methode, um die Verteilung der Wachstumsintensität in wachsenden Teilen zu bestimmen. Verh Naturh Medic Ver 2: 70–153

Avanzi S, d'Amato F (1967) New evidence on the organization of the root apex in leptosporangiate ferns. Caryologia 20: 257–264

Bailey IW (1919) Phenomena of cell division in the cambium of arborescent gymnosperms and their cytological significance. Proc Natl Acad Sci USA 5: 283–285

Bailey IW (1920) The cambium and its derivative tissues. II. Size variations of cambial initials in Gymnosperms and Angiosperms. Am J Bot 7: 355–367

Bailey IW (1930) The cambium and its derivative tissues. V. A reconnaissance of the vacuome in living cells. Z Zellforsch Mikrosk Anat 10: 651–682

Bailey IW (1954) Contributions to plant anatomy. Chron Bot 1: 3–8, Waltham, Mass USA

Bannan MW (1950) The frequency of anticlinal divisions in fusiform cambial cells of *Chamaecyparis*. Am J Bot 37: 511–519

Bannan MW (1951) The reduction of fusiform cambial cells in *Chamaecyparis* and *Thuya*. Can J Bot 29: 57–67

Bannan MW (1953) Further observations on the reduction of fusiform cambial cells in *Thuya occidentalis*. Can J Bot 31: 63–74

Bannan MW (1957) The relative frequency of the different types of anticlinal divisions in Conifer cambium. Can J Bot 35: 875–884

Bernier G (1964) Etude histophysiologique et histochimique de l'évolution du méristème apical de *Sinapis alba* L., cultivé en milieu conditionné et en diverses durées de jours favorables ou défavorables à la mise à fleurs. Mém Acad R Belg Sci 16: 1–150 (cf 119 et seq)

Bernier G, Bronchart R (1963) Application de la technique d'histoautoradiographie à l'étude de l'incorporation de Thymidine tritiée dans les méristèmes caulinaires. Bull Soc R Sci Liège 32: 269–283

Bersillon G (1951) Sur le point végétatif de *Papaver somniferum* L. Structure et fonctionnement. CR Acad Sci Paris 232: 2470–2472

Besnard-Wibaut C (1970) Evolution des synthèses d'ADN dans le méristème caulinaire et la moelle de l'*Arabidopsis thaliana,* race Stockholm, soumis à un traitement vernalisant. CR Acad Sci Paris 270: 2932–2935

Besnard-Wibaut C (1977) Histoautoradiographic analysis of the thermoinductive processes in the shoot apex of *Arabidopsis thaliana* L. Heynh, vernalized at different stages of development. Plant and Cell Physiol 18: 949–962

Brachet J (1940) La détection histochimique des acides pentoxe-nucléiques. C R Soc Biol 133: 88–90

Brossard D (1973) Le bourgeonnement épiphylle chez le *Bryophyllum daigremontianum* Berger (Crassulacées). tude cytochimique, cytophotométrique et ultrastructurale. Ann Sci Nat Bot Paris 12e sér 14: 93–214

Brumfield RT (1943) Cell lineage studies in root meristems by means of chromosome rearrangements induced by X-rays. Am J Bot 30: 101–110

Buvat R (1952) Structure, évolution et fonctionnement du méristème apical de quelques Dicotylédones. Ann Sci Nat Bot 11e sér 13: 199–300

Buvat R (1955) Sur la structure et le fonctionnement du point végétatif de *Selaginella caulescens* Spring., var. *amoena.* CR Acad Sci Paris 241: 1833–1835

Buvat H (1958) Recherches sur les infrastructures du cytoplasme, dans les cellules du méristème apical, des ébauches foliaires et feuilles développées d'*Elodea canadensis.* Ann sci Nat Bot 11e sér 19: 121–162

Buvat R (1977) Origine golgienne et lytique des vacuoles dans les cellules méristématiques des racines d'Orge *(Hordeum sativum).* CR Acad Sci Paris 284 D: 167–170

Buvat R, Genevès L (1951) Sur l'inexistence des initiales axiales dans la racine d'*Allium cepa* L. (Liliacées). CR Acad Sci Paris 232: 1579–1581

Buvat R, Liard O (1951) Nouvelle constatation de l'inertie des soi-disant initiales axiales dans le méristème radiculaire de *Triticum vulgare.* CR Acad Sci Paris 233: 813–814

Buvat R, Robert G (1979) Vacuole formation in the actively growing root meristem of barley *(Hordeum sativum).* Am J Bot 66: 1219–1237

Buvat R, Roger-Liard O (1954) La prolifération cellulaire dans le méristème radiculaire d'*Equisetum arvense* L. CR Acad Sci Paris 238: 1257–1258

Camefort H (1950) Structure du point végétatif de *Picea excelsa.* CR Acad Sci Paris 231: 65–66

Camefort H (1951) Structure du point végétatif de *Ginkgo biloba* en période d'activité (initiation foliaire). CR Acad Sci Paris 233: 88–90

Camefort H (1954) Présence et localisation de l'acide ribonucléique dans le point végétatif de quelques Gymnospermes. CR Acad Sci Paris 238: 922–924

Camefort H (1956a) Etude de la structure du point végétatif et des variations phyllotaxiques chez quelques Gymnospermes. Thèse Doctorat ès Sci Paris, Masson, Paris

Camefort H (1956b) Structure de l'apex caulinaire des Gymnospermes. Année Biol 32: 401–416 (Bibliographie)

Caspary R (1859) *Aldrovandia vesiculose* Monti. Bot Zeit 17: 117–124

Catesson AM (1953) Structure, évolution et fonctionnement du point végétatif d'une Monocotylédone: *Luzula pedemontana* Boiss. et Reut. (Joncacées). Ann Sci Nat Bot 11e sér 14: 253–291

Catesson AM (1964) Origine, fonctionnement et variations saisonnières du cambium d'*Acer Pseudoplatanus* L. (Acéracées). Ann Sci Nat Bot 12e sér 5: 229–456

Catesson AM (1970) Evolution des plastes de pomme au cours de la maturation du fruit. Modifications ultrastructurales et accumulation de ferritine. J Microsc 9: 949–974

Catesson AM (1974) Cambial cells. In: Robarts AW (ed) Dynamic aspects of plant ultrastructure. Mc Graw Hill, New York, pp 358–390

Catesson AM (1984) La dynamique cambiale. Ann Sci Nat Bot 13ème sér 6: 23–43

Catesson AM, Roland JC (1981) Sequential changes associated with cell wall formation in the vascular cambium. IAWA Bull n.s. 2: 151–162

Cheadle VI (1937) Secondary growth by means of a thickening ring in certain monocotyledons. Bot Gaz 98: 535–555

Clowes FAL (1954) The promeristem and the minimal construction centre in grass root apices. New Phytol 53: 108-116

Clowes FAL (1961a) Duration of the mitotic cycle in a meristem. J Exp Bot 12: 283-293

Clowes FAL (1961b) Apical meristems. In: Botanical Monographs. Blackwell Scientific, Oxford, 217 p

Clowes FAL (1963) The quiescent centre in meristems and its behaviour after irradiation. Brookhaven Symp Biol 15: 46-58

Clowes FAL (1965a) The duration of the G1 phase of the mitotic cycle and its relation to radiosensitivity. New Phytol 64: 355-359

Clowes FAL (1965b) Meristems and the effect of radiation on cells. Endeavour 24: 8-12

Cottignies A (1974) Définition cytophysiologique de la dormance du méristème chez le *Fraxinus excelsior* L. Etude de l'activité mitotique et de la teneur en DNA nucléaire dans le point végétatif apical dormant. CR Acad Sci Paris 278 D: 2763-2766

Cottignies A (1979) The blockage in the G1 phase of the cell cycle in the dormant shoot apex of Ash. Planta 147: 15-19

Cottignies A (1984) Visualisation des centres générateurs selon la théorie phyllotaxique de *Plantefol* dans le point végétatif du Frêne (*Fraxinus excelsion* L.). Can J Bot 62: 2636-2643

Cottignies A (1985) Dormance et croissance active chez le Frêne (*Fraxinus excelsior* L.). Thèse Doct ès Sci Nat, Univ P & M Curie, Paris

D'Amato F, Avanzi S (1965) DNA content, DNA synthesis and mitosis in the root apical cell of *Marsilea strigosa*. Caryologia 18: 383-394

D'Amato F, Avanzi S (1968) The shoot apical cell of *Equisetum arvense*, a quiescent cell. Caryologia 21: 83-89

Douliot H (1890) Recherches sur la croissance terminale de la tige des Phanérogrames. Ann Sci Nat Bot 7 ème sér 11: 283-350

Emberger L (1968) Les plantes fossiles dans leurs rapports avec les végétaux vivants. Masson, Paris

Eriksson J (1878) Über das Urmeristem der Dikotyledonen Wurzeln. Jahrb Wiss Bot 11: 380-436

Esau K (1942) Vascular differentiation in the vegetative shoot of *Linum*. I. The procambium. Am J Bot 29: 738-747

Esau K (1948) Phloem structure in the grapevine, and its seasonal changes. Hilgardia 18: 217-296

Flahault C (1878) Recherches sur l'accroissement terminal de la racine chez les Phanérogames. Ann Sci Nat Bot 6e sér 6: 1-168

Foster AS (1938) Structure and growth of the shoot apex in *Ginkgo biloba*. Bull Torrey Bot Club 65: 531-536

Foster AS (1941a) Zonal structure of the shoot apex of *Dioon edule Lindl*. Am J Bot 28: 557-564

Foster AS (1941b) Comparative studies on the structure of the shoot apex in seed plants. Bull Torrey Bot Club 68: 339-350

Grégoire V (1935a) Données nouvelles sur la morphogenèse de l'axe feuillé dans les Dicotylées. CR Acad Sci Paris 200: 1127-1129

Grégoire V (1935b) Les liens morphogénétiques entre la feuille et la tige dans les Dicotylées. CR Acad Sci Paris 200: 1349-1351

Grégoire V (1938) La morphogenèse et l'autonomie morphologique de l'appareil floral. Cellule 47: 285-452

Guttenberg H von, Burmeister J, Brosell HJ (1955) Studien über die Entwicklung des Wurzelvegetationspunktes der Dikotyledonen. Planta 46: 179-222

Hanstein J (1868) Die Scheitelzellgruppe im Vegetationspunkt der Phanerogamen. Abh aus dem Gebiet der Naturwiss Math und Med Bonn, oder: Festschr Niederrhein Ges Natur und Heilkunde, pp 109-143

Hanstein J (1870) Die Entwicklung des Keimes der Monokotylen und der Dikotylen. Bot Abh 1: 1-112

Heijnowicz Z (1971) Upward movement of the domain pattern in the cambium producing wavy grain in *Picea excelsa*. Acta Soc Bot Pol 40: 499-512

Heijnowicz Z (1973) Morphogenetic waves in cambia of trees. Plant Sci Lett 1: 359-366

Heijnowicz Z (1974) Pulsation of domain length as support for the hypothesis of morphogenetic waves in the cambium. Acta Soc Bot Pol 43: 261-271

Hofmeister W (1851) Vergleichende Untersuchung der Keimung, Entfaltung und Fruchtbildung der höheren Kryptogamen und der Samenbildung der Coniferen. Leipzig

Holle HG (1876) Über den Vegetationspunkt der Angiospermen-Wurzeln, insbesondere die Haubenbildung. Bot Zeit 34: 241–263

Huber B (1952) Tree physiology. Annu Rev Plant Physiol 3: 333–346

Janczewski E de (1874) Recherches sur l'accroissement terminal des racines dans les Phanérogames. Ann Sci Nat Bot 5ème sér 20: 162–201

Kaplan R (1937) Über die Bildung der Stele aus dem Urmeristem von Pteridophyten und Spermatophyten. Planta 27: 224–268

Koch L (1891) Über Bau und Wachstum der Sproßspitze der Phanerogamen. I. Die Gymnospermen. Jahrb Wiss Bot 22: 491–680

Korody E (1937) Studien am Sproßvegetationspunkt von *Abies concolor, Picea excelsa* und *Pinus montana*. Beitr Biol Pflanz 25: 23–59

Lance A (1954) Répartition de l'acide ribonucléique dans les méristèmes apicaux de deux Composées. CR Acad Sci Paris 239: 1238–1239

Lance A (1957) Recherches cytologiques sur l'évolution de quelques méristèmes apicaux et sur ses variations provoquées par des traitements photopériodiques. Ann Sci Nat Bot 11ème sér 18: 91–421

Lance-Nougarède A, Loiseau JE (1960) Sur la structure et le fonctionnement du méristème végétatif de quelques Angiospermes aquatiques ou semi-aquatiques dépourvues de moelle. CR Acad Sci Paris 250: 4438–4440

Larson PR (1976) Procambium vs cambium and protoxylem vs metaxylem in *Populus deltoides* seedlings. Am J Bot 63: 1332–1348

Lawalrée A (1948) Histogenèse florale et végétative chez quelques Composées. Cellule 52: 215–294

Lier FG (1955) The origin and development of cork cambium cells in the stem of *Pelargonium hortorum*. Am J Bot 42: 929–936

Loiseau JE (1959) Observations et expérimentation sur la phyllotaxie et le fonctionnement du sommet végétatif chez quelques Balsaminacées. Ann Sci Nat Bot 11ème sér 20: 1–214

Louis G (1935) L'ontogenèse du système conducteur dans la pousse feuillée des Dicotylées et des Gymnospermes. Cellule 44: 87–172

Majumdar GP (1942) The organization of the shoot in *Heracleum* in the light of development. Ann Bot NS 6: 49–81

Marty D (1973) Aspects particuliers de l'ontogenèse des thylacoïdes dans les feuilles panachées de *Coleus blumei* Benth. CR Acad Sci Paris 277 D: 45–48

Marty F (1973) Mise en évidence d'un appareil provacuolaire et de son rôle dans l'autophagie cellulaire et l'origine des vacuoles. CR Acad Sci Paris 273 D: 1549–1552

Marty F (1978) Cytological studies on GERL, provacuoles and vacuoles in root meristematic cells of *Euphorbia*. Proc Natl Acad Sci USA 75: 852–856

Michaux N (1966) Structure et fonctionnement du méristème apical de l'*Isoetes setacea* Lam. CR Acad Sci Paris 263: 501–504

Michaux N (1967) Le méristème apical adulte de l'*Isoetes setacea* Lam. Etude histoautoradiographique: essai de détermination de la longueur du cycle mitotique. Rev Cytol Biol Vég 30: 353–372

Michaux N (1968) tude cytologique du méristème apical du *Pteris cretica* L. CR Acad Sci Paris 267 D: 1442–1444

Michaux N (1970a) Détermination, par cytophotométrie, de la quantité d'ADN contenue dans le noyau de la cellule apicale des méristèmes jeunes et adultes du *Pteris cretica* L. CR Acad Sci Paris 271 D: 656–659

Michaux N (1970b) tude comparée de la structure et du fonctionnement du méristème apical adulte de l'*Isoetes setacea* Lam et du *Pteris cretica* L. Bull Soc Bot Fr Mém 117: 83–101

Michaux N (1971) Durée du cycle mitotique dans le méristème apical du jeune sporophyte du *Pteris cretica* L. CR Acad Sci Paris 273 D: 336–339

Michaux N (1972) tude cytophotométrique du méristème caulinaire de l'*Isoetes setacea* Lam. au cours de son cycle annuel. CR Acad Sci Paris 274 D: 3453–3456

Naegeli C (1845) Wachstumgeschichte der Laub und Lebermoose. Z Wissensch Bot 2: 138–210

Naegeli C, Leitgeb H (1868) Entstehung und Wachstum der Wurzeln. Beitr Wiss Bot 4: 73–160

Nougarède A (1965) Organisation et fonctionnement du méristème apical des Végétaux vasculaires. Travaux dédiés à Lucien Plantefol. Masson, Paris, pp 171–340

Nougarède A (1967) Experimental cytology of the shoot apical cells during vegetative growth and flowering. Int Rev Cytol 21: 203–351

Nougarède A, Loiseau JE (1963) tude morphologique des rameaux du *Lycopodium selago* L.; structure et fonctionnement de l'apex. CR Acad Sci Paris 257: 2698–2701

Nougarède A, Pilet PE (1965) Distribution des ribosomes le long de la racine du *Lens culinaris* L. CR Acad Sci Paris 260: 2899–2902

Nougarède A, Rembur J (1977) Determination of cell cycle and DNA synthesis duration in the shoot apex of *Chrysanthemum segetum* L. by double labelling autoradiographic techniques. Z Pflanzenphysiol 85: 283–295

Nougarède A, Rembur J (1978) Variations of the cell cycle phases in the shoot apex of *Chrysanthemum segetum* L. Z Pflanzenphysiol 90: 379–389

Nougarède A, Rembur J (1985) Le point végétatif, un modèle pour l'étude du cycle cellulaire et de ses contrôles. Bull Soc Bot Fr, Actual Bot 132 (1): 9–34

Nougarède A, Rondet P (1973) Un modèle original d'organisation de la tige: étude du fonctionnement plastochronique chez le *Pisum sativum* L, var. nain hâtif d'Annonay. CR Acad Sci Paris 277 D: 997–1000

Nougarède A, Rondet P (1975) Index mitotique et teneurs en DNA nucléaire du méristème axillaire de la feuille de rang 6 et du point végétatif terminal, chez le *Pisum sativum* var. nain hâtif d'Annonay. CR Acad Sci Paris 280 D: 709–712

Nougarède A, Rondet (1976) Durée des cycles cellulaires du méristème terminal et des méristèmes axillaires du *Pisum sativum* L. CR Acad Sci Paris 282 D: 715–718

Nougarède A, Gifford EM Jr, Rondet P (1965) Cytohistological studies of the apical meristem of *Amaranthus retroflexus* under various photoperiodic regimes. Bot Gaz 126: 248–298

Nougarède A, Tepfer S, Tepfer M (1973) Les bourgeons au cours de l'état pérennant chez le *Phytolacca decandra* L. Modalités de conservation du plant durant l'hiver et reprise de l'activité printanière. CR Acad Sci Paris 276 D: 957–960

Perbal G (1974) L'action des statolithes dans la réponse géotropique des racines de *Lens culinaris*. Planta (Berl) 116: 153–171

Philipson WR (1949) The ontogeny of the shoot apex in Dicotyledons. Biol Rev 24: 21–50

Plantefol L (1947) Hélices foliaires, point végétatif et stèle chez les Dicotylédones. La notion d'anneau initial. Rev Gén Bot 54: 49–80

Plantefol L (1948) Fondements d'une théorie phyllotaxique nouvelle. La théorie des hélices foliaires multiples. Masson, Paris

Popham RA (1951) Principle types of vegetative shoot apex organization in vascular plants. Ohio J Sci 51: 249–270

Prantl K (1874) Untersuchungen über die Regeneration des Vegetationspunktes an Angiospermen-Wurzeln. Arb Bot Inst Würzburg 1: 546–562

Reeve RM (1948) Late embryogeny and histogenesis in *Pisum*. Am J Bot 35: 591–601

Reinke J (1871) Untersuchungen über Wachstumsgeschichte und Morphologie der Phanerogamen-Wurzel. Hanstein's Bot Abh 1: 1–50

Rembur J, Nougarède A (1977) Duration of cell cycles in the shoot apex of *Chrysanthemum segetum* L. Z Pflanzenphysiol 81: 173–179

Rougier J (1955) Sur le point végétatif de l'*Aquilegia vulgaris* L. CR Acad Sci Paris 240: 654–656

Russow E (1872) Vergleichende Untersuchungen betreffend die Histiologie (Histographie und Histiogenie) der vegetativen und sporenbildenden Organe und die Entwicklung der Sporen der Leitbündel-Kryptogamen, mit Berücksichtigung der Histiologie der Phanerogamen, ausgehend von der Betrachtung der Marsiliaceen. Mém Acad Imp Sci St Pétersbourg VII Vol 19, 1: 1–207

Sachs J v (1868) Lehrbuch der Botanik fig 111

Sachs J v (1874) Traité de Botanique. Traduction française by Van Thieghem. (See fig 112 p 190). F Savy ed Paris

Saint Côme R (1966) Application des techniques histoautoradiographiques et des méthodes statistiques à l'étude du fonctionnement apical chez le *Coleus blumei* Benth. Rev Gén Bot 73: 241–323

Sanio C (1860) Vergleichende Untersuchungen über den Bau und die Entwicklung des Korkes. Jahrb Wiss Bot 2: 39–108

Schmidt A (1924) Histologische Studien an Phanerogamen Vegetationspunkten. Bot Arch 9: 345–404

Srivastava LM (1966) On the fine structure of the cambium of *Fraxinus americana* L. J Cell Biol 31: 79–93

Strasburger E (1872) Die Coniferen und die Gnetaceen. Ambr Abel, Leipzig, Bermühler, Berlin

Taillandier J (1969) Structure et fonctionnement du méristème terminal de l'*Anagallis arvensis* L. et localisation des synthèses d'ADN au cours des diverses phases du développement. CR Acad Sci Paris 268 D: 676–679

Treub M (1875) Le méristème primitif de la racine dans les Monocotylédones. Mus Bot Leyde 2: 17–98

Wardlaw W (1955) Experimental and analytical studies of Pteridophytes. XXVIII. Leaf symmetry and orientation of ferns. Ann Bot 19: 389–399

Wareing PF, Phillips IDJ (1970) The control of growth and differentiation in plants. Pergamon, Oxford

Wilson BF (1964) A model for cell production by the cambium of conifers. In: Zimmermann MH (ed) The formation of wood in forest trees. Academic Press, London, pp 19–36

Wodzicki TJ (1980) Control of cambial activity. In: Little CHA (ed) Control of shoot growth in trees. Proc IUFRO Workshop, Fredericion, Canada, pp 173–183

Yarbrough JA (1949) *Arachis hypogaea*. The seedling, its cotyledons, hypocotyl and roots. Am J Bot 36: 758–772

Cytology of the Processes of Differentiation and Dedifferentiation During the Ontogeny of Vascular Plants

3.1 The Plasticity of Plants Cells

The study of the origin of the various types of meristems and their derivatives suggests that the ontogenesis of the vegetative organs of vascular plants concerns, on the one hand, the diversification of cells derived from meristematic proliferation, i.e. the cells concerned with the formation of all the various tissues of the plant and, on the other hand, with the production of more or less specialized cells that regain the condition of meristematic cells. This occurs after the proliferating activity has been restored in these cells.

Thus, cells derived from primary meristems develop through a *differentiation* process, then some of them are involved again in a histogenetic mechanism and return to the meristematic condition, i.e. they *dedifferentiate*.

The first process, the differentiation, is a common phenomenon of development which occurs in all beings with tissues. In contrast, the dedifferentiation does not seem to occur in the normal ontogenesis of animals in general.

In plants one can see that a cell, differentiated for instance into medullary or phloem parenchyma, may lose this specificity to produce vascular or cork cells after a phase of proliferation. Such an evolution shows that the characteristics of already differentiated plant cells are not necessarily irreversible.

The regression may even become more accentuated when branching processes generate primary meristems derived from more or less differentiated cells. In normal ontogenesis, this can occur in the formation of roots, rootlets and adventitious buds or in the axillary meristems of leaves.

Experimentally, it is possible to induce the complete dedifferentiation of very varied differentiated cells: epidermal cells, cortical parenchyma, medullary, phloem, collenchyma cells etc. Such processes take place particularly in many cases after the cutting of stems, roots or leaves.

For example in a Liliaceae close to *Hyacinthus, Brimeura amethystina,* a few epidermal cells having completed their differentiation, may regress when a fragment of leaf cut out off a *completely developed* lamina is cultured in vitro: cells derived from the epidermal ones produce *primary meristems* and are capable of giving rise to a *whole plant* (Fig. 3.1). In the same leaf, chlorophyllous cells of assimilating parenchyma can also regress in producing a perfectly functional root meristem (Fig. 3.2).

Thus plant cells, even in advanced plants, maintain a considerable plasticity that does not exist in animal cells. They can lose their primary specificity and, in proliferating again, give rise to a cell lineage of multiple tissue potentialities.

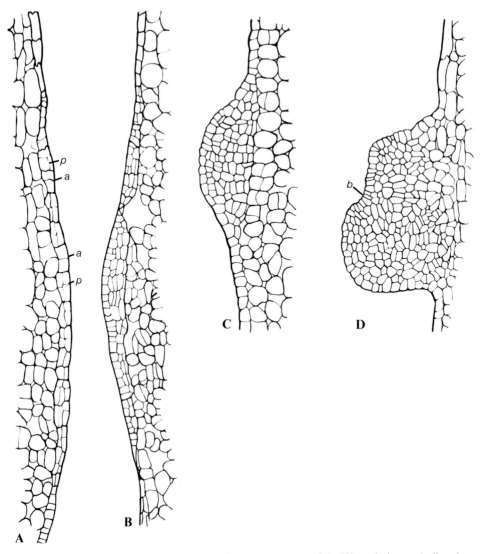

Fig. 3.1 A–D. Leaf cuttings of *Brimeura amethystina* L. Process of dedifferentiation preluding the regeneration. Dedifferentiation of epidermal cells. **A** Recurrence of an anticlinal *(a)* and periclinal *(p)* mitotic activity; **B, C** edification of a body of meristematic cells, becoming organized in a bulbil *b* (**D**)

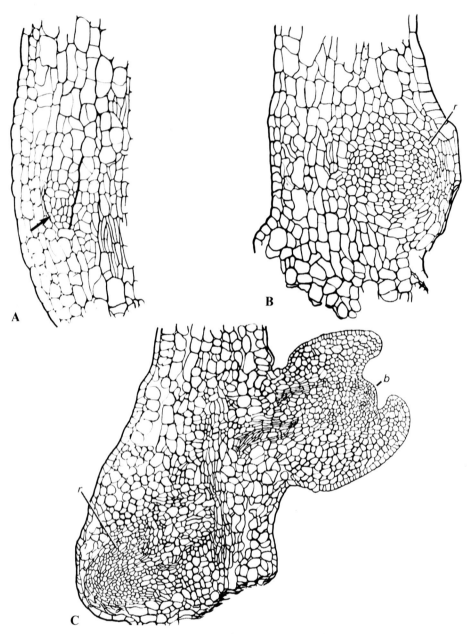

Fig. 3.2 A–C. Dedifferentiation of green cells of the *Brimeura amethystina* mesophyll (**A**, *arrow*). The recurrence of activity of these cells gives rise to a meristematic body (**B**, *r*) being arranged in a root meristem (**C**, *r*). **C** shows the simultaneous formation of a root *(r)* and a bulbil *(b)* provided with an apical meristem. Reactivation of mesophyll internal cells gives rise to proconducting strands which link up to the conducting system of the mother leaf

190

The cytological study of this cellular "rejuvenescence" and of the potentialities of cells which regress in such a way allows the determination of stages which, by symmetry and on the whole correspond to the stages of differentiation.

3.2 General Cytological Scheme of Differentiation

The study of meristems shows that the *most meristematic* cells, i.e. having, in normal ontogenesis, the most numerous and varied filiation, are primary meristematic cells generated by the initiating ring or by the proliferating area of the root meristem.

From such a cell, the most general scheme of its evolution is as follows: after successive mitoses, its derivatives are finally arranged in one or several differentiated tissues.

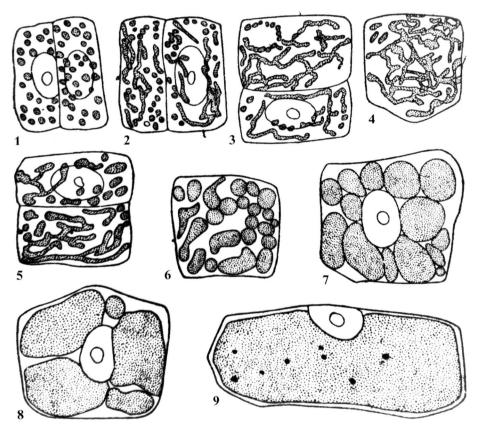

Fig. 3.3. Evolution of the vacuolar apparatus in the course of differentiation of the cells derived from the barley apical root cells. **1-5** Still meristematic cells; **6,7** cells at the beginning of differentiation, hydration and hypertrophy of vacuoles; **8** onset of confluence of hypertrophic vacuoles; **9** differentiated cell of the cortical parenchyma containing a single large vacuole (vital staining with neutral red); (after GUILLIERMOND 1938)

For simplification, we shall deal separately with (1) the unit: nucleus – cytoplasm - vacuoles, (2) the chondriosomes and plastids; and (3) the cell wall for which we shall give some details of structure to clarify the terminology which is still confusing.

3.2.1 Development of the Nucleus, the Cytoplasm and Vacuoles

The classical works of GUILLIERMOND (cf. 1934 a, b, 1938) elucidated the stages of this evolution. In meristematic cells, the nucleus is central, voluminous and the vacuoles are small and numerous with a concentrated content (Fig. 3.3/1, 2).

The differentiation begins as soon as the proliferation is less active, but long before it ceases. This onset is characterized by the enlargement of the vacuoles which become rich in water. Then the cytoplasm shows a spongious aspect (Fig. 3.3/3, 4). The cell often grows in length and the vacuoles occupy more and more space in it. Soon they are squashed together and surround the nucleus (Fig. 3.3/5-7), so that the cytoplasm which separates them is reduced to very thin films. Then, the vacuoles become united two by two, their number decreases, whereas the volume increases (Fig. 3.3/8) and finally the cell contains a single vacuole only, which fills almost the whole volume (Fig. 3.3/9). The nucleus is thus shifted against one of the cell wall faces and its spherical aspect is changed to a lenticular or flat ovoid shape. The large vacuole remains clearly enclosed in the cytoplasm which is reduced to a very thin parietal film, applied against the cell wall and also around the nucleus (Fig. 3.3/9). Only some cytoplasmic *trabeculae* can be found in the vacuolar space.

This development can be observed particularly in the cortex cells of the roots of various germinations, the vacuoles of which are easily stained with neutral red (Plate 3.1), it is quite generalized and early. It is completed by the development of the chondriome and the cell wall.

3.2.2 Development of the Chondriome

At the beginning of this century, under this heading, cytologists referred to a typically double structure in differentiated plant cells, including the chondriosomes and the plastids. Authors generally agreed to recognize that these two types of inclusions represent genetically distinct filiations of intracellular organelles.

Plate 3.1 A-D. Evolution of vacuoles during the differentiation of wheat cortical root cells. Vital staining with neutral red.

A Cells of the meristematic proliferation zone; exiguous vacuolar apparatus in the shape of excessively slender filaments and very small globular vacuoles.

B Id, beginning of swellings localized along vacuoles which are still filamentous.

C Farther from the meristem, into already larger cells, modification of the vacuolar apparatus into spherical vacuoles, at first independent, then joining together, *(lower cells)* in the shape of more voluminous enclaves, which press themselves together.

D Cortical cells near the end of differentiation; they grow longer while vacuoles, in the process of confluence, fill up the greater part of the cell volume; × 1250

192

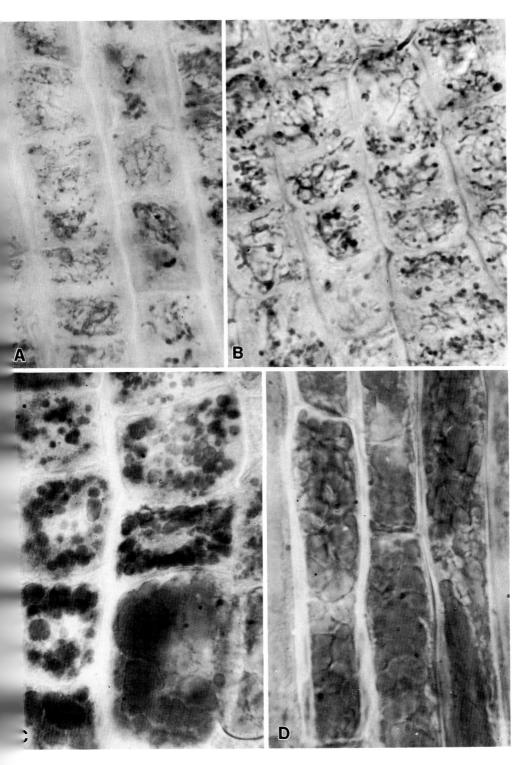

193

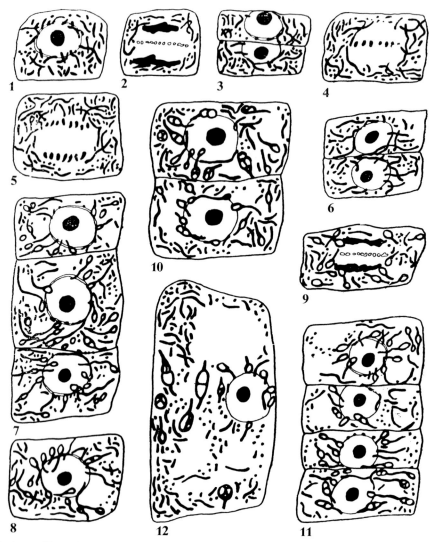

Fig. 3.4. Differentiation of leucoplasts in the *Ricinus* root. **1-6** Meristematic cells, leucoplasts rougly identical to chondriosomes; however, the filamentous formations represent, probably but not exclusively, undifferentiated proplastids; **7-11** cells in the course of differentiation; elaboration of starch into filamentous elements, i.e. leucoplasts; **12** differentiated cell, from the central cylinder parenchyma, showing leucoplasts elaborating compound starch grains; × 1500. RE-GAUD's technique, after GUILLIERMOND 1938 in GUILLIERMOND et al. 1933)

In meristematic cells, these two categories of elements have approximately the *same morphology,* which makes their distinction difficult with a light microscope; they also present the same *biochemical features.* Chondriosomes (mitochondria) and plastids are *living* cellular elements chemically *related* before they become differentiated; but the distinction of the two categories appears quickly when the *differentiation* occurs and this fact can be confirmed with electron microscopy.

Thus, in the primary meristematic cells, all the chondriome elements are granules, rods or generally short filaments. During the process of differentiation, some of these elements develop into plastids: they increase and elaborate substances characteristic of the tissue which develops: carotenoids and chlorophyllous pigments, starch. The casual pigmentary production allows the distinction of three types of plastids: leucoplasts (Fig. 3.4), without pigments, chromoplasts and chloroplasts (Fig. 3.5) depending on the predominating pigments.

The remaining elements of the meristematic chondriome develop into chondriosomes which have characteristics of granules (mitochondria) of short rods or of more or less elongated filaments (chondriocontes) (Figs. 3.4 and 3.5). The produc-

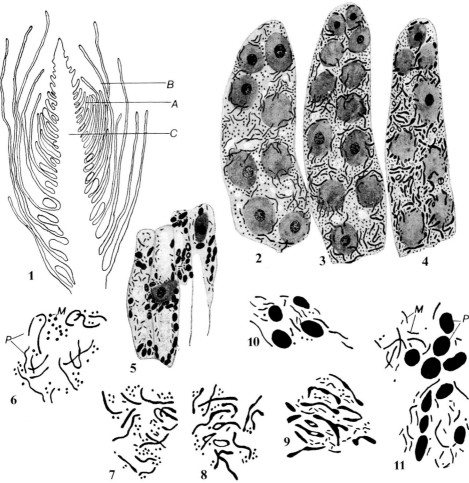

Fig. 3.5. Differentiation of chloroplasts in the *Elodea canadensis* bud. **1** Schema of position of the differentiation sites; *A* leaf primordium at the stage of greening appearance; *B* leaf containing differentiated chloroplasts; *C* place of the stem where chlorophyll begins to appear. **2-4** Chloroplast differentiation states, starting from filamentous chondriosomes. **5** Mature leaf cells, completed chloroplasts, short rod-like chondriosomes. **6-11** Successive aspects of differentiation of chloroplasts *(P)* and mitochondria *(M);* (after GUILLIERMOND 1922)

tions resulting from their activity are generally not visible in morphological study (GUILLIERMOND 1920).

During the differentiation of the chondriome, in most cases, and among other features, chondriosomes remain as fragile as the meristematic chondriome to fixatives including alcohol or acids. This fact can also be noticed in leucoplasts and chromoplasts; in contrast, chloroplasts become generally more resistant to these fixatives when they differentiate. It even seems that cytologists preferred the use of acid fixatives to carry out some particular studies of their structure.

When both structures of the chondriome are differentiated, their shape or their content may be modified in *a reversible way* under the influence of the physiological activity of the cells (EMBERGER 1925, 1927). Thus, in a chlorophyllous cell exposed to light, starch grains may be elaborated into chloroplasts which become distended. During the night, the starch is depolymerized and disappears from the chloroplasts: it becomes a soluble glucid (saccharose) transported in the plant.

In the same way, if chondriosomes tend to be short (very divided state) in young cells, and long (less divided state) in advanced cells, they can also *transitorily* go from one state to the other under the influence of the cellular activity variations. Thus, the evolution of chondriosomes does not exclude some changes in their morphological features.

Since the end of the 1950s, many descriptions of the ultrastructure of mitochondria and proplastids performed with electron microscopy have completed these classical data, for example in the bud of *Elodea canadensis* (BUVAT 1958). In the most meristematic cells (Plate 3.2 A) undifferentiated proplastids resemble meristematic mitochondria. The granulofibrillar stroma includes clear areas in both categories of organelles; these areas are presently acknowledged as being the site of mitochondrial and chloroplastic DNA (cf. NASS and NASS 1963; RIS and PLAUT 1962). The more internal membrane of the double envelope generates a small number of scattered cristae, generally shorter in mitochondria than in proplastids (Plate 3.2 A). However, plastids are slightly more voluminous and begin to elaborate starch early. During the differentiation, the internal cristae of proplastids grow considerably in length and begin to join in one part of their surface (Plate 3.2 B–D). They become "thylakoids", flat cisternae, the transection of which is parallel to the outline of the envelope.

Plate 3.2 A–D. Differentiation of chloroplasts in the *Elodea canadensis* bud (beginning) (osmic fixation).

A Highly meristematic cell of the apical meristem. Poorly structured mitochondria *m*, as well as proplastid *pl*, yet little different from mitochondria, and in the course of division; *e* endoplasmic reticulum; *g* plastoglobules; *i* invaginations of the plasmalemma; *p* plasmodesmata; the *arrows* indicate the formation of a crista due to a fold of the proplastid internal membrane; × 30000.
B Beginning of differentiation of a proplastid: lengthening of cristae that tend to become parallel when forming thylakoids. *g* Plastoglobules; × 82500.
C Young chloroplasts in the shape of very elongated chondriosomes, corresponding to the filamentous chondriosomes of Fig. 3.5 (**2, 3, 6** and **7**) and containing long thylakoids partly associated; × 24000.
D Young chloroplast into which thylakoids begin to join in grana; × 70000.

(After BUVAT 1958)

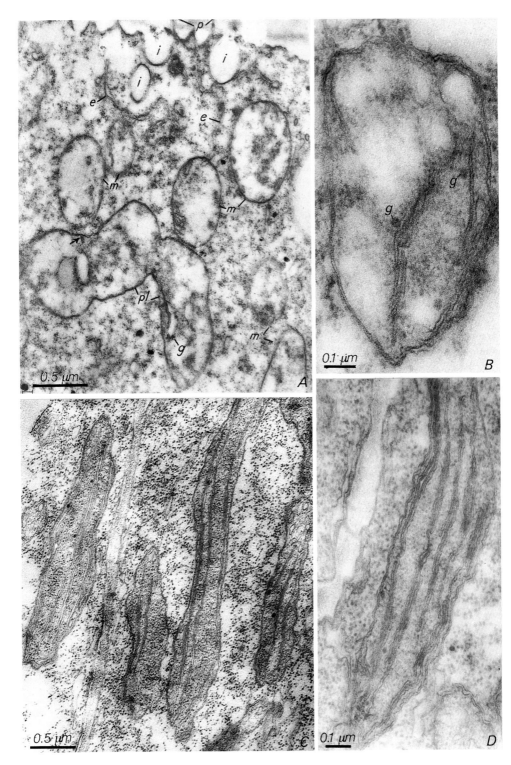

Plastoglobules appear within the stroma. The multiplication and growth of thylacoids result in the well-known structure of typical granar chloroplasts (Plate 3.3 A).

In chlorophyllous cells with a high metabolic activity, the differentiation of mitochondria does not perceptibly change their shape or size, but it is mainly characterized by a large increase in the number of cristae (Plate 3.3 B) and therefore in their density (BUVAT and LANCE 1958).

The cytoplasm and nucleus are also the sites of modifications related to differentiation. For example, in the cytoplasm, the ribosome density decreases (NOUGAREDE and PILET 1965), which is indicated by a lower rate of basophilia. In the nucleus, one can often observe process of partial condensation of the chromatin and a reduction of the nucleolar volumes. The Golgi apparatus demonstrates its activity related to cellular differentiation either by participating in the genesis of new vacuoles (Plate 2.4 B) (MARTY 1973; BUVAT and ROBERT 1979) and supplying the vacuolar system with acid hydrolases (COULOMB and COULOMB 1973) or by providing the plasmalemma with vesicles loaded with precursors of parietal material (various polysaccharides). In fusing with the plasmalemma, these vesicles deliver their content in the parietal space after a phase of restructuration (cf. Sect. 3.2.3) (Plate 2.4 A).

Here, we do not intend to deal with the details of these phenomena which are presently studied with the up-to-date methods of ultrastructural cytology, cytochemistry and molecular biology.

3.2.3 Development of the Cell Wall

The skeletal wall alone was noted by the first histologists and suggested the term "cells". But, after having studied only this skeleton, histologists devoted themselves to the study of the cellular content which results in the cytology. As a consequence, for some time, histologists were no longer involved in the study of cell wall in plants. However, the research carried out on this plant element has been recently stimulated and yielded important data.

▷

Plate 3.3 A, B. Completion of the differentiation of chloroplasts and mitochondria in *Elodea canadensis* leaves (osmic fixations).

A Ultrastructure of a differentiated chloroplast; *e* envelope, composed of two membranes; *gl* plastoglobules; *gr* grana; *pv* peripheral vesicles; *iz* intergranar zones; × 39 500.
B Mitochondria in a differentiated chlorophyllous cell; particularly numerous cristae (cf. **A**, Plate 3.2); *arrows* indicate the mitochondrial DNA sites; × 48 000

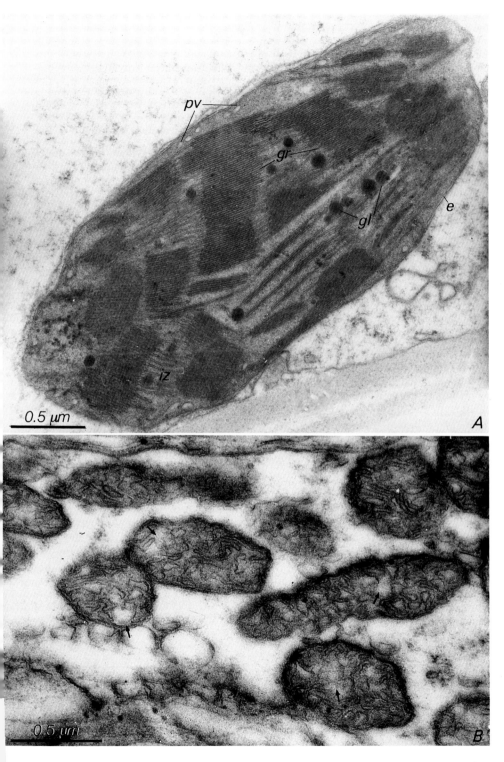

3.2.3.1 Relations Between Cytoplasm and Cell Wall

Cytologists considered the skeletal wall as a paraplasmic derivative, exterior to the protoplasm by which it is generated. In fact, it has long been acknowledged that thin trabeculae of cytoplasm go through the wall, along canaliculi called plasmodesmata (Plate 3.4). Moreover, the close association of the pectocellulosic wall with the ectoplasmic pellicle seems to affect the exchange processes of material between the cell and the internal medium of plants. The cell wall and the cytoplasm are found to be in a close and complex relationship. Several authors (FREY-WYSSLING) even suggested that the wall was living. It is more accurate to consider that it is composed of *paraplasmic material,* therefore passive, with cytoplasm extensions and that the cytoplasm governs the formation and the growth of the cell wall.

The infrastructure of plasmodesmata has been studied with electron microscopy. For a long time, the question was posed as to how did the ectoplasmic pellicles, i.e. the plasmalemma, become deposited along the canaliculi of plasmodesmata. More particularly, one wondered if these plasmalemma were connected in the same way as a synapse, like in nervous cells. It was easy to observe that the plasmodesmatal canaliculus was smoothly coated with the plasmalemma (Plate 3.4 A–D). It extends from one cell to the other *without interruption* along the tubule. Besides, the lumen of the plasmodesmatal central space includes a thin tubule which connects the endoplasmic reticulums from one cell to the other. This "desmotubule" ensures a continuity between the endoplasmic systems of the cells of a tissue which are supplied with plasmodesmata (Plate 3.4 A, B; Plate 3.5 A–C). In addition, between the desmotubule and the plasmalemma, there is a space left in the canaliculus which ensures a hyaloplasmic continuity between cells (Plate 3.4 A–D).

These data led to the review of the conventional concepts of the cellular structure of plant tissues. The term cell per se (the "cell" of R. HOOCKE 1664), which means prison, supposes that the content of each cellular unit is encompassed in a closed space which in fact does not exclude the metabolic exchanges. In general, this seems to occur in animal tissues, where each cell is completely individualized. In contrast, plasmodesmata ensure a *cytoplasmic continuity* between the cells of plant tissues, which resembles organisms deprived of cells and with coenocytic structure. Since the mean diameter of plasmodesmata is about 30 to 50 µm, it seems to be justified to minimize the role of these communications. In fact, it has been estimated that about 10 000 to 20 000 plasmodesmata connect each meristematic cell with the neighbouring ones (STRUGGER 1957) and about 6000 to 24 000 connect the differentiated cells of the parenchyma between themselves (KRULL

$\triangleright$

Plate 3.4 A–D. Intercellular connections; ultrastructure of plasmodesmata.

A,B Longitudinal sections; **C,D** transversal sections. *P* Pectocellulosic cell wall; *er* endoplasmic reticulum; *pl* plasmalemma; *d* axial tubule (desmotubule) in continuity with the endoplasmic reticulum (see **A** and **B**).

Fixation: O_sO_4; **A,B,C** $\times 200000$; **D** $\times 87000$

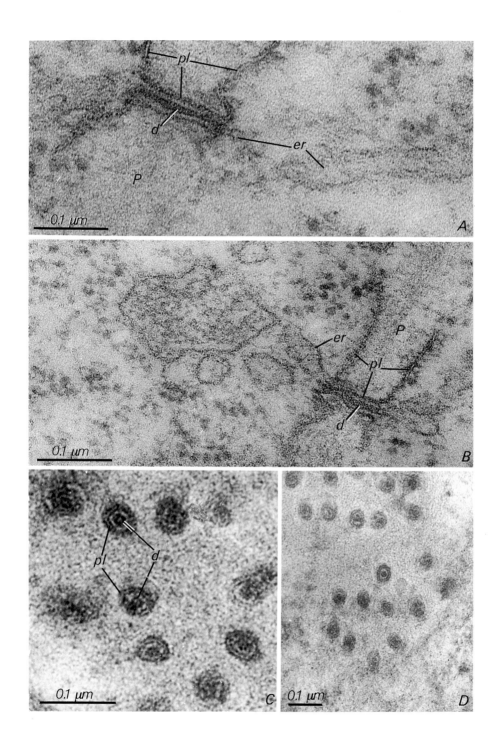

1960). The continuity of the cytoplasm in the tissues of the superior plants is therefore an important quantitative factor. It is also manifested in these tissues by the dissemination of infectious agents, the viruses; it is well known that these viruses are unable to pass through the plant cell walls and the dead cells; plasmodesmata constitute the only passage from one cell to the other for viruses.

The cytoplasmic continuity of the plant tissues increases the physiological role which must be recognized for cell walls. They are obviously the site of material transport through the apoplastic route and the comparison between animal tissues and plant tissues suggests that the cell walls represent an equivalent feature and support the internal medium of plants, forming a network in the plant protoplasm like the blood capillaries between animal cells.

The enumeration of the chemical components of the cell wall should therefore include the numerous metabolites, i.e. the organic and inorganic molecules and ions which progress along the apoplastic route. Moreover, cyto-enzymologic research shows that the walls contain enzymes which play a part in the differentiation process, mainly hydrolases, such as acid phosphatase (GIORDANI 1980, with numerous references) and peroxidases (LIBERMAN-MAXE 1974; CZANINSKI 1978; CZANINSKI and CATESSON 1969). The latter are active in the differentiation of xylem cells, among other tissues, and take part in the lignification process.

Finally, LAMPORT's works (1970, references to previous publications) revealed the presence of a parietal hydroxyproline-rich protein which influences the stretching property of walls, called *extensine* by the author.

Plate 3.5 A–F. Intercellular connections.

A–C Relations between the endoplasmic reticulum and plasmodesmata. **D–F** Origin of plasmodesmata, at the telophase, when the "cell plate" is forming. Tubules of endoplasmic reticulum (desmotubules) go across either the cell wall or the cell plate, in this case between the pectic vesicles *(arrows)*.

A × 15000; **B** × 43000; **C** × 35000; **D–F** × 31000; **A, F** barley roots; **B–E** wheat roots. Fixation: $KMnO_4$

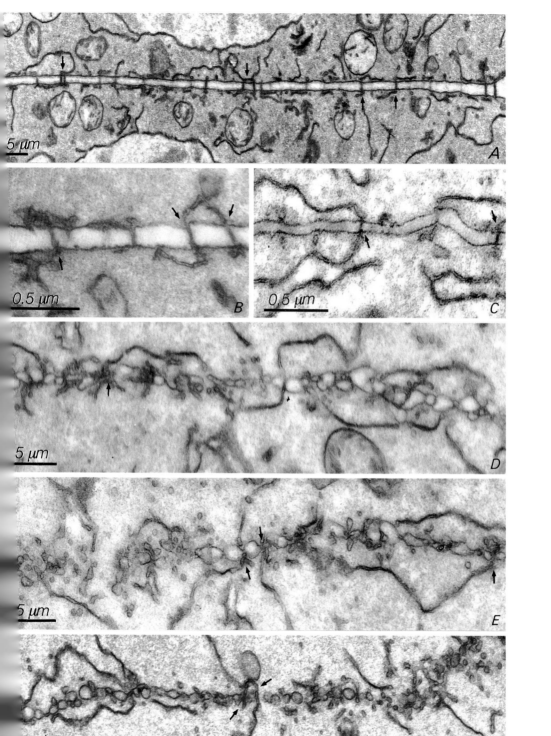

5 μm

0.5 μm   *B*

0.5 μm   *C*

5 μm   *D*

5 μm   *E*

5 μm   *F*

A

3.2.3.2 Heterogeneity of the Cell Wall Structure: Cell Wall Layers

The wall of a differentiated cell is generally heterogeneous. It is composed of several layers in close contact (Fig. 3.6). In the median area of the parietal wall, which separates two cells, there is a median substance which is the *middle lamella*. On each side, layers derived from one or the other cell are juxtaposed. The first is called the primary layer or *primary cell wall:* it is applied against the middle lamella (Fig. 3.6). Towards the interior of the cell cavity, other layers can be found, the whole structure producing the *secondary cell wall*. This one is produced only in those cells in which the cell walls are most differentiated (vessels, sclerenchyma), it is frequently composed of three layers (Fig. 3.6).

3.2.3.3 Origin of the Cell Wall and the Middle Lamella

When the karyokinesis ends, the phragmoplast edificates a new wall. Some vesicles occur at the equator and join together to form the cell plate, the first draft of the future wall. This lamella soon becomes rich in pectose material and grows until it comes in contact with the wall of the mother cell. It will constitute the *middle lamella* of the new cell wall.

The vesicles of the cellular plate are derived from the dictyosomes of the Golgi apparatus, thus demonstrating their role in the parietal productions as from the initiation of cell walls. In their progression to the equator these vesicles seem to be guided by the phagmoplast microtubules (BAJER 1968).

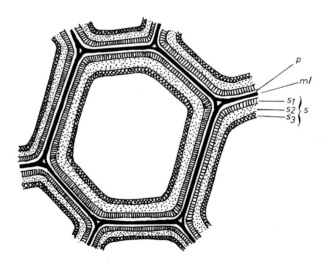

Fig. 3.6. Scheme of the general structure of the cell wall. Only some plant cell types differentiate secondary wall layer (xylem, fibres). For clarity, the thickness of the layers has been amplified. *ml* Middle lamella; *p* primary cell wall; *s* secondary cell wall, here constituted of three layers S_1, S_2, S_3

The middle lamella is mainly pectic. It mainly contains calcium and magnesium pectates (BONNER 1950). The use of electron microscopy demonstrated that, when they join together, the granules which occur at the equator of the phragmoplast leave pores in which the cytoplasm of the mother cell is continuous. This shows the cytoplasm *continuity* between daughter cells. The continuity seems to appear before the formation of the middle lamella by means of tractus, perpendicular to the cell plate, in relation with profiles of *endoplasmic reticulum* (Plate 3.5 D–F). This is in agreement with PORTER and MACHADO (1960) that the distribution of the endoplasmic reticulum on both faces of the cellular plate determines the distribution of plasmodesmata. So, along with the cytodieresis, a relationship of continuity is established within the endoplasmic reticulum, in addition to the continuity of the plasmalemma which lines the pectocellulosic wall, along plasmodesmata, between neighbouring cells (Plate 3.4 A, B; Plate 3.5 D-F).

In addition to the endoplasmic reticulum tubules, the phragmoplast microtubules have been considered as playing a part in the formation of plasmodesmata. The microtubules which go through the equatorial plane, not only in the karyokinesis spindle, but also in the surrounding cytoplasm, would maintain a cytoplasmic continuity between daughter cells (BAJER 1968; HEPLER and JACKSON 1968; ROBARDS 1968).

When it is in contact with the cell wall of the mother cell, the young middle lamella is separated from the middle lamella of the former wall by the so-called primary pectocellulosic layers of this cell wall ("primary wall"). The connection of these middle lamellae, a process disregarded by histologists, was particularly studied by JUNGERS (1937) and MARTENS (1937). In cells in which walls are not very slender (cambium, secondary walls of parenchymatous cells) MARTENS saw the edge of the new lamella resting against the former, primary cell wall. At the same time, on both sides, the cell plate is coated with primary parietal deposits which spread on old walls to which they are added (Fig. 3.7 a, d–h). The edge of the middle lamella becomes thickened and forms an "intraparietal" cavity (Fig. 3.7 b). Between the cavity and the primary middle lamella a "primary cell wall" layer, pertaining to the mother cell, can be observed. The further cell growth enlarges the cavity, stretches the fragment of cell wall which separates it from the old lamella, until it tears up, thus creating a triangular lacuna. In some cases this process is accompanied by destruction of the former lamella opposite the new intraparietal cavity (Fig. 3.7 c, d, e′–h′). Two intraparietal cavities can also join together, forming a quadrangular lacuna, or join with a pre-existing meatus or lacuna (Fig. 3.7 d″, e″–h″).

These mechanisms suggest that the setting of a new cell wall is linked to the formation of new meatus or to the enlargement of a pre-existing meatus. But there are primary meristematic tissues with very slender-walled cells lacking lacunae. However, it is quite probable that the processes described by MARTENS are common but one can understand that if the lamella of the mother cell wall is still only coated by thin cellulosic deposits the separating trabecula is very small, easily eliminated and that the meatus is reduced to an unrecognizable formation. Thus, it is only further on, during the processes of growth and differentiation that, subsequently to the turgescence, the destruction of the middle lamella can be increased resulting in a meatus visible with light microscopy.

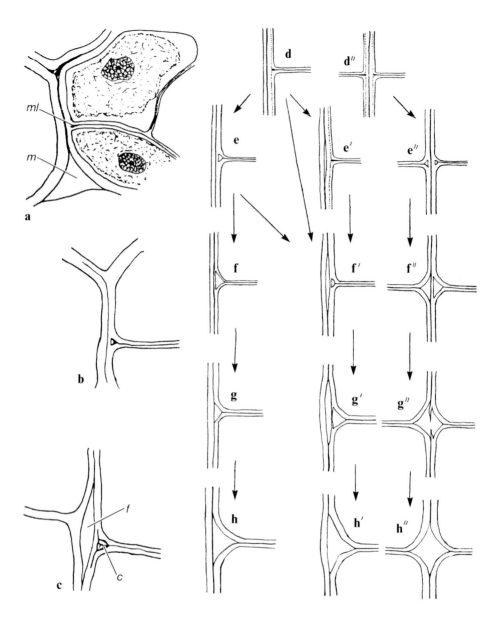

Fig. 3.7 a–h. Origin of the intercellular spaces. **a** Parenchyma cell of the stem of *Sambucus nigra;* the edge of the middle lamella *ml* abuts against the primary cell wall of the mother cell and thickens at this contact; *m* meatus. **b** Cortical parenchyma of *Hoya carnosa;* beginning of the formation of an "intramembranous cavity" in the pad which edges the middle lamella. **c** Id, in the case of formation of an intercellular split *f* facing the intramembranous cavity *c;* **d–h** formation scheme of a triangular gap, involving the distension of the primary cell wall, its resorption, opposite the cavity of the middle lamella, and the consecutive joining of the last one with the mother cell middle lamella. **e′–h′** Similar process, but accompanied by an intercellular split. **d″,e″–h″** Formation of a quadrangular lacuna. Two intramembranous cavities join together. In addition to the elimination of the primary cell wall, this case implicates a distension and a tearing of the middle lamella; (after MARTENS 1937)

In addition, MARTENS showed the very early formation of the gaps in the cortical area of root meristems. An ultrastructural and cyto-enzymologic study would be useful in elucidating the connection mechanisms between the middle lamellae and the mechanisms which cause the meatus formation.

3.2.3.4 The Primary Cell Wall

The parietal layer deposited on the cell plate constitutes the "primary cell wall" which is at first poor in cellulose and rich in "pectic compounds" (holopentosides + pectates) in polysaccharides and hemicelluloses.

From the beginning, the cellulose deposit is irregular and elaborates and inframicroscopic reticulated structure (Plate 8.1 c). Some strands of the reticulum thicken more than others encompassing areas of the thinner cell wall which constitute the *"primary pits"* (see Plate 8.1 c and Fig. 3.8 a). Plasmodesmata are numerous in these areas. However, others interpenetrate the primary layer outside of these pits (Fig. 3.8 a). The primary wall arises in *growing* young cells. This reticulated structure gives the cell wall enough plasticity to grow, even if it is already quite

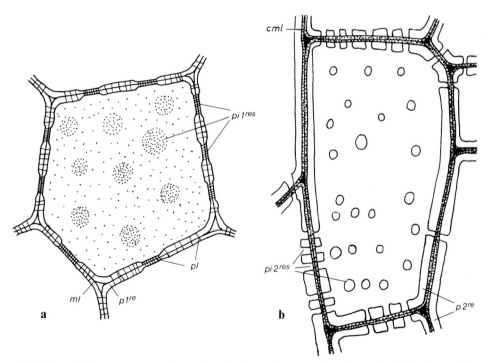

Fig. 3.8 a, b. Primary and secondary pits. **a** Scheme of a parenchymatous cell with primary cell wall only *(p1^{re})*. Primary pits *(pi1^{res})* are areas of thinner cell walls, with numerous plasmodesmata *(pl)*. Other plasmodesmata remain outside the pits; *ml* middle lamella. **b** Scheme of a secondary cell-walled cell *(p2^{re})*; cell of a secondary xylem ray of *Castanea. pi2^{res}* Non-bordered secondary pits; *cml* compound middle lamella (middle lamella + primary cell wall)

thick, as is the case with some tissues. On transections, in profile view and with polarized light the primary wall appears anisotropic.

In fact, in this case the average orientation of the cellulosic fibrils follows the outline of the cell wall. Conversely, in front view, fibrils are oriented in all directions in young cells or in those cells in which the cell wall growth arises with no preferential direction; in that case the primary cell wall appears isotropic. But, when the cell wall extends according to a selective orientation, the cellulose fibrils are firstly transversally deposited, then affected by another orientation and tend to be parallel to this direction. In this case a more or less marked anisotropy is demonstrated with the polarizing microscope. ROELOFSEN (1959) presented a suggestive image of these modifications in comparing the cell wall with overlapping fish nets stretched out in the same direction. The collenchyma, studied by ROLAND (1966) is a good example of this phenomenon (see p.453–455).

During differentiation, the primary wall is thickened by the addition of amorphous polysaccharide material and cellulose fibrils, first deposited on the middle lamella.

VIAN and ROLAND (1972) elucidated the origin of this material and the mechanisms which elaborate the cell wall, using cytochemical labelling of the plasmalemma and the delimitation membranes of the Golgi vesicles (phosphotungstic acid at low pH: APT, thiocarbohydrazide-Ag proteinate or PATAg). The authors demonstrated that dictyosomes generate clear vesicles, the membrane and the content of which cannot be labelled at the origin (vesicles *V1;* Plate 3.6). They mature as they progress into the cytoplasm. The content of polysaccharides positive to the APT or PATAg tests increases. The membrane labelling is first localized (vesicle *V2*), then it extends to the whole surface reflecting the restructuralization which allows the membrane to become similar to the plasmalemma with which anastomosis can occur (vesicles *V3* and *V4,* Scheme 3.1). The content, itself also restructured, obtains a reticulated aspect and each tractus of secreted material is in contact with a point of the membrane. On opening through anastomosis with the plasmalemma, the content of the vesicles is released in the periplasmic space, where it produced the precursor material of the parietal organization. Schemes 3.1 and 3.2 summarize these mechanisms.

Plate 3.6 A, B. Root cap cells of *Pisum sativum,* ultrastructural evolution of polysaccharide Golgi vesicles.

A Dictyosomes *D,* emitting *v1* vesicles, not reactive with the phosphotungstic acid (PTA), neither their membrane nor their content. This reactivity appears as they progress into the cytoplasm, first in a discontinuous manner *(arrows)* on the membrane *(v2),* whereas their content is marked in the form of a reticulated weft. Subsequently, the membrane of vesicles reacts on its whole surface *(v3)* and becomes, in this respect, similar to the plasmalemma *(pm)* with which it will fuse.
B Id, but more clearly showing that the reactive fibrils, a secretion product, are in continuity with the restructured sites of the vesicle membrane *(double arrows).* *m* Mitochondria, the membranes of which are not reactive; *re* endoplasmic reticulum, also non-reactive, as well as the tonoplast *t; V* vacuole; *simple arrows* indicate the site of the onset of membrane restructuralization of *v2* vesicles.

× 73 000. Fixation: glutaraldehyde – O_sO_4; reagent: phosphotungstic acid at low pH (PTA); (after VIAN and ROLAND 1972)

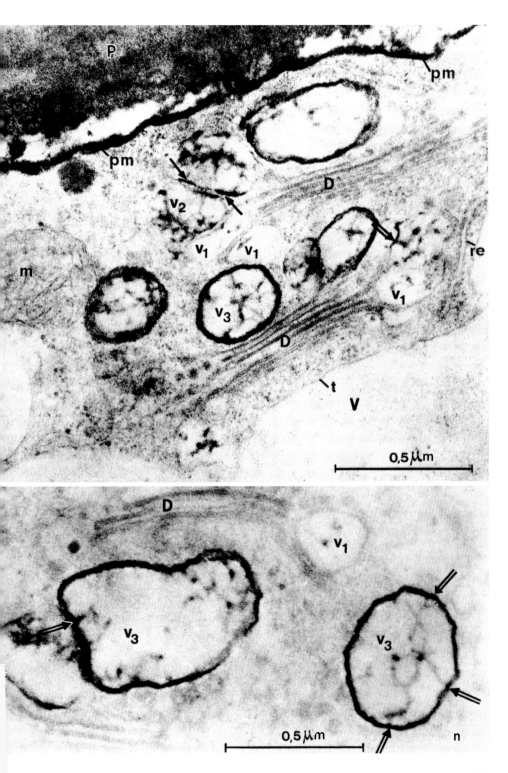

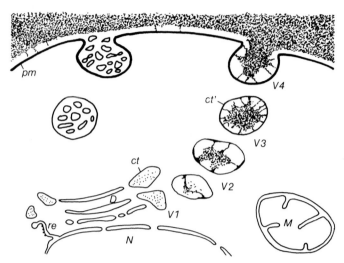

Scheme 3.1. Modalities of membrane differentiation, preceding anastomosis with the plasmalemma, using cytochemical tests. *V1* young Golgi vesicle with poorly reactive membrane, *ct,* similar to the other cytomembranes; *V2* localized restructuralization in the membrane of a vesicle; *V3* vesicle of precursor concentration with a thoroughly modified membrane, *ct',* showing the same positivity to tests as the plasmalemma, *pm; V4* open vesicle at the surface of the cytoplasm in the periplasmic space. The anastomosed membrane, prior to fusion, has the same reactive features. The multivesicular bodies, *cm,* within the cytoplasm and the plasmalemmasomes located on the surface, are entirely composed of membranes cytochemically similar to the plasmalemma (after VIAN and ROLAND 1972)

This organization, as yet inadequately known, involves all together the peripheral skeleton of the cells and the screen constituted by the middle lamella or the pre-existing layers of the primary wall (PRAT and ROLAND 1971). This screen is the support for the distribution of further deposits of parietal material. Several observations have led to the belief that microtubules associated with the plasmalemma orientate the cellulose fibrils released in the paraplasmic space, following the opening of Golgi vesicles and including the precursors of the fibrils (VIAN and RO-LAND 1972). Generally, both structures are parallel, mainly in young cells in the process of elongation in which, at first, they are arranged in short pitch helices, and in the secondary parietal derivatives in which the orientation of these structures is rigorous and which varies from one layer to the other. The evidence of parallelism has shown that the microtubules may either reach a length superior to the circumference of the elongating cells, or become associated into peripheral and helical units (LLOYD 1984).

3.2.3.5 The Secondary Cell Wall

The most rational distinction between primary and secondary cell walls has raised doubts and discussions. A number of histologists stated that the primary cell wall is only a very thin, not very cellulosic deposit which covers the middle lamella.

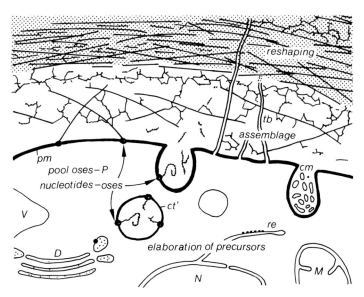

Scheme 3.2. Secretion phenomena playing a part in the building of growing cell walls.

The elaboration of precursors is related to certain cytomembranes which present common cytochemical features: plasmalemma, *pm,* and restructured membranes, *ct',* of cytoplasmic vesicles.

A coordinate synthetic activity manifested at these levels gives rise to complex molecules: fibrillar elements on the plasmalemma, *pm,* heteropolymeres with specific sequences, matrix precursors within the vesicles of concentration. (However, there are probably no clear delimitations at the sites where the components occur.)

The precursors are stored in the periplasmic space where, by means of different chemical bindings, particularly of divalent calcium bridges, they edificate three-dimensional superstructures.

The structure remains plastic. It will be modified and reoriented during cellular growth and secondarily arranged.

tb Plasmic tubule penetrating into the developing wall. For the remaining part of the scheme same legend as in Scheme 3.1 (after VIAN and ROLAND 1972)

However, it seems easier to consider that the primary cell wall contains all the material elaborated *during the cell growth* (KERR and BAILEY 1934).

With reference to this concept, the thickness of the primary cell wall is variable and, moreover, the cells, in which no parietal differentiations occur *subsequent to their growth,* have no secondary cell wall. In this work, we shall accept this way of thinking which seems the most consistent with the study of plant tissues.

One of the main features of the secondary cell wall is the fact that it develops on a surface which will not perceptibly change since the cell no longer grows. At least when it arises, the secondary cell wall is mainly composed of cellulose with, in addition, non-cellulosic osides and hemicelluloses. The successive layers (often three) of this wall can be particularly distinguished by differences in orientation of the fibrous molecules of cellulose. As a consequence, the cell wall is *anisotropic* and on transections, the different layers appear variably luminous with the use of polarized light in suitable orientations.

211

The thickening caused by these layers spares pores called secondary pits where no secondary deposit occurs (Figs. 3.8 b and 7.5). At the level of these pits, the cytoplasm is thus in contact with the *primary cell wall*. The secondary pits may occur above primary pits. Several secondary pits are sometimes constituted in the place of one former primary pit. But there can be other pits above the thicker parts of the primary cell wall. Conversely, primary pits can be covered by the secondary cell wall.

In fact, in the cell walls of the cells which generate secondary layers, it is difficult, even impossible, to distinguish the primary wall from the middle lamella; the whole forms a "compound middle lamella". Vascular cell pits are secondary pits, some of which are complicated with marginal differentiations (bordered or areoled pits). On the contrary, the pores found in the sieve areas of the phloem cells are due to the evolution of primary pits, with a few exceptions (see p. 352). The formation of secondary walls characterizes *the end of differentiation* and it is generally considered that the cells involved in this formation can no longer dedifferentiate. Their development would become *irreversible*. Very often, this extreme differentiation entails cell death, as is the case of vessels or sclerenchyma fibres. In fact, the secondary cell wall appears as a differentiation of mechanical justification.

3.2.3.6 Changes in the Secondary Cell Wall

In the course of elaboration, the cell wall increases mainly by the addition of pectocellulosic substances or some polyosides, but further on it can be chemically modified by impregnation of the pectocellulosic material. Among the substances added to this material, *lignine* is most often found. They can also be mineral substances; for example, epidermal and peripheral tissues of Graminaceae, Cyperaceae, Equisetaceae have silica-impregnated cell walls.

The modifications due to the impregnation of the cell wall frequently obliterate the delimitations between the successive wall layers. Sometimes the "primary cell wall", one or several layers of the "secondary cell wall" together with the middle lamella form a common area in which three or five layers integrade. As previously, with regards to the secondary pectocellulosic productions, this zone can be denoted as a "compound middle lamella".

These modifications of the cell wall will be further described with more details with respect to the various types of tissues. Let us only underline that *all* the layers which form the wall may change and in particular that all of them, even the middle lamella, may become lignified.

3.3 The Course of the Differentiation

In the section 3.2 we described the evolution of the cell components derived from primary meristematic cells in the process of differentiation. The evolution is not always simultaneous for all the components and it is possible to demonstrate a chronology which cannot be rigorous, as cases are numerous, but which is most frequently encountered. In most cases two phases are acknowledged.

3.3.1 The First Phase

As soon as they begin to develop, cells derived from the primary meristems increase in size and the main cytological modification, which is also the first to occur, is the enlargement, then the confluence of the vacuoles. The morphological consequence of cell growth is the shifting of the cytoplasm and its nucleus against the cell wall.

The centre of the cells is soon filled with a large vacuole. Between the vacuole and the still thin and pectocellulosic cell wall, there is a cytoplasmic film, thicker where the cytoplasm includes the nucleus, enclosing chondriosomes and plastids still undifferentiated or only slightly differentiated.

These cytological features are similar to those described in cambial cells and the medullary meristem cell (Figs. 2.29 and 2.50 B, C).

To summarize this onset of differentiation, one can see that it corresponds to the passage from the primary meristematic condition to the secondary meristematic condition. This evolution is variable depending on cell types: it is, for example, slower in the procambial strands than in the ventral or dorsal parenchyma of the foliar buttress, but it is generally common in all cellular types in the process of differentiation.

Of course, this method of expression is figurative, as all cells do not develop into functional secondary meristematic cells. In many plants, they do not organize themselves into cambium. But, experimentation as well as observations still show that this cytological condition characterizes quite well the potentialities of the cells in such a condition. In fact, whereas the *primary meristematic cells* of the initiating ring are organogenic since they can initiate leaves and leaf segments, the secondary meristematic cells proliferate only to produce one or two determined tissues and do not edificate any organ; they are histogen. For example, the cambial generating zone exclusively produces xylem and phloem and the medullary meristem only produces medullary parenchyma. When, experimentally, organs are obtained from cells including a large vacuole, it is because, previously, some of their derivatives have regained the primary meristematic features existing in cells with small vacuoles, and have generated an apical meristem. This is the case with the formation of buds in the cambial zone of root cuttings of *Cichorium* or *Tragopogon* (BU-VAT 1943).

3.3.2 The Second Phase

When they reach the state of a large vacuole, cells either keep on differentiating without proliferating perceptibly, or remain inactive for a while before resuming a proliferating activity which is then oriented; in such a case they constitute a *secondary meristem.*

In the first case, the differentiation takes place in situ after the first phase, resulting in *primary* tissues. This is the only existing mode in most Monocotyledons and in living Pteridophytes.

In the second case, a proliferating phase *occurs between* the two phases of differentiation, resulting in secondary tissues.

In any case, the second phase of the differentiation involves mainly the components *different from* the unit cytoplasm-vacuoles which is affected by the first phase. Shifted by the vacuole, the nucleus becomes lenticular or variably flattened as it appears in the cells which no longer divide.

The proplastids of the young cells differentiate, except for particular cases, into chloroplasts, leucoplasts or into chromoplasts, amyliferous or not, according to the specificity of the cell involved. Finally, the paraplasmic products are elaborated: storage material, secondary products of the elaborating activity and parietal products. The latter have mainly been used in the past to determine the main tissues, sometimes too exclusively.

One should keep in mind that the parietal differentiations, valuable criteria of cellular differentiation, are only an element of it. Let us also consider that, whereas the first phase is quite general and regular, the second one, on the contrary, is the phase of *specialization* and *diversification* of cells.

This process of evolution seem to be the most common but, it is obvious that the factors of differentiation are numerous and varied enough to give particular features to each case. It should be noted that the most variable cellular components with regards to morphology, plastids and vacuoles, in addition to the differentiating evolution of the cell, are influenced by many factors such as the time of the day or the season. The morphological aspects described previously do not exclude variations which are either species cases, or temporary modifications.

Plate 3.7 a–k. *Lycopersicum esculentum,* stem cuttings. First phase of dedifferentiation of periphloem chorophyllous cells.

a Normal periphloem cell; chondriosomes in the shape of granules and short rods; lentil-shaped choroloplasts, relatively small nucleolus.
b First division of chloroplasts: constrictions followed by stretching; distension and tendency to vesiculization of chondriosomes.
c Cells coming from first redivisions of initial cells; chloroplasts divide actively and begin to regress.
d–i Gradual reduction of chloroplasts, following repeated partitions; tendency of transitory vesiculization of chondriosomes; enlargement of nucleoli.
k First phase of dedifferentiation nearly completed; cells becoming more or less isodiametric; plastids still in process of division, now only slightly larger than mitochondria.

× 1200. Fixation: REGAUD; staining: hematoxylin

214

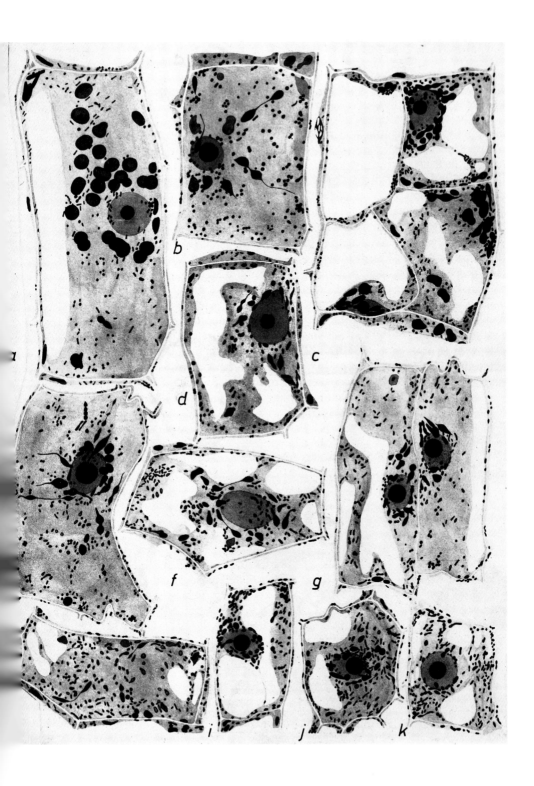

3.4 The Main Types of Tissues

In a first approach, it is possible to classify the tissues elaborated during the second phase with respect to the most striking cytological modifications of cells.

If the differentiation mainly involves the protoplasm and, above all, at least apparently, mitochondria and plastids, it produces parenchymatous or secretory tissues. Generally, cell walls do not develop much: they get thicker and remain cellulosic. Such are for example the amyliferous parenchyma (Fig. 4.4), the assimilating palisade parenchyma (Fig. 4.2), the tannin cells of the rose medulla, cells containing soluble reserves of the beat root (saccharose) or Jerusalem artichoke (inuline), soluble material which is stored in the vacuole.

If the differentiation mainly involves the cell wall, it can induce the genesis of supporting tissues: collenchyma (cellulosic thickenings) (Fig. 8.3), sclerenchyma (lignification) (Fig. 8.25) or protecting tissues: epidermal (cuticle differentiation); cork (impermeability due to suberin) (Figs. 5.3 and 5.21).

Finally, in other cases, the differentiation associates the development of cell walls closely with that of living matter. Such is the case in conducting tissues: xylem and phloem tissues. In these tissues the differentiation is so pronounced that it results in cell death. But these tissues have to keep some activity which could not exist with an arrangement of dead cells. This activity is maintained by the juxtaposition of different cellular types with some living elements. They represent complex tissues, the activity of which relies upon the association of living cells and senescent or dead cells, led to this extreme condition by differentiation: the total or partial removal of the living matter.

Plate 3.8 a–g. Stem cuttings of *Lycopersicum esculentum*. Second phase of dedifferentiation of periphloem chlorophyllous cells.

a, b Cells reaching the end of the first phase of dedifferentiation; single vacuole, sometimes crossed by cytoplasmic strands. **a,** *(right upper cell)* plastids only slightly distinct or indistinct from mitochondria; nucleus applied against the cell wall.

c, d Smaller cells; the cytoplasm begins to divide the vacuole into separate organelles; the process is more advanced in the *left cell* of **d**; the nucleus becomes spherical and moves apart from the cell wall; nucleoli larger and larger.

e Cells of a young root primordium, already recognizable; nucleus at the cell centre; numerous small globular vacuoles.

f Meristematic cells of a neoformed root; considerable reduction of vacuoles, some of which are stretched-shaped; relatively voluminous nuclei, with large nucleolus; plastids practically indistinct, chondriome divided into short elements, mitochondria and short rods.

g Reproduction of the *three left cells* of **f** at the same scale as in Plate 3.7;

× 1750 (**a–f**) and × 1200 (**g**). Fixation: REGAUD; staining: hematoxylin

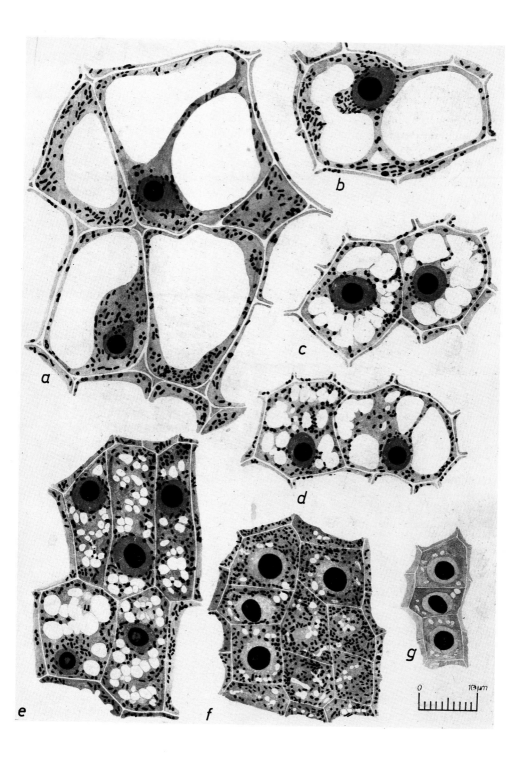

a

b

c

d

e

f

g

0   10 μm

217

3.5 Processes of Dedifferentiation in the Course of Ontogeny

3.5.1 Circumstances of Dedifferentiation

In their most general features, we have just considered the two phases through which cells, derived from primary meristems, can reach a state of completed maturity.

But we have also mentioned situations in which all the derivatives of primary meristematic cells do not exclusively develop progressively. Thus, the elaboration of cambiums often involves already differentiated cells inducing their regression to the *"secondary meristematic"* stage.

Even spontaneously, the arising of adventitious roots, secondary roots and axillary buds, especially when they are late, may lead cells to dedifferentiate until the primary meristematic state.

3.5.2 The Phases of Dedifferentiation

The complete regression, from cells of secondary tissues or primary tissues having completed their differentiation, can be experimentally obtained in many cases (cutting, culture of plant tissues) (Figs. 3.1 and 3.2).

The dedifferentiation begins by the recurrence of proliferation and the loss of the structural attributes issued from the second phase of differentiation (plastids, reserves, etc.). The cells return to the secondary meristematic stage: this represents a *first phase* of dedifferentiation (Plates 3.7, 3.9 and 3.10). The regression of the large vacuole, the migration of the nucleus to the centre of the cell, occur only after the cells have resumed the primary meristematic condition: this represents a *second phase* of dedifferentiation (Plates 3.8, 3.11).

Plate 3.9 a–h. *Brimeura amethystina* L., leaf cutting. Onset of dedifferentiation of chlorophyllous mesophyll cells.

a Normal middle cell, peripheral cytoplasm, relatively short chondriosomes, lentil-shaped chloroplasts containing starch grains.
b–d Phase of amylogenesis, with reduction of the chromophil substance of plastids, chondriosomes in the shape of long filaments (chondriocontes).
e,f Condensation of the elaborating substance of plastids, which have lost their chlorophyll and have obtained a typical aspect of amyloplasts. The chondriome mostly divided into granular mitochondria. The vacuole is still single and voluminous.
g,h After the phase of amylogenesis: regression of amyloplasts, decrease of their size and polarization of the plastidal substance, which forms a cupule covering the starch grain (**g**), then taking the shape of granules or short rods (**h**).

After the resumption of the starch, the cells have reached the end of the first phase of dedifferentiation. The second phase is then similar to that illustrated in Plate 3.8 for *Lycopersicum esculentum* cuttings

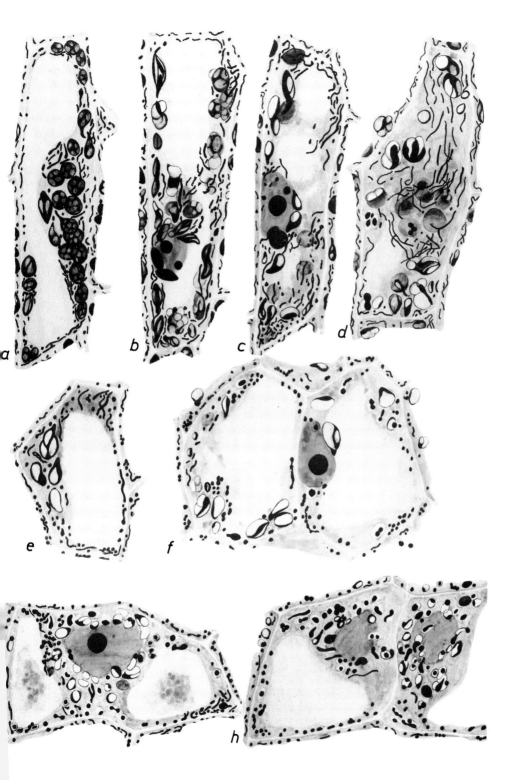

219

Apparently, the recurrent proliferating activity which is a prelude to dedifferentiation is generally preceded by a phase of metabolite increase, essentially of a marked amylogenesis for example. It is observed in leaf samples (*Brimeura amethystina;* Plate 3.8/2–4), but also in several examples of elaboration or activation of primary meristem, either induced (in vitro culture of *Tobacco* medulla, BROSARD 1970) or spontaneous (foliar bulbil of *Bryophyllum,* BROSSARD 1973); suppression of the inhibition of axillary meristem of *Pisum* (NOUGAREDE and RONDET 1975 etc.).

The first phase does not produce new organs, the cells dedifferentiated up to the secondary meristematic type are at the most *histogenic.* The cells which have gone through the second phase, contrarily tend to edificate apical meristem: they have become *organogenic.*

This shows that during the course of differentiation as well as in the course of dedifferentiation, *the cytological aspect of cells is characteristic of their potentialities.*

These experimental data are consistent with the phenomena of normal ontogeny of vascular plants. During ontogeny, dedifferentiation process occur spontaneously: the cells, which are in the first phase of regression, generate *cambiums* totally or partially and are thus *histogenic;* the cells which return to the primary meristematic condition generate apices which add new organs to the plant and are thus *organogenic* (adventitious and secondary roots, axillary buds).

These regressions are more or less extended depending on species and on the selected organs. Thus, the pericyclic cells from which secondary roots arise are generally poorly differentiated, as well as the axillary cells of leaves when the formation of the axillary meristem is early, however, this is not always the case.

Finally, the thorough cytological study of the initiating ring itself shows a constant variation of the meristematic features of the cells: each foliar initiation is accompanied by a notable accentuation of these features (see p.150; Fig.2.32; Plate 2.2/2, 3).

Plate 3.10. Phloem parenchyma of *Daucus carota,* cultivated in vitro. Beginning of the first phase of dedifferentiation.

1 Ordinary phloem parenchyma cell; chromoplastids loaded with compound starch grains and carotene crystals in the shape of flexuous and angular lamellae; cytoplasm reduced to a thin film applied against the cell wall, and partially seen in front view; a few chondriosomes; single vacuole filling the major cell volume.

2 After 28 days of in vitro culture; still undivided cell, reaching the end of a preliminary phase: unfinished hydrolysis of starch; the cytoplasmic film, always very thin, is seen from a front view and, laterally, it is seen in profile; the resorption of starch initiates the dedifferentiation of chromoplasts; some of them, in *a,* have released their carotene crystals *(d)* into the cytoplasm and have now taken a ramified form; others condense their substance on part of these crystals *(b)* or continue to surround them *(c).*

A,B,C Morphological aspect of the dedifferentiation of chromoplasts. **A** Structure of normal chromoplasts; the plastidal substance entirely surrounds the carotene crystals and the starch grains, even if the pellicle is only very thin. **B** The resorption of starch is accompanied by the condensation of the plastidal substance, which partly uncovers the grains *(1–6),* and then resolves in discontinuous elements, resembling chondriosomes *(7)* and releases the crystal into the cytoplasm *(8).* **C** The starch has disappeared and the chromoplasts have taken the form of ramified filaments, either continuous *(1)* or discontinuous *(2,4,5,8).* Sometimes these plastids still encompass carotene crystals *(4)* or keep their mark for a moment

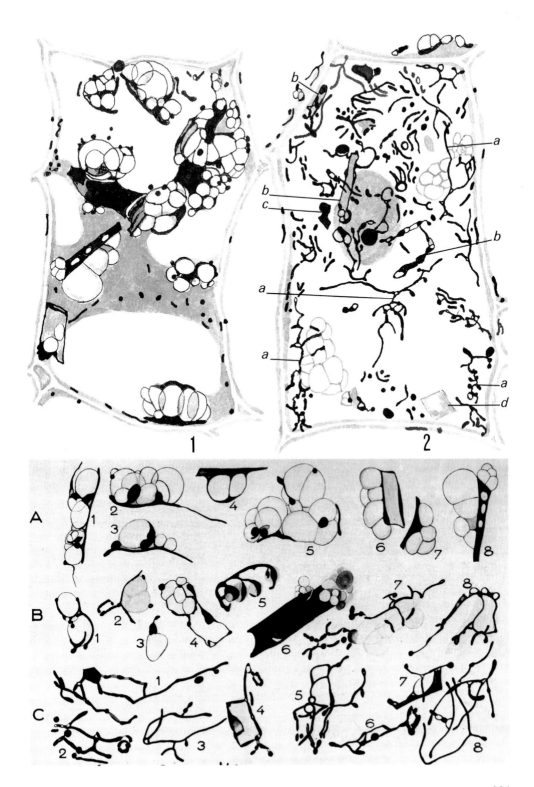

1

2

b

a

b

c

b

a

a

a

d

A

1
2
3
4
5
6
7
8

B

1
2
3
4
5
6
7
8
7

C

1
2
3
4
5
6
7
8

3.5.3 Dedifferentiation and Mode of Growth

Thus, it seems that all young plant cells tend to differentiate and to pass through the phases of all living structures, youth and growth, maturity, senescence, death. The cell which rises in the initiating ring is not spared: if it is not concerned with another activity, the cell begins to differentiate within the initiating ring itself. If it participates in the production of a leaf, the cell ceases to differentiate, it regresses in proliferating and its derivatives, when they no longer proliferate, if not earlier, will resume the course of differentiation. The same thing could be said about an axillary cell which becomes parenchymatous prior to being involved in bud formation. In fact, the ontogeny of vascular plants is achieved by means of the association of processes of differentiation and dedifferentiation. Only the first process seems to take place in animal ontogeny. The dedifferentiation processes which occur, in addition, to plant ontogeny represent the manifestation of a physiological mechanism, opposed to differentiation, *which could aim to maintain in plants the meristematic tissues which ensure their continued growth.*

The concept of indefinite growth, which contrasts plants to animals, seems to be linked to dedifferentiation processes in plants, and to their absence in *animals,* in which growth is limited, i.e. "determined", and in which the differentiation is irreversible, at least with respect to the specificity which appears during development.

Plate 3.11. Rhizogenesis in in vitro cultures of *Daucus carota* phloem, last phase of dedifferentiation of cells derived from the phloem parenchyma.

1 Neoplastic tissues at the end of the first phase of dedifferentiation, chiefly characterized by the regression of chromoplasts, no longer clearly distinct from chondriosomes. Some cells have elaborated tracheiform cell wall thickenings.

2 Neoplastic cambial zone, arising at the contact of tracheiform cells of a neoformed nodule. These cells illustrate the transition from the parenchymatous state to the secondary meristematic state, typical of the first dedifferentiation phase.

3 Beginning of the second dedifferentiation phase, redivision of cells and growth of the cytoplasm; chondriome broken up into short elements.

4 Activation of the proliferation, more and more reduced cells; nuclei become subspherical, but vacuoles still voluminous.

5 Accentuation of the preceding features during proliferation: plastids resemble chondriosomes, but are slightly thicker; more and more abundant mitochondria; beginning of vacuole fragmentation; nuclei roughly spherical with voluminous nucleoli.

6 Cells of a young neoformed root meristem, reaching the end of their dedifferentiation; nuclei at the centre of the cells; voluminous nucleoli; abundant cytoplasm, containing only small vacuoles, most of them filamentous and branched, the content of which is often precipitated by the fixative; plastids only slightly distinct from mitochondria or rodlike chondriosomes, which form a particularly abundant chondriome.

× 1500 (**1–5**); × 2400 (**6**). Fixation: REGAUD; staining: hematoxylin; (BUVAT 1944)

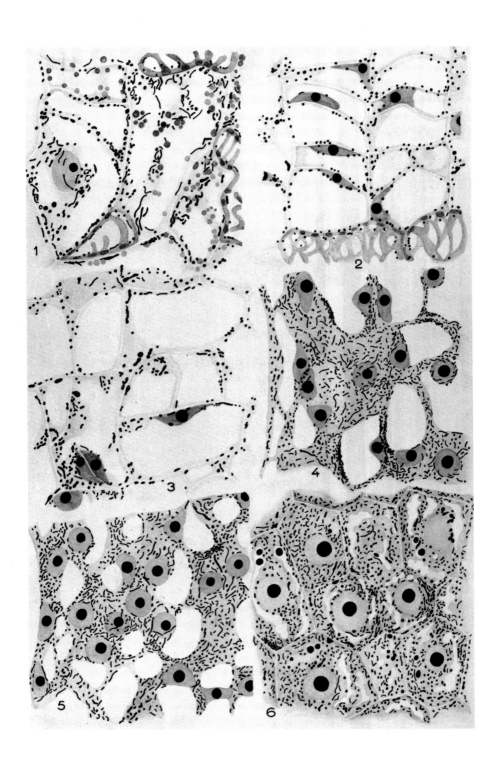

In other terms, it can occur that an animal cell loses the attributes acquired through its differentiation and starts proliferating again, but it will only generate cells of the same category as itself. Conversely, certain differentiated plant cells may, after they have been reactivated, produce all the cellular types of an organ differing even from the organ from which they have derived, they prove to be "pluripotential" and even "omnipotential".

References

Bajer A (1968) Fine structure studies on phragmoplast and cell plate formation. Chromosome (Berl) 24: 383–417

Bonner J (1950) Plant biochemistry. Academic Press, London

Brossard D (1970) Evolution de la synthèse d'amidon durant l'organisation des nodules, puis des méristèmes primaires obtenus in vitro, à partir d'explants de moelle de Tabac. Comparaison avec le point végétatif de la plante entière. CR Acad Sci Paris 271 D: 56–59

Brossard D (1973) Le bourgeonnement épiphylle chez le Bryophyllum daigremontianum Berger (Crassulacées). Etude cytochimique, cytophotométrique et ultrastructurale. Ann Sci Nat Bot 12e sér 14: 93–214

Buvat R (1943) Phénomènes de dédifférenciation dans le bourgeonnement des tissus du tubercule de Chicorée à café cultivés in vitro. CR Acad Sci Paris 216: 574–576

Buvat R (1944–45) Recherches sur la dédifférenciation des cellules végétales. Ann Sci Nat Bot 11e sér 5: 1–130, 6: 1–119

Buvat R (1958) Recherches sur les infrastructures du cytoplasme, dans les cellules du méristème apical, des ébauches foliaires et des feuilles développés d'Elodea canadensis. Ann Sci Nat Bot 11e sér 19: 121–162

Buvat R, Lance A (1958) Evolution des infrastructures de mitochondries au cours de la différenciation cellulaire. CR Acad Sci Paris 247: 1130–1132

Buvat R, Robert G (1979) Vacuole formation in the actively growing root meristem of barley (Hordeum sativum). Am J Bot 66: 1219–1237

Coulomb C, Coulomb P (1973) Participation des structures golgiennes à la formation des vacuoles autolytiques et à leur approvisionnement enzymatique dans les cellules du méristème radiculaire de la Courge. CR Acad Sci Paris 277 D: 2685–2688

Czaninski Y (1978) Localisation ultrastructurale d'activités peroxydasiques dans les parois du xylème du Blé pendant leur différenciation. CR Acad Sci Paris 286 D: 957–959

Czaninski Y, Catesson AM (1969) Localisation ultrastructurale d'activités peroxydasiques dans les tissus conducteurs végétaux au cours du cycle annuel. J Microsc 8: 875–888

Emberger L (1925) Sur la réversion des plastes chez les Végétaux. CR Acad Sci Paris 181: 879–880

Emberger L (1927) Nouvelles recherches sur le chondriome de la cellule végétale. Rev Gén Bot 39: 341–363, 420–448

Giordani R (1980) Dislocation du plasmalemme et libération de vésicules pariétales lors de la dégradation des parois terminales durant la différenciation des laticifères articulés. Biol Cell 38: 231–236

Guilliermond A (1920) Sur l'évolution du chondriome dans la cellule végétale. CR Acad Sci Paris 170: 194

Guilliermond A (1922) Remarques sur la formation des chloroplastes dans le bourgeon d'Elodea canadensis. CR Acad Sci Paris 175: 286

Guilliermond A (1934a) Les constituants morphologiques du cytoplasme: le chondriome. Actualités scient et indust. Hermann, Paris, no 170

Guilliermond A (1934b) Les constituants morphologiques du cytoplasme; le système vacuolaire ou vacuome. Actualités scient et indust. Hermann, Paris, no 171

Guilliermond A (1938) Introduction à l'étude de la Cytologie. Actualités scient et indust. Hermann, Paris, no 741, p 59, nos 741, 743

Guilliermond A, Mangenot G, Plantefol L (1933) Traité de Cytologie végétale. Le François, Paris

Hepler PK, Jackson WT (1968) Microtubules and early stages of cell plate formation in the endosperm of *Haemanthus katherinae* Baker. J Cell Biol 38: 437–446

Jungers V (1937) L'origine des méats chez le *Viscum album*. Cellule 46: 111

Kerr T, Bailey IW (1934) The cambium and its derivative tissues. X. Structure, optical properties and chemical composition of the so-called middle lamella. J Arnold Arbor Harv Univ 15: 327–349

Krull R (1960) Untersuchungen über den Bau und die Entwicklung der Plasmodesmen im Rindenparenchym von *Viscum album*. Planta (Berl) 55: 598–629

Lamport DTA (1970) Cell wall metabolism. Annu Rev Plant Physiol 21: 235–270

Liberman-Maxe M (1974) Localisation ultrastructurale d'activités peroxydasiques dans la stèle de *Polypodium vulgare* (Polypodiacées). J Microsc 19: 169–182

Lloyd CW (1984) Toward a dynamic helical model for the influence of microtubules on wall patterns in plants. Int Rev Cytol 86: 1–51

Martens P (1937) L'origine des espaces intercellulaires. Cellule 46: 355–388

Marty F (1973) Mise en évidence d'un appareil provacuolaire et de son rôle dans l'autophagie cellulaire et l'origine des vacuoles. CR Acad Sci Paris 273 D: 1549–1552

Nass MMK, Nass S (1963) Intramitochondrial fibers with DNA characteristics. I, II. J Cell Biol 19: 593–611, 613–629

Nougarède A, Pilet PE (1965) Distribution des ribosomes le long de la racine du *Lens culinaris*. L. CR Acad Sci Paris 260: 2899–2902

Nougarède A, Rondet P (1975) Synthèse et utilisation de l'amidon dans les axillaires du *Pisum sativum* L. (var. nain hâtif d'Annonay) après la levée de dominance apicale. Rev Cyt Biol Vég 38: 197–215

Porter KR, Machado RD (1960) Studies on the endoplasmique reticulum. IV. Its form and distribution during mitosis in cells of onion root tip. J Biophys Biochem Cytol 7: 167–180

Prat R, Roland JC (1971) Etude ultrastructurale des premiers stades de néoformation d'une enveloppe par des protoplastes végétaux séparés mécaniquement de leur paroi. CR Acad Sci Paris 273D: 165–168

Ris H, Plaut W (1962) Ultrastructure of DNA-containing areas in the chloroplast of *Chlamydomonas*. J Cell Biol 13: 383–391

Robards AW (1968) A new interpretation of plasmodesmatal ultrastructure. Planta (Berl) 82: 200–210

Roelofsen PA (1959) The plant cell wall. Handbuch der Pflanzenanatomie. Borntraeger, Berlin

Roland JC (1966) Organisation de la membrane paraplasmique du collenchyme. J Microsc 5: 323–348

Strugger S (1957) Der elektronenmikroskopische Nachweis von Plasmodesmen mit Hilfe der Uranylimprägnierung an Wurzelmeristemen. Protoplasma 48: 231–236

Vian B, Roland JC (1972) Différenciation des cytomembranes et renouvellement du plasmalemme dans les phénomènes de sécrétions végétales. J Microsc 13: 119–136

PART II

Histological Differentiation of Vascular Plants

The final differentiation of the cells derived from meristems results in a great variety of cellular types in the organism of any vascular plant. The most specialized cells are not always the most "alive" but they are generally the most characteristic of the great taxonomic groups (e.g. vascular cells and vessels). On the contrary, the cells which appear most important due to the vital activity, those in which the living protoplasm is most abundant, are maybe the most similar in the whole series of plants considered.

CHAPTER 4

Parenchyma

4.1 General Characteristics

The differentiation of parenchyma is mainly concerned with the living substance and the *non-parietal* products of elaboration. These tissues occur in all places where the plant development did not induce more specialized differentiations or, to use another expression, they constitute the *fundamental tissue* of all organs, at least when they are most active. Due to this distribution, parenchyma have been compared to the *conjunctive tissue* of animals. This comparison is quite inadequate if we consider that parenchyma along with meristems are the main sites of the *biochemical activity* of the plant. Therefore, they are the first to be discussed.

The future parenchymatous cells, when they are derived from meristems, first differentiate by considerably growing; the growth is an element of differentiation. The linear dimensions can be two to ten times higher than those of the initial meristematic cells. Thus, parenchymatous cells are *large cells* about a few tens of μm in diameter or length.

The increase of size is mainly due to the increase of vacuoles which soon join in a single large enclave, almost filling up the whole cellular volume. Then the cytoplasm constitutes a thin film applied against the wall which generally remains quite thin and cellulosic. The cell wall develops only a *primary* layer and shows primary plasmodesmata-rich pits. Besides, there are other plasmodesmata outside of the pits (see Fig. 3.8 a).

The nucleus is shifted against one of the cell walls flattening more or less when the vacuoles become united. Then in the parietal cytoplasmic film, differentiations characteristic of the second phase occur (p. 214). They are mainly concerned with mitochondria and plastids.

Usually the *aspect of plastids is characteristic* of the type of parenchyma. Again according to this type, there are differences in the *paraplasmic content*. A few examples are given here.

4.2 Main Types of Parenchyma

Parenchyma are often considered as *food storage tissues*. Depending on the *physiological* significance of their biochemical activity, it is possible to distinguish *assimilatory parenchyma* which are the principal sites of *photosynthesis* or its sequels and *storage parenchyma* preferential storage sites of excess assimilation material.

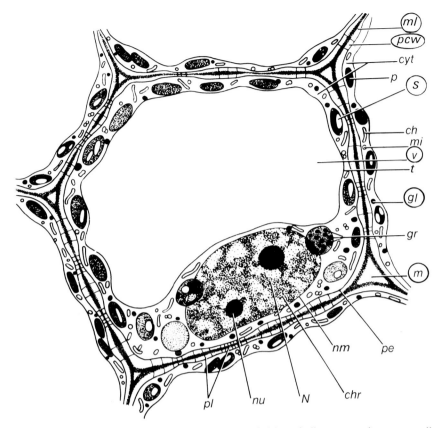

Fig. 4.1. Leaf cell of *Elodea canadensis:* organization of chlorophyllous parenchymatous cell, as seen with the light microscope. The *encircled letters* indicate the paraplasmic constituents. *s* Starch (in the chloroplasts); *ch* filamentous chondriosomes; *chr* chromatin; *cyt* cytoplasm; *gl* lipophil globules; *gr* grana (in the chloroplasts); *ml* middle lamella of the cell wall; *m* meatus; *mi* mitochondria; *nm* nuclear membrane; *pcw* pectocellulosic primary cell wall; *N* nucleus; *nu* nucleolus; *p* lentil-shaped chloroplasts; *pe* plasmalemma; *pl* plasmodesmata; *t* tonoplast; *v* vacuole (after Buvat 1965)

4.2.1 Assimilatory Parenchyma

They mainly correspond to chlorophyllous parenchyma but it should be noted that proteogenesis also takes place in the non-chlorophyllous superficial cells of the root: thus, here there are other assimilatory cells; they use substances provided by the leaves and resulting from the activity of chlorophyllous parenchyma. The latter are distributed in areas that can be reached by light in the stems, the leaves and the reproductive organs. The plastids of such parenchyma have differentiated into chloroplasts and their chondriome is usually divided into short mitochondria. In leaves, the assimilating parenchyma constitute the mesophyll which may be homogeneous as in the leaves of *Elodea canadensis* (Fig. 4.1) or heterogeneous when the upper leaf face is the most exposed to light; the upper half of the mesophyll

229

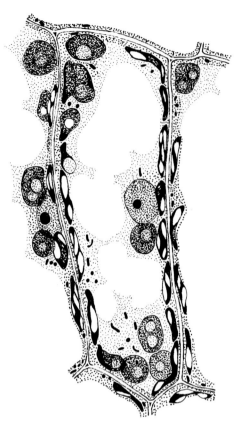

Fig. 4.2. *Nerium oleander,* leaf palisade parenchyma. Numerous chloroplasts, generally containing starch grains, applied against the cell wall and enclosed in the thin cytoplasm pellicle encircling one large and single vacuole; nucleus equally enclosed in a thickening of cytoplasm, surrounded by chloroplasts, some of them in front view, in parts of the cytoplasm of the upper face of the cell: granular or very short mitochondria. Fixation: REGAUD; staining: hematoxylin

is then differentiated into *palisade parenchyma* (Fig. 4.2; Plate 4.1 A), whereas the lower half becomes a *lacunar or spongy parenchyma* with intercellular spaces which facilitate the gaseous exchanges (Plate 4.1 B and Plate 4.2).

The *palisade parenchyma,* the most differentiated, is composed of cells huddled one against the other in ordered regular layers, including numerous chloroplasts, generally lentil-shaped and thicker than the cytoplasmic pellicle that they distend. This encompassing pellicle maintains them against the membrane and around the nucleus.

▷

Plate 4.1 A, B. *Nerium oleander,* mesophyll.

A Palisade parenchyma comprising two layers of cells, beneath the upper epidermis *ep* and the two-layered hypodermis *hy.* **B** Spongy parenchyma, filling up the dorsal (lower) half part of the mesophyll. Chloroplasts of the two parenchymas contain starch grains *(arows).*

Fixation: REGAUD; staining: hematoxylin

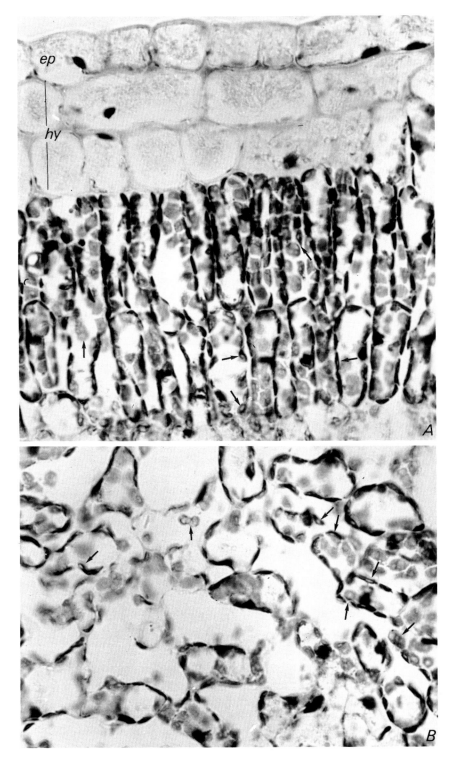

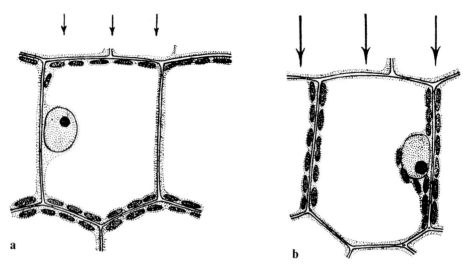

Fig. 4.3. Schema of chloroplast distribution in the leaf cells related to the light intensity: weak light, **a,** *short arrows;* strong light, **b,** *long arrows.* In the first case, the larger surface of chloroplasts is exposed to light; in the second, they present their profile to the light rays and shade one another along anticlinal cell sides

Such a cytological structure contributes to the regulation of the functioning of the palisade cells normally exposed to light from which they use the energy required for photosynthesis. If the light is too intense, active photosynthesis may produce a secondary release of heat higher than that suitable for the cell. However, the plastids line the walls parallel to light rays (Fig. 4.3 b). Thus, only their profile is exposed to light (minimal surface) and, moreover, to a certain extent, they provide shade for each other; the photosynthesis is then limited. If the light is weak, the plastids preferentially line the walls perpendicular to light rays, so exposing a larger surface to light (Fig. 4.3 a) for better use of the deficient light energy.

Temporarily, the assimilatory cells accumulate protids, glucids and lipids. Among the glucids, the most apparent is starch which forms temporary enclaves in chloroplasts (Fig. 4.2, Plates 4.1 and 4.2).

The productions issued from the activity of assmiliatory parenchyma are set into motion by hydrolysis and migrate, mainly during the night, to the reserve parenchyma.

Plate 4.2. *Cucurbita maxima,* ultrastructure of lacunar (spongy) parenchyma cells.

Poorly developed chloroplast grana, large vacuole, cytoplasm with a strong density of ribosomes. *p* Chloroplasts; *s* starch; *m* mitochondria; *px* peroxisomes; × 18 500. Fixation: glutaraldehyde-O$_s$O$_4$; contrast according to REYNOLDS (1963)

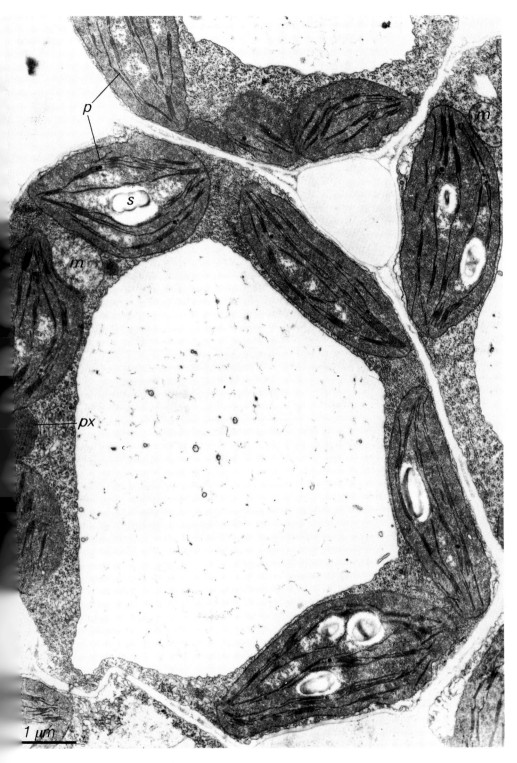

1 μm

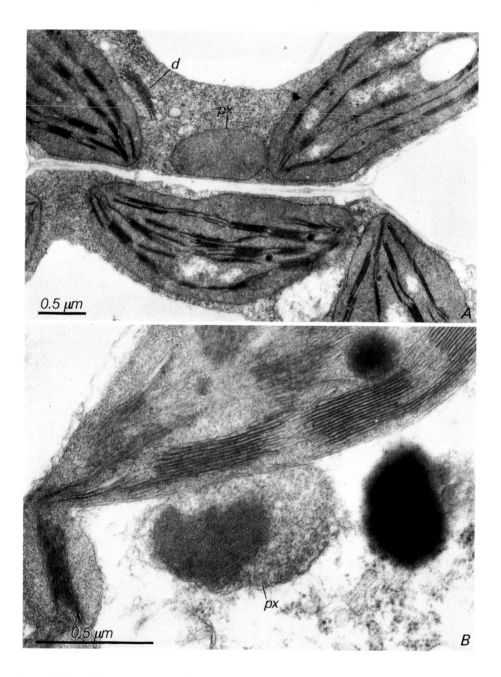

0.5 μm

A

0.5 μm

B

Plate 4.3 A, B. Peroxisomes in leaf parenchyma.

A *Curcubita maxima,* spongy parenchyma; *px* peroxisome with homogeneous content beside a chloroplast; *d* dictyosome; ×26000.

B *Ruta chalepensis,* leaf parenchyma; *px* peroxisome, against a chloroplast, containing a crystalline "nucleoid"; ×63000.

Fixation: glutaraldehyde-O_SO_4; contrast according to Reynolds (1963)

The infrastructures of assimilatory parenchyma cells demonstrate their bio-chemical activity through the development of active surfaces. Those of chloroplasts have often been observed (Plate 3.3 A); moreover, chondriosomes are generally richer in cristae (Plate 3.3 B) which is linked to their high oxidative and phospho-rylative activity. Probably related to this activity, which generates hydrogen perox-ide, peroxisomes are frequently found in the vicinity of chloroplasts, as osmophilic corpuscles with homogeneous content (Plate 4.3 A) or with a crystalline nucleus (Plate 4.3 B) on ultrathin sections.

4.2.2 Storage Parenchyma

In the organs (roots, rhizomes, tuberized or not, albumen or cotyledon parenchy-ma of seeds) in which the products of anabolism are stored, the storage of such products occurs in parenchymatous cells.

In many cases, glucid storage is materialized as starch within very differentiat-ed leucoplasts (potato, bulb scales, floury albumens, etc.; Fig. 4.4). In other cases, glucids remain soluble and accumulate within the vacuole (beet root: saccharose;

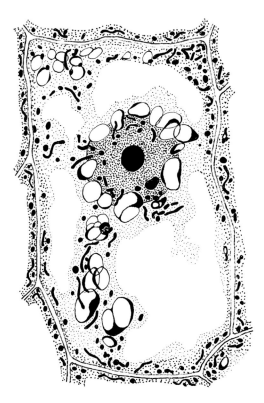

Fig. 4.4. Bulbil of *Brimeura amethystina* L. Example of starch storage parenchyma cell, not chlo-rophyllous. Each amyloplast contains one or several starch grains, distending the plastidal sub-stance (after BUVAT 1944)

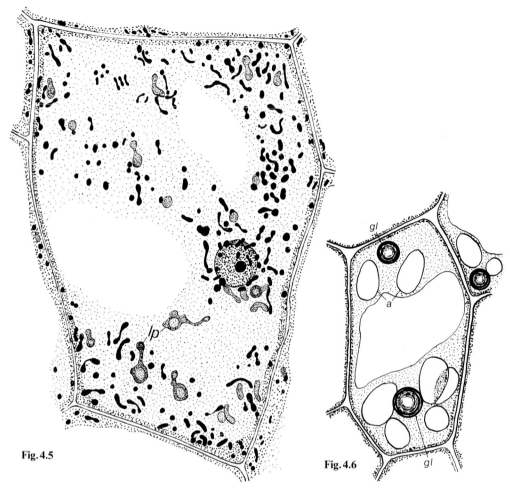

Fig. 4.5

Fig. 4.6

Fig. 4.5. *Cichorium intybus* L., var. "chicorée à café". Phloem parenchyma cell of tuberized root. An example of non-materialized saccharide storage cell of parenchyma (inulin, dissolved in vacuoles). In this tubercle, cortical tissues have been exfoliated and the essential part of the storage is localized in the phloem parenchyma. In the absence of starch, the leucoplasts *lp* remain discrete and little differentiated. The cytoplasm is reduced to a thin peripheral film, in front view on the parts preserved by the section (after BUVAT 1944)

Fig. 4.6. Young scale of *Lilium candidum* bulb. Lipid globules *gl* scattered in the cytoplasm between amyloplasts *a*

rhizome of Jerusalem artichoke or endive: inulin), in these cases plastids are poorly differentiated, they are simple *leucoplasts* (Fig. 4.5).

Lipid reserves (glycerides) accumulate within the cytoplasm itself, in oilseeds *(Ricinus)* or in storage organs such as *Lilium* bulbs (Figs. 4.6) where they constitute large refringent granules, improperly called "oleoplasts".

Protein reserves are not always materialized but sometimes produce enclaves of vacuolar origin, the "aleurone grains", resulting from the dehydration of vacuoles enriched in protids during seed maturation. They can also form "crystalloids"

236

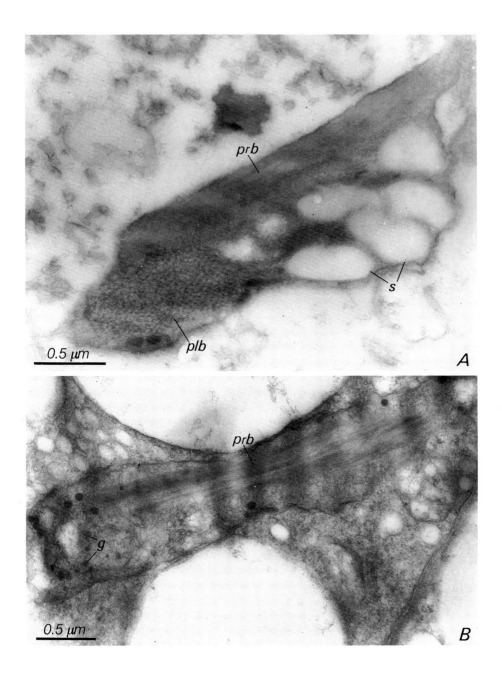

Plate 4.4 A, B. *Phajus wallichii* (Orchidaceae), root proteoplasts.

A Amyliferous proteoplast enclosing a "prolamellar body"; *plb; prb* fibrillar proteinaceous bundle; *s* starch grains; ×38 000. **B** Section devoid of starch and prolamellar body; the fibrillar texture of the proteinaceous bundle is well apparent; *g* plastoglobules; ×32 000.

Osmic fixations

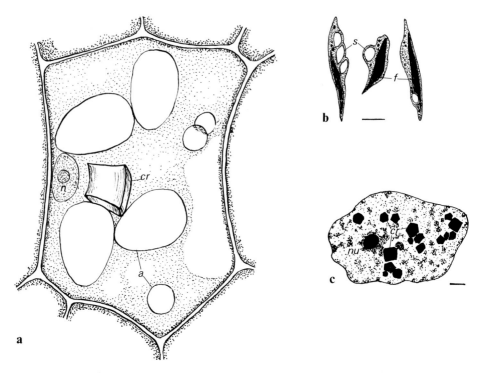

Fig. 4.7 a–c. Proteinaceous crystalloids in parenchyma cells. **a** Young potato tuber cell: *cr* crystalloid in the cytoplasm, between voluminous amyloplasts *a; n* nucleus. **b** Proteoplasts of *Phajus wallichii* (Orchidaceae), enclosing, in addition to starch grains *s*, a fusiform proteinaceous inclusion, *f,* the fibrous texture of which is perceptible with the electron microscope (see Plate 4.4) (drawn from an electron micrograph). **c** *Asplenium fontanum,* frond, parenchyma cell nucleus, containing proteinaceous crystalloids *cr; nu* nucleolus (drawn from an electron micrograph, courtesy of ARSANTO 1973). The *bar* in **b** and **c** represents 1 μm

in the cytoplasm (potato, Fig. 4.7 a), in the plastids (*Phajus,* BUVAT 1959, Fig. 4.7 b and Plate 4.4) and even in the nucleus (*Campanula trachelium, Asplenium fontanum,* Fig. 4.7 c).

In the next section, another more particular type of parenchyma will be mentioned, the parenchyma with cellulose reserves.

4.3 Intercellular Relations and Morphology of Cells

The parenchyma cells have a typical polyhedral shape (often 12 to 14 faces) and are isodiametric, but the angles are rounded off during differentiation and this fact creates a network of *gaps* allowing the gas circulation in the plant.

These intercellular spaces may have large volumes in certain parenchyma, e. g. the lacunar spongy parenchyma of the leaves in relation with the substomatic chambers (Plate 4.1 B; Plate 4.2).

238

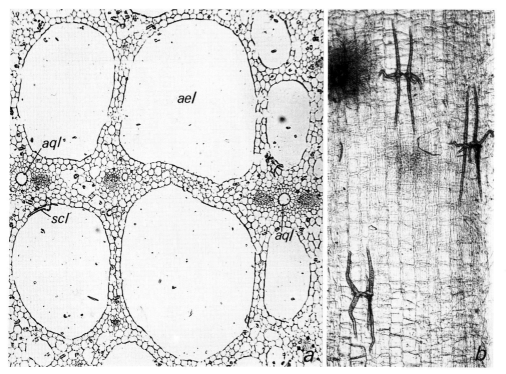

Fig. 4.8 a, b. *Nuphar* sp., petiole, aerenchyma. **a** Septa of about two or three layers of parenchymatous cells in thickness separate aeriferous spacious lacunae *(ael)*; into the conducting strands, an aquiferous lacuna *(aql)* takes the place of the xylem; *scl* sclereid, partially preserved by the cutting, one arm of which is applied against the lacuna edge. **b** Entire sclereids in longitudinal view; the parenchymatous septum is in front view (cf. also Chap. 8, B Fig. 8.19)

Finally in aquatic plants, parenchyma constitute septa of a single or of few layers, separating cavities filled with gas (Fig. 4.8) (aeriferous parenchyma or aerenchyma).

A very particular type of parenchyma is encountered is some seeds with polysaccharide storage. The cell walls are thickened by accumulation of parietal substances in which hemicellulose prevails (MEIER 1958), but they stay alive for a long time and remain connected by means of long plasmodesmata (e. g.: albumen of date, Fig. 4.9). On germination, the walls of the parenchyma become thinner, and this is evidence of the role of the enclosed polysaccharides acting as storage material.

The cell wall resulting from this reversible cellulose thickening is considered as exclusively "primary" by the authors, who believe that the secondary cell wall is an irreversible differentiation.

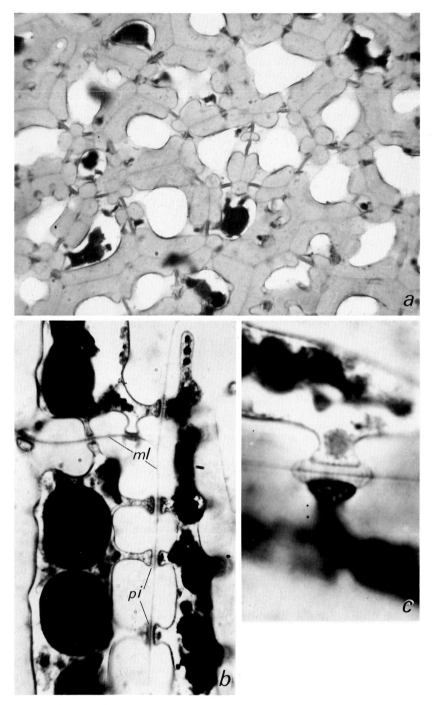

Fig. 4.9 a–c. Albumen of date *(Phoenix dactylifera).* **a** Thickening of primary walls, sparing the pits. **b** Pits *pi* between thickened walled cells; *ml* middle lamellae. **c** Detail of a pit, crossed by numerous plasmodesmata

4.4 Secondary Parenchyma

The subero-phellodermic generating layer, the phellogen, produces one or a few cell layers towards the interior, which develop into *secondary parenchymatous cells.* Their cell walls are more or less thickened but only by pectocellulosic *primary* parietal formations. Their content has the typical aspect of that of the cells of relatively superficial parenchyma: well-developed chloroplasts can be found in particular, when the light is not completely occluded by the periderm which covers them.

4.5 Parenchyma in the Series of Vascular Plants

The histological features of parenchyma tissues are probably similar in the series of vascular plants. Unlike other tissues described further, they have not been considered for the characterization of the large groups. These characteristics are mainly related to the mode of life or to the habit of the plant or organ to which they belong.

Thus, the leaves with more or less vertical lamina have no palisade parenchyma or sometimes a poorly differentiated one. In most cases, the mesophyll is wholly lacunar.

References

Arsanto JP (1973) Nature protéique des structures paracristallines intranucléaires dans les tissus des jeunes frondes de l'*Asplenium fontanum* (Polypodiacée). CR Acad Sci Paris 276 D: 1345–1348

Buvat R (1944) Recherches sur la dédifférenciation des cellules végétales. Ann Sci Nat Bot 11e sér 5 (6): 244 p

Buvat R (1959) Infrastructures des protéoplastes de la racine de *Phajus wallichii* (Orchidacées). C R Acad Sci Paris 249: 289–291

Buvat R (1965) Le cytoplasme végétal. In: Travaux dédiés à Lucien Plantefol. Masson, Paris, pp 81–124

Meier H (1958) On the structure of cell walls and cell wall mannans from ivory nuts from dates. Biochem Biophys Acta 28: 229–240

Reynolds ES (1963) The use of lead citrate at high pH as an electron opaque stain in electron microscopy. J Cell Biol 17: 208–212

CHAPTER 5

Protective Tissues

A. Epidermis

In some cases, one can consider that all the tissues described in the following sections correspond to differentiations necessary to the functioning of parenchyma. Conducting tissues supply the parenchyma with the substances they need, protective tissues spare them the casual troubles of the medium.

In fact, the plant organism is a whole in which each tissue is necessary to the others and the complexity of vascular plants can be interpreted as a consequence of the adaptation to life in *aerial* and *underground* mediums, mediums which have taken the place of the ancient aquatic one.

Above all, protective tissues are most differentiated in organs which are in direct contact with these two mediums. There are two kinds of protective tissues: the epidermis and the suber or cork. This chapter is devoted to the study of *epidermis*.

5.1 Histological, Ontogenic and Chemical Characteristics

The cytological structure of epidermal cells is quite variable, but this fact does not alter the histological and histochemical uniformity of the epidermal tissue.

True epidermis is only found in organs *derived from a cauline vegetative apex* (LINSBAUER 1930), with no marked secondary increase (stems of herbaceous Monocotyledons, young stems of Dicotyledons, leaves, etc.). Generally, epidermal cells form a continuous layer on the surface of these organs, without any *gap* between their anticlinal walls, except in places where particularly differentiated cells constitute a "stoma" (Fig. 5.2).

When the apical meristem includes an independent superficial tunical layer, as in Dicotyledons, in general, epidermal cells are derived from this layer exclusively by means of anticlinal mitoses which harmonize them with the development of underlying tissues. The epidermis differentiation does not depend directly on these tissues. The origin of epidermis is therefore consistent with the concept of "dermatogen" (HANSTEIN 1868, 1870).

However, in some vascular plants the superficial initial cells of the meristem undergo both anticlinal and periclinal cleavages (Pteridophytes, Gymnosperms, Graminaceae). The epidermis is individualized later, after the arrangement of the cells which will generate the more internal tissues or after the initiation of the first leaf buttresses.

The cells which cover the gemmule and the hypocotyl of the "primary body" of the plant (see p. 100) during embryogeny, together with the meristematic cells initiating the epidermis, form the *protoderm* of HABERLANDT (1914).

In addition to the embryonic and ontogenic criteria, epidermal cells are well characterized by a histochemical criterion.

The histochemical characteristic of epidermis is the secretion of an impermeable lipid material called *cutin* in the external cell wall and on this cell wall. In accumulating on the wall, the cutin forms a continuous pellicle, called the "cuticle", perforated only in the stoma areas (Fig. 2.46).

The cutin is composed of numerous mono-, di- or trihydroxylated fatty acids of relatively high molecular weight (14 to 18 carbon atoms); more than 20 were singled out and identified by means of thin-film chromatography by EGLINTON and HUNNEMAN (1968). Most of them are also found in suberin (see below KOLATTUKUDY 1981).

The *cutin* differentiates mainly the epidermis of aerial plants from the superficial layers of roots such as the hair layer, which accumulate *suberin* within their walls but do not form a coating similar to the cuticle.

5.2 Epidermal Cells

Very varied cellular types can be differentiated in the epidermal tissue. The least differentiated cells form the main part of the tissue, but in addition there are *stomatal cells* much more specialized and sometimes very varied appendages: the "trichomes" which may be simple protective hairs or hairs composed in many

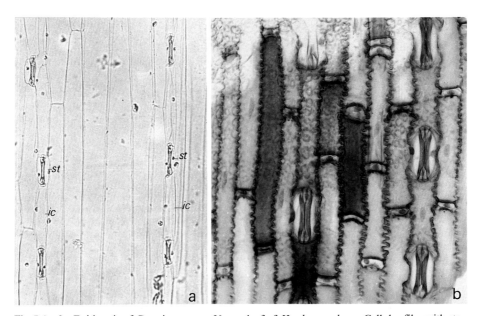

Fig. 5.1 a, b. Epidermis of Graminaceae. **a** Young leaf of *Hordeum vulgare*. Cellular files with stomata alternate with rows of long cells, without stomata; *st* stomata, provided with their two subsidiary cells; *ic* intervenant cells. **b** *Agropyrum glaucum* leaf. Wrinkled cell walls in the files of cells, mainly those without stomata; long cells alternate with short cells which differentiate into "silica cells"

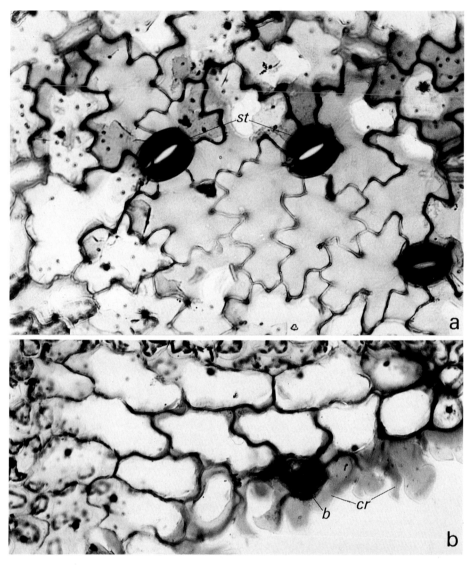

Fig. 5.2 a, b. *Ranunculus acris,* leaf epidermis, paradermal sections. **a** Lower epidermis (dorsal) stomatiferous; sinuous angular cells; *st* stomata. **b** Upper epidermis (ventral); flexuous outline of cells, different from dorsal epidermal cells; *cr* cuticular structure, in transversal section; *b* base of a protective trichome

different ways, or glandular hairs. Besides, epidermis can include suberified cells, secretory cells and mineralized cells, lithocysts and silica cells (see H Prat 1948, 1951).

5.2.1 Common Epidermal Cells

The shapes of common epidermal cells are variable, often in relation to the shape (therefore the growth) of the organ that they cover. Thus, the leaves of Monocotyledons have epidermal cells elongated in the same direction as the leaf (Fig. 5.1). It is the same with stems or petiole or with the vein coating of leaf lamina.

In lamina of horizontal leaves the upper epidermis is frequently very different from the lower one (Fig. 5.2). The last one is often composed of widely sinuous cells (Fig. 5.2 a). In Graminaceae particularly, one can frequently observe epidermal cells with an undulate outline (Fig. 5.1 b). The reasons for such a shape have long been discussed but not acknowledged.

Common functional epidermal cells are alive. They include a large vacuole and their cytoplasm forms a thin pellicle applied against the wall, distended by the nucleus enclosed within. In most cases, it seems that common epidermal cells *have no chloroplasts*. In addition to common chondriosomes, there are only more or less differentiated leucoplasts or leucoplasts similar to chondriosomes (*Hyacinthus,* Fig. 5.3 a). However, it is not a general fact.

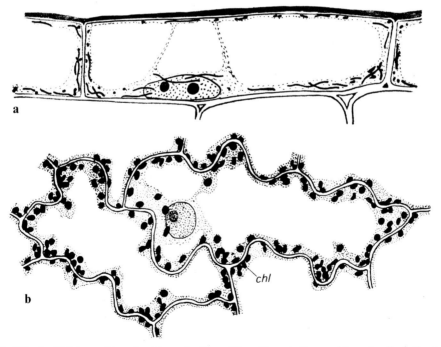

Fig. 5.3 a, b. Cytology of banal epidermal cells. **a** Non-chlorophyllous epidermal cell of *Hyacinthus;* large unique vacuole, undifferentiated plastids, nucleus against the internal cell wall. **b** Chlorophyllous epidermal cells: upper epidermis of *Polypodium vulgare; chl* chloroplasts

Several species belonging to all the great groups of vascular plants have chlorophyllous epidermis. This fact is often the rule in Filicineae, it is frequent in Angiosperms, especially in *shade plants* living in forest underwood (Fig. 5.3 b).

The cutin production being acknowledged as the chemical characteristic of epidermis, we shall now discuss the cytological characteristics of this production.

The cutin is partially excreted at the exterior of the superficial cell wall, in some cases in the anticlinal walls also, by a process still unknown. The cutin forms an impermeable pellicle applied against the wall, i. e. the cuticle. Moreover, the external wall of epidermal cells becomes impregnated with cutin and sometimes the other walls also. The cutinization of the wall is achieved by apposition of more of less cutinized pectocellulosic layers alternating with layers richer in pectin giving the membrane a lamellar aspect.

The cuticle itself, formed by the secretion out of the wall, also often has a laminated structure, due to the discontinuous elaboration. The cuticle surface is sometimes smooth and uniform, or almost so, but in other cases, it is ornamented and rough (Fig. 5.4). It often shows ripples, the origin and role of which have been related to the growth modalities (see Fig. 9.5 b, d). Such ripples could be produced by an early active secretion, especially on flower petals, allowing the enlargement of cells. In other cases, they could be the expression of a growth oriented in the cell growth direction as it could be the case in the staminal hairs of *Tradescantia* (MARTENS 1934).

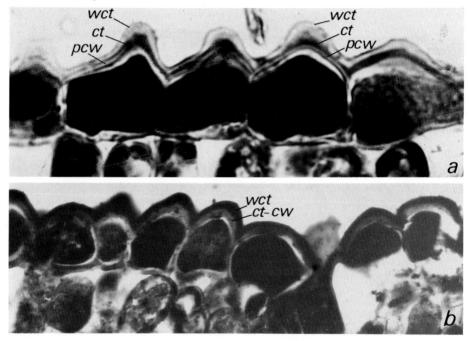

Fig. 5.4 a, b. *Tamarix gallica,* epidermal coating of young leaves. **a** Fixation: Regaud – hematoxylin; *pcw* pectocellulosic cell wall; *ct* cuticle; *wct* coating mostly composed of waxes. **b** Id, but after action of Sudan Black B, which stains lipid substances; the waxy coat *wct* is strongly stained in black; the whole cuticle-cell wall *(ct–pcw)* appears homogeneous and less reactive. The epidermal cells are tanniniferous and their vacuole is contrasted by the bichromate of the fixative

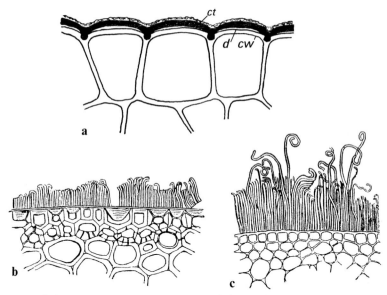

Fig. 5.5. a Leaf of *Musa* (banana tree). Detachment *d* of the cuticle *ct* from the external epidermal cell wall *cw*. **b, c** Extracuticular waxy secretions in the shape of rods or filaments perpendicular to the epidermal surface of sugarcane at an internode (**b**) and at a node (**c**) (in GUILLIERMOND et al. 1933, after DE BARY 1884)

On the internal side, the cuticle often extends through the middle lamellae into the anticlinal walls which separate the cells (Fig. 2.46). It has been suggested that these extensions could correspond to the filling of the tears made in the first cuticle layers, tears related to the growth, which would be selectively found above the anticlinal walls.

The cuticle formation supposes that the precursors of this membrane pass through the external wall and sometimes the radial walls of the epidermal cell. In the radial and internal walls of these cells, there are plasmodesmata, but in the external wall their existence is doubtful, although SIEVERS (1959) described them under the heading of *ectodesmata*. However, such ectodesmata, if they constitute the passage of the cutin constituents, do not grow through the already completed layers of the cuticle; the constituents are therefore released in the internal layers most recently completed and one can speculate whether they can circulate towards the periphery (FRANKE 1961). It has been considered that such precursors are more or less fluid-saturated fatty acids, which can become solid by polymerization when they are in contact with air (FREY WYSSLING and MUHLETHALER 1959).

In fact, the cuticle is not strictly lipophilic; it is slightly hydrophilic and sensitive to the atmospheric humidity due to the polar radicals of its fatty acids.

The cuticle is sometimes partially detached from the cellulosic wall (leaf of banana tree), remaining adherent due to the laminae which penetrate into the anticlinal wall (Fig. 5.5 a). It is the air layer located between the cuticle and the cell wall which gives the surface of certain leaves their silvery appearance. A similar detachment can be observed in some epidermis which have become secretory.

Finally, some productions with comb shapes, scales or digitiform appendages have been erroneously attributed to the cuticle, they are in fact folds of the external cell walls, lined by the cuticle.

The cuticle has a very variable thickness; it is non-existent in the immerged epidermis of aquatic vascular plants, very thin in the young organs and in the protoderm but it can become thicker than the epidermal cells which secrete it. For example, this is the case of the conifer leaves. The thickening typically consists of successive cutin layers, but, in addition, the cuticle is often covered by various secretions, particularly *waxes* which make it more hydrophobic and impermeable. This waxy coating may severely impair the efficiency of phytosanitary treatments. Generally, waxes are found both in the cuticle and on the surface as well as in the cutinized wall. They frequently produce variable structures on the organ periphery: granules, intermingled filaments, cristae or fingerlike productions provided with hooks (Fig. 5.5b, c). The waxy layer may develop in excessive length as on the palm tree leaves with "carnauba wax", *Copernicia cerifera* of Brazil on which it reaches several millimetres.

Other secretions may accumulate on the cuticle: sodium chloride on the leaves of salt plants, resins, gums or mucilages which make the organ surface sticky *(Silene nutans)* or even polyterpenes generating rubber *(Eucalyptus)*.

In numerous plants (*Equisetum,* Graminaceae, Cyperaceae, palm trees. etc.) the walls of the epidermal cells become impregnated with *silica*.

Finally, the walls of epidermal cells may lignify, especially on conifer needles and on the leaves of vascular plants of ancient origin, considered more or less "primitive" (Cycadales, Joncaceae, Cyperaceae). In Graminaceae, the lignification of vascular sheaths frequently involve the epidermis.

5.2.2 Stomatal Cells (Guard Cells)

Stomata are apertures in the epidermis, facilitating the gas exchange of the plant with the external medium. They are typically composed of two "stomatal cells" or "guard cells", kidney-shaped, with an aperture left between them or "ostiole" (Fig. 5.6, Plates 5.2 A and 5.4). Through this pore, the internal atmosphere within the intercellular spaces of parenchymas communicates with the exterior. Generally, a larger space under the ostiole constitutes the "substomatal chamber" (Fig. 5.6).

Fig. 5.6 a–h. Examples of stomata. **a, b** *Agropyrum glaucum;* **a** front view of a stoma; **b** transversal section in the narrow median part of the stoma cells *st,* according to *arrows* of **a'**, where the cell walls are strongly thickened. **a', b'** Drawings after **a** and **b**; *n* nuclei, stretched in the median part of the stomatal cells, but lentil-shaped in the subsidiary cells *su; sch* substomatal chamber. **c, d** *Persica vulgaris;* lips formed by the cuticle, thickenings of the stomatal cell walls on the ostiole face. **e** *Linum usitatissimum;* stem stoma, roughly superficial, *col* collenchyma. **f** *Salvia pratensis;* stem stoma, distinctly prominent over the epidermis. **g** *Nerium oleander;* stomata localized into a stomatiferous crypt *cry* on the dorsal side of the leaf. **h** *Pinus silvestris;* stoma sunk into the mesophyll of a young needle. *ct* Cuticle; *ep* epidermis; *o* ostiole; *tr* protective trichomes (in the stomatiferous crypt), other abbroviations as in **a, a', b, b'**

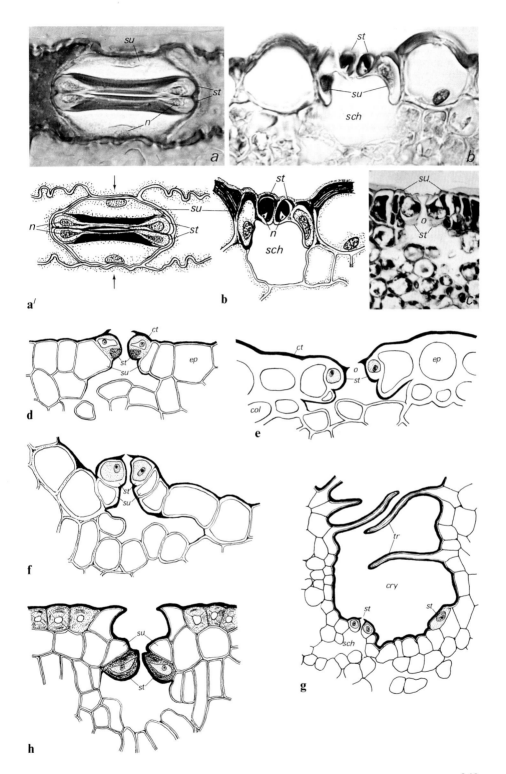

5.2.2.1 Cell Wall Differentiation of Stomatal Cells

Stomatal cells are specially differentiated epidermal cells. Besides their particular shape, they differ from common epidermal cells by their parietal and protoplasmic structures. Firstly, they always have well-differentiated chloroplasts generally loaded with starch, even when common cells are colourless. Moreover, the wall thickenings of the guard cells are localized in such a manner that the turgor variations change their shape. In the simplest cases, the wall is thicker on the side of the ostiole than on the opposite side (Fig. 5.6), with the result that the thin wall bulges away all the more as the turgor increases, and that the whole cell bends, whereas the ostiole increases.

In other cases (some Graminaceae) stomatal cells appear to be elongated with bulging extremities. The middle part has thickened walls, whereas the ends have thin walls sometimes incomplete so that the cytoplasms of the two cells are confluent (BROWN and JOHNSON 1962). Increase in turgor causes a swelling in the ends and parts the rigid middle walls away, enlarging the ostiole (Fig. 5.6). Other wall differentiations probably control the stomatal aperture. Thus, in conifers, stomatal cells in which the cavity is also widened at the end, are sunk into the mesophyll by subsidiary cells and in both walls are partly lignified (rigid) and partly cellulosic (elastic) so that the turgor changes operate the opening and closing of the ostiole (Fig. 5.6h; FLORIN 1931).

Like the other epidermal cells, stomatal cells are lined with cuticle which spreads down into the ostiole and lines the external wall of the substomatal chamber. Added to the strictly parietal differentiations, the cuticle can produce cristae or lamellae on the front side of the stomatal cells which modify the shape of the ostiole (Fig. 5.6).

Stomata can form small prominences on the epidermis (*Nicotiana,* cf. ESAU 1965) or, on the contrary, be sunk into the underlying parenchyma, as on the needles of conifers (Fig. 5.6h) and on the leaves of many species adapted to dryness and to strong light (*Nerium oleander;* Fig. 5.6g). The relationship of stomatal cells to the "subsidiary cells" (see below) and to the common epidermal cells is progressively established during the differentiation of the epidermis.

5.2.2.2 Ontogenesis and Classification of Stomata

Stomatal cells arise through particular divisions of undifferentiated cells of the protoderm or the young epidermis. Both "guard cells" arise through the *symmetrical* division of a "stomatal mother cells". This may be a young epidermal cell similar to common cells which initiates both stomatal cells through a single symmetrical mitosis. The two "guard cells" are then surrounded by common epidermal cells, all similar, which are not derived from the same *initial cell* as the guard cell. This ontogenic type of stomata is said to be *perigenous* (FLORIN 1958; PANT and MEHRA 1965). In Gymnosperms, the completed stomatal systems of *haplocheile* type are related to this category (FLORIN 1931, 1933).

In other ontogenic modes, the *stomatal mother cell* is derived from an *initial cell,* also called *meristemoid cell* (BÜNNING 1952, PANT and MEHRA 1965) which divides *asymmetrically* a determined number of times and finally produces a small

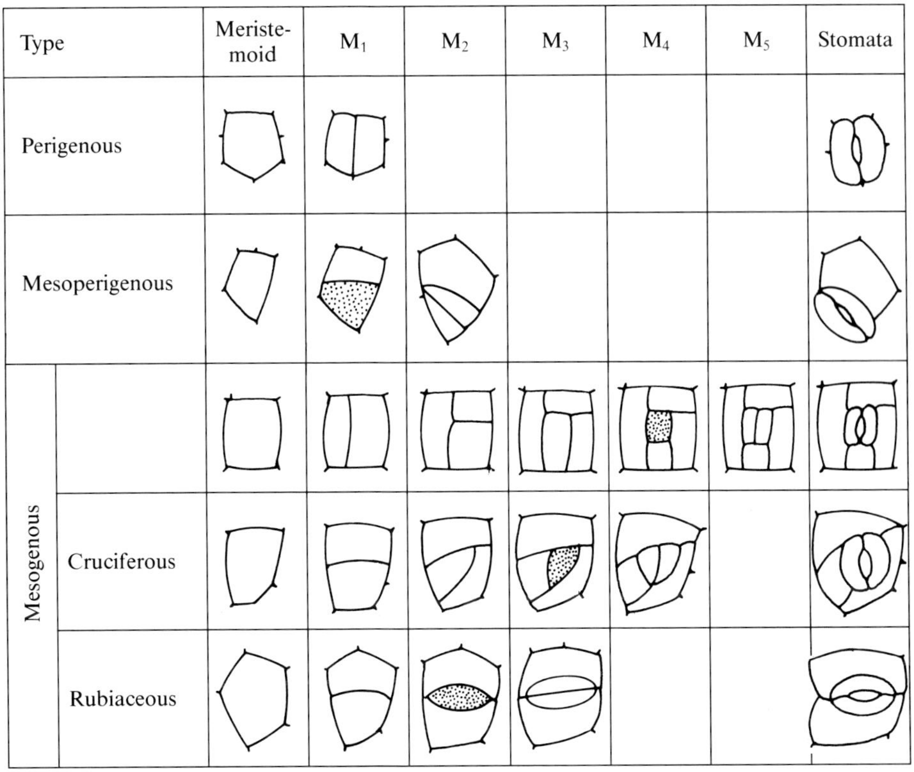

Type		Meristemoid	M₁	M₂	M₃	M₄	M₅	Stomata
Perigenous								
Mesoperigenous								
Mesogenous								
	Cruciferous							
	Rubiaceous							

Scheme 5.1. Ontogenic types of stomata derived from the so-called meristemoid cell (BÜNNING 1952). This cell can generate the stomatal cells directly through symmetrical mitosis (perigenous types) or can divide several times to produce the mother cell of the stoma *(grey)* which gives rise to the guard cells through symmetrical mitosis (mesoperigenous and mesogenous types).

M_1 to M_5 Successive mitoses of the meristemoid cell, only the last one is symmetrical (modified and simplified after PANT 1965)

cell which is the stomatal mother cell, which *symmetrically* divides into two similar stomatal cells. In such cases, the *subsidiary* cells are the sister cells of the stomatal cells that they surround. This type is called *mesogenous* and it seems that the *syndetocheile* type of FLORIN is related to it. In this last type the meristemoid cell produces the stomatal mother cell and two subsidiary sister cells of the mother cell (Scheme 5.1).

Other stomatal systems include both subsidiary cells, sisters of the stomatal mother cell and epidermal cells which are not derived from the meristemoid initial; they are said to be of *mesoperigenous* type (PANT and MEHRA 1965).

The main categories have themselves been divided into a great number of types according to the ontogeny and the elaboration of the mature apparatus. The number of the subsidiary cells has been noted particularly (VESQUE 1889; METCALFE and CHALK 1950). No subsidiary cells are present in the *ranunculaceous* or *anomocytic* type; however, there are three subsidiary cells, one smaller than the others in the *cruciferous* or *anisocytic* type. Moreover, the orientation of the sub-

251

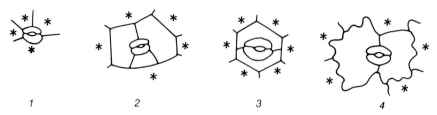

1   2   3   4

Scheme 5.2. Main types of stomata according to:

VESQUE (1889)	METCALFE and CHALK (1950)
1. Ranunculaceous	anomocytic
2. Cruciferous	anisocytic
3. Rubiaceous	paracytic
4. Caryophyllaceous	diacytic

Asterisks indicate the common epidermal cells which surround the cells derived from the meristemoid initial (after LANDRÉ 1972)

sidiary cell walls compared to the orientation of the ostiole is also taken into account: in the *rubiaceous* or *paracytic* type, the cell walls and ostiole are parallel, in the *caryophyllaceous* or *diacytic* type, they are perpendicular (Scheme 5.2).

From extensive ontogenic study several types within the former categories can be distinguished: about 10 according to PANT and MEHRA (1965) and up to 26 according to FRYNS-CLAESSENS and VAN COTTHEN (1973). This is a matter for specialists and is not discussed further here. One of the main interests of these studies concerns the systematics and phylogeny. Thus, in Umbellifers, GUYOT (1966) showed that the types of stomata detect a phylogenic evolution which fits well with the comparative study of the other productions such as the pollen structures.

In Monocotyledons, STEBBINS and KUSH (1961) acknowledged phylogenic indications in the variations of composition of the stomatal complexes found on leaf epidermis. Species with four subsidiary cells or more (e.g. Commelinaceae) are regarded as most primitive compared to Graminaceae in which the stomatal apparatus includes two subsidiary cells, and to the *Allium* genus which has none.

In Prephanerogams and mesozoic Gymnosperms, the leaf stomata of these morphologically similar fossils enabled FLORIN (1933) to distinguish the leaves of Cycadales in which stomata are *haplocheile* from those of Bennettitales which have a *syndetocheile* stomatal apparatus.

▷

Plate 5.1 A, B. *Sinapis alba.*
A young stoma in the process of differentiation. The two stomatal cells have a dense cytoplasm, due to the abundance of ribosomes. The ostiole is not yet differentiated. Common oganelles are well represented: *chl* granar chloroplasts, devoid of starch; *d* dictyosomes; *m* very structured mitochondria; *mt* microtubules, in great number around the whole periphery of the cytoplasm; *n* nucleus, central and relatively voluminous; *er* endoplasmic reticulum; *v* vacuoles; × 15500. Paradermal section; fixation: glutaraldehyde-O_5O_4; contrast: KMnO$_4$; (courtesy of LANDRÉ 1969 a).

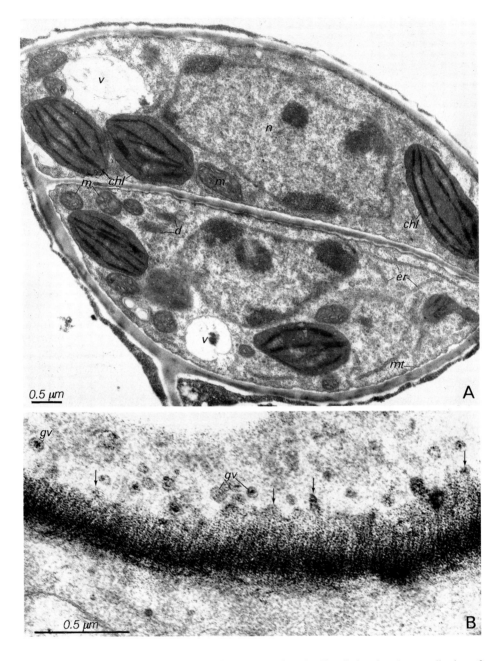

B Stomatal cell at the stage of ostiole differentiation; dorsal cell wall showing the contribution of polysaccharide Golgi vesicles, *gv,* to the edification of the cell wall *(arrows);* the fibrous texture of this cell wall is well visible; × 50000. Technique PATAg according to THIÉRY (1967) for the detection of acid polysaccharides (courtesy of LANDRÉ 1970).

Schematically, one can consider that the evolution leads from perigenous types to mesogenous types, which are numerous.

In many species, the same organ may include several different types of stomata (*Solanum nigrum,* LANDRÉ 1976) which appear successively during its development but remain in the epidermal tissues of adult organs. This is a difficult question posed to systematists.

The first stomata differentiate early, especially on leaf buttresses; they do not arise all at once but successively over a long period of growth. In the reticulated, veined leaves, stomata are mixed at different stages of development (most Dicotyledons, a few Monocotyledons).

In leaves with parallel veins, stomata are more and more differentiated, *from the end to the base;* they are arranged in longitudinal rows (most Monocotyledons, a few Dicotyledons).

5.2.2.3 Ultrastructure of Stomatal Cells

Few publications describe the ultrastructure of the stomatal cells and their development through stomata differentiation.

On the cotyledon epidermis of *Sinapis alba* (LANDRÉ 1969a, b, 1972), which has a stomatal apparatus of anisocytic type, before the ostiole formation, the two cells of a young stoma have a dense cytoplasm due to the abundance of ribosomes. Within, common organelles are well represented: rough endoplasmic reticulum, densely structured mitochondria, not very active dictyosomes, numerous granar chloroplasts still starch-free and finally, the vacuolar apparatus reduced to a few vacuoles of small size; microtubules line the plasmalemma on the hyaloplasm side. The relatively voluminous and dense nucleus is at the centre (Plate 5.1 A).

The ostiole aperture is indicated by the thickening of the middle lamella of the cell wall common to both cells, then of the primary cell wall (Plate 5.2 B). From this moment, chloroplasts accumulate starch grains, whereas the stroma becomes clearer, suggesting a hydration process, confirmed by the enlargement of vacuoles and some type of mitochondria hypertrophy (Plate 5.2 A). Membrane remnants in vacuoles and myelinic bodies in mitochondria indicate that the differentiation of guard cells is accompanied by lytic processes.

Plate 5.2 A, B. *Sinapis alba.*

A Mature differentiated stoma, open ostiole. Differences from the young stoma: clearing of the cytoplasm, extensive development of the vacuolar apparatus and storage of numerous starch grains in chloroplasts. V_1, V_2, V_3 vacuoles of the three cells which surround the stoma; $\times 5500$. Fixation: glutaraldehyde-OsO_4, contrast: $KMnO_4$; (courtesy of LANDRÉ 1972).

B Young stoma; beginning of lysis of the cell wall middle lamella, at the origin of the ostiole *(arrow); s* starch grains, still of small size, in chloroplasts; *n* nuclei; *v* vacuoles, contrasted with the fixative; $\times 12300$. Fixation: $KMnO_4$ (courtesy of LANDRÉ 1969a)

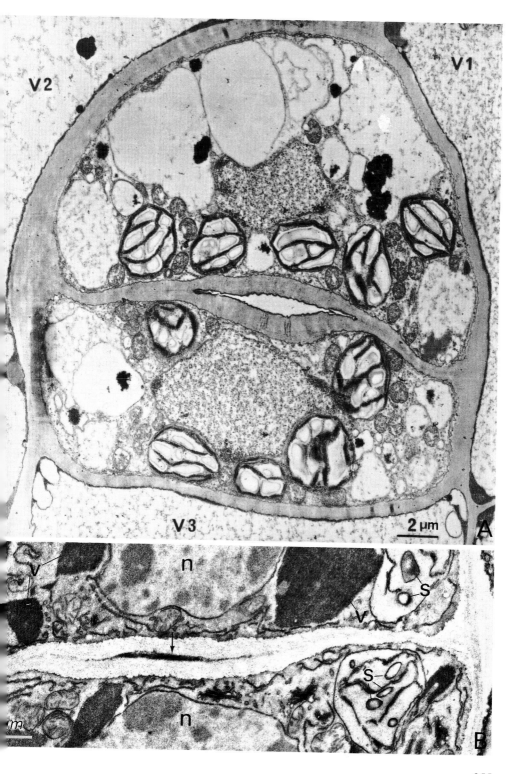

V2

V1

2 µm

A

V3

v

n

s

V

m

s

n

B

255

Conversely, the onset of wall differentiation coincides with the high activity of dictyosomes and a marked development of microtubules (LANDRÉ 1969 *b*, 1970, 1972). The dictyosomes generate numerous vesicles of three types: (1) small and dense, (2) medium-sized with a clear content, (3) larger, irregular with a transparent content also (Plate 5.3 A, B).

The PATAg technique used for the detection of polysaccharides with vic-glycol groups (THIÉRY 1967) stains the small vesicles uniformly and shows a nucleoid precipitate in the larger vesicles which are confluent. The vesicles are integrated into the plasmalemma and their content is released into the periplasmic space, thus thickened and stained like the vesicles, but in which reactive polysaccharides become fibrillar (Plate 5.1 B). Thus, the Golgi apparatus plays an obvious role in the elaboration of stomatal cell walls, but the amount of endoplasmic reticulum diverticles neighbouring these walls suggests that it is also involved in such processes (LANDRÉ 1969 *a*, 1972).

The large amount of microtubules shown during the development of stomatal cell walls is characteristic; it does not exist in subsidiary cells or in common epidermal cells. They progress along the plasmalemma at about 200 Å for the most internal ones, but their tips are probably attached to the plasmalemma, but this is usually difficult to see on electron plates. Between the bundles or the layer of parallel microtubules, there are numerous vesicles similar to those derived from dictyosomes, vesicles apparently guided by the microtubules towards the plasmalemma. The plasmalemma surface is, in fact, very rough, because of anastomoses with the vesicles and the multivesicular cytoplasmic bodies and because of the indentations which seem to correspond to the fixing points of the microtubules (LANDRÉ 1969 *b*). Microtubules, parallel to the wall fibrils, were observed in several examples (LLOYD 1984).

Finally, the statistical use of cytophotometric measurements revealed a particular development of stomatal cell nuclei and epidermal cells of *Solanum nigrum* (LANDRÉ 1973 *a, b*, 1976). This epidermis is composed of common diploid cells, tetraploid cells and heteroploid cells which no longer segment, the nuclei of which include a DNA excess added to values 2 C and 4 C. These *amplification* phenomena occur in several animal and plant tissues in which they appear as an element of differentiation.

According to species and stomatal types, the cytological evolution of the "guard cells" during their development is obviously variable, for example with regards to the earliness of amylogenesis. However, it seems that the main features, concerning particularly plastids and the vacuolar apparatus, are generalized if we consider the *physiological states of the stomata..*

Plate 5.3

A Stoma of *Sinapis alba*. Cytoplasmic areas of two young stomatal cells, still at the onset of differentiation. Cytoplasm rich in ribosomes, partly associated with the endoplasmic reticulum, *er*, itself abundant; dictyosomes *d* equally numerous, actively producing three kinds of vesicles: small ones, with dense content *(thin arrows)*, slightly larger ones, with clear content *(arrow heads)* and still larger ones *v*; little differentiated chloroplasts devoid of starch. Ostiole, *ost*, not yet differentiated, in the cell wall separating the two cells. *Asterisks* indicate plasmalemma winding resulting from the exocytosis of clear Golgi vesicles (see Plate 2.4 A).

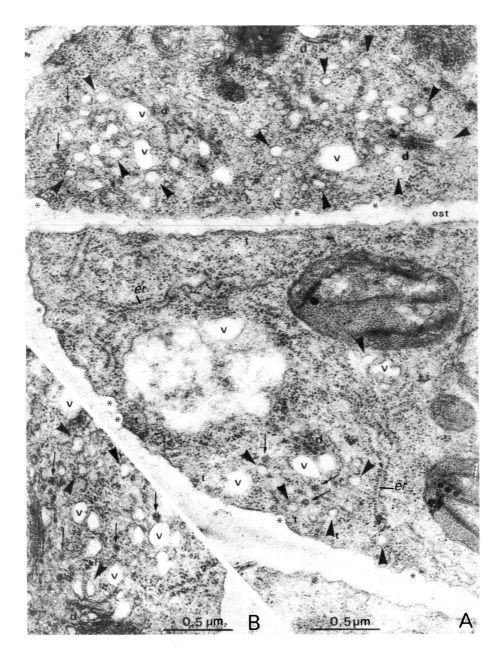

B Detail of an area presenting two dictyosomes, surrounded by three kinds of vesicles, as in **A**; × 37 000. Fixation: glutaraldehyde-O₅O₄; contrast: KMnO₄ (courtesy of LANDRÉ 1970)

5.2.2.4 Cytophysiology of Stomata

The closing and opening mechanisms in stomata are still poorly known in spite of the extensive research on this subject (see LOUGUET 1962, 1974). This book excludes a discussion of the numerous hypotheses proposed. However, turgor, K^+ ions, carbon anhydride and light seem to play an important part in such mechanisms. The review of the literature yields a few cytophysiological facts apparently well established.

In light microscopy, the in vitro observations either directly or with neutral red staining yields data consistent with the description of ultrastructural cytology summarized below (MOURAVIEFF 1955; GUYOT and HUMBERT 1970; HUMBERT 1976).

1. Closed stomata have a reduced vacuolar apparatus divided into numerous small vacuoles either globular or reticulated.
2. Open stomata include one or two vacuoles per cell, which fill the main part of the cellular volume.
3. Half-open stomata have an intermediate but very inconstant vacuome which changes continuously in shape (HUMBERT 1976).

These results obtained from direct observations on several species are not always consistent with the vital stainings; in fact, neutral red staining causes fragmentations or confluences between vacuoles, as well as precipitates which do not reflect the outline of vacuoles.

From stomata in which the aperture or closure was physiologically controlled (HUMBERT et al. 1975, 1978) the ultrastructural study with electron microscopy (HUMBERT 1976) confirmed these variations (Plate 5.4 A–D). Moreover, the same author, under light microscopy, obtained the progressive aperture of stomata of *Anemia rotundifolia* (Schizeaceae) in 40 to 60 min, under the influence of K^+ ions in the form of potassium nitrate. The phenomenon occurs along with the disappearance of the initial vacuolar pattern and the appearance of large vacuoles.

The role of K^+ ions is very debated, but from what precedes, one can imagine that the stomatal changes occur together with water circulation either among the constituents of the stomatal cells or between these cells and the subsidiary cells or the neighbouring epidermal cells. In fact, the hyaloplasm and the mitochondrial stoma are denser in open stomata, whereas mitochondria cristae are more dilated (HUMBERT 1976). The other cellular organelles, particularly plastids, are slightly but *not symmetrically* modified on the aperture and closure of the ostiole.

Plate 5.4 A–D. *Pelargonium hortorum;* cytological modifications of guard cells, in relation to opening and closing movements of stomata.

A Stoma which has been closed for more than 8 h; vacuolar apparatus divided into numerous vacuoles *v,* with a finely granular content; *p* plastids enclosing voluminous starch grains *s;* nucleus showing little condensed chromatin; × 3550.
B Beginning of opening movement; expansion of vacuoles *v* and suggestive images of anastomosis; × 4250.

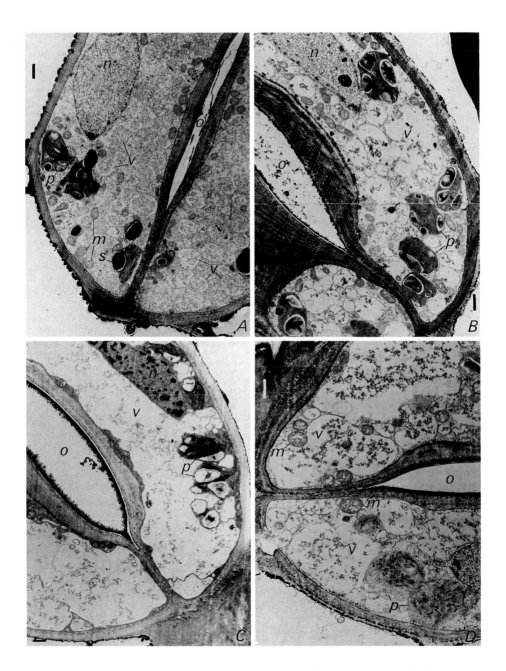

C Open stoma; each cell encloses a unique vacuole, the precipitated content of which is very scattered; nucleus enriched in condensed chromatin, forming disseminated amounts in the nucleoplasm; ×3550.

D Stoma in the course of closing movement, beginning of division of vacuolar apparatus into multiple vacuoles, with still diluted content; mitochondria *m* and plastids *p* clearly dilated (compare with *A*); the granar ultrastructure of the plastids is more apparent; *o* ostiole; ×5000.

(Courtesy of HUMBERT et al. 1978)

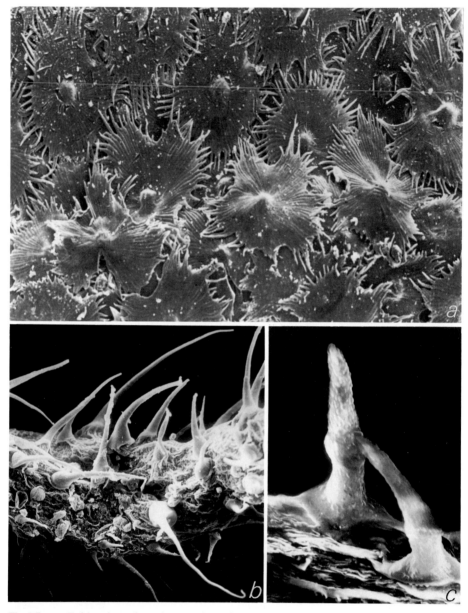

Fig. 5.7 a–c. Epidermis surfaces in scanning microscopy. **a** *Elaeagnus angustifolius:* peltate trichomes. **b** *Thymus vulgaris,* protective trichomes on a young leaf epidermis. **c** *Thymus vulgaris,* calyx trichomes, showing cuticular structure of the apical cell

Fig. 5.8 a–e. Examples of protective leaf trichomes. **a, b** Dorsal epidermis of *Platanus orientalis* leaf. **a** Morphology of a "candelabrum" trichome; **b** front view of the dorsal epidermis of a young leaf showing the density of the protective coat. **c** Dorsal epidermis of *Olea europaea* leaf;

260

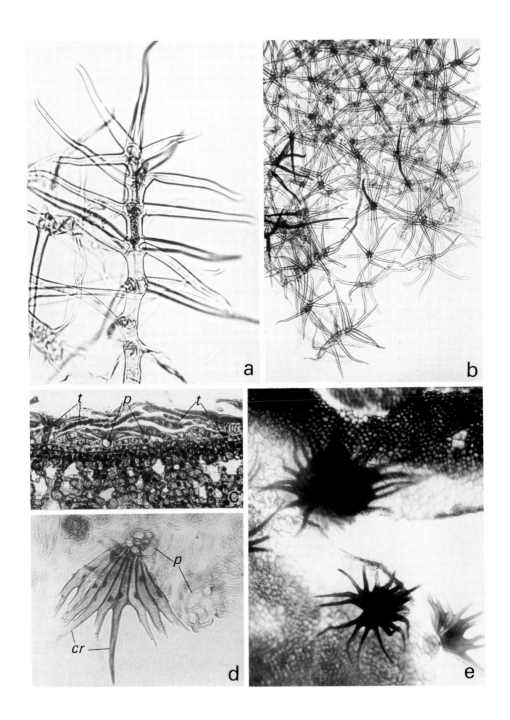

section perpendicular to the leaf lamina: strong density of peltate trichomes *t,* forming a thick covering; *p* monocellular pedicels. **d, e** *Elaeagnus angustifolius,* paradermal sections of leaf laminae. **d** Partial section of a peltate trichome, part of the crown *cr* and two pluricellular pedicels *p.* **e** Aspect of trichomes in front view

5.2.3 Trichomes

The increase of cells derived from the apical protoderm can be followed or not by periclinal cleavages which produce epidermal appendages of extremely varied shapes. These productions can be divided into two groups: the "protective" trichomes without secretory cells and the "glandular" trichomes with secretory cells.

5.2.3.1 Protective Trichomes

Protective trichomes are very diverse, they can be unicellular or multicellular. The first ones are simple or branched. They sometimes look like the absorbing hairs of roots, but are not as developed (Fig. 5.7b, c; Fig. 5.14a).

Multicellular hairs may be composed of a plain file of cells, but they are often much more complex: starred trichomes, peltate, branched trichomes etc. (Figs. 5.7a, 5.8).

The cytological aspect of these hairs is similar to that of common epidermal cells. They are lined with a cuticle which is in continuity with the cuticle of the epidermal surface.

Such "protective" trichomes develop often very early on the outline of vegetative apices (Fig. 5.9). Initiatied on the upper layer of the meristem, i.e., the protoderm, in places where it is still undifferentiated, they sometimes include only one cell or a few cells, but which become giant cells (Fig. 5.9). Of course, these tri-

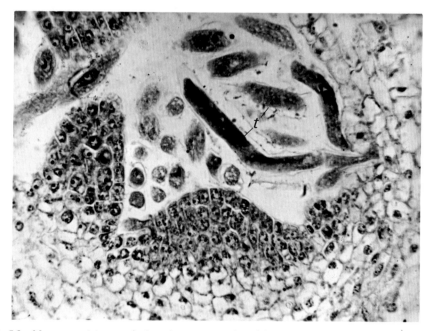

Fig. 5.9. *Myosurus minimus,* apical meristem; protective trichomes *t* constituted of giant cells, surmounting the apex, ×500. Fixation: REGAUD; staining: hematoxylin

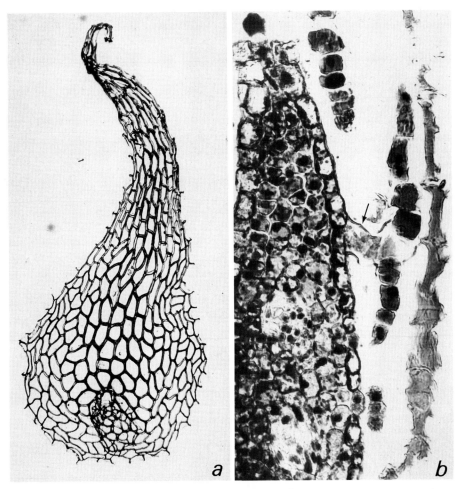

Fig. 5.10 a, b. *Ceterach officinarum,* epidermal leaf-shaped scale of this fern leaf; **a** front view; **b** profile, on a section of the leaf showing the pedicel *(arrow)* inserted on the lower (dorsal) epidermis, and the lamellar extension, constituted of a single layer of cells. Sections of the other scales are visible, one of them being superposed on the first one. Fixation: NAVASCHIN; staining: hematoxylin

chomes protect the apical meristem that they cover. They not only provide shelter from light, which is unfavourable to the meristematic state, but also from the dryness of the aerial medium. They create a relatively constant atmosphere above the apex.

The protecting role of trichomes is particularly obvious on young organs in development, the tissues of which are more sensitive to atmospheric and light variations. (Fig. 5.8 b)

Frequently, the trichomes grow old earlier than the other epidermal cells, they die soon and seem to disappear on adult organs. The temporary hairiness of the young leaves of the beech tree is a good example.

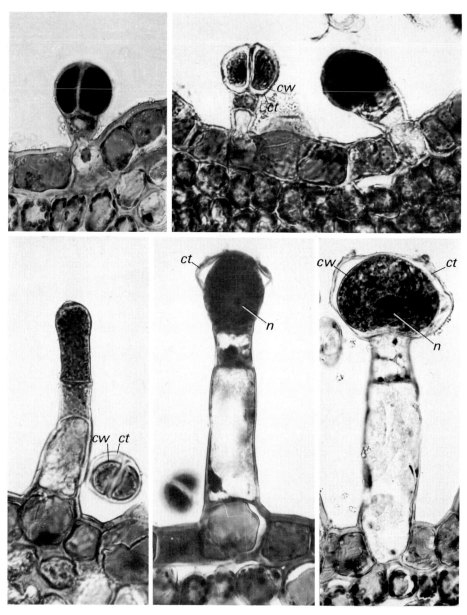

Fig.5.11. Various types of secretory trichomes of the *Salvia pratensis* stem. *ct* cuticle, becoming unstuck from the cell wall *cw; n* nucleus in the secretory head cells, strongly chromophil. Fixation: REGAUD, staining: hematoxylin

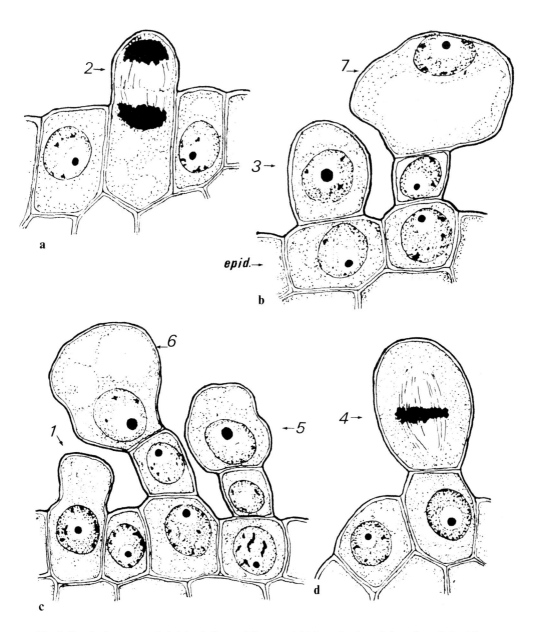

Fig. 5.12 a–d. Ontogeny of *Atriplex halimus* trichomes. *1* **(c)** Lengthening of the primordial epidermal cell. *2* **(a)** Division of this cell, giving birth to the trichome initial cell *3* **(b**). *4* **(d)** Division of the initial cell, producing the pedicel cell and the apical cell *5* **(c)** which differentiates later (*6, 7* **b, c**); (after SMAOUI 1975)

The epidermal formations sometimes attain a significant size and develop foliated scales. In particular, this is the case on the younger parts of the sporophytes of several ferns, but such productions are also found on the epidermis of a few Angiosperms *(Hippophae rhammoides)*. In various cases, in which these organs are very developed, they arise on a protuberance built with the participation of *subepidermal cells:* they are no longer plain epidermis productions (ferns; Fig. 5.10).

5.2.3.2 Secretory or Glandular Trichomes

The formation of numerous trichomes involves the differentiation of part of their cells (distal cells) into secretory cells. Secretory trichomes are generally composed of a foot, a body or pedicel and a head including the secretory cells (Fig. 5.11). Moreover, all of these cells secrete a cuticle in continuity with the epidermis cuticle. The shapes of these small organs are also very variable. Most often, secretions correspond to terpene substances (essential oils, resins, etc.), that accumulate within the cytoplasm, taking the shape of refringent globules, and they can also be stored between the cuticle and the cell wall (*Ballota;* Fig. 5.11). Cytologically, the foot and body cells of these trichomes are comparable to common epidermal cells. Secretory cells have a denser, opaque content and are poorly vacuolated.

Secretory trichomes will be discussed in more detail in Chapter 9 which is devoted to secretory tissues.

5.2.3.3 Development of Trichomes

When they are multicellular, the trichomes arise from the pericline division of an epidermal cell (Fig. 5.12/1, 2). A basal cell is thus generated forming the foot. The upper cell divides again, when the trichome has more than two cells, to produce the pedicel cell and the apical cell (Fig. 5.12/3-5).

The cleavages of the upper cell and its derivatives are anticline or pericline depending on the species and the time at which the trichome is produced.

5.2.4 Secretory Epidermis

The differentiation into secretory cells may occur in epidermal cells of a common morphological type. Generally, the cells of the epidermal surface become secretory all together. Thus, there are secretory epidermal tissues on the leaf ventral face and on the stipules of *Salix* in which the epidermis has a "palisade aspect". The epidermis of rose petals also secrete essential oils which produce the fragrance. These oils may be stored between the distended cuticle and the upper wall. Secretory surfaces develop sometimes parenchymatous protuberances derived from subepidermal tissues, thus producing some types of glandular outgrowths (*Nerium* petioles) (Fig. 5.13). Nectaries which have a sweet secretion belong to this type. Secretory epidermis will be discussed in more detail in Chap. 9.

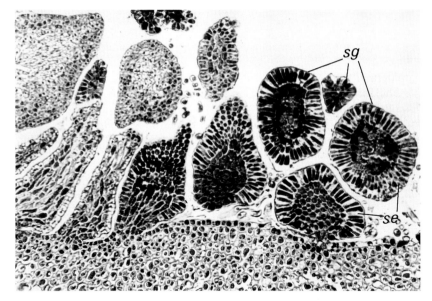

Fig. 5.13. Secretory glands *sg*, situated at the base of *Nerium oleander* petioles. *se* Secretory epidermis surrounding the glandular fingerlike protusions

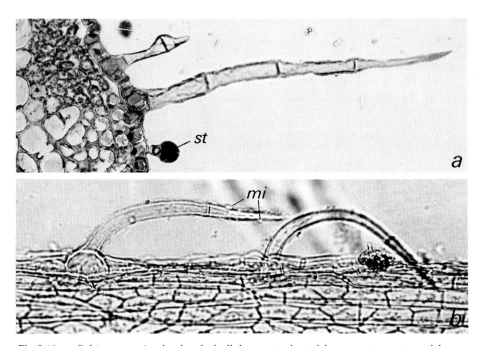

Fig. 5.14. a *Salvia pratensis*, simple pluricellular protective trichome; *st* secretory trichome. **b** *Cannabis sativa*, mineralized trichomes; *mi* mineral incrustations

267

5.2.5 Lithocysts

Epidermal cells may be loaded with mineralized concretions and may develop into "stony cells" or *lithocysts*. According to species, lithocysts are generated by basic epidermal cells *(Ficus)* or by epidermal hairs (*Humulus, Cannabis*, etc., Fig. 5.14b). The formation of lithocysts is generally a very important differentiation process of the epidermal cell.

As an example, we shall describe the formation of lithocysts in the upper epidermis of *Ficus* leaves studied in details by HILTZ (1951); HILTZ and POBEGUIN (1949).

The young epidermal cell which is going to differentiate into a lithocyst is first individualized by the thickening of its superficial cell wall (Fig. 5.15b). While the cell wall increases, the thickening is completed by a pectocellulosic appendage which grows in length in the cell and will become the support, the *pedicel*, of the

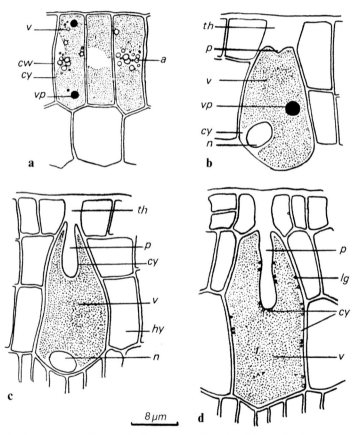

Fig. 5.15a–d. Ontogeny of lithocysts in the upper epidermis of *Ficus elastica* leaves. Neutral red vital staining (I). **a** Upper epidermis of a very young leaf; **b** thickening *th* of the superficial cell wall of a future lithocyst, and the appearance of the pedicel *p*; **c, d** development of the pedicel. *a* Amyloplasts; *cy* cytoplasm; *hy* hypodermal cell; *lg* lipid granules; *cw* pectocellulosic cell wall; *n* nucleus; *vp* vacuolar precipitates; *v* vacuole; x ≈ 1600; (after HILTZ 1951)

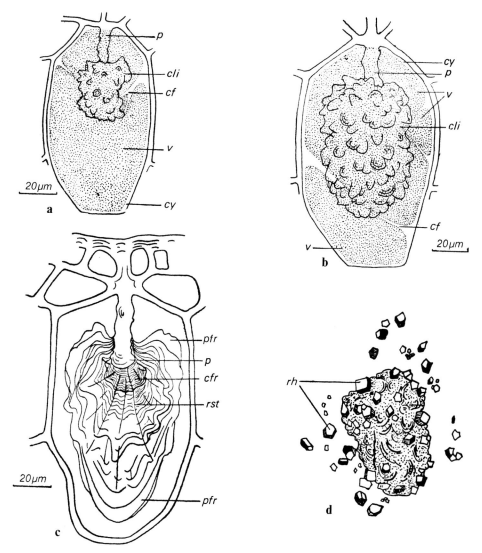

Fig. 5.16 a–d. Ontogeny of lithocysts in the upper epidermis of *Ficus elastica* leaves (II). **a, b** Growth of the cystolith *cli* around the pedicel *p; v* large vacuole with cytoplasmic filaments *cf* across; *cy* cytoplasm, almost exclusively peripheral (neutral red vital staining). **c** Decalcification of a cystolith, which reveals the cellulosic and callosic framework with a central part, *cfr,* and a peripheral part, *pfr;* the whole structure is striated by radial striae, *rst.* **d** Isolated cystolith maintained in water. The amorphous calcium carbonate, unstable in water, crystallizes into calcite rhombohedra, *rh;* x ≈ 500 (after HILTZ 1951)

mineralized concretion, the whole set forming the "cystolith". The pedicel is therefore parietal and is entirely surrounded by the cytoplasm which forms it (Fig. 5.15).

After the pedicel formation, the cytoplasm elaborates around its end a fuzzy pectocellulosic weft, thinly striated, which is soon loaded with silica and calcium silicate (Fig. 5.16).

269

At the same time, silica forms a deposit on the remaining part of the pedicel. This weft will constitute the support for the mineralized material of the cystolith which will be elaborated further on by the cytoplasm. It is mainly composed of calcium carbonate which forms a deposit associated with some silicate, so that the mineralized concretion is amorphous, a fact which is not common. The amorphous calcium carbonate is, in fact, rarely found and is probably stabilized by the associated with a small quantity of calcium silicate in lithocysts. Moreover, isolated cystoliths maintained in water are unstable and release the calcium carbonate which is precipitated into *calcite* crystals (Fig. 5.16 d).

One should keep in mind that lithocysts are living cells, the cystolith of which is a very particular parietal differentiation elaborated by the cytoplasm.

5.2.6 Other Types of Epidermal Cells

The epidermal tissues of most plants differentiate other types of particular cells; their specificity and distribution is of taxonomic interest. This is the case in Graminaceae (METCALFE 1960): the epidermal tissues of stems and both faces of leaf lamina and sheaths have specialized and variable structures according to the area considered in each organ (PRAT 1948, 1951). Common epidermal cells are elongat-

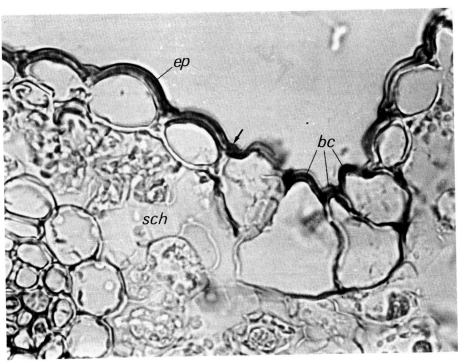

Fig. 5.17. *Agropyrum glaucum,* bulliform cells *bc* at the bottom of longitudinal leaf folds. *ep* Banal epidermal cells; *sch* substomatal chamber; the *arrow* indicates the site of a stoma which was not in the plane of the section

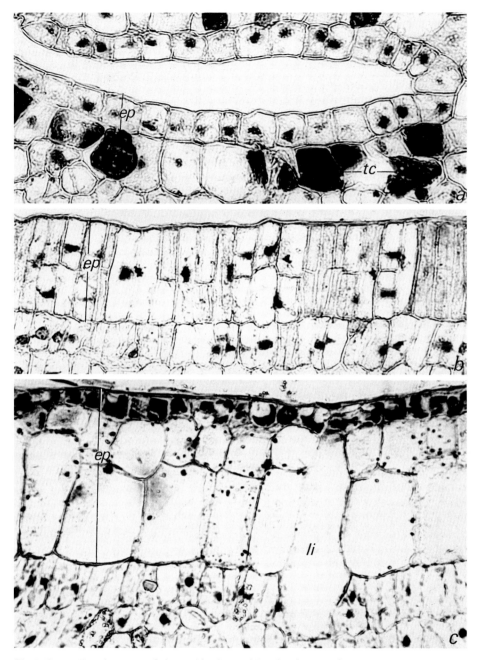

Fig. 5.18 a–c. Development of the multiseriate epidermis of *Ficus elastica*. **a** Upper epidermis *ep* (adaxial) of a very young leaf, composed of a single layer of roughly cubic cells. *tc* Tanniniferous cells. **b** More advanced stage; higher and narrower cells, issued from anticlinal cleavages, and onset of splitting of the epidermal layer into two layers through periclinal cleavages. **c** Still young leaf epidermis, but composed of three layers; *li* lithocyst in the process of differentiation

ed in the same direction as the organ and arranged in lines. According to their localization, for example with respect to leaf veins, monocellular or multicellular lines of common cells are or are not broken off by shorter cells. The latter are regularly grouped in a pair of cells, one of them is loaded with silica, the other is suberified (Fig. 5.1). Other epidermal cells differentiate papilla or trichomes through unequal divisions (see H PRAT 1948, 1951).

Im Graminaceae and in other Monocotyledons adapted to dryness such as the oyats *(Psamma arenaria)* of sandy littorals, the epidermal cells of the upper face of the leaves, which fold longitudinally, are differentiated into water storage cells, the "bulliform cells" either on the whole epidermis, or in deep-lying folds (Fig. 5.17). They are globular cells, which may be poorly chlorophyllous or not chlorophyllous, the volume of which is mostly occupied by a large vacuole. The radial walls (anticline) are thin and apparently elastic, whereas the superficial wall has the same cutinized thickness and is lined with the same cuticle as common cells. These cells seem to play a major role in the folding and unfolding of foliar lamina due to the turgor variations. Moreover, they could also participate in the lamina enlargement during ontogeny.

5.3 Multiseriate Epidermis

Many leaves include, beneath the upper layer, between this layer and the mesophyll, chlorophyll-free cells enclosing a large vacuole. The tissue is generally called *hypodermis,* but evolution study shows that, in some cases, the hypodermis is derived from the division of epidermal cells; in such a case it is not a true hypodermis, but a *multiseriate epidermis* which is produced through pericline divisions during the differentiation of the cells generated by the "protoderm" (Fig. 5.18).

In particular, multiseriate epidermis can be found in Moraceae *(Ficus),* Piperaceae, Malvaceae, Begoniaceae and Pittosporaceae. They are also present in Monocotyledons (*Tradescantia,* palm trees) and even in ferns (some species of *Polypodium,* LINSBAUER 1930). According to the species and to the age of the organ, the multiseriate epidermis may be composed of 2 to 15 layers of cells. Besides, while young cells increase in thickness, their pericline divisions may be inconstant, so that the tissue is not always arranged in regular or continued layers (Fig. 5.18).

▷

Fig. 5.19. Scheme of differentiation stages of stomatal complexes in the leaf epidermis of *Hordeum vulgare.* **1** Undifferentiated young epidermis; *fst* file of cells which will produce stomata; *bec* banal epidermal cells. **2** Unequal mitosis, *umi,* for the cytoplasm, which becomes polarized, giving birth to a stoma mother cell, *stmc,* the cytoplasm of which is denser, and to an "intervenant cell" *ic,* the cytoplasm of which is less dense. Thus formed, the stoma mother cells *stmc* induce a polarization of the adjacent banal cells *(arrows)* and the division of their nucleus. **3** The adjacent banal cell gives birth to a dense cytoplasmic cell, which becomes a subsidiary cell, *subs,* of the future stoma, and to a new banal cell in which the polarization fades out. **4, 5** The mother cell of each stoma, associated to two subsidiary cells, divides into two stomatal cells *stc,* by a mitosis, *emi,* equal and symmetrical with regards to the cytoplasm. **6** The stomatal cells complete their differentiation; (after STEBBINS and JAIN 1960; STEBBINS and SHAH 1960)

Like the hypodermis (*Nerium* leaves), the multiseriate epidermis generally indicates adaptability to dryness (it is richer in water) and to strong light (protection of underlying chlorophyllous cells of the mesophyll).

5.4 Cytophysiological Problems
Related to the Differentiation of Epidermis

The epidermis tissue is complex. Its differentiation from the protoderm results in the elaboration of several types of very different cells, each one being often isolated among the neighbouring cells which belong to other types. Such cells are comparable to "idioblasts".

The individual development of epidermal cells, which does not prevent their perfectly ordered distribution, raises subtle problems concerning the determinism of such a development. STEBBINS and co-workers (STEBBINS and JAIN 1960; STEBBINS and SHAH 1960) investigated these problems in relation to cellular differentiation in the epidermis of Graminaceae and other more or less advanced Monocotyledons (*Allium,* Commelinaceae).

For example, in barley, in which stomata are arranged in monocellular files, which are separated by lines of common epidermal cells (Figs. 5.1 and 5.19), the mother cells of stomata result from an unequal mitosis with regards to the cytoplasm, so that each file is composed of a mother cell regularly alternating with an intermediary cell, the *intervenant cell* (Fig. 5.19/2).

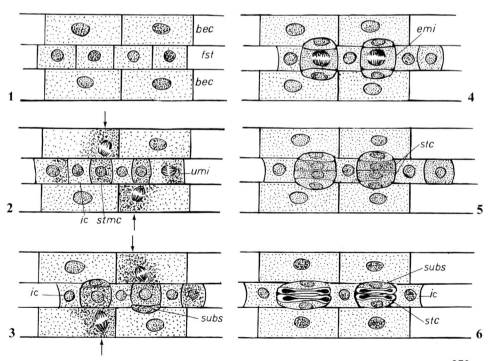

With a careful and complex study of several abnormalities, the authors presented various arguments showing that it is the mother cell of the stoma, the cytoplasm and the nucleus of which are denser than those of the intervenant cell, which induces an unequal mitosis in the adjacent cell on each side of the stomata lines (Fig. 5.19/2, 3). Thus, each lateral cell, sometimes adjacent to two or several mother cells of stomata, produces a *subsidiary cell,* the cytoplasm and the nucleus of which are denser than in the sister cell (Fig. 5.19/**3, 4**) and a new lateral cell. The differentiation of the complex ends through the mitosis of the mother cell of the stoma, this time equal as to the cytoplasm (Fig. 5.19/4, 5), which produces both stomatal cells (stc), and by the differentiation of the latter (Fig. 5.19/6).

This "polarization-asymmetry" sequence seems to occur in all idioblastic productions of epidermis. Their determinism is not established, but it may be researched in the physiological modifications initiated by the cells themselves rather than in the hormonal or other influences of the whole set of the organs. The polarization process occurs generally (STEBBINS and SHAH 1960) when the cells proceed from one stage of development to another, which is basically different. However, the order of the epidermal differentiations is certainly related to the whole development of the plant.

From the detailed studies of LANDRÉ (1976) on the development of the epidermis during the growth of the leaf of *Solanum nigrum,* one can see that when the protodermal cells, which generate the first stomata, are still rather meristematic in the dorsal epidermis of the top of the leaf buttress, the cytoplasmic polarization is capable of easily producing the mother cells of stomata through a single asymmetrical division: in that case stomata are of an anomocytic type. When the young leaf grows, common epidermal cells differentiate and the cytoplasmic condensation of polarization is slower and several asymmetrical divisions are necessary to produce the stomatal mother cell: in that case stomata are of an anisocytic type.

5.5 Epidermal Productions in Vascular Plants

The epidermal tissue is a protective tissue generally found on all the sporophytes of vascular plants and Bryophytes. Its ecological variations are very constant in all these plants. The cuticle is all the more developed when the plant lives in a drier or more enlightened medium. It is almost inexistent on the submerged organs of aquatic plants.

The common epidermal cells of plants or organs exposed to strong light have generally no differentiated chloroplasts; they are sometimes multiseriate. Conversely, plants living in the shade and aquatic plants have generally chlorophyllous epidermis. Chloroplasts are particularly developed in the epidermal cells of fern leaves. It would be interesting to elucidate the generalization of such facts and the influence of this determinism through systematic and experimental research.

In addition to its protecting role, the epidermis may have an important excretory role. This is the case in many so-called *aromatic* plants, Labiatae for example. The epidermal excretion was paralleled with the growth of organs, i.e. the metabolism of growing organs releasing secondary products such as essential oils (FREY WYSSLING).

From a phylogenic standpoint, we have already mentioned that the mode of formation of *stomata*, excluding epidermal organs, was the basis for important phylogenic considerations. Thus, in *Cycadales*, the mother cell of a stoma divides only once to produce both stomatal cells: stomatal cells are not the direct sister cells of the surrounding common epidermal cells. In *Bennettitales*, the sterile leaves of which are similar to those of Cycadales, the two cells derived from the mother cell divide again to produce both stomatal cells and subsidiary cells (FLORIN 1933).

Subsidiary cells are not always different from common epidermal cells, even when the study of the stomatal development shows their transitory formation.

The presence of foliated dermal productions particularly developed on the young fronds of ferns (Fig. 5.10) retained the attention of phylogenetists. Superficial subepidermal tissues often participate in the formation of such organs, suggesting some comparisons between these generally deciduous productions and the small leaves of Psilotinae or Lycopodineae (BOWER 1935, 1967).

Finally, the epidermal productions and in particular the trichomes are often used in the recognition of species due to the diversity and the specificity of their shapes (Figs. 5.7, 5.8, 5.11).

B. Cork (Phellem)

Cork or suber is the protective tissue which replaces epidermis on long-standing aerial organs. It is also found on the surface of underground organs. It constitutes the main part of the *periderm*.

5.6 Origin of Cork: Primary and Secondary Cork

On young root, as soon as the hair layer becomes passive, the underlying cortical cells change into suberous cells without dividing themselves. Thus, the cork resulting from this modification is called *primary* cork. The same occurs in many stems of Monocotyledons when they produce cork cells from subepidermal parenchymatous cells.

On advanced stems and roots of Dicotyledons and Gymnosperms, among other tissues, the periderm includes a *subero-phellodermal* layer which generates *secondary* cork (phellem) layers towards the outside (see p. 169).

Thick stems of arborescent Monocotyledons have a particular periderm without a well-defined secondary meristematic layer. We shall examine these various cases.

5.7 Differentiation of Cork Cells

Regardless of the cork tissue selected, it consists, wholly or in part, of cells which have been submitted to a characteristic differentiation, i.e. *the suberization*. This process may start before the end of cell growth. It consists in adding to the cellulosic cell wall of the young cell or the parenchyma cell, a membranous (coat), a kind of secondary wall including an impermeable substance, the suberin. This impermeable lamella, the cell of which is lined through apposition, isolates its content and prevents the exchanges with the environment and with the neighbouring cells, so that when the differentiation is completed, the cell dies, the protoplasm degenerates and disappears and the cavity is soon filled with gas. Thus, cork is in fact a dead tissue in opposition to the epidermis. It is poor in water and the primary pectocellulosic cell wall often becomes very thin and difficult to single out.

Generally, cork cells are derived from the phellogen layer or from parenchyma cells which do not have very thick walls (Fig. 5.20). This characteristic allows particularly their distinction from supporting cells and lignified cells. However, suberization may occur in cells primarily differentiated into collenchyma of lignified cells, thus producing suberous thick-walled cells, but the parietal thickness is only partly caused by suberization.

Suberin is a lipid complex of organic acids of high molecular weight, similar to those of the cutin but probably arranged in a different way, including 14 to 18 carbon atoms, 1 or 2 acid functions and 0 to 3 hydroxyls. In addition, it has another component which does not occur in the cutin, the *friedelin* with 30 carbon atoms.

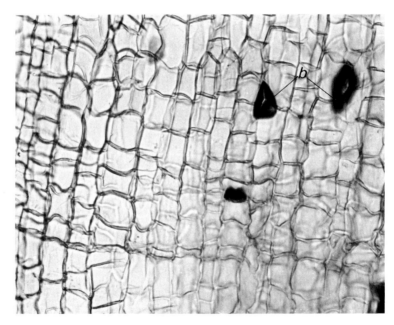

Fig. 5.20. *Quercus suber,* rhytidome. Transversal section in the cork (secondary suber). Radial files of cells, the protoplasmic content of which has disappeared and which are air filled (some air bubbles, *b,* remain in the preparation)

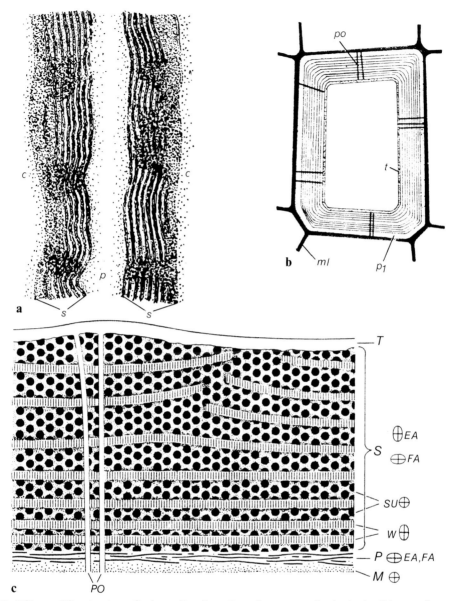

Fig. 5.21 a–c. Ultrastructure of suber cell walls. **a** Lamellar texture of suberized cell layers of potato tuber. *c* Cytoplasmic faces; *p* primary pectocellulosic cell wall (+ middle lamella); *s* suberized layers of the cell wall, where suberin lamellae *(dark)* alternate with waxy lamellae *(clear)*; × 100 000 (after an electronic micrograph of Sitte 1962 b).

b Scheme of cork cell profile; the lamellar cell wall (in general, much thinner) determines an anisotropy termed "of shape" (Formanisotropie); *ml* middle lamella; *p₁* pectocellulosic primary cell wall; *po* pores, resulting from obturated former plasmodesmata; *t* tertiary layer, secreted in extremis by the cell; (after Sitte 1961).

c Interpretation scheme of the suberized cell wall. *M* Middle lamella; *P* primary pectocellulosic cell wall; *S* suberized layers; *SU* suberin (isotropic); *T* tertiary cell wall layer; *W* waxes [linear molecules oriented perpendicular to lamellae, responsible for a distinct (or proper) anisotropy (Eigenanisotropie)]. *FA Formanisotropie; EA* Eigenanisotropie (after Sitte 1962 a)

277

Suberin is selectively stained by Sudan III and Sudan Black B, i.e. stainings of fatty bodies. According to the global analysis of cork, about 30% fatty acids have been detected.

In cork cells, suberin is associated to other substances, particularly waxes, terpene polymers (resins) and very often to many tannins. Pure suberin is colourless. The yellow or brown colour of cork is mainly due to tannins or resins, but such substances are mainly found within the cellular cavities, rather than in the wall itself.

Conversely, suberin and waxes are closely associated within the thickness of the "secondary" wall of cork cells. By means of polarized light and electron microscopy, FALK and HADIDI (1961) and SITTE (1961, 1962a, b) showed that the suberized wall is lamellated, granting it an anisotropy related to the cellular outline *(Formanisotropie)* (Fig. 5.21a, b). This structure results from the alternation of suberin and waxes layers (Figs. 5.21c). The first layers are apparently isotropic, whereas the second layers show a different anisotropy, to which it is orthogonal. This anisotropy *(Eigenanisotropie)* could result from the orientation of the fatty acid chains of waxes perpendicularly to parietal layers (Fig. 5.21c).

The future cells of cork sometimes secrete cellulosic parietal layers towards the inside, after the deposits of suberized layers and prior to the complete degeneration of their protoplasm. Cellulosic parts of these cell walls could lignify. Finally, poorly recognizable formations sometimes line the inside of cork cells. They have been considered as a "tertiary layer" in the cork of *Quercus suber* (SITTE 1962a) but they may be the only protoplasmic and paraplasmic remnants after the death of cells.

The degeneration of cork cells does not always occur quickly. Suberin alone is not as strongly impermeable as waxes, due to polar groupings and to the lower molecular weight of its fatty acids. The paraffins and fatty acids of high molecular weight (C_{16} to C_{32}) of waxes probably increase the impermeability of cork considerably.

5.8 Primary Cork

The primary cork results from the delayed suberization of already differentiated cells, particularly of *cortical parenchymatous cells*. It is frequently observed on young roots above the hair area. It also occurs in the stems of various Monocotyledons (*Typha, Phoenix,* numerous Graminaceae), or it often forms when the epidermis is torn by stem growth. Although this cork tissue is elaborated at the expense of cells of primary tissues, it still results from a secondary differentiation of these cells.

Fig. 5.22 a, b. *Prunus domestica,* rhytidome of a 2-year-old stem. **a** Transversal section; *phd* phelloderm; *s* secondary suber (cork) composed of numerous layers of flattened cells. **b** Longitudinal-radial section, showing the moderate elongation of cork cells

5.9 Secondary Cork (Peridermal Cork)

This term refers to the cork tissue derived from the phellogen (subero-phellodermal) layer. At the same time, this layer generates, towards the inside, secondary parenchymatous cells which constitute the phelloderm. These tissues, together with the generating layer, form the *periderm*.

Periderm occurs mainly on long-standing organs, having a continued growth in thickness, in Dicotyledons and Gymnosperms. It is exceptional in arborescent Monocotyledons. Such is the case with palm trees of *Roystonia* genus, the trunk of which is covered by a smooth white periderm. Some species of *Cocos* and *Aloe* also show continued and successive peridermal formations.

Secondary cork cells are arranged in radial lines generally distorted after the death of cells (turgor disappearance) and after increase of the circumference on which they stand (transversal stretching; Fig. 5.20). Their walls are generally thin.

Periderm does not last long, it is progressively exfoliated along with more or less underlying tissues by the new periderm which is more deeply elaborated.

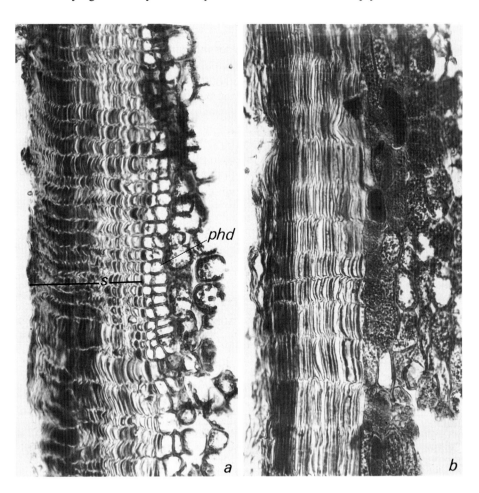

All the tissues which will be exfoliated by the newly formed periderm constitute the *rhytidome*.

The generating layer is frequently interrupted and in such cases produces cork plates. Rhytidome is then "scaly" (e.g. *Quercus, Ulmus*). Trunks with smooth cortex have a continued generating peridermal tissue (e.g. *Fagus, Prunus;* Fig. 5.22).

The renewal of the generating subero-phellodermal layer is frequently produced from the ancient, no longer functional phloem, which has a few living parenchymatous cells left. The more external phloem cells thus separated from the internal phloem are added to the internal periderm and generally become sclerified; in that case *rhytidome* consists of both peridermal cells and cells derived from the phloem (e.g. *Juglans regia, Cephalanthus occidentalis,* ESAU 1964).

5.10 Storied Cork of Arborescent Monocotyledons

This type occurs in many Monocotyledons having secondary increases. Its origin and development are particular.

It is initiated by cortical cells (therefore primary) which tangentially divide again once or several times, each division resulting in a radial file of two to eight cells. But the cortical initials are not regularly distributed and do not form any definite generating layer (Fig. 5.23 a). The short cellular files are only associated side by side in a discontinuous way, forming tangential rows connected at certain points and overlapping elsewhere, while being separated by layers of non-redivided cells.

In this way, several discontinuous layers of this particular cork may occur, separated by non-suberized cortical cells which are finally crushed. The resulting dissociation is different from the desquamation of rhytidome in Dicotyledons and Gymnosperms (e.g. *Dracaena, Cordyline,* etc.) (Fig. 5.23 b, c).

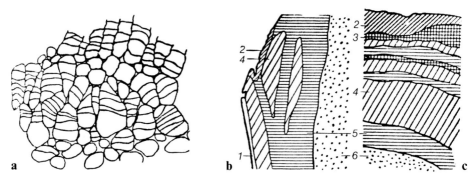

Fig. 5.23 a–c. Lignous monocotyledon cork. **a** Storied cork of *Curcuma longa* cortex; repeated tangential divisions of cortical cells, but without alignment in regular layers. **b** *Cordyline australis,* radial section of the stem. **c** *Cordyline indivisa,* transversal section of the stem. **b** and **c** Zonation of the periderm, where storied suberized cells *(5),* suberized non-redivided cells *(4)* and non-suberized cortical cells *(3)* alternate, the whole structure lying beneath a coat of crushed superficial cells *(2),* where an epidermis is sometimes recognizable *(1)* (after PHILIPP 1923)

5.11 Lenticels

We have mentioned that the epidermis, i.e. the impermeable coating of young aerial organs, is interrupted by stomata through which gas exchanges are carried out with the external medium. In the same way, the secondary cork, which often constitutes a thick and impermeable coating on advanced organs, is interrupted by lenticel organs which allow the gas exchanges (HABERLANDT).

Lenticels are found on almost all long-standing organs, roots and stems, having a periderm. However, continued peridermis *without lenticels* is also found (e.g., *Philadelphus, Vitis*). Such periderms are exfoliated each year and replaced by new and deeper ones.

Contrarily to the compact suberized tissue, lenticel tissue consists of round cells, separated by marked meatus. These meatus ensure a communication between the atmosphere and the intercellular spaces of the deeper cells. In Gymnosperms, the so-called filling cells of lenticels are suberized early. In Dicotyledons three types of lenticels have been distinguished (ESAU 1965) related to the localization and time of suberization. (1) Lenticels of relatively primitive species *(Magnolia, Liriodendron)* exhibit an annual alternation of loose areas of thin-walled cells and compactly arranged layers having thicker walls. (2) In *Fraxinus, Quercus, Tilia*, etc. filling cells are not suberized, contrarily to "closing cells" (Fig. 5.24b) arising at the end of the season. (3) This alternation of loosely arranged non-suberized layers with suberized layers of "closing cells" occurs several times a year in other species (*Betula, Fagus, Robinia*, etc.) and these strata are exfoliated by the new growth of the phellogen.

When the generating layer arises from external cortical layers, the first lenticels result from a particular and localized activity of subepidermal parenchymatous cells which first divide irregularly. In the mass which is thus produced the deeply seated cells undergo oriented divisions yielding a loose meristematic zone which takes the place of the subero-phellodermal layer (Fig. 5.25 c). Towards the outside this zone generates some files of cells which round out and, towards the inside, some parenchymatous cells which will be connected to the phelloderm issued from the peridermal cambium (Fig. 5.24a).

Fig. 5.24. a Completed lenticel of *Sambucus nigra*. Exfoliation of suberized epidermal cells *(ep. sub); a–b, c–d, e–f* cellular rows comprising the phellogen (of blurred localization) which has produced (1) towards the exterior, non-suberized filling cells *(a–b, c–d)* or suberized cells *(e–f)*, and (2) towards the interior, some phelloderm cells *(b, d, f)*. Deeper cleavages thicken the lenticel *(g, g', g'')*. On the *left* side, the phellogen of the lenticel is connected with the periderm phellogen; × 225 (after DEVAUX 1900).
b Mature lenticel of *Coriaria myrtifolia*. Formation of layers of closing cells *(ff, f', f')* which become suberized, alternating with filling cells *("cellules comblantes", cc, cc')* infrequently suberized *(s)*. The *f'–f'* layer is in the course of setting and is still only partly suberized *(a, b, c). phd* Phelloderm of the lenticel, much more developed than the periderm lenticel, visible on the sides, and into which deep cleavages *(d, d')* still occur; *p, p'* lacunae, below the lenticel edges; × 125; (after DEVAUX 1900; original author's lettering)

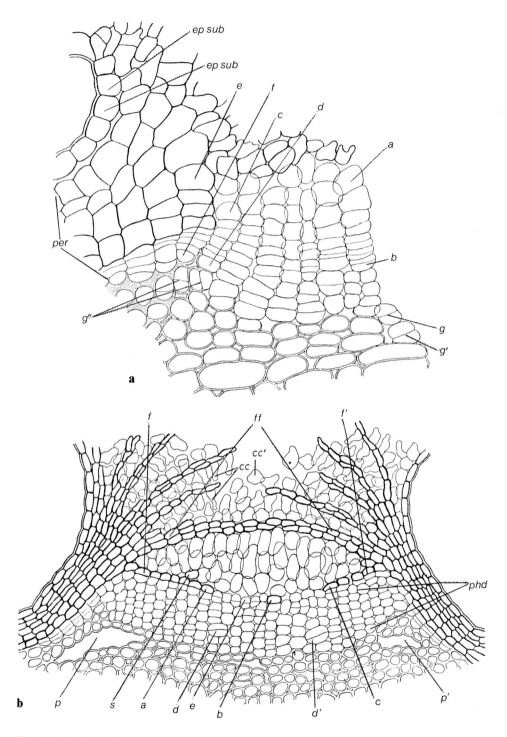

Fig. 5.24a, b. Legend see p. 281

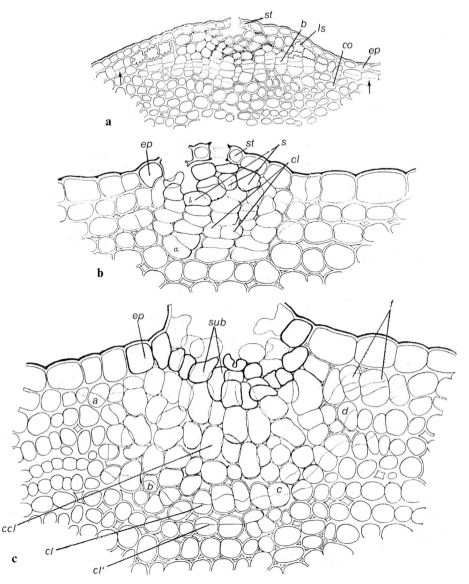

Fig. 5.25. a *Rhamnus frangula,* young lenticel in the process of differentiation below a stoma *(st),* within a stem still covered with epidermis *(ep)* and in which the periderm is also at its beginning (subepidermal cleavages are marked by *arrows).* Only some secondary cork cells *(ls)* have been produced. The cortical cells underlying the substomatal chamber *(b)* divide periclinally; *co* collenchyma; × 240.

b *Sambucus nigra,* young lenticel. Breaking of the epidermis next to a stoma *(st);* the neighbouring epidermal cell *(ep)* and the cells bordering the substomatal chamber *(s)* are suberized. The underlying cortical cells undergo periclinal cleavages *(cl)* and produce cellular rows *(a–b);* × 225.

c *Sambucus nigra,* more advanced lenticel. Accentuation of the epidermis breaking *(ep)* and necrosis of cells outside the cupule of suberized cells *(sub);* hypertrophy of the cupule underlying cells *(c cl)* and periclinal cleavages of deeper cortical cells *(cl, cl')* forming a concave generating layer *(a b c d); f* beginning of the formation of a periderm in the subepidermal cells; × 225 (after Devaux 1900; original author's lettering)

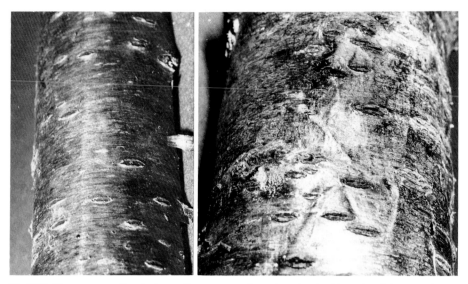

Fig. 5.26. Two aspects of lenticels on *Cerasus avium* branches, the rhytidome is of the smooth type

The first lenticels may rise very soon on a young organ still covered with undamaged epidermis, and before the formation of the remaining part of the periderm. Generally, they arise beneath stomata and in this case, the cells which surround the substomatal chamber develop into a generating zone (Fig. 5.25 a, b). This development is accompanied by the regression of chloroplasts among other processes; it starts with a dedifferentiation like the birth of the phellogen layer.

Late arising lenticels together with lenticels of successive periderms no longer have a relationship with stomata. They may arise simultaneously with the suberophellodermal layer and in some cases later, beneath it. In this case lenticels break off the periderm.

Lenticels may increase or are distended by the thickening of the stem. Their lenticular shape (hence their name) comes from this stretching (Fig. 5.26). If the bark is fissured and rough (as in *Quercus, Robinia*) lenticels are found deep down in the furrows.

As already mentioned with respect to the suberization of lenticel cells, the functioning of the generating zone of lenticels can show seasonal variations. In fall, it often produces more compactly arranged layer which form a kind of protective operculum but which is ruptured as secondary growth occurs. The generating area of lenticel can also produce alternations of loosely arranged and tightly arranged tissues (*"cellules de fermeture"* DEVAUX 1900; "closing cells" of English authors) not related to seasons (*Prunus avium, Coriaria myrtifolia*, Fig. 5.24 b).

References

Bary A de (1884) Comparative anatomy of the vegetative organs of the Phanerogams and Ferns. Clarendon, Oxford

Bower FO (1935, 1967) Primitive land plants. Macmillan, Londres (1935); Hafner, New York, Londres (1967)

Brown WV, Johnson SC (1962) The fine structure of the grass guard cell. Am J Bot 49: 110–115

Bünning E (1952) Morphogenesis in plants. Surv Biol Prog 2: 105–140

Devaux H (1900) Recherches sur les lenticelles. Ann Sci Nat Bot sér 8 (12): 1–240

Eglinton G, Hunneman DH (1968) Gas chromatographic – mass spectrometric studies of long chain hydroxy-acids. I. The constituent cutin acid of apple cuticle. Phytochemistry 7: 313–322

Esau K (1964) Structure and development of the bark in Dicotyledons. In: Formation of wood in forest trees. Academic Press, London, pp 37–50

Esau K (1965) Plant anatomy, 2nd edn. Wiley, New York

Falk H, Nabil el Hadidi M (1961) Der Feinbau der Suberinschichten verkorkter Zellwände. Z Naturforsch Teil B Anorg Chem Org Chem 16: 134–137

Florin R (1931) Untersuchungen zur Stammesgeschichte der Coniferales und Cordaitales. Kungl Svenska Vetenskapsakad Handl Ser 3 10: 1–588

Florin R (1933) Studien über die Cycadales des Mesozoikums nebst Erörterungen über die Spaltöffnungsapparate der Bennettitales. Kungl Svenska Vetenskapsakad Handl 12: 1–134

Florin R (1951) Evolution in Cordaites and Conifers. Acta Horti Bergiani 15: 285–388

Florin R (1958) Notes on the systematics of the Podocarpaceae. Acta Horti Bergiani 17: 403–411

Franke W (1961) Tröpfchenausscheidung und Ektodesmen – Verteilung in Zwiebelschuppenepidermen. Ein Beitrag zur Frage der Ektodesmenfunktion. Planta (Berl) 57: 266–283

Frey-Wyssling A, Mühlethaler K (1959) Über das submikroskopische Geschehen bei der Kutinisierung pflanzlicher Zellwände. Naturf Ges Zürich Vrtljschr 104: 294–299

Fryns-Claessens E, Cotthen W van (1973) A new classification of the ontogenic types of stomata. Bot Rev 39: 71–138

Guilliermond A, Mangenot G, Plantefol L (1933) Traité de Cytologie végétale. Le François (ed) Paris

Guyot M (1966) Les stomates des Obellifères. Bull Soc Bot Fr 113: 244–273

Guyot M, Humbert C (1970) Les modifications du vacuome des cellules stomatiques d'*Anemia rotundifolia* schrad. CR Acad Sci Paris 270 D: 2787–2790

Haberlandt G (1914) Physiological plant anatomy. Macmillan, London

Hanstein J (1868) Die Scheitelzellgruppe in Vegetationspunkt der Phanerogamen. Abh aus dem Gebiet der Naturwiss Math und Med Bonn, oder Festschr Niederrhein Ges Natur und Heilkunde: pp 109–143

Hanstein J (1870) Die Entwicklung des Keimes der Monokotylen und der Dicotylen. Bot Abh 1: 1–112

Hiltz P (1951) Contribution à l'étude des lithocystes et des cystolithes de *Ficus elastica*. Dipl Et Sup Paris, librairie gén de l'Enseignement, 27 p

Hiltz P, Pobéguin T (1949) Sur la constition des cystolithes de *Ficus elastica*. CR Acad Sci Paris 228: 1049–1051

Humbert C (1976) Recherches sur la différenciation et la cytophysiologie des stomates. Thèse Doct Sci, Dijon

Humbert C, Louguet P, Guyot M (1975) Etude ultrastructurale comparée des cellules stomatiques de *Pelargonium X hortorum* en relation avec un état d'ouverture ou de fermeture des stomates physiologiquement défini. CR Acad Sci, Paris 280 D: 1373–1376

Humbert C, Louguet P, Guyot M (1978) Modifications ultrastructurales des cellules de garde et mouvements des stomates chez le *Pelargonium X hortorum*. Rev Cytol Biol Vég Bot 1: 233–257

Kolattukudy PE (1981) Structure, biosynthesis and biodegradation of cutin and suberin. Annu Rev Plant Physiol 32: 539–567

Landré P (1969a) Premières observations sur l'évolution infrastructurale des cellules stomatiques de la Moutarde (*Sinapis alba* L.) depuis leur mise en place jusqu'à l'ouverture de l'ostiole. CR Acad Sci Paris 269 D: 943–946

Landré P (1969b) Quelques aspects infrastructuraux des stomates des cotylédons de la Moutarde (*Sinapis alba* L.). CR Acad Sci Paris 269 D: 990–992

Landré P (1970) Activité golgienne en liaison avec celle du plasmalemme dans les cellules stomatiques de la Moutarde (Sinapis alba L) lors de la formation de l'ostiole. CR Acad Sci Paris 271 D: 904–907

Landré P (1972) Origine et développement des épidermes cotylédonaires et foliaires de la Moutarde (*Sinapis alba* L.). Différenciation ultrastructurale des stomates. Ann Sci Nat Bot 12e sér 13: 247–322

Landré P (1973a) Étude cytophotométrique des types cellulaires d'un épiderme adulte. Distributions comparées des teneurs en DNA nucléaire des cellules épidermiques et des cellules stomatiques du *Solanum nigrum* L. CR Acad Sci Paris 276 D: 2055–2058

Landré P (1973b) Etude cytophotométrique des types cellulaires d'un épiderme adulte. Pluralité des valeurs en DNA nucléaire des cellules épidermiques et des cellules stomatiques du *Solanum nigrum* L. CR Acad Sci Paris 276 D: 2673–2676

Landré P (1976) Teneurs en DNA nucléaire de quelques types cellulaires de l'épiderme de la Morelle noire (*Solanum nigrum* L.) au cours du développement de la feuille. Etude histologique et cytophotométrique. Ann Sci Nat Bot 12ème ser 17: 5–104

Linsbauer K (1930) Die Epidermis. In: Linsbauer, Handbuch der Pflanzenanatomie, Band 4, Lief 27. Borntraeger, Berlin

Lloyd CW (1984) Toward a dynamic helical model for the influence of microtubules on wall patterns in plants. Int Rev Cytol 86: 1–51

Louguet P (1962) Sur une méthode d'étude du mouvement des stomates utilisant la diffusion de l'hydrogène à travers les feuilles. Actes Coll Intl Méthodologie de l'Écophysiologie végétale, Montpellier, pp 307–316

Louguet P (1974) Les mécanismes du mouvement des stomates: étude critique des principales théories classiques et modernes et analyse des effets du gaz carbonique sur le mouvement des stomates du *Pelargonium X hortorum* à l'obscurité. Physiol Vég 12: 53–81

Martens P (1934) Recherches sur la cuticule. IV. Le relief cuticulaire et la différenciation épidermique des organes floraux. Cellule 43: 289–320

Metcalfe CR (1960) Anatomy of the monocotyledons. I. Gramineae. Clarendon, Oxford

Metcalfe CR, Chalk L (1950) Anatomy of the dicotyledons, 2 vol. Clarendon, Oxford

Mouravieff I (1955) Recherches sur la cytologie et la physiologie des cellules stomatiques chez quelques Angiospermes. Thèse d'Université, Lyon

Pant DD (1965) On the ontogeny of stomata and other homologous structures. Plant Sci Ser 1: 1–24 (cited by Humbert 1976)

Pant DD, Mehra B (1965) Ontogeny of stomata in some Rubiaceae. Phytomorphology 15: 300–310

Philipp M (1923) Über die Verkorkten Abschlussgewebe der Monokotylen. Bibl Bot 92: 1–28

Prat H (1948) Histophysiological gradients and plant organogenesis. Part I. Bot Rev 14: 603–643

Prat H (1951) d° Part II. Ibid 17: 693–746

Sievers A (1959) Untersuchungen über die Darstellbarkeit der Ectodesmen und ihre Beeinflussung durch physikalische Faktoren. Flora (Jena) 147: 263–316

Sitte P (1961) Die submikroskopische Organisation der Pflanzenzelle. Ber Dtsch Bot Ges 74: 177–206

Sitte P (1962a) Zum Feinbau der Suberinschichten im Flaschenkork. Protoplasma 54: 555–559

Sitte P (1962b) Feinbau und Funktion der Pflanzenzellwand. Umsch Wiss Tech 9: 273–276

Smaoui A (1975) Les trichomes vésiculeux d'*Atriplex halimus* L. - Modalités de sécrétion saline d'une plante halophile. Thèse spécialité, Aix-Marseille II - 1975

Stebbins GL, Jain SK (1960) Developmental studies of cell differentiation in the epidermis of monocotyledons. I. *Allium, Rheo* and *Commelina*. Dev Biol 2: 409–426

Stebbins GL, Kush GS (1961) Variation in the organization of the stomatal complex in the leaf epidermis of monocotyledons and its bearing on their phylogeny. Am J Bot 48: 51–59

Stebbins GL, Shah SS (1960) Developmental studies of cell differentiation in the epidermis of monocotyledons. II. Cytological features of stomatal development in the Gramineae. Dev Biol 2: 477–500

Thiéry JP (1967) Mise en évidence des polysaccharides sur coupes fines en microscopie électronique. J Microsc 6: 987–1018

Vesque J (1889) De l'emploi des caractères anatomiques dans la classification des végétaux. Bull Soc Bot Fr 36: 41–89

CHAPTER 6

Phloem

The development of tissues concerned with the relationship and material exchange between the various parts of the plant may be considered as a consequence or as a prerequisite of the increasing complexity and size of the organs of ground plants growing in the atmosphere. Tissues most particularly differentiated for this function or "conducting tissues" are divided into two types: *phloem* and *xylem* or *vascular tissue*. The terms phloem and xylem, introduced by NAEGELI (1858), appear to be the most convenient and their generalized use is justified.

By means of conventional experiments (annular barking, polarity of tissues) and chemical studies, it has been assessed that vascular tissues carry along the "primary sap", i.e. the mineral solutions issued from the ground, whereas phloem tissues mainly carry "elaborated sap", i.e. the solutions of organic substances synthesized within the plant or issued from such synthesis.

Of course, the conducting functions related to mineral and organic material transport are not so clearly shared between both types of tissues, but they do occur. Thus, it is easy to understand that the conducting function of phloem is of importance. In spite of the abundance of anatomical, cytological and physiological studies devoted to this subject, information on this tissue is incomplete.

6.1 Characteristic Cell Element: The "Sieve Cell"

Phloem is a complex tissue composed of several different types of cells. However, it is characterized by a constant element, although more or less differentiated, depending on groups: *the sieve cell*, thus termed due to the particular differentiation of the wall structures. The sieve cell is first described by a basic example: a developed Angiosperm, *Cucurbita pepo*, in which phloem has been the subject of many studies. Sieve cells are elongated in the same orientation as the organ involved. The transversal size is commonly larger than that of neighbouring cells (Fig. 6.1). In the phloem tissue of *Cucurbita*, sieve cells are arranged end to end.

6.1.1 Cell Wall Differentiations

The walls of sieve cells are *pectocellulosic*. They are slightly thicker than those of neighbouring cells but they have been considered as *primary*, i.e. walls are initiated during the sieve cell differentiation in a single phase of pectocellulosic elaboration applied against the middle lamella. This differs from that which occurs in the cells in which the protoplasm elaborates *secondary strata*, after the deposit of

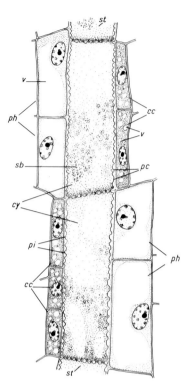

Fig. 6.1. Scheme of the main types of cells of *Cucurbita pepo* phloem. *cc* Companion cells; *sb* diffluent slime bodies; *cy* diluted cytoplasm after disappearance of vacuoles (mictoplasm); *pi* lateral cell wall pits of sieve cells; *ph* phloem parenchyma; *st* sieve tube; *v* vacuoles. Nuclei are of the euchromocentric type (after BUVAT 1963 b)

this primary layer. However, in many species of Angiosperms, lateral walls of sieve cells are lined with an internal layer, distinct in its characteristic nacreous aspect. The "nacreous wall" can occur in primary and secondary phloem. It is absent in many species and inconsistently found in others (e.g. sieve cells with or without nacreous walls in apple tree phloem; *Betula alba* individuals having nacreous walls, others not, according to ESAU and CHEADLE 1958). With a prospective study concerning 240 plants, divided into 142 species and 121 genera, ESAU and CHEADLE observed nacreous walls in the secondary phloem of 45 species, whereas 97

▷

Plate 6.1 A–C. Differentiated cells of *Cucurbita*, sieve cell walls.

A Functional sieve plate, the pores have only a slight callose coating *ca; mf* mictoplasm filaments; *cc* companion cells; *m* mitochondria. × 5100.

B Sieve plate, the pores are widely lined with callose *ca*, and crossed by the finely filamentous mictoplasm; *b* bars of the sieve areas; × 8350.

C Oblique section passing through a sieve cell wall. Although narrowed by callose sheaths *ca*, the pores, with osmiophilic content, allow some mictoplasm lineaments to subsist; × 19000

Osmic fixations

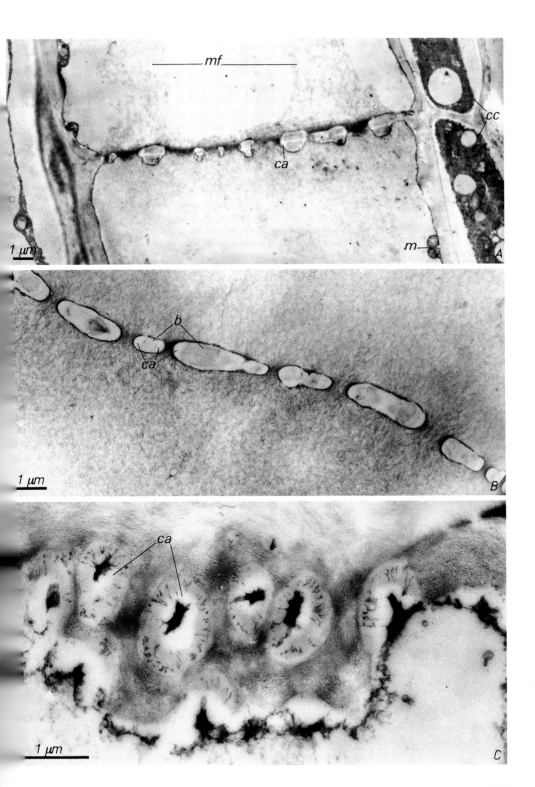

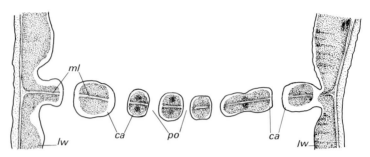

Fig. 6.2. Scheme of a section of *Cucurbita* sieve plate. *ca* Callose; *ml* middle lamella; *lw* lateral cell wall; *po* pores

species were devoid of it. Nacreous walls vary in thickness: very thin *(Prunus, Quercus);* middle thickness or very thick *(Magnolia);* they give a positive reaction to tests for cellulose and pectins; they sometimes occlude the cellular cavity completely and can shrink and fold into more aged, no longer functional cells.

In the following section, "end cell walls" refer to cell walls which delimit the long cell at both ends, and "lateral cell walls" are the faces obtained through median transversal section. With regards to polarity, it is established that the conduction occurs from the one end considered to be *proximal* to the other, considered to be *distal.*

Sieve cells of *Cucurbita* have characteristic end cell walls. Such walls are sieve areas with clusters of pores through which the cellular content progresses. From one cell to the next one, pores are longitudinally lined up in a series of canalicula through which successive cells are connected. These connections will be discussed later. Pores constitute *sieve areas,* or more briefly, a sieve (Figs. 6.1 and 6.2).

Each pore is comparable to a plasmodesma but differs by a larger size and by the cytological and cytochemical features of the protoplasm enclosed.

Each canaliculus is commonly lined with *callose,* forming a hollow cylinder, with a variable thickness according to the age of the sieve cell (ESAU 1961, ULLRICH 1962) (Plate 6.1, and Fig. 6.2). The existence of the callose sheath was denied by SALMON (1946). In fact, the formation of callose increases when the sieve area ceases to function but the presence of callose in functional sieve elements is not doubtful.

However, the former observations of MANGIN (1890) are still valuable. MANGIN gave the name of callose to the substance which is deposited around phloem pits and finally occludes the sieve elements with a *callus.*

Callose, formerly described by MANGIN as distinct from petic compounds is mainly composed of glucose polymers, whereas through hydrolysis pectic compounds release a majority of osides derived from *arabinose* (in C_5). Pectic acids include derivatives of this pentose and galactose. The binding mode β-1-3 (KESSLER 1958) and the spiral shape of the molecular linkage allow the distinction between callose and cellulose. In cellulose, elements are linked in β-1-4 and the molecular linkage is in a straight line. Callose stains a clear blue with resorcin blue and anilin blue, and above all, even in small amounts, may be detected by its characteristic fluorescence (CURRIER 1957).

290

In a possible general way, in numerous cells different from phloem cells, callose may occur almost immediately, above all opposite plasmodesmata, after a wound or even after the slight modification entailed by plant sampling for microscope study (ESCHRICH 1956, CURRIER 1957). Besides it is as quickly destroyed through a *callase*. The question was raised whether the deposit of *callose* in the pores of sieve areas is not induced by the sample preparation. Such a localization is, however, very characteristic and, moreover, it changes as the plant ages; thus, it can be considered as obviously specific of phloem conducting cells.

Outside the sieve areas, these cells show other depressions along the lateral walls, comparable to the primary pit fields of parenchyma. The accurate study of each pit field shows that it encompasses thin tubules. These pits constitute "sieve areas" less differentiated than the "sieve plates" of "end walls" (Figs. 6.33 and 6.34).

6.1.2 The Content of Sieve Cells

The structure and distribution of the contents of sieve cells are very difficult to study. In fact, in "living" organs, there is a certain pressure within the cell. The section of the organ which must be fixed generally causes a depression which can disorganize the cell content or at least modifies it greatly. Fortunately, this effect would occur only in the vicinity of the section (SALMON). However, one must be very careful in the cytological study of sieve cells.

6.1.2.1 The Nucleus

These cells characteristically lose their nucleus during differentiation. The nucleus disappears progressively (Figs. 6.3 and 6.4). During cell differentiation, the nucleus grows in length passively, sometimes constricted, simulating an amitosis and above all shows chromatin degeneration. In the secondary phloem of *Cucurbita,* the nucleus swells progressively with the formation of alveoles, and the chromatin structures (in this case prochromosomes) after a transitory tendency to be diluted into the nucleoplasm are less and less reactive to the Feulgen staining technique which selectively stains the chromatin. The nucleus *loses its chromatin* through a process of dechromatinization: it becomes swollen, alveolar (Fig. 6.3) and is reduced to some type of non-reactive bladder, devoid of nucleoli.

This degeneration process through *dechromatinization* seems characteristic of *secondary* sieve cells. In the primary phloem, SALMON described a different degenerating process, through *"pycnose"* briefly discussed in this book. It was not observed with *Cucurbita pepo,* but with several different Dicotyledons and some Monocotyledons *(Convallaria).* In such cases, the nucleus increases considerably, becoming *amiboid,* and lengthens significantly, whereas chromatin spreads uniformly into the nucleoplasm. Finally, the nuclear material is scatterd in the cytoplasm as droplets. The nucleolus, which is also susceptible to break up, disappears together with the nucleus (Fig. 6.4).

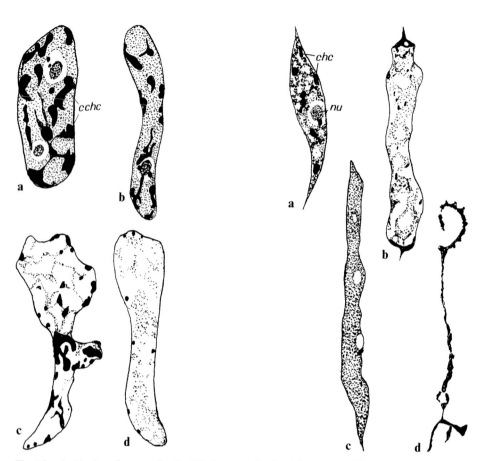

Fig. 6.3 a–d. Nuclear degeneration by "dechromatinization" during the differentiation of *Bryonia dioica* secondary sieve cells. **a** Normal nucleus, type "areticulated", with compound chromocentres *(cchc)*; **b** beginning of nucleus hypertrophy; **c** nucleus hypertrophy and chromatin reduction in a relatively differentiated cell; **d** almost entirely dechromatinized nucleus (after SALMON 1946)

Fig. 6.4 a–d. Pycnotic degeneration of the nucleus during the differentiation of *Solanum tuberosum* sieve cells. **a** Normal nucleus of the "semi-reticulated" type with chromocentres *chc;* **b** nucleus hypertrophy and alveolization, the chromatin is temporarily scattered into thin granules; **c** pycnotic degeneration following the preceding granules diffusion, resulting in a uniform colouration of the nucleus, after FEULGEN technique; **d** after nuclear involution, chromatin in the shape of tracts and troplets, more or less discontinuous. *nu* Nucleolus (after SALMON 1946)

▷

Plate 6.2 A–C. Nuclear degeneration in the course of sieve cell differentiation.

A Stem of *Cucurbita pepo*, immature sieve cell; lobate and swollen nucleus with a clearer content: case of degeneration by dechromatinization. *pch* Prochromosomes; *d* dictyosomes; *m* mitochondria. × 4700.

B, C Pycnotic degeneration of the nucleus *n* in sieve cells of *Polypodium vulgare*, stage more advanced in **C**; *er* endoplasmic reticulum amounts; *p* plastids; *dg* dense globule; × 5500 (**B**) and 17 000 (**C**).

Fixations: glutaraldehyde – O_sO_4; contrast: $KMnO_4$ (after Liberman-Max 1983)

292

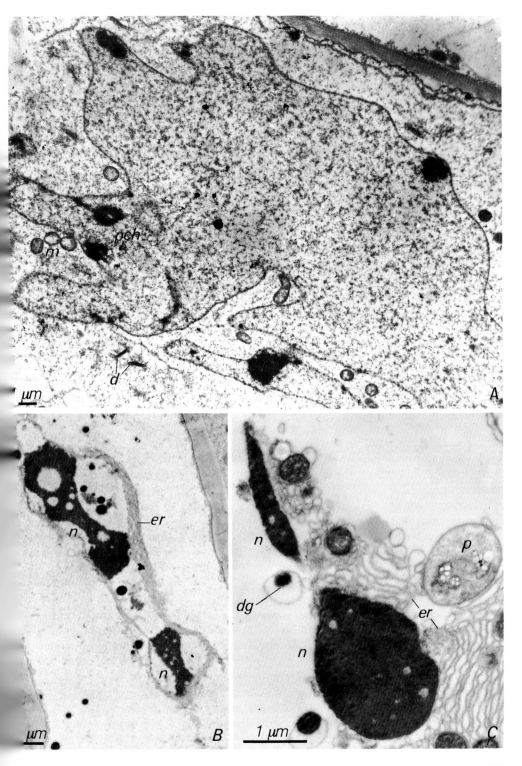

A number of authors have observed the nuclear degeneration with electron microscopy (ESAU and CHEADLE 1965; ENGLEMAN 1965; MAXE 1966; ARSANTO 1970, etc.; cf. ARSANTO 1979). These observations confirm the elongation and the hypertrophy of the nuclear vesicle along with a deeply lobed surface. The content becomes clearer and the chromatin disappears either before *(Cucurbita)* or after the nucleolus (sycamore, NORTHCOTE and WOODING 1966; *Pisum,* ZEE and CHAMBERS 1968) (Plate 6.2 A). The nuclear membrane hangs loosely in the cellular cavity, like an empty goat bottle, then it divides and disappears. In Polypode (MAXE 1966) the nucleus is divided into granules through pycnotic degeneration, granules which may remain or disappear (Plate 6.2 B, C).

In differentiated sieve cells (e.g. *Eucalyptus,* ESAU 1947) chromophil bodies were reported, compared to nucleoli rejected from the nucleus. It would be useful to study these bodies with electron microscopy to acknowledge their ultrastructure.

6.1.2.2 Plastids and Mitochondria

Very young cells, before developing into sieve cells, have a chondriome comparable to that of cambial cells, composed of granulous mitochondria along with undifferentiated plastids which are difficult to distinguish from mitochondria in light microscopy (Fig.6.5 a). Later, plastids grow moderately while remaining more or less spherical (Fig.6.5 b).

Small pastids of sieve cells have long been acknowledged (STRASBURGER 1891) but their interpretation is now more accurate (SALMON 1946). They are present in all sieve cells until the end of their development. Their most remarkable feature, although not observed in all species, is the production of tiny young starch grains, stained *red* by iodine, but digested by amylase. This kind of starch is not produced in all species, or in all types of phloem in the same species. For instance, in *Cucurbita,* sieve cells of the secondary phloem sometimes have amyliferous plastids, whereas those of the primary phloem are devoid of starch (Plate 6.3 A).

During the differentiation of sieve cells, plastids change only moderately. They remain quite similar to mitochondria from which they differ, except for amylaceous, proteinaceous or lipid inclusions, by a slightly larger size and a lower development or the absence of cristae together with a marked clearing of stroma (ESAU and CRONSHAW 1968a) (Plate 6.3 A).

Phenomena resulting from plastid degeneration were observed in differentiated sieve cells as well in Dicotyledons (Plate 6.3 B) and Monocotyledons in which plastids seem to become very fragile so that they burst on fixation or inclusion processes, while their content is scattered in the cell lumen (Plate 6.3 C).

In addition, sieve cell plastids of Monocotyledons and of quite numerous Dicotyledons (Magnoliaceae, Aristolochiaceae, Caryophyllaceae, certain leguminous plants, etc.) produce proteinaceous siderophil granules. Since the former observations carried out with light microscopy (SALMON 1946), the plastidal inclusions of sieve cells were studied with electron microscopy in a number of species. In particular, the extensive research of BEHNKE (cf. BEHNKE 1972, 1973, with references to numerous previous publications) enabled the author to separate type S plastids

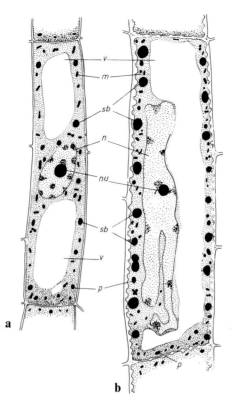

Fig. 6.5 a, b. Scheme of differentiation of *Cucurbita pepo* sieve cells. **a** Young cell, beside the cambium; *n* nucleus in median situation; *p* poorly differentiated plastids, slightly larger than mitochondria *m; sb* onset of formation of slime bodies; two or a few vacuoles. **b** Cell in the course of differentiation. Degenerating, hypertrophied and amoeboid nucleus; only one great vacuole; cytoplasm mainly peripheral; mitochondria and plastids fairly unchanged; thickening of lateral cell walls, with pits; beginning of formation of sieve plates; (after BUVAT 1963 b)

(S: starch) devoid of proteinaceous inclusions, typical of most Dicotyledons, from type P plastids (P: proteinaceous) which can also include starch and occur in Monocotyledons and in quite a number of Dicotyledons. In the phloem of Monocotyledons, proteinaceous inclusions of plastids are commonly cone – shaped and their structure is partially crystalloid (Plate 6.4). From the data obtained through the observation of the group of chosen species, BEHNKE infered some facts of phylogenetic interest. Thus, *type P* plastids occur only in Monocotyledons and in the families which are classified on the basis of certain Dicotyledon phylums. Such a distribution suggests that both types of plastids were perhaps present in ancient Angiosperms and that type P has survived only in Monocotyledons which in that case have developed independently from Dicotyledons.

With respect to this interpretation, it is interesting to note that proteinaceous inclusions were acknowledged in the sieve cell plastids of several Gymnosperms: *Pinus strobus* (SRIVASTAVA and O'BRIEN 1966; MURMANIS and EVERT 1966); *Pinus pinea* (WOODING 1966, 1968); *Pinus silvestris* (PARAMESWARAN 1971); *Picea abies* (BEHNKE 1972).

Mitochondria, commonly found as granules in young sieve cells, do not change very much during cell differentiation. They are less numerous in completed cells, but survive as long as sieve cells are functional (BUVAT 1963a, b, Plate 6.3 A–C). The infrastructure of mitochondria does not change much; cristae are more or less swollen, a fact which may be due to a high fragility towards fixatives which do not all yield the same figures (ESAU and CRONSHAW 1968a). Alteration of the internal membrane and cristae can also be noted possibly as a result of degeneration processes, but practically normal mitochondria survive in all cases (BUVAT 1963; ESAU and CHEADLE 1965).

These organelles probably remain functional in differentiated sieve cells in which CATESSON and LIBERMAN MAXE (1974) showed the presence of cytochrome – oxydase and cytochrome c, suggesting that they may participate in the energetic processes of assimilate transfer.

6.1.2.3 The "Slime Bodies" or P-Proteins

Two categories of inert inclusions are often formed in abundance in sieve cells: some of them of a very particular, proteinaceous nature, are the "slime bodies", the others are lipid granulations and are not specific for sieve cells and tend to disappear when differentiation ceases. Conversely, slime bodies are elaborated through the differentiation of sieve cells and are very characteristic of this tpye of cell. They have long been singled out (WILHELM 1880; STRASBURGER 1891) and referred to as *Schleimkörper* by German authors. Slime bodies (of English authors), quite common in Dicotyledons, are generally missing in Monocotyledons. They occur in the still young cell, within the cytoplasm, without any contact with organelles. They grow and sometimes join together; in microscope preparations, they are frequently carried along towards the ends of the cell, most of them towards the distal sieve area where they accumulate much strongly stained material which spreads in the sieve area and penetrates into the pores, thus representing the sieve accumulation (Fig. 6.6).

In the *Cucurbita* phloem, as in numerous species, slime bodies become more fluid and disorganized as the differentiation of sieve cells ends, and they become dispersed in the vacuolar content no longer delimited by the tonoplast (see Fig. 6.1). In other species *(Robinia),* slime bodies remain intact in functional sieve cells.

Plate 6.3 A–C. Plastid evolution in sieve cells.

A *Cucurbita* sieve cell. Little differentiated plastids *pl,* showing only a few sinuous lamellae; *m* mitochondria; × 31000.

B *Id* degeneration of part of plastids *pd,* which become empty; in contrast, mitochondria *m,* rich in cristae and partly swollen, remain in the cytoplasm; × 31000.

C *Hordeum vulgare,* germination, sieve cell of root at the end of differentiation. The membrane of plastids *pd* has disappeared and their content has been spread in the mictoplasm. This fact is not constant and may be an artefact, but it seems to indicate that plastids become fragile, a fact which does not occur with mitochondria *(m); po* pores of the sieve area; × 17500.

Fixation: glutaraldehyde O_8O_4; contrast: $KMnO_4$

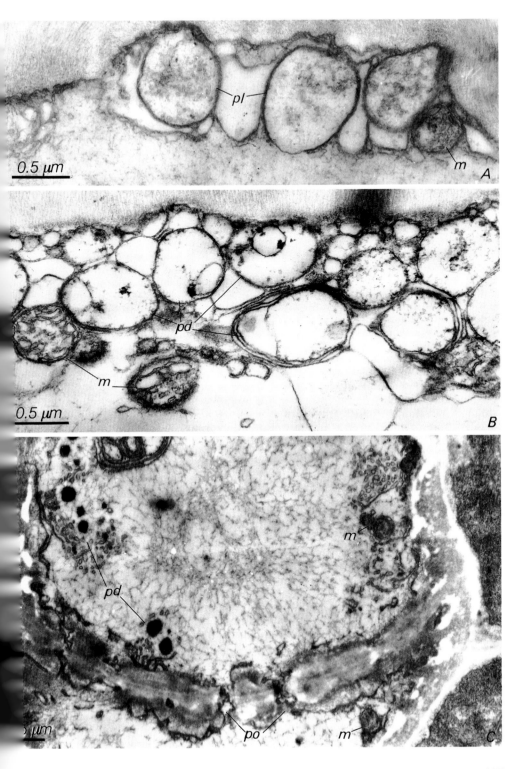

0.5 µm

pl

m

A

0.5 µm

pd

m

B

pd

m

po

m

µm

C

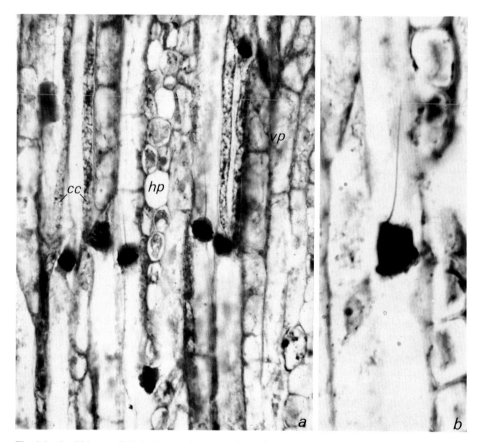

Fig. 6.6 a, b. Phloem of *Robinia pseudoacacia;* sieve plate accumulations. **a** Tangential section in functional phloem; as it frequently occurs, the slime bodies have gathered against the sieve, which is often interpreted as an artefact; *cc* companion cells, with numerous granular mitochondria; *hp* horizontal parenchyma; *vp* vertical parenchyma. **b** Sieve accumulation having kept one of its "flagellae". Fixation: REGAUD – hematoxylin

Plate 6.4 A–C. Plastids of Monocotyledon sieve cells, enclosing cuneiform and cristalline proteinaceous inclusions *(arrows).*

A, B Root protophloem of barley germinations.
C Protophloem of mature seed, dehydrated embryos;

× 30000 (**A**), 50000 (**B**), 45000 (**C**). Fixations: glutaraldehyde – OsO_4; contrasts: $KMnO_4$ (**A** and **B**), uranyl acetate-lead citrate (**C**)

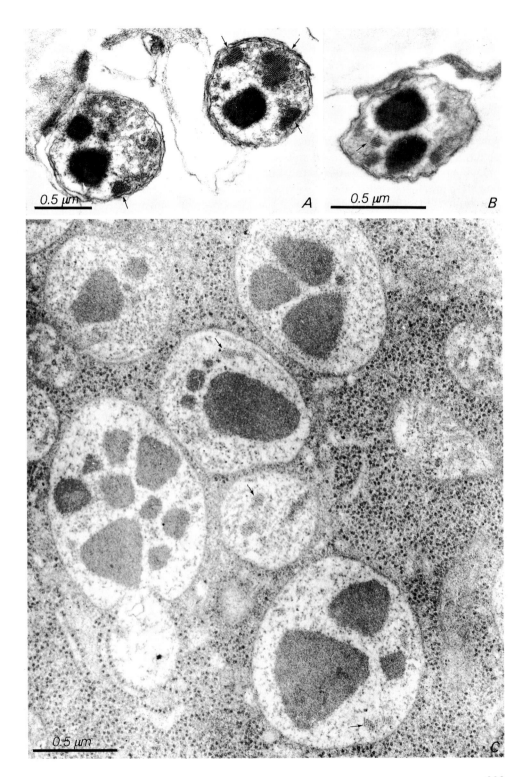

0.5 μm

A

0.5 μm

B

0.5 μm

C

299

According to ESAU (1965) sieve accumulations would correspond to artefacts caused by the translocation of the cellular content when the material is prepared and fixed. This movement would result from the fall in pressure which occurs along with the processes of material preparation. However, at least in some cases, it seems that such artefacts could magnify an existing structure, indicating the polarity of sieve cells.

Slime bodies have characteristic shapes depending on the species (Fig.6.7): spheroidal as in *Cucurbita* (Fig.6.5) or spindle shaped, twisted as in several species of different families or genera (*Lamium, Tussilago, Polygonum, Lonicera, Mahonia, Sambucus, Lappa, Clematis,* etc.; Fig.6.7) and in some cases they resemble flagellate organisms, particularly in most leguminous plants (bean, *Robinia pseudoacacia, Cicer arietinum;* Fig.6.8). Such flagellate bodies have suggested the presence of parasitic protozoa in sieve cells. They are in fact proteinaceous formations, i.e. "P-proteins" and their ultrastructure was revealed with electron microscopy (ESAU and CRONSHAW 1967). ESAU and CRONSHAW proposed the term P-protein (= Phloem protein) to designate the content of the phloem slime bodies which present cytochemical reactions to proteins (BAILEY 1923; LAFLECHE 1966; CRONSHAW and ESAU 1967 among other authors).

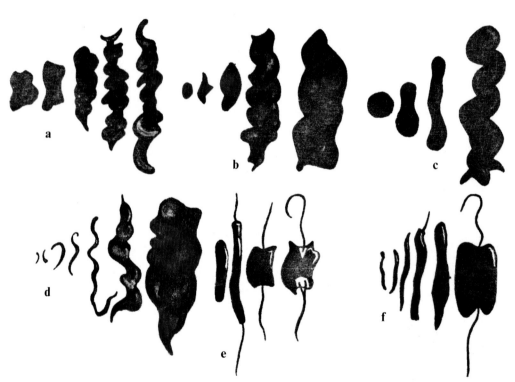

Fig.6.7a-f. Some specific shapes of slime bodies, and their evolution during the sieve cell differentiation. **a** *Mahonia;* **b** *Sambucus;* **c** *Lappa:* "turriculated" inclusions, formed from globular bodies; **d** *Clematis:* id, but first stages filamentous; **e** *Cytisus;* **f** *Phaseolus:* inclusion becoming "flagellate" more or less early (after SALMON 1946)

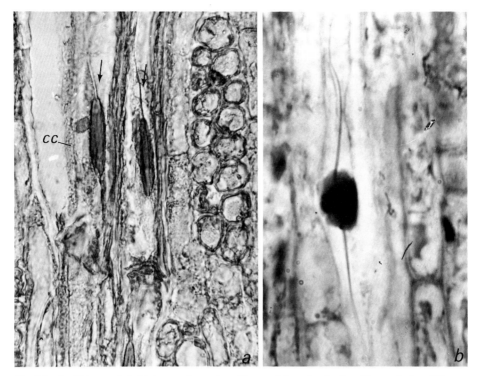

Fig. 6.8 a, b. Phloem of *Robinia pseudoacacia*. **a** Tangential section in the phloem during differentiation; young flagellate slime bodies incompletely developed *(arrows)*; *cc* cytoplasmic dense companion cells. REGAUD fixation the 7th of May, iodine-iodide reactive. **b** Flagellate slime body in a differentiated sieve cell. Fixation: REGAUD – hematoxylin

With electron microscopy, P-protein inclusions sometimes seem to be amorphous, but most commonly, they show a granulous (on thin sections), fibrous or tubular structure, according to the state of differentiation of sieve cells and to the fixation techniques (Plate 6.5 A). The most reliable material prepared with glutaraldehyde – osmic fixation and the inclusion in epoxyresin (araldite-epon) allows the distinction of four different types of structure (CRONSHAW and ESAU 1967, 1968).

1. In the *Nicotiana* phloem, differentiating young cells enclose P-proteinaceous bodies composed of assembling of tubules about 200–240 Å in diameter. Such tubules form the P_1-protein.
2. Further on, the corpuscles are disorganized and become dispersed as striated fibrils about 100–150 Å in diameter. The striated fibrils are the P_2-protein.
3. In the *Cucurbita* phloem, slime bodies are commonly composed of a larger element of thinly fibrillar features together with a smaller element of tubular structure (Plate 6.5 B). The first element results from the collection of thin, non-striated fibrils which arise in the cytoplasm and are named P_3-proteins. Tubular structures of the small element are similar to P_1-protein tubules.

301

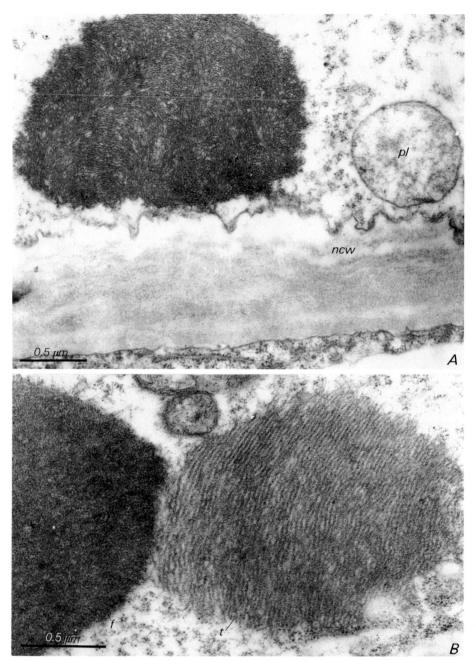

Plate 6.5 A, B. *Cucurbita*, slime bodies (P-proteins); sieve cells in the process of differentiation.

A Homogeneous body, finely fibrous (P₃-protein) arranged in tubules of small diameter (about 180 Å = P₄-protein); *ncw* nacreous cell wall; *pl* plastid. × 35 000.

B Heterogeneous body constituted of a fibrous element *f* (P₃-protein becoming P₄) and a tubular element *t*, composed of relatively large tubules (diameter: 200 to 250 Å = P₁-protein). × 45 000

Fixations: **A** glutaraldehyde – O_SO₄; **B** O_SO₄

302

4. In the larger element, P_3-protein is locally arranged in thinner tubules than those of P_1, about 180 Å in diameter, producing P_4-protein (CRONSHAW and ESAU 1968).

Inframicroscopic structural variations of P-proteins, systematized in two examples, *Nicotiana* and *Cucurbita*, demonstrate a macromolecular evolution of P-proteins during the differentiation of sieve cells. A number of scientists observed different structures, either fibrillar, tubular or both in various species (examples in KOLLMANN 1964; NORTHCOTE and WOODING 1966; LAFLECHE 1966; etc.). At first it was not possible to obtain generalized considerations from the specific variations observed, which are comparable to those allowing the distinction of genera *Nicotiana* and *Cucurbita*.

ARSANTO's works (1977, 1979, 1982) gave evidence of the relationship yielding a fundamental unity to the variously described structures depending on the species and the authors.

Since 1969, EVERT and DESHPANDE, in underlining the casually striated aspect of P_1-tubules and P_3-fibrils of *Ulmus americana* sieve cells, suggested that such components are arranged in a helical shape. In the same way, using the "image enhancement" technique of MARKHAM et al. (1963), PARTHASARATHY and MÜHLETHALER (1969) considered that P_1-*tubules* were composed of two helical-shaped subfibrils, twisted together in *Cucurbita maxima*. KOLLMANN et al. (1970) observed the same duality in dissociating in vitro exsudate filaments of *Cucurbita* and *Nicotiana* phloem. Other scientists reported similar results (see ARSANTO 1979); finally, using freeze etching, LAWTON and JOHNSON (1976) elucidated that each subfibril consists of a linear association of spheroidal subelements about 35 Å in diameter.

Using a method allowing the high magnification of electronic figures (500 000 to 2 000 000), ARSANTO (1977), in sieve cells of *Ecballium*, showed the existence of the two helices constituting the P_1-tubules, but demonstrated that each helix is itself composed of two elementary helices, similarly twisted one around the other. He defined the P_1-tubule as a *double helix of order 1 (DH 1)*, each helix being constituted of a *double helix of order 2 (DH 2)* resulting from the twisting of two elementary helices (Plate 6.6 C, D).

On transversal sections of P_1-tubules, the enhancement of the image (MARKHAM et al. 1963) revealed nine to ten elements with the same diameter (35 Å) as the elementary filaments (Plate 6.7 c series *1* and *2*) and in some cases, the arrangement in pairs of the subunits of DH 2 (Plate 6.7 c series *3*, No. *5*).

With regards to the striate fibrils (P_2-protein), which occur in particular in *Polygonum fagopyrum* when sieve cells are differentiated enough and when small proteinaceous bodies become diluted in the cell lumen, ARSANTO showed that they only correspond to helices of order 1 (DH 1) with a loose pitch produced by the stretching of P_1-tubules (ARSANTO 1977, 1979, 1982) (Plate 6.8 a, c; compared with Fig. b and d).

P_3-fibrils have been compared by the author to elementary filaments when they are associated in pairs forming the double helices of order 2 (DH 2), whereas structure P_4 would be formed through the association of *DH 2* into *DH 1*, i.e. an incomplete structure of P_1-tubules, the pitch of which is not as tight as in completed P_1-tubules.

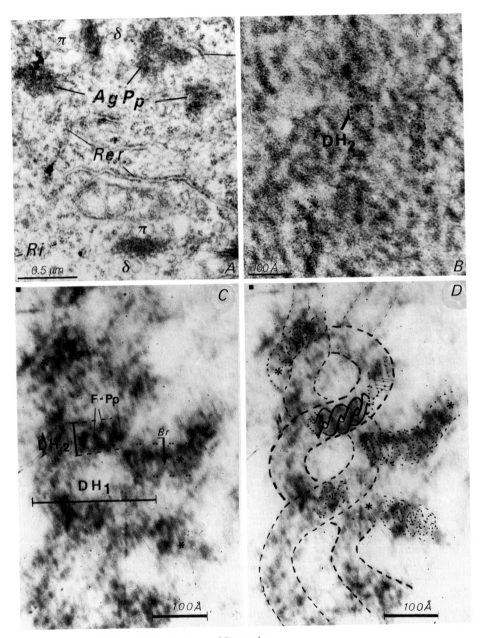

Plate 6.6A–D. Origin and ultrastructure of P-proteins.

A Rising of P-protein aggregates *(Ag Pp)* in a young sieve cell of *Ecballium elaterium* in the neighbourhood of rough endoplasmic reticulum profiles *(Rer)*, and numerous polysomes *(Ri)*; π proximal and δ distal faces of dictyosomes; × 32 500.

B Ultrastructure of P-protein aggregates in a sieve cell of *Ecballium* in the course of differentiation: elementary double helices *(DH₂,* see text), 30 to 80 Å in diameter, according to their distended or contracted state. Diameter of constitutive helices: about 15 to 30 Å; × 1 180 000.

C,D Ultrastructure of P-protein at the time of association of the double helices into "super double helices" *DH₁,* resulting in the P₁-protein type through a contraction process. **D** reproduces **C** with its explanation; × 1 450 000.

(after ARSANTO 1979); author's original lettering

Flagellate inclusions of most leguminous plants correspond to particularly compact assembling of double helices of order 1 (ARSANTO 1979, 1982), the well-ordered arrangement determining their paracrystalline structure (e.g. *Cicer arietinum*). Such a pattern is more easily observed in the flagellate or teased ends of these inclusions in its fibrillar form (Plate 6.9 A, B). In beans (LAFLÈCHE 1966) the paracrystalline structure of these "hairy" areas is clearly shown in differentiated sieve celle (Plate 6.9 C).

In proteinaceous bodies, resulting from their association, P_1-tubules are regularly arranged in bundles, either hexagonal, pentagonal or quadrangular (Plate 6.7 a). Moreover, they are connected together through transversal connections about 40 to 90 Å in diameter, depending on the author (WOODING 1969; LAWTON and JOHNSON 1976; ARSANTO 1977, 1979, 1982) (Plate 6.7 b, asterisks; Plate 6.8 d).

The origin of P-proteins is not clearly established. They arise within the hyaloplasm as fibrils comparable to P_3-proteins which soon become associated in aggregates in the shape of type DH 2 double helices (ARSANTO 1979) (Plate 6.6 A, B). Besides, some P_3-proteins issued in many species *(Ecballium)* from the disorganization of DH 1 helices of P_1-proteins have a similar structure. The cytoplasmic areas where such fibrils arise are rich in ribosomes, endoplasmic reticulum and dictyosomes, but their participation in the genesis of these proteins has not been demonstrated.

Plate 6.7 a–c. Ultrastructure of bundles of P_1-tubules.

a Transversal section through a bundle of P_1-protein tubules *FP₁* in the stem metaphloem of *Ecballium,* adjacent to a P-protein aggregate *Ag Pp;* × 57 500.

b Enlarged detail of the *framed part* of *a;* quadrangular or hexagonal arrangement of P_1-tubules (diameter: 140 to 180 Å) distant of 200 to 250 Å. Intertubular connections maintain this organization *(asterisks);* × ≈ 470 000.

C Application of the image rotation technique (MARKHAM et al. 1963) to detect the number of constitutive elementary fibrils of a P_1-tubule. The original section is marked *0* in the three series. Rotations have been made according to 360°/n. In series *1* and *2,* the reinforcement occurs for n = 9 or 10, which indicates about ten subunits on the thin section. In series *3,* the strenghtening preferentially suggests five subunits, but these subunits seem to be double *(arrows):* they are composed of paired filaments forming the double elementary helix *DH₂;* × 1 400 000

(after ARSANTO 1979)

Plate 6.8 a–d. Compared ultrastructures of P_1 and P_2 protein bundles in the *Polygonum fagopyrum* root phloem.

a P_2-protein bundle *(FP₂)* in a protophloem differentiated sieve tube; striated aspect of P_2-fibrils; × 65 000.

b P_1-protein bundle *(FP₁)* of a sieve tube in the course of differentiation; tubular aspect, × 65 000.

c Structure of a P_2-striated fibril, at high magnification; × 1 300 000.

d Structure of a "P_1-tubule" at the same magnification as in *c.* Obviously, P_2-fibrils arise from a simple stretching of double helices *DH₁,* constitutive of P_1-tubules. *DH₂* elementary double helices; *Br* arms securing connection between double helices *(asterisks);* × 1 300 000.

(after ARSANTO 1979)

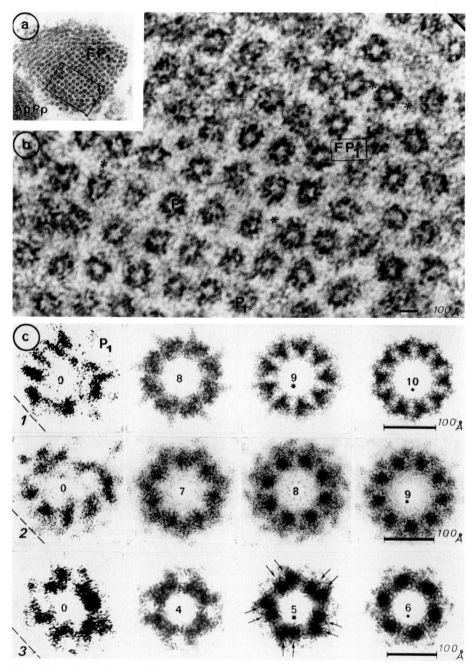

Plate 6.7 a–c. Legend see p. 305

306

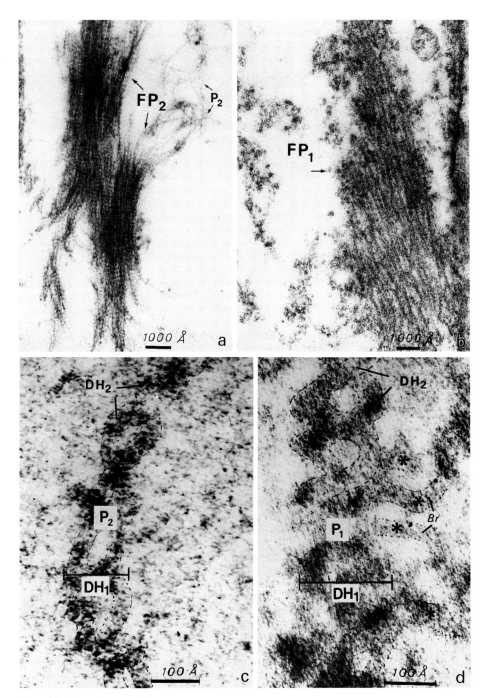

Plate 6.8. Legend see p. 305

307

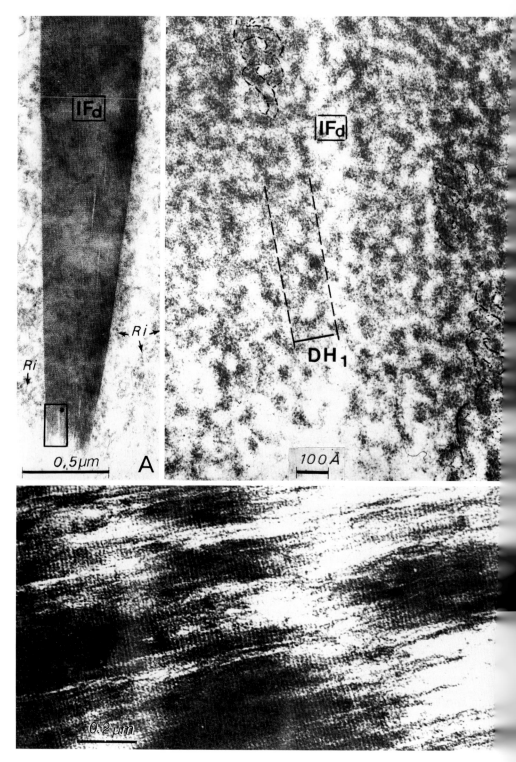

In the same areas, at different points, dictyosomes give rise to spiny or alveo-late vesicles (CRONSHAM and ESAU 1967, 1968; NEWCOMB 1967; DESHPANDE 1974). Such vesicles allow the translocation of proteins in various materials (ROTH and PORTER 1964; FRIEND and FARQUHAR 1967; cf. BUVAT 1981). One can there-fore speculate as to whether such vesicles do or do not participate in the genesis and the translocation of the materials required to eloborate the DH 1 of P_1-pro-teins. In fact, ARSANTO (1977) demonstrated that the "spines" of vesicles consist of an elementary type "DH 2" double helix. Moreover, according to LAWTON and JOHNSON (1976), they could result from the assemblage of globular units about 35 Å in diameter, like the elementary fibrils of DH 2 P-protein.

The physiological significance of P-proteins remains completely conjectural.

In spite of many cytochemical and biochemical studies, the number and fea-tures of P-proteins are not very accurate. Most investigations, most often carried out with exsudates, detect a "major protein" of about 116 000 daltons mol. wt. and another one of 30 000 daltons but they are probably not the only ones, particularly as exsudates not only include P-proteins (BEYENBACH et al. 1974).

Due to of the content in amino acids, these molecules are in some way compa-rable to actin and tubulin, however, they are all different.

However, P-protein filaments can be decorated with "arrowheads" by heavy meromyosin, as actin. Besides in vitro, the 116 000 dalton protein has an aspect re-sembling that of actin filaments. The presence of a protein comparable to actin was demonstrated in the phloem of *Heracleum sosnowskyi* by KURSANOV et al. (1983), but the authors hesitated to localize this protein in sieve tubes, in samples including substances also derived from companion cells and phloem parenchyma.

Therefore, P-protein may have features in common with "contractile" proteins of an actinomyosin type. In view of these facts, attempts have been made to local-ize ATPase in sieve cells and thus led ARSANTO (1979) to detect a labelling in the P-protein bodies of differentiated cells which agreed with the numerous controls required to single out an ATPase (cf. ARSANTO 1979), although such a character-ization is very difficult.

If an ATPase is really present in slime bodies, it is possible to compare the or-ganization of P-proteins in associated tubules with that of the axostyle of various flagellates, or with the flagellae of Ciliae in which the *dynein,* which supplies the energy required for movements, should be replaced by lateral connections be-tween P_1-tubules (ARSANTO 1979).

◁——————————————————————————————————

Plate 6.9 A–C. Flagellate inclusions of *Cicer arietinum* (**A, B**) and *Phaseolus vulgaris* (**C**).

A Aspect of the flagellate inclusion *IF_d* in a condensed state; a transversal striation is apparent; *Ri* ribosomes; × 47 000.

B Enlargement of the *framed part* of **A,** showing that the inclusion is composed of the close asso-ciation of countless type *DH_1* helices; × 850 000; (after ARSANTO 1979).

C *Phaseolus vulgaris,* ultrastructure of a flagellate inclusion in a differentiated sieve cell. Longitu-dinal alignment of fibrils, 25 Å in diameter, well matched with a transversal periodic striation, at a period of 150 Å; × 82 000; (after LAFLÈCHE 1966).

Fixations: glutaraldehyde – O_sO_4; contrast: $KMnO_4$

Although hypothetical, these considerations suggest that P-proteins have a peristaltic role in the circulation of the sap. This role was also considered by FENSOM (1972) following "microperistaltic contractions" observed and filmed by LEE et al. (1970, 1971) in living and functional phloem of *Heracleum*.

Different hypotheses have been formulated to explain the significance or the physiological role of P-proteins. WEBER et al. (1974) postulated a lytic origin related to differentiation; MILBURN (1971) considered that they participate in the occlusion of sieve areas. Several arguments oppose these hypotheses, in particular the presence of P-protein in companion cells or in phloem parenchyma cells contiguous to sieve tubes, a fact which cannot be explained with regards to the previous hypotheses.

To conclude, microperistaltism together with the fact that P_1-tubules are double helices with variable pitch (type P_2 resulting from their loosening) are in favour of an accelerating and regulating role of P-proteins in the circulation of assimilates in sieve tubes, but this is only hypothetical, particularly as such proteins are absent in numerous Angiosperms, including Monocotyledons in which, however, translocation occurs.

6.1.2.4 Cytoplasm and Vacuoles

When they are young, the future sieve cells have a large amount of cytoplasm, are strongly basophil and have numerous small vacuoles which, in some cases and temporarily, form a reticulated system similar to the system produced by the *vacuolar aggregation* (MANGENOT 1929). The vacuoles enlarge through differentiation and join to produce two, then a single vacuole which is no longer stained by neutral red (SALMON 1946). This large vacuole shifts the protoplasm against the walls where it is reduced to a film, only thickened against sieve areas.

In spite of their fragility, the development of vacuoles and cytoplasm has been observed in electron microscopy. During development, new vacuoles, produced through autophagy (Plates 6.10 and 6.11), are added to those derived from initial, procambial or cambial cells (BUVAT and ROBERT 1979). In particular, procambial cells, which generate the first sieve cells of protophloem, generally vacuolated in stems, maintain a very meristematic structure in roots where they may be devoid of vacuoles (Plate 6.10 A). The last ones are produced through autophagy from

--▷

Plate 6.10 A, B. Autophagic processes during the differentiation of barley root sieve cells.

A Onset of differentiation of a procambial cell into a protophloem cell. Cytoplasmic areas are sequestrated by double membranes (sequestration vacuoles *sv*), which will become vacuoles; lysis of the sequestrated cytoplasm in these autophagic vacuoles *av;* × 15 500.
B Id, autophagic vacuoles in which some organelles are still not undoubtedly individualized (dictyosomes), or already quite unrecognizable; *d* dictyosomes; *m* mitochondria, some of them during sequestration *(arrows); p* plastids; *nw* "nacreous" wall; *er* endoplasmic reticulum; × 18 000.
Fixations: glutaraldehyde – O_sO_4; contrast: $KMnO_4$

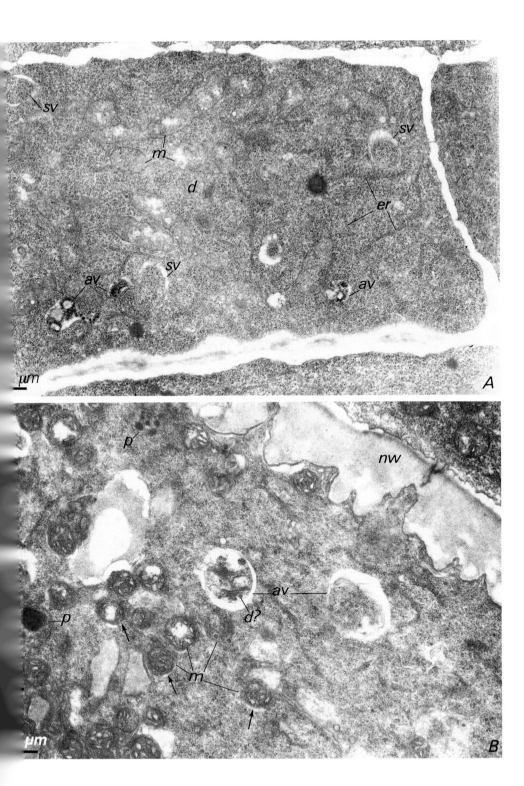

311

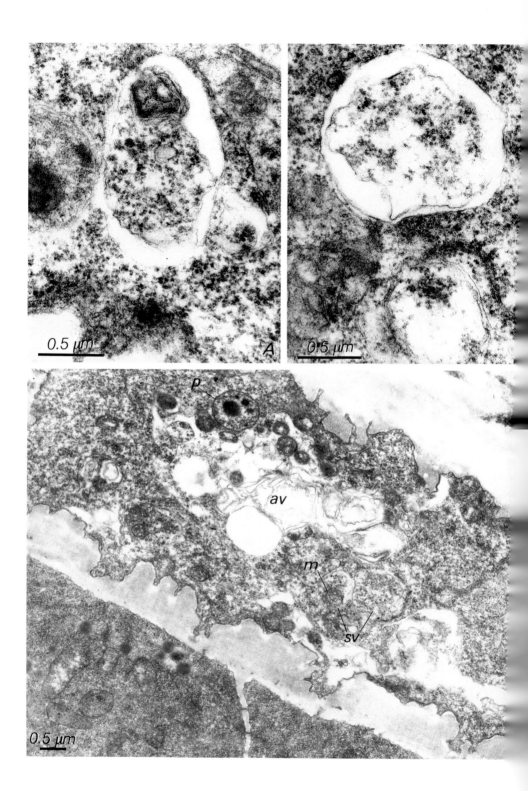

provacuolar Golgi formations, as in primary meristems (cf. p. 144). In most species, the tonoplast separates from the peripheral cytoplasm, floating freely in the cellular cavity, then it breaks off and disappears (BUVAT 1963a, b; BEHNKE 1965; BOUCK and CRONSHAW 1965; LAFLÈCHE 1966, etc.; Plate 6.12). The tonoplast disappears sooner or later, sometimes before the nucleus (ZEE and CHAMBERS 1968) sometimes after. Rarely, vacuoles may even be found until the cell ceases to function, in the protophloem in which differentiation occurs within a few hours and in which sieve cells do not live long, about 24 h (BUVAT and ROBERT 1979).

The consequence of this disorganization is that the vacuole content is mixed with the cytoplasm, the structure of which becomes diluted. The whole content of sieve cells was called *mictoplasm* by ENGLEMAN (1965).

The *endoplasmic reticulum* of sieve cells develops originally through differentiation. At the beginning, it shows a common configuration, i.e. anastomosed saccules distributed in the whole cytoplasm. During differentiation, in the mictoplasm, it is divided into two sets (ESAU and CRONSHAW 1968b), one part of the cisternae are closely applied against the plasmalemma as a perforated network, the other part remains in the mictoplasm for some time but tends to produce the piling up of membranes, parallel or anastomosed in networks, or in spheres mainly localized against lateral walls. Such associations occur in numerous Dicotyled-

---

Plate 6.11 A–C. Onset of differentiation in a barley protophloem sieve cell, derived from procambial cells. Lytic origin of vacuoles.

A, B Sequestration vacuoles, the two membranes persist and the content, already cleared, is not yet digested; × 34 800 (**A**), 33 400 (**B**).

C Lysis completion of the content imprisoned in autophagic vacuoles *av.* In the most advanced ones, only remnants of membranes subsist, no longer allowing the identification of sequestrated organelles; *sv* more recent sequestration vacuoles, one of them enclosing a mitochondrion *m*, still recognizable; *p* plastid; × 13 500.

Fixation: glutaraldehyde – O_SO_4; contrast: $KMnO_4$

Plate 6.12 A, B. *Cucurbita* sieve cells. Release of tonoplasts *t* during their differentiation.

A On the *left* cell, beginning of the release of the tonoplast *t*, when, on the *right* cell, the tonoplast has disappeared. In this cell, the content is a "mictoplasm" *(mic)*; × 5800.

B Advanced phase of tonoplast release; the tonoplast *t* becomes partly free in the cell cavity; × 5100.

Fixation: O_SO_4; contrast: $KMnO_4$

Plate 6.13 A–E. Branched plasmodesmata between sieve cells and companion cells.

A–C *Hordeum* root protophloem. **D** stem *Cucurbita* phloem. The small canal which crosses the sieve cell wall ramifies at the middle lamella level *(arrows)* into several branches crossing the companion cell wall *cc; sc* sieve cell; × 30 000 (**A**), 42 000 (**B**), 36 000 (**C**), 41 000 (**D**).

E Differentiated sieve cell of *Cucurbita*; endoplasmic reticulum coiled against the plasmalemma *pl* and encircling some mitochondria *m; mic* mictoplasm; *lw* lateral wall of the sieve cell; × 20 000

Fixation: glutaraldehyde – O_SO_4

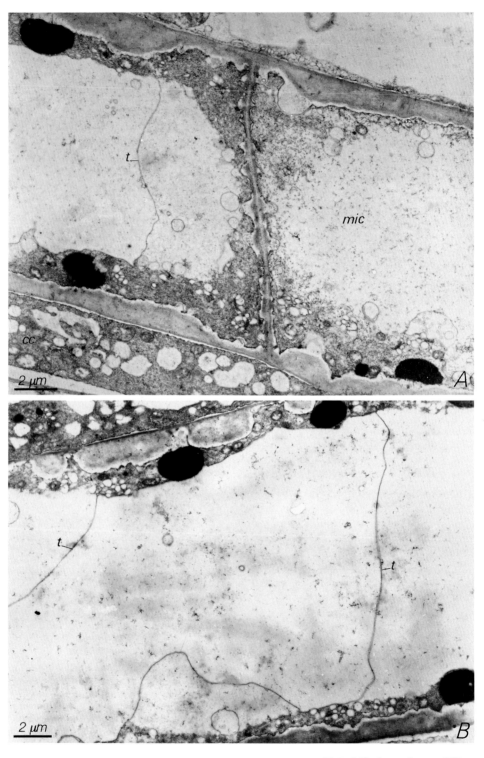

Plate 6.12. Legend see p. 313

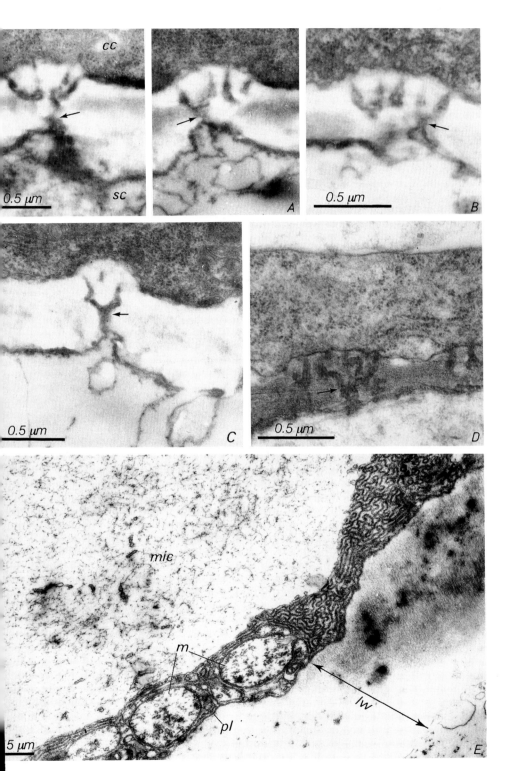

Plate 6.13. Legend see p. 313

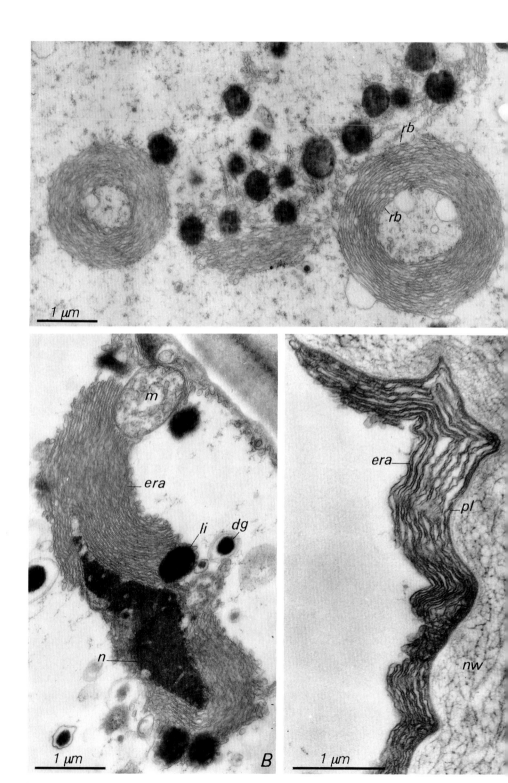

ons, e.g. the *Cucurbita* genus in which they are not very developed (cf. ESAU and GILL 1971, for previous references), in Monocotyledon (BEHNKE 1968; PARTHA-SARATHY 1974 a, b), in conifers (cf. SRIVASTAVA and O'BRIEN 1966) and also in Pteridophytes (LIBERMAN MAXE 1971; BURR and EVERT 1973) (Plates 6.13 E and 6.14). Such piling up can also be applied agains the degenerating nucleus (LIBER-MAN-MAXE 1971). The endoplasmic reticulum of young cells is mainly of a rough type and the cytoplasm is dense, rich in ribosomes. Through differentiation, ribosomes are more and more scattered and the endoplasmic reticulum becomes almost entirely smooth. However, some ribosomes may survive on the externally of such formations (ESAU and GILL 1971). However, the characteristics of the endoplasmic reticulum of sieve cells suggested the specific terms of *sieve tube reticulum* (BOUCK and CRONSHAW 1965) or *sieve element reticulum* (SRIVASTAVA and O'BRIEN 1966).

In section 6.2.1 we shall see that the endoplasmic reticulum of sieve cells participates in the elaboration of the pores of these elements.

Young sieve cells include numerous dictyosomes (Plate 6.15) which manifest an intense activy during their differentiation, then disappear in completed sieve cells. Already in 1963 (BUVAT 1963 c) such an activity was demonstrated by histo-autoradiography (NORTHCOTE and WOODING 1966, and other authors) and by means of cytochemical techniques (CATESSON 1973; ARSANTO and COULON 1974); it mainly concerns the formation of nacreous walls when they are present, and probably the thickening of all walls. Saccules of the distal face of dictyosomes, in particular, produce vesicles including polysaccharide precursors of cell walls (see below). The proximal saccules give rise to another type of vesicle, smaller and darker (probably alveolate) which is positive to acid phosphatase and belongs to the lysosomal apparatus (ARSANTO 1979). In particular, it plays a part in the formation of autophagic vacuoles which participate in the differentiation of sieve cells (BUVAT and ROBERT 1979). Although the publications on these dictyosomes describe only Golgi vesicles, it is probably peripheral tubuloreticulated systems, possibly connecting dictyosomes together, which produce such vesicles (ARSANTO 1979).

Sieve cells of Pteridophytes, with the exception of Lycopodineae, show characteristic inclusions, the "refringent globules" or "dense globules" (MAXE 1964), strongly contrasted with osmic fixatives (Plate 6.16 A–C). In the Polypode, these

◁——

Plate 6.14 A–C. *Polypodium vulgare,* sieve cells; piles of endoplasmic reticulum cisternae.

A Endoplasmic reticulum associations in spheric bodies or *Nebenkerne.* The most external and most internal membranes still bear some ribosomes *rb;* × 16 000.

B Endoplasmic reticulum amounts *(era)* around the pycnotic remnants of degenerating nucleus *(n); dg* dense globule; *li* lipid granule; *m* mitochondrion; × 19 000.

C Endoplasmic reticulum amounts *(era)* against the plasmalemma *pl; nw* "nacreous" cell wall; × 24 300.

Fixation: glutaraldehyde – OsO_4; contrast: $KMnO_4$; [after LIBERMAN – MAXE 1971 (**A** and **B**); 1983, (**C**)]

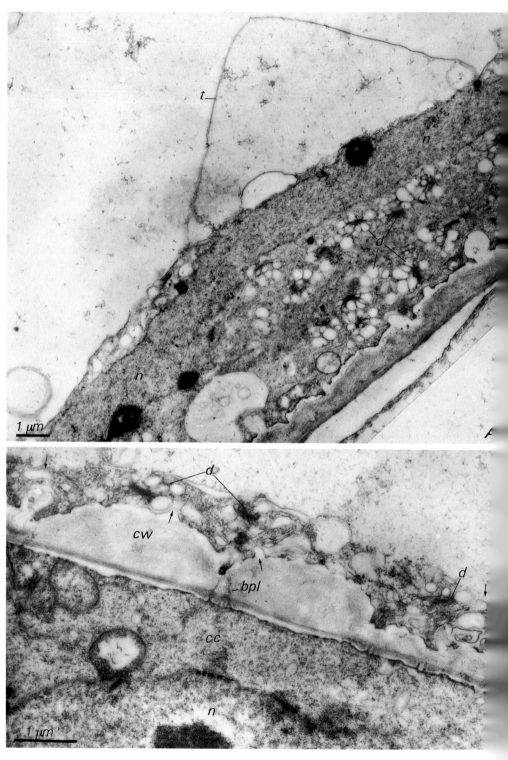

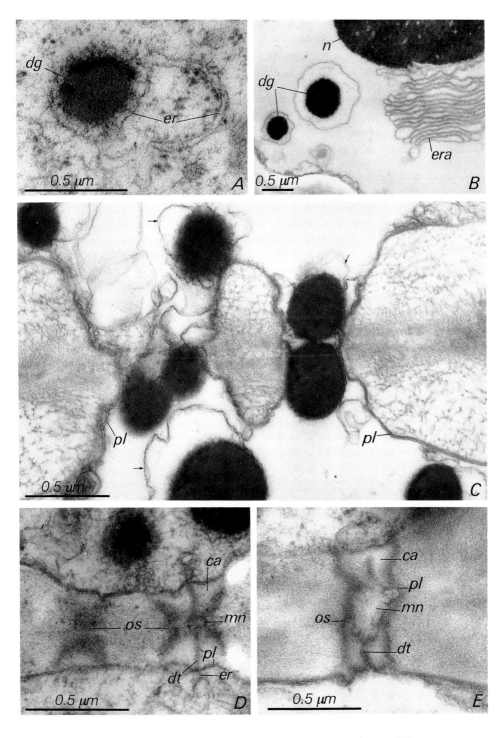

◁ **Plate 6.15.** Legend see p. 320 △ **Plate 6.16.** Legend see p. 320

319

granules occur in the endoplasmic reticulum, swelling the end of cisternae (LIBER-
MAN-MAXE 1971); in other genera *(Platycerium, Phlebodium)* they occur in dictyosomes (EVERT and EICHHORN 1974) or *(Davallia)* in smooth areas of the endoplasmic reticulum, which are probably related to Golgi saccules (FISHER and EVERT
1979). These granules are typically surrounded by reticular or Golgi membranes in
which they occurred (Plate 6.16 A, B). Commonly, they consist of a very osmiophilic centre and a less opaque peripheral zone, sometimes granulous, particularly
in young sieve cells.

Dense globules mainly consist of protein, they are stained by mercuric brom-
ophenol blue (MAXE 1964) but negative to lipid staining (Sudan Black B, Nil Blue)
and to polysaccharide tests (PATAg, according to THIÉRY 1967) (LIBERMAN MAXE
1978). Comparison with peroxisomes is suggested, but they are never found to be
crystalline and are apparently negative to the peroxidase test (LIBERMAN MAXE
1974). They increase in number during the differentiation of Polypode sieve cells
and are frequently located at the pore apertures (Plate 6.16 C). Their role has not
been established.

◁

Plate 6.15. A, B. Golgi apparatus activities during the differentiation of lateral cell walls of sieve
cells in *Cucurbita.*

A Cell at the phase of tonoplast release *(t);* numerous dictyosomes *d* producing a profusion of
transparent vesicles. It is known that these vesicles enrich the cell wall with polysaccharide material by an exocytosis process; *n* highly elongated nucleus.
B Continuation of Golgi cell wall elaborations after the disappearance of the tonoplast; *d* dictyosomes surrounded with transparent Golgi vesicles, similar to the periplasmic space of the cell
wall, *cw; arrows* indicate Golgi contribution to the periplasmic space; *bpl* branched plasmodesma
between the companion cell *cc* and the sieve cell; *n* companion cell nucleus; × 8800 **(A)** and
16 500 **(B).**

Fixation: O_SO_4; contrast: $KMnO_4$

Plate 6.16 A–E. *Polypodium vulgare,* phloem, "dense globules" **(A–C)** and formation of sieve cell
pores **(D, E).**

A Reticular origin of dense globules: the globule *dg* is surrounded by a membrane which is in
continuity with a rough endoplasmic reticulum profile *er;* × 55 000.
B Dense globules surrounded by a membrane, probably derived from the endoplasmic reticulum;
era endoplasmic reticulum amount; × 17 000.
C Two completed pores between two sieve cells, in which dense globules have become stuck, still
enclosed in a membrane *(arrows); pl* continuous plasmalemma, along the pores, from one sieve
cell to another; × 45 600.
D Beginning of pore formation; *mn* median nodule enlarging the plasmodesma at the level of the
middle lamella; *ca* egg timer-shaped callose deposit, parts the edges of the sheath of osmiophilic
substances *os* aside, around the plasmodesma; *pl* plamalemma; *dt* desmotubule; *er* endoplasmic
reticulum in continuity with the desmotubule; × 55 000.
E More advanced state of pore differentiation, increasing of the median nodule *mn,* pushing away
the callose cap and the osmiophilic substance *os* up to open a cylindrical cavity which becomes
the future pore; *dt* desmotubule; × 55 000.

Fixation: glutaraldehyde – O_SO_4; contrast: $KMnO_4$; [after LIBERMAN-MAXE 1968 **(D, E)**, 1971 **(A,
B)**, 1983 **(C)**]

6.2 Relations Between Sieve Cells, and Between Sieve Cells and Adjacent Cells

Sieve cells are connected through pores of sieve areas. The fine structure of these areas, at first controversial in the absence of electron microscopy, has been acknowledged with the use of this microscope, which also enables the investigation of the their ontogeny.

6.2.1 Differentiation of Sieve Plates

At the origin, the end wall which separates two successive sieve cells is thin and includes common plasmodesmata. In the areas which will develop into sieve fields, each pore develops from a single plasmodesma (Plate 6.17 a, b). In various species, the endoplasmic reticulum of these elements, in the process of differentiation, is partially scattered into the cytoplasm and partially applied against the plasmalemma. In particular, endoplasmic reticulum saccules progress over both faces of the wall around plasmodesmata, whereas the areas which will form the sieve "bars" are not systematically lined with reticular material (Plate 6.17 b, e).

Plate 6.17 a–e. Differentiation of the metaphloem sieve cell wall.

a, b *Cicer arietinum*, stem metaphloem. Callose discs *ca* reaching the middle lamella; **a** at the beginning of widening of plasmodesma *Pd;* **b** at the time of formation of the median cavity *cm*. Callose discs are covered by an endoplasmic reticulum cisterna *Re,* and are not reactive.

c–e *Ecballium elaterium*, stem metaphloem. **c** Stage of fusion of callose discs entailed by the disappearance of the middle lamella *Lm;* **d** recently opened pore *Po,* surrounded with callose; **e** another image of the sieve plate in the course of differentiation, showing the plasmalemma *Pl,* more reactive against the bars *B* of the future sieve plate than at the sites of the future pores, where it is emcompassed between the callose discs and the endoplasmic reticulum. *Di* dictyosome; *π* proximal face; *δ* distal face; *V₃* polysaccharide Golgi vesicles, yielding material for the setting of bars; *V₄* large Golgi vesicles, not reactive.

Fixations: glutaraldehyde – O$_S$O$_4$; PATAg test for acid polysaccharide detection; (after ARSANTO 1979); author's original lettering

Plate 6.18 a–c. Secondary phloem, sieve plate differentiation.

a *Robinia pseudoacacia,* sieve area during differentiation. Detection of acid phosphatase activities (GOMORI's technique). The lead phosphate precipitate is localized on callose discs, between the bars *b;* *er* endoplasmic reticulum; × 20 000.

b *Acer pseudoplatanus,* sieve area at the onset of differentiation. At this stage, callose discs *ca* are reactive to the PATAg test, according to THIÉRY (1967), for the detection of acid polysaccharides; *d* dictyosomes, producing polysaccharide vesicles *gv,* which supply the bars with thickening material *(arrows); pl* plasmalemma, facing callose discs less reactive than at the level of bars; *er* endoplasmic reticulum, closely applied against the plasmalemma at the callose disc location; × 40 000 (PATAg reaction, according to THIÉRY 1967).

c More advanced state of sieve plate differentiation. Callose less and less reactive; differences of plasmalemma reactivity *(pl)* facing the bars and facing the callose discs are more pronounced than in **b;** × 40 000; same technique as in **b.**

(after CATESSON 1973)

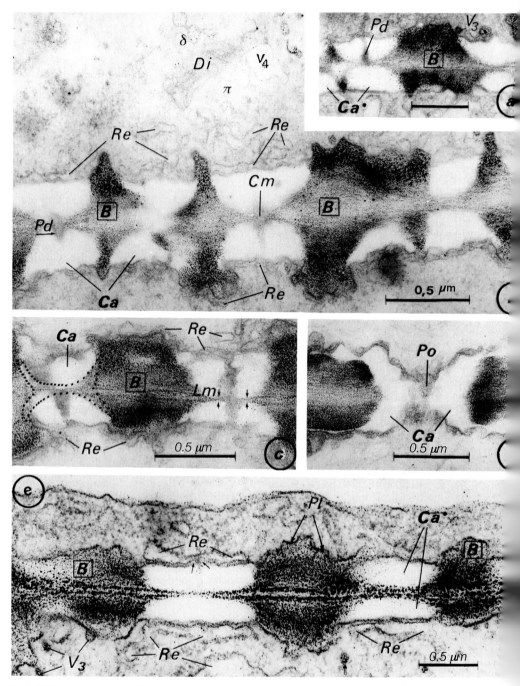

Plate 6.17. Legend see p. 321

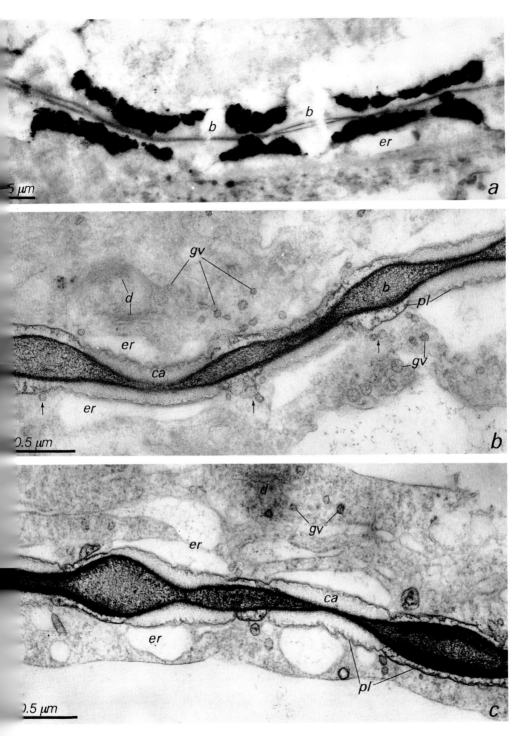

Plate 6.18. Legend see p. 321

On each side of plasmodesmata a callose disc occurs, surmounted by a saccule of endoplasmic reticulum. It seems that the association of reticulum with callose platelets impairs the movement of Golgi vesicles which, in different places, contribute to the wall thickening. Between the two callose discs, the thickening process ceases and, moreover, lytic processes enlarge the middle of the plasmodesmal canal and disorganize the wall which becomes thinner and then disappears. Both callose discs are brought into contact, and fuse, the callose is arranged in a sheath around the newly opened pore (Plate 6.17c, d) (ESAU and CHEADLE 1965; ESAU and CRONSHAW 1968b; LIBERMAN MAXE 1968; ARSANTO 1979). The existence of lytic phenomena was demonstrated by CATESSON (1973), who observed a high acid phosphatase activity at the level of callose discs, at the end of pore formation (Plate 6.18a). These facts suggest that the endoplasmic reticulum is concerned with the transport of hydrolase enzymes which perforate the sieve areas, and also maybe with the synthesis and carrying of callose precursors, whereas Golgi vesicles carry the cellulose precursors away to the plasmalemma which covers the future bars of sieve areas (Plate 6.18b) (CATESSON 1973; ARSANTO and COULON 1975). The acid polysaccharide test (PATAg, THIERY 1967) shows labelling differences between the areas of the plasmalemma, whether they line the bars (clear labelling) or the sites of future pores (poor or absent labelling). This plasmalemma differentiation could entail the distribution of both categories of precursors (CATESSON 1973; Plate 6.18b, c).

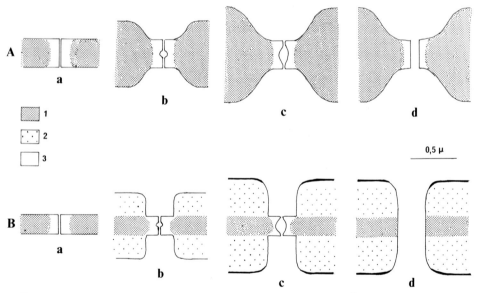

Fig. 6.9 A, B. *Polypodium vulgare,* differentiation of sieve cell pores. **A** In the protophloem, constitution of a median nodule from a plasmodesma *(a),* at the level of the middle lamella *(b);* this nodule enlarges *(c)* and the gap extends to the ancient plasmodesma extremities *(d).* The pore is then open and is bordered with a sheath of callose *(white).* **B** In the metaphloem, a similar beginning, but accompanied by a thickening of the cell wall "nacreous" coat, and by the formation of a larger pore *(d).* In addition, the callose sheath disappears in the late metaphloem. *1* Dense polysaccharide cell wall; *2* "nacreous" thickening; *3* callose (after LIBERMAN-MAXE 1968, 1978)

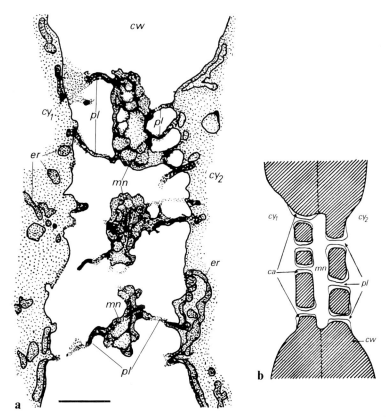

Fig. 6.10 a, b. Structure of pores, between two Gymnosperm sieve cells *(Metasequoia glyptostroboides)*. **a** Tangential longitudinal section; *cy* cytoplasm; *mn* elongated median nodules, on which several plasmodesmata *(pl)* converge on each side; *cw* cell wall; *er* endoplasmic reticulum; ×38000; drawing after an electron micrography from KOLLMANN and SCHUMACHER (1963). **b** Scheme of interpretation; *ca* callose, other abbreviations as in **a** (KOLLMANN and SCHUMACHER 1963)

The enlargement of plasmodesmata gives rise to a "median nodule" generally poorly apparent in Angiosperms, but well characterized in Pteridophytes such as Polypode (LIBERMAN MAXE 1968), in which each pore is also derived from a single plasmodesma (Plate 6.16 D, E and Fig. 6.9), and much more developed in Gymnosperms (KOLLMANN and SCHUMACHER 1963; KOLLMANN 1964; MURMANIS and EVERT 1966; WOODING 1966). Sieve cell pores of these Gymnosperms are formed in primary pits including *several* plasmodesmata. The middle lamella is lysed under the pit-field surface and becomes a median nodule extending to several plasmodesmata on either side (Fig. 6.10 a, b). The plasmodesmata are enlarged in their turn, and become coated with plasmalemma sometimes including remaining tracts of endoplasmic reticulum. It should be recalled that there are no P-proteins in the phloem of Gymnosperms or in that of Pteridophytes.

Along with the formation of pores, the Golgi system is strongly activated at the level of sieve bars, which become thickened with polysaccharide-loaded vesicles

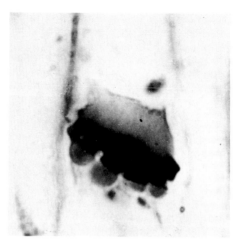

Fig. 6.11. Secondary phloem of *Robinia pseudoacacia;* so-called *gouttelettes de Lecomte,* a structure considered an artefact. The sieve plate accumulation penetrates into the sieve pores and forms droplets hanging in the lower sieve cell. Fixation: REGAUD – hematoxylin

generated by dictyosomes and labelled using the silver proteinate technique, according to THIERY (1967). The maturation of the vesicles, demonstrated in other tissues by VIAN and ROLAND (1972) is achieved during their progression into the sieve cell, maturation which enables them to fuse with plasmalemma and to release their content in the periplasmic space (ARSANTO and COULON 1975). These authors also assessed the determining contribution of Golgi productions in the formation of nacreous walls.

At the end of differentiation, the pores of sieves areas are invaded with mictoplasm and particularly P-proteinaceous filaments (BEHNKE 1971). Other cytoplasmic components can be found in pores, but the doubtful fixation of sieve tubes does not allow an accurate interpretation of figures. However, it is certain that cell plasmalemmas which, as previously mentioned (p. 313), survive closely applied against walls, are connected together in pores and therefore in continuity (Plate 6.19 A).

With regards to the endoplasmic reticulum, saccules which surmount the callose platelets, when they are present, survive only temporarily and disappear when pores are perforated (ESAU and CRONSHAW 1968 b).

▷

Plate 6.19 A, B. *Cucurbita pepo* phloem.

A Sieve plate bar, entirely covered by the plasmalemma *pl,* which is continuous from one cell to the other; *ca* callose ring; *ml* middle lamella; *pw* primary cell walls; *er* endoplasmic reticulum cisternae, applied against the plasmalemma; *mic* mictoplasm, penetrating into the pores and in continuity between sieve cells; × 60 500.
B Typical companion cell; dense cytoplasm, rich in ribosomes; vacuolar apparatus divided into small vacuoles *v; p* poorly differentiated plastids; *m* short or spherical mitochondria; *d* numerous dictyosomes; *n* nucleus, provided with a large nucleolus, and in a central position; *sc* sieve cells; *er* endoplasmic reticulum; × 5170.

Fixations: OsO_4; contrast: $KMnO_4$

326

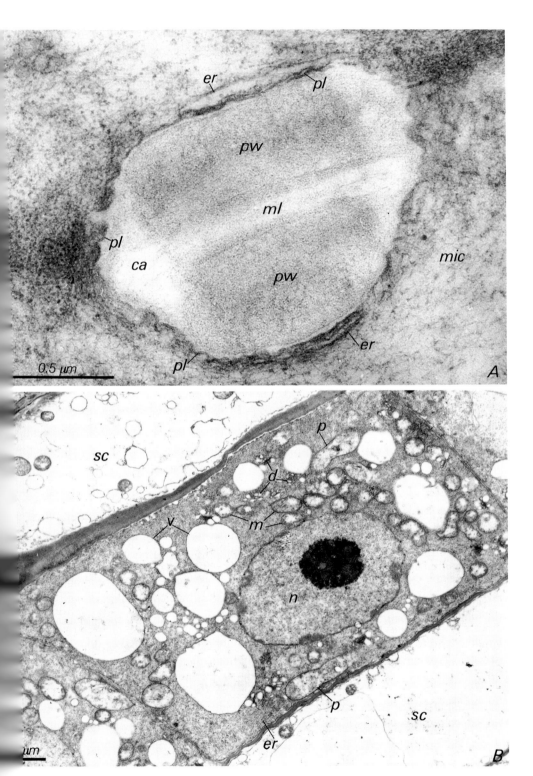

327

Endoplasmic reticulum tracts may occur in pores, in different places, mainly in species devoid of P-proteins. However, the pore content has no significant occluding structure (BEHNKE 1971; JOHNSON 1973).

In several preparation, sieve cells show a polarity underlined by the global accumulation of their content at one of their ends, particularly P-proteins which constitute the *sieve accumulations* mentioned earlier (p. 296). Such accumulations have sometimes been considered as occluding structures of pores, possibly confounded with callose deposits. But LECOMTE (1889) showed that the material accumulated at the surface of sieve areas penetrates into pores and can be found in the "following" cell as hanging droplets beneath the sieve wall. For this author "LECOMTE droplets" materialized the circulation of sap (Fig. 6.11). Presently, such figures are known to be artefacts caused by pressure variations related to the sampling and fixation of phloem tissues.

6.2.2 Differentiations in Lateral Walls

Lateral walls of sieve cells have plasmodesmata, partly concentrated in thinner areas which form primary pits. The plasmodesmata through which sieve cells and companion cells communicate, are of a particular type; on the sieve cell wall side a simple tubule reaches the middle lamella and becomes ramified in several branches which progress through the companion cell wall side (Plate 6.13 A–D). Usually the ramified plasmodesmata are lined with plasmalemma which is thus continuous between these elements, and they typically enclose an axial tubule issued from the endoplasmic reticulum of each cell (ESAU and CRONSHAW 1968 b; ARSANTO 1979).

6.3 Companion Cells

In the phloem of Angiosperms, sieve cells are commonly associated with smaller cells different from the other cells of phloem parenchyma. Closely applied against sieve cells, they have been named "companion cells".

Companion cells arise along with the last mitoses preceding the differentiation of sieve cells. A longitudinal mitosis divides a cell relatively large in diameter (Plate 6.20), resulting in the sieve cells and a narrower cell which will develop into a companion cell. The narrow cell can divide in turn, in both directions, but more commonly only transversally. Derivatives are series of small cells which can or cannot be arranged in line when going from one sieve cell to the next (Fig. 6.1).

Such a continuity occurs mainly in Monocotyledons and in some herbaceous Dicotyledons, plants in which the phloem is poor in parenchyma cells.

▷

Plate 6.20. *Cucurbita pepo,* phloem

Origin of sieve cells and companion cells: longitudinal cleavage of long cambial initials, unequal in the cytoplasm. The future sieve cell is wider and with large vacuoles, the future companion cell

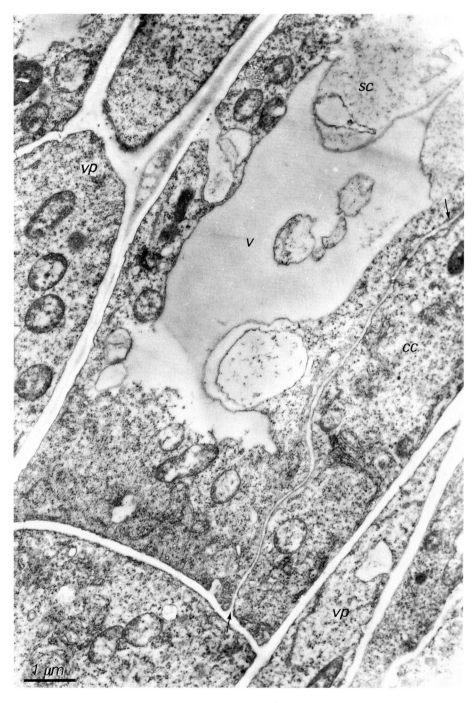

cc is narrower and shows no vacuole in the present section; *vp* vertical phloem parenchyma; *arrows* indicate the separation cell wall between the two cells, following a recent cleavage; × 14000.

Fixation: glutaraldehyde – O_sO_4; contrast: $KMnO_4$

The cleavage wall between companion cells and sieve cells generally shows particularly developed primary pits with ramified plasmodesmata. In developed species, these pit fields may be comparable to sieve areas less differentiated than terminal sieve plates, and similar to those which occur between contiguous sieve cells (Figs.6.33b; 6.34a, c).

The cytoplasm allows a clear distinction between companion cells and neighbouring cells of phloem parenchyma because of its high density and its vacuolar system which is divided into numerous, more or less spherical vacuoles (Fig.6.1 and Plate 6.19 B).

The marked basophilia of the cytoplasm is caused by the large amount of ribosomes. In case of secondary phloem, the basophilia is higher than in the initial cambial cell, a largely vacuolated cell. It seems that, after the cleavage, unequally with regards to the cytoplasm which separates the future sieve cell from the companion cell, the latter, in addition to the nucleus, is mainly loaded with cytoplasm and relatively few vacuoles (Plate 6.20) but in various quantities, depending on organs and species, e.g. more in trees according to A.M.CATESSON (personal communication). The vacuolar system of the companion cell develops later, through an autophagic process similar to that occurring in procambial cells.

The nucleus is rather voluminous and central (Plate 6.19 B); the chondriome commonly consists of granular mitochondria together with short elements (Fig.6.12).

Plastids are poorly differentiated and slightly larger than mitochondria. They rarely include chlorophyll; exceptionally, phytoferritin deposits (CATESSON 1966) and a few small starch grains may be observed within.

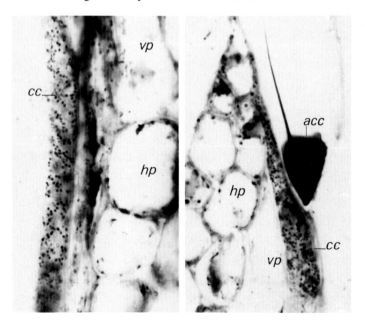

Fig.6.12. Granular mitochondria of companion cells *cc;* tangential sections in the secondary phloem of *Robinia pseudoacacia. acc* Flagellate sieve cell accumulation; *vp* vertical parenchyma; *hp* horizontal parenchyma. Fixation: REGAUD – hematoxylin, May 25th

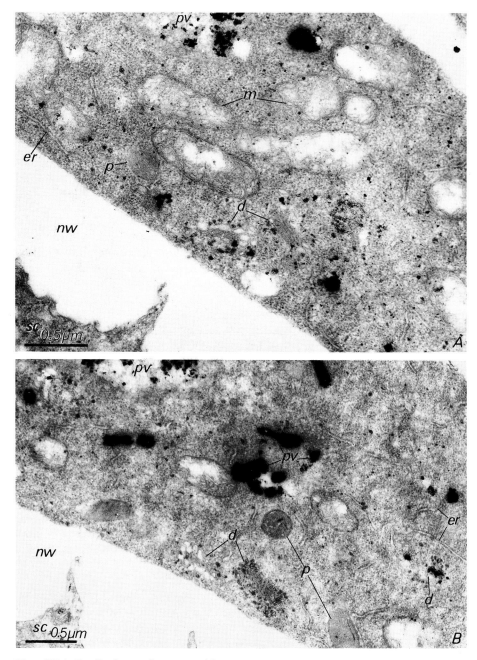

Plate 6.21 A, B. *Hordeum vulgare,* root phloem.

Ultrastructural cytology of companion cell differentiation; acid phosphatase activities. The lead phosphate precipitates are localized in some Golgi vesicles, dictyosome saccules (especially in **A**) and provacuoles *pv* (especially in **B**); *sc* sieve cells; *d* dictyosomes; *m* mitochondria; *p* plastids, poorly differentiated; *nw* "nacreous" cell wall; *er* endoplasmic reticulum; × 34 000 (**A**), 27 000 (**B**)

(GOMORI's technique 1952)

331

Companions cells show a normal endoplasmic reticulum and numerous dicty-osomes. Several publications reported acid phosphatase acitivity in the endoplas-mic reticulum, an uncommon fact (see CATESSON and CZANINSKI 1967). On the contrary, in the dictyosomes and vacuoles of companion cells associated with young sieve elements, hydrolase has a high activity as assessed by BUVAT and RO-BERT (1979) (Plate 6.21). Peroxidases were also detected in vacuoles, on the tono-plast and the plasmalemma (CZANINSKI and CATESSON 1970; CATESSON 1980).

This cytological state is that of companion cells applied against functional sieve tubes. The evolution of such cells is discussed. ESAU (1965 cf. p. 288) consid-ered that they are *physiologically* and *anatomically* associated to sieve cells and that they die when the latter degenerate. SALMON (1946), on the contrary, suggested that companion cells could normally develop into functional sieve cells which re-place the cells which no longer function.

This matter should be separately studied in Dicotyledons in which the func-tional phloem is constantly restored through cambium activity, this, however, does not necessarily imply the contribution of companion cells, and in long-living Monocotyledons without secondary conductive tissues in which the phloem func-tions more than a year without being restored.

In the first case, observations carried out in the phloem of *Robinia* showed that the functioning and death of sieve cells and their companion cells occur simulta-neously, suggesting that both types of cells constitute an indivisible anatomical and physiological association (Fig. 6.13).

In the second case, the example of Vanilla, a liana, in which death occurs at the 2-year-old base of the stems, was studied by DUNOYER DE SEGONZAC (1958). The author reported the very slow and progressive differentiation of companion cells, but did not note *any replacement* of sieve cells.

Thus, the common evolution of companion cells and sieve cells seems to be confirmed.

Companion cells occur only in the phloem of Angiosperms but they are some-times missing in the first phase of differentiation of primary phloem (proto-phloem) and in the phloem of certain primitive, lignified Dicotyledons (BAILEY and SWAMY 1949).

However, it seems that they are always present in the extremities of conducting bundles of thin leaf veins, in which they are concerned with the loading of sieve tubes (FISCHER and EVERT 1982). Sometimes they become comparable to transfer cells.

Sieve cells of Gymnosperms and Pteridophytes have no companion cells. However, in conifers, cells of phloem parenchyma and phloem ray cells can be-come closely associated with sieve cells and show a strongly stained cytoplasm, like that of companion cells of Angiosperms. Observed already in 1891 by STRAS-BURGER, these cells have been designated as "albuminous cells". Their strong posi-tivity was interpreted as an indication of a large protein amount, but recent works performed with electron microscopy in particular (SAUTER 1974, 1980; SAUTER et al. 1976), assessed that these cells rather include a high density of polysomes and organelles such as mitochondria; moreover, the vacuome is divided into numerous small vacuoles which give them a denser aspect than common parenchymatous cells (Fig. 6.14a). The ontogenic origin allows their differentiation with companion

Fig. 6.13. *Robinia pseudoacacia,* ancient phloem of the stem, tangential section passing outside the fibre layers. Simultaneous necrosis of sieve tubes and their companion cells *(asterisks)*. The phloem parenchyma only remains alive. *vp* Vertical phloem parenchyma. Fixation: REGAUD - hematoxylin

cells of Angiosperms, as they are not produced by the division of mother cells of sieve cells, but represent particular differentiations of certain cells of vertical phloem parenchyma or, more frequently, of phloem rays.

Conversely, a number of cytological and physiological features allow a comparison with companion cells: (1) they are starch-free, at least in summer when they become associated with activated sieve cells; (2) they are related to sieve cells through dissymmetric areas, sieve areas on the sieve cell side, areas with numerous plasmodesmata on the STRASBURGER cell side; (3) the cytoplasm density is similar to that of companion cells; (4) they die together with the sieve cells with which they are associated (Fig. 6.14 b).

Cytophysiological studies indicate that the similarity is also functional: the activity of various dehydrogenases and acid phosphatases is greater than in common parenchmytous cells, when the functional phloem is most activated (SAUTER and BRAUN 1972).

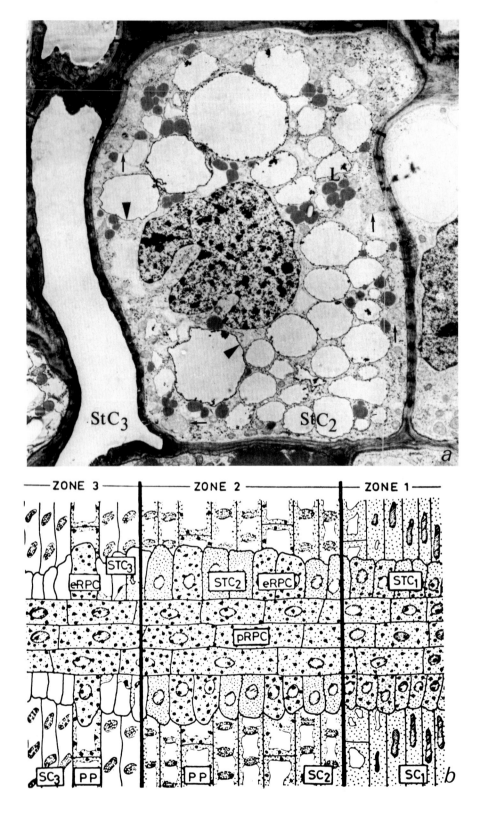

ZONE 3   ZONE 2   ZONE 1

6.4 The Phloem Parenchyma

In addition to sieve cells and companion cells or STRASBURGER cells, the phloem is commonly composed of living cells with parenchymatous features which often correspond to storage cells. Starch, lipids or secretion products such as tannins and resins are stored within. Some of them can even become clearly secretory and, in some cases, become irregularly anastomosed resulting in a network of "pseudo-laticifers" or "articulated laticifers" (Chicoraceae).

In the *primary* phloem, there are only parenchymatous cells elongated parallel with sieve tubes.

In the *secondary* phloem, in addition to the former system, there are radially elongated cells forming the phloem rays. Here, elongated cells parallel to sieve tubes are designated as *"vertical phloem parenchyma"* and "phloem rays" as *horizontal phloem parenchyma*. The secondary vascular tissue shows comparable systems.

Cells of phloem parenchymas, whether vertical or horizontal, are active and play an important role in the seasonal variations of the phloem physiology as well as, casually, in the regeneration of rhytidomes. They are the only cells to survive in the old phloem which is no longer functional.

In the functional phloem, they show a parietal cytoplasm, surrounding a generally single vacuole, contrarily to the companion cells. The nucleus shows the common structure of the species concerned, it is applied against the cell wall.

In young aerial organs, in which light is not stopped by the rhytidome, the cells include developed chloroplasts but these chloroplasts are deficient in photosystem I (JUPIN et al. 1975) (Plate 6.22 A, B). They can actively produce starch within the leucoplasts or chloroplasts and phytoferritin deposits are frequently found within (CATESSON, 1966; Plate 6.22 C, D; Fig. 6.15).

◁──

Fig. 6.14 a, b. Strasburger cells. **a** Cytology of a Strasburger cell, from a phloem ray of *Pinus nigra*. Relatively dense cytoplasm, numerous small, globular vacuoles, well outlined by a regular tonoplast *(arrowheads)*, voluminous and lobed nucleus. StC_2 Live Strasburger cell (in *zone 2* of **b**); StC_3 dead Strasburger cell (in *zone 3* of **b**); *L* lipid globules; *arrows* indicate plastids devoid of starch and poorly differentiated; courtesy of SAUTER et al. 1976). **b** Schematic development of a *Larix* phloem ray from the cambium *(zone 1)* to the advanced wood *(zone 3)*. Strasburger cells are *stippled* and are devoid of starch (denoted as *big dots* in other parenchyma cells). SC_1, SC_2, SC_3 sieve cells, immature in *zone 1*, mature in *zone 2*, dead in *zone 3*. STC_1, STC_2, STC_3 Strasburger cells, alive and adjacent to sieve cells in *zones 1* and *2*; in *zone 3*, STC_3 *(white)* have died *at the same time* as the associated sieve cells SC_3. *pRPC* procumbant ray parenchyma cells; *eRPC* upright ray parenchyma cells; *PP* vertical parenchyma cells. These three kinds of cells remain alive and contain starch (courtesy of SAUTER 1980)

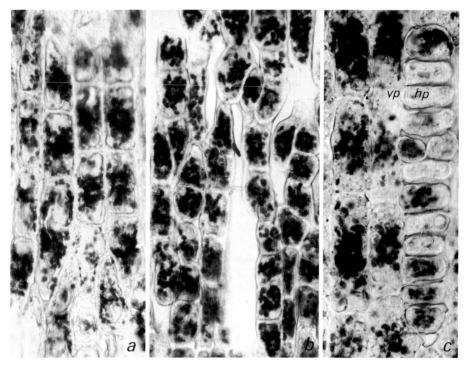

Fig.6.15a–c. *Robinia pseudoacacia,* phloem parenchyma, amylogenesis. **a** In April; **b** in September; **c** in November. *vp* Vertical parenchyma; *hp* horizontal parenchyma. Tangential sections prepared with lugol (iodine-iodide reactive); plastids are strongly coloured due to the starch grains

In functional phloem, the chondriome is abundant, often broken off in mitochondria, indicating a great activity, or it is associated into long chondriocontes. Sometimes, cells of phloem parenchymas soon become tanniniferous (Fig.6.16).

6.4.1 "Vertical" Phloem Parenchyma (Axial Parenchyma)

Derived from procambial cells or from "long" cambial cells, vertical parenchyma cells can remain very elongated and often fusiform or can divide transversally resulting in scattered, short longitudinal cords composed of almost isodiametric small cells (Fig.6.26 C). These small cells may have a different biochemical activity. Thus, they can elaborate calcium oxalate crystals in a form which does not occur in fusiform cells. This can be noted in the secondary phloem of *Castanea* (Fig.6.26).

6.4.2 "Horizontal" Phloem Parenchyma (Ray Parenchyma)

"Phloem rays" are derived from short initials of the cambium (Fig.2.51). Relatively short when they arise, ray cells lengthen radially, this orientation corre-

336

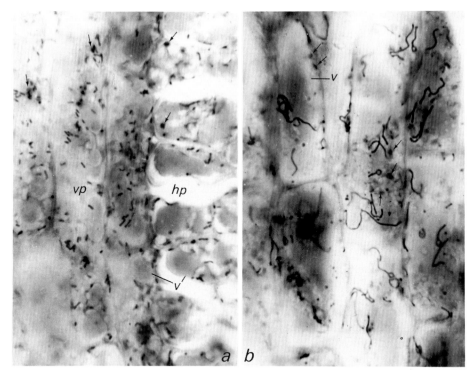

Fig. 6.16 a, b. *Robinia pseudoacacia,* phloem parenchyma, chondriosomes. **a** Young functional phloem, chondriosomes in the shape of globular or short rodlike mitochondria. *vp* Vertical parenchyma; *hp* horizontal parenchyma. **b** Ancient phloem: filamentous chondriosomes. *v* Tanniniferous vacuoles; *arrows* indicate plastids. Fixation: REGAUD – hematoxylin, in June

sponding to their larger size. Commonly, they have a large amount of cytoplasm and accumulate more storage material than axial parenchyma cells.

6.4.3 Phloem Parenchyma Pits

The cell walls of these two parenchyma systems have numerous, well-observed *primary pits,* sometimes occurring only on the radial faces (Fig. 2.51 A, B). Such primary pits occur also between parenchymatous cells and companion cells or sieve cells. In such a case, on the side of the conducting cell, a slightly differentiated "sieve area" takes the place of the primary pit field, like between companion cells and sieve tubes.

6.4.4 Some Cytochemical Data

In phloem parenchyma cells, the cytoplasm is less rich in ribosomes hence less chromophil than in companion cells. The endoplasmic reticulum is well developed particularly in the shape of long saccules parallel to the cell walls. Dictyosomes are rather scattered when cell differentiation is completed.

In the secondary phloem of *Robinia* and sycamore, the endoplasmic reticulum, in spring, has an acid phosphatase activity similar to that existing in companion cells (CATESSON and CZANINSKI 1967). It is exceptional and disappears in winter.

Dictyosomes and sometimes vacuoles are positive with the same test (technique of Gomori). The authors suggested that the high metabolic activity characteristic of the spring reactivation increases the hydrolase syntheses in the reticulum and that these proteins are transferred into vacuoles through the Golgi apparatus.

Various nucleoside phosphatases were detected in the phloem of several species (cf. CRONSHAW 1980); their infrastructural localization allows interesting suggestions on the mechanisms of assimilate translocation. The nucleoside phosphatase activites, particularly of ATPase, are more pronounced in the plasmalemma of companion cells and frequently of sieve cells. They also occur in the plasmalemma of phloem parenchyma cells. Exceptionally, they are absent in the plasmalemma of sieve cells in *Pisum,* a fact which seems to be related to the existence of "transfer cells" in the phloem of this species. The plasmalemma of these cells has a particularly pronounced ATPase activity when parietal excrescences are differentiated.

After being scattered in the mictoplasm, mitochondria, dictyosomes and, in some cases, vacuoles and P-proteins have ATPase activity. A potassium-dependent ATPase was acknowledged in the sieve cells of *Ecballium elaterium* (ARSANTO 1979). The physiological interest of these cyto-enzymologic tests will be analyzed in Section 6.9 (cf. p.363).

Phloem parenchyma cells, mainly when they are chlorophyllous, include peroxisomes in which catalase has been detected (CZANINSKI and CATESSON 1970). In addition to these corpuscles, peroxidase activities generally involve only the walls and the plasmalemma in Angiosperms. With regards to Pteridophytes, such activities are absent in the phloem parenchyma of the Polypode, except in peroxisomes, but they occur in the narrow strip of cells adjacent to both phloem and xylem (LIBERMAN MAXE 1974).

Plate 6.22 A–D. Cytology of the phloem parenchyma.

A,B *Cucurbita pepo,* young stem phloem (first internode): moderately differentiated chloroplasts: little developed grana; clear areas containing the chloroplastic DNA in a dense stroma; in **B,** the chloroplast is in the process of division; *d* dictyosome; *m* mitochondria; *px* peroxisome; × 34000.

C,D *Acer pseudoplatanus* (after CATESSON 1966). **C** Amyliferous leucoplast in a phloem derivative cell close to fusiform cambial initials (young vertical parenchyma); *s* starch; *fe* phytoferritin cluster; *m* mitochondrion; × 23000. **D** Chloroplast of a young horizontal parenchyma cell; outside of the granar structure, numerous phytoferritin clusters *(fe);* × 96000.

Fixations: glutaraldehyde – OsO_4; contrast: $KMnO_4$

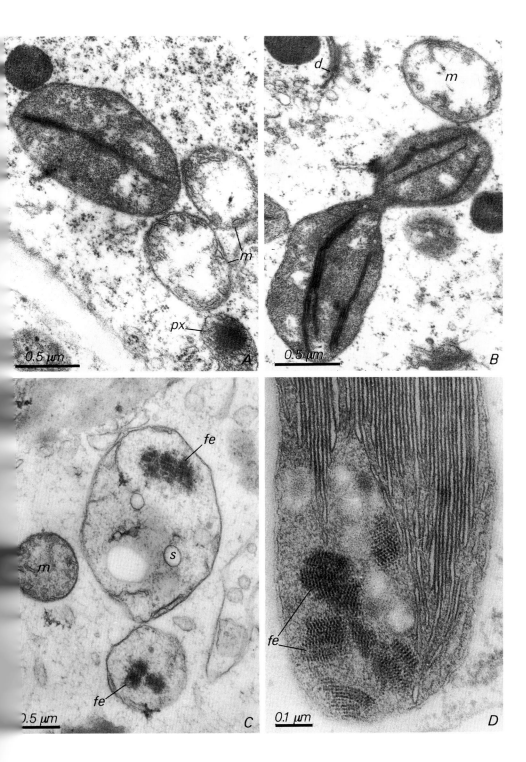

339

6.5 Phloem Fibres

When the phloem ceases to function, conducting cells together with parenchyma cells remain either globally cellulosic and are then crushed during growth or become sclerified. Other cells, intercalated in the functional phloem or at the periphery of this phloem, are more or less sclerified during differentiation and become phloem fibres.

They can be found in primary phloem as well as in secondary phloem. Some fibres begin to differentiate in organs which are still in the lengthening phase (particularly in the primary phloem); these fibres may become very long. Those which, in primary or secondary phloem, arise after the growth in length of organs, may also become elongated through intrusive growth, but they are shorter than the primary fibres of the same plants.

Primary fibres occur in the phloem of growing organs. The first differentiation corresponds to the end of the transversal divisions which continue to affect the neighbouring cells. This results in long, thin-walled cells. But the elongation of the future fibres is not only related to the organ growth in length. The fibres can be-

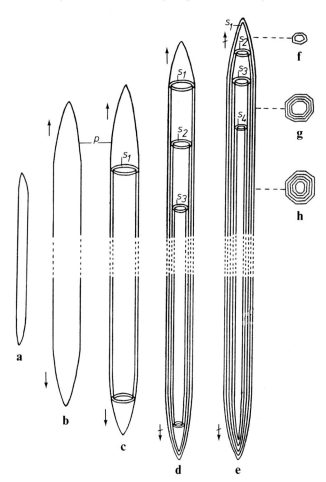

come longer than the areas where they have arisen, through a growth localized to their ends, a growth process which give them a fusiform aspect. By means of this double-growth process, primary phloem fibres sometimes reach considerable length. For example, primary phloem fibres of *Boemeria nivea* reach 550 mm (AL-DABA 1927), those of hemp are about 10 to 15 mm in length.

When the growth in length of the organ is completed, the walls in the median portion of the fibre become thickened by the addition of successive secondary cylindric encased lamellae (Fig. 6.17). Meanwhile, the ends where the wall remains thin, grow in length, progressing between the vertical phloem parenchyma cells in the middle lamellae (KUNDU 1942). In this case, new localized tripartite associations occur composed of the former primary layer of the parenchyma cell, the divided middle lamella and the newly elaborated portion of the fibre wall. The elements of this new structure are involved in a common growth process (SCHOCH-BODMER and HUBER 1951).

The ending growth generally lasts longer towards the apex than towards the base, which is easily understandable if we consider that the base tissues are completed whereas, towards the apex, the lengthening process progresses towards partially differentiated tissues.

When the whole lengthening ceases, some secondary lateral layers extend to the ends, while new layers are produced in the whole cell wall. The thickening is sometimes very large, reaching about 90% of the fibre diameter in some cases *(Linum usitatissimum)* (Fig. 6.18).

At least during their elaboration, secondary parietal layers are very slightly applied to the primary wall from which they become easily detached in microscope preparations; they become folded and even dissociated from one another in such preparations (Fig. 6.18).

Secondary fibres are derived from long cambial cells, like those which generate the vertical parenchyma and sieve cells. Of course, such fibres arise in organs with completed length and extend only through "intrusive growth" of their ends. In this way they become fusiform cells generally shorter than the primary fibres of the same organ. For example, in the hemp stem, secondary fibres are about 2 mm in length, whereas the average length in primary fibres is 12 mm.

The evolution of the cellular content of fibres is not well established and rather difficult to study. However, is should be noted that the growth and differentiation

---

Fig. 6.17 a–h. Growth and differentiation of primary phloem fibre cell walls. **a** Young fibre at the moment of its individualization in the procambium; **b** onset of growth in length and diameter, involving the whole cell, the cell wall is still entirely primary; **c** end of growth and addition of a secondary coat S_1 applied on cell walls along the median part of the future fibre. The extremities, where the primary cell wall p only subsists, lengthens along by a mode of intrusive growth *(arrows);* **d** the S_1 coat spreads towards the extremities and reaches the lower one, where the intrusive growth stops *(crossed arrow);* secondary coats S_2 and S_3 are deposited in the median part and progress towards the extremities; **e** the secondary coats reach the two extremities, and the intrusive growth is then stopped *(crossed arrows);* an S_4 coat is finally added (in this example). **f–h** Transversal sections at the levels indicated in **e**.

The scheme takes into account the fact that intrusive growth is more important and more prolonged at the upper extremity, where it occurs in younger parts of the organ, which is still growing (after ESAU 1965)

which occur in walls are performed by the living cell material. If, at the end of the evolution, when the cellular cavity is reduced to a narrow canaliculus, the protoplasm disappears, one should consider that fibres are living cells up to the end of their evolution, as well as sieve cells. Young fibres enclose plasmodesmata in the median part and later the wall thickening spares secondary pits, simple or slightly bordered. Thus, they preserve possible exchanges with neighbouring cells.

The cytology of phloem fibres is not established. It seems that plastids are poorly differentiated and are of small size, however, they store starch mainly in fall and winter. Mitochondria are numerous. As it occurs in certain xylem fibres, the nucleus may divide independently of the fibre division. Phloem fibres can remain cellulosic (flax) or become only slightly impregnated with lignim (hemp, ramie), or become highly lignified (jute), a generalized fact in Monocotyledons.

In the secondary phloem, in such cases where fibre bundles are in contact with phloem radial cells, the latter sometimes show markedly thickened walls and become sclereids but remain living. They are connected to the neighbouring cells, through tubular pits (Fig. 6.28; cf. sect. 6.7).

6.6 Primary Phloem

6.6.1 Protophloem

The first differentiations which appear on one side of the proconducting strands generated by the apical meristems, the phloem pole, form the *protophloem*. During the processes of parenchymatization, which occur on the periphery of the proconducting strands, the protophloem cells remain highly meristematic; ribosomes are particularly numerous in these cells, which causes a pronounced basophilia within the cytoplasm.

The vacuolar system is often non-existent or reduced to the provacuoles (tubulo-vesicular system) from which true vacuoles will evolve, resulting from sequestration and autophagy processes (Plate 6.10 A, B).

The protophloem is characterized by sieve cells with a narrow diameter, the sieve areas being hardly visible. It is not even known whether they exist within the protophloem of Gymnospersm. Companion cells often lack on the side of these first sieve tubes even in Angiosperms, in the stem and in the root, but they can probably be found within the leaf veins.

However, between the sieve tubes, the protophloem includes vertical phloem parenchyma (Plate 6.23). The protophloem is the essential conducting tissue of young growing organs. After the first sieve cells have lost their nuclei, they stop following the elongation of the organ; they find themselves passively overstretched and squashed. Therefore, the sieve tubes of the protophloem of the stems and the roots function for a brief period only. What remains of them is flattened up, can be seen for some time and then disappears. In contrast, within the fine leaf veins, they function during the whole life of the leaf.

Very often at this point, the remaining cells of the protophloem, which are up till then parenchyma cells, differentiate into *primary phloem fibres*. In the older or-

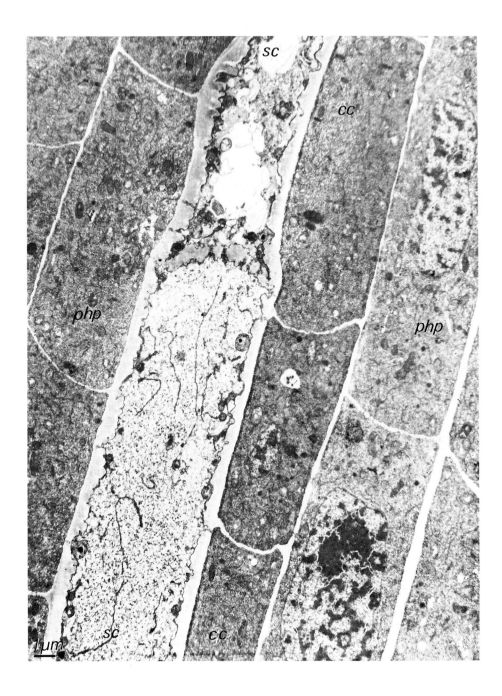

Plate 6.23. *Hordeum vulgare*, young root protophloem.

The sieve cells *sc* complete their differentiation earlier compared to the companion cells *cc* and to the phloem parenchyma *php*, very abundant in this protophloem and still meristematic; × 5000. Fixation: glutaraldehyde – O_SO_4; contrast: $KMnO_4$

343

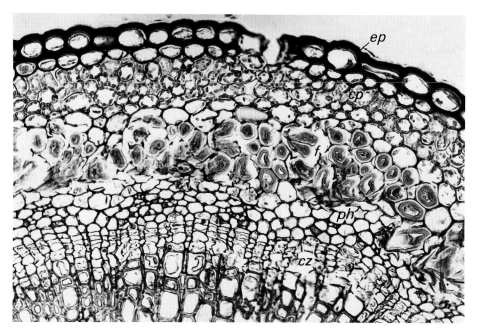

Fig. 6.18. *Linum usitatissimum,* young stem, transversal section showing the periphloem fibre ring *f,* surrounding the primary phloem *ph; ep* epidermis underlined by a subepidermal layer, both being collenchymatous; *cp* cortical parenchyma; *cz* cambial zone

gan, these fibres appear, *surrounding the phloem.* Such periphloem fibres have often, but erroneously, been described as "pericyclic fibres". This is especially the case for aerial stems for which the histogenesis does not allow, like for the roots, to define a true pericycle. Among the textile fibres of flax, the first to be differentiated, at least, are such a case (Fig. 6.18) (ESAU 1943).

The fibres of the primary phloem are then put together in bundles forming, at least in the stem, a network resulting from numerous junctions, similarly to the conducting tissue. These fibres are economically highly interesting. Flax has always been raised even in ancient times, especially in the Mediterranean areas as well as hemp in the Far East for textile industries. The length of the fibres is important for weaving, which is an advantage for protophloem fibres. The separation of reticulated bundles is achieved through steeping. This process enables pectic enzymes secreted by bacteria and fungi to attack the middle lamellae then the primary walls of the supple tissues surrounding the bundles of fibres. The latter are then mechanically isolated.

▷

Fig. 6.19. *Helianthemum tuberosum* (ornamental variety), transversal section of a young stem; conducting bundle typical of young Dicotyledons. The major part of the phloem is constituted of metaphloem *(mph),* comprising sieve tubes, companion cells (small diameter) and parenchyma, but no fibres. These last cells will form periphloem strands, still undifferentiated *(f); cb* cambium; *col* collenchyma, with intercellular spaces and lacunae; *ep* epidermis; *sc* secretory canals; *vb* vascular bundles

6.6.2 Metaphloem

The primary phloem which differentiates after *the growth in length* of the organ is *the metaphloem*. It can be slightly developed, as for instance in some stems of Dicotyledons or, on the contrary, constitute the only conducting tissue of the elaborated sap during its whole life (Monocotyledons, Pteridophytes).

The sieve cells of the metaphloem are wider and longer than those of the protophloem. Their sieve areas are also more visible and distinct. Companion cells, which do not exist except in Angiosperms, are associated with the sieve cells of the metaphloem in Dictotyledons and Monocotyledons.

In the metaphloem of Dictoyledons, the sieve tubes and their companion cells are surrounded by vertical phloem parenchyma (Fig. 6.19). In the Monocoty-

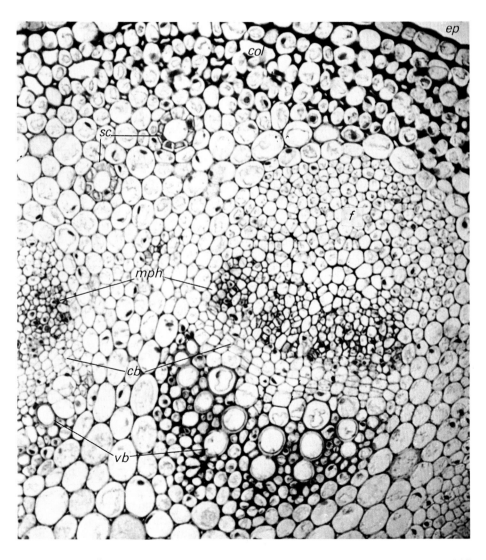

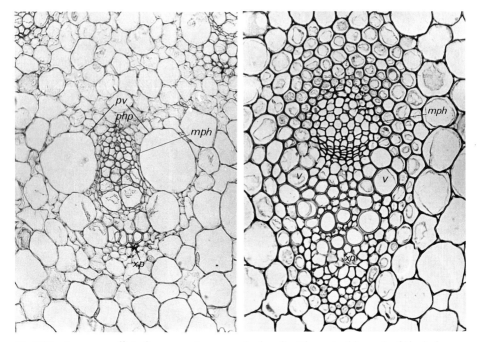

Fig. 6.20. *Asparagus officinalis,* young stem, vascular bundle. The metaphloem *(mph)* includes on-ly sieve cells (large diameters) and companion cells (small diameters), the arrangement of which is similar to a tiled floor. *php* Phloem pole; *xp* xylem pole; *pv* peripheral vessels, typical of Mono-cotyledons, still incompletely differentiated

Fig. 6.21. *Ranunculus acris,* vascular bundle. Like in Monocotyledons (Fig. 6.20), the metaphloem *mph* is selectively composed of sieve cells (polygonal cells of large diameter) and their companion cells (small diameter, most of them rectangular in section). As in Fig. 6.20, the whole system has an aspect of a more or less constant regular tiling. The differences are represented by the occurrence of a cambial zone and the absence of peripheral vessels. *v* "Superposed" vessels (the more recent ones); *xp* xylem pole

ledons, the sieve tubes and companion cells often form, *alone,* the main part of the phloem strands. The phloem parenchyma cells are exclusively present on the periphery (Fig. 6.20). The metaphloem thus, generally allows one to determine whether an organ comes from a Monocotyledon or from a Dicotyledon. However, there are Dicotyledons, the metaphloem of which is of a Monocotyledon type: they are called the "V-shaped bundles" Dicotyledons (Renonculaceae) (Fig. 6.21).

The metaphloem of Dicotyledons *usually lacks fibres,* which always arise in the protophloem. It can, however, become later sclerified after it ceases to function.

In Monocotyledons, fibres can sometimes be found in the metaphloem, but usually a sclerenchyma sheath develops, enclosing the totality of the conducting bundles, in addition to the strand which is possibly formed where the proto-phloem developed (Fig. 6.22).

The delimitation between protophloem and metaphloem can be uncertain and can be related only to the distinction of the moment at which the elongation of the organ ceases. However, it can also be quite clear when the protophloem alone

346

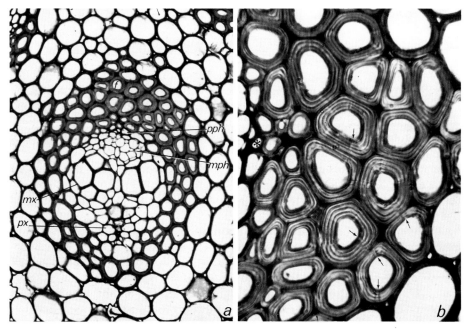

Fig. 6.22 a, b. *Dracaena* sp., conducting bundle of the stem. **a** General view in transversal section; *f* phloem fibres covering the phloem bundle and extending laterally through fibres with thinner walls surrounding the xylem; *mph* metaphloem; *pph* protophloem (phloem pole); *mx* metaxylem; *px* protoxylem (xylem pole). the metaphloem is only composed of sieve cells (large diameter) and companion cells (smaller diameter). The basic parenchyma is changed into sclerenchyma with cell walls thickened only slightly. **b** Enlargement of the region of phloem fibres developed around the phloem pole *(asterisk)*. The middle lamellae *(very dark)* are visible, like the multiple coats of the secondary cell walls, crossed by intercellular communications *(arrows)*

does not include any companion cell (aerial organs of some Monocotyledons but not the roots).

In persisting plants whith absence of secondary growth (Monocotyledons such as palms, bamboos, ferns etc.) the question can be raised as to whether the sieve cells remain functional for many years, despite the fact that they are enucleated. This seems possible but uncertain. We recall that, according to SALMON (1946), the first sieve cells could be replaced by others which would be differentiated later from companion cells. This transformation, as seen previously, does not seem to occur in the metaphoem of the persisting stem of Vanilla (DUNOYER DE SEGONZAC 1958).

6.7 Secondary Phloem

Derived, towards the outside, from the generating zone, the secondary phloem is also distinct from the metaphloem because of the interpenetration of the *horizontal parenchyma* (phloem rays) in the whole of the "vertical" cells. This horizontal parenchyma comes from *short* cells of the cambium, whereas all the cells of the "vertical" system, i.e. sieve cells, companion cells, cells of the vertical parenchyma, fibres, etc., arise from *long* cells ("fusiform" initials).

The proliferating activity of the cambium and the growth rate of the cells of the phloem it produces are seasonal. The result is a seasonal zonation in the secondary phloem, but it is usually less stressed than in the xylem. This is, on the one hand, due to the destruction more or less accentuated of the older layers and, on the other hand, due to the differences in the transversal growths of the various cell layers.

However, many Dicotyledons and Gymnosperms form tangential bands of fibres which stratify the secondary phloem (liber). The word "liber" was given because of the resulting foliated aspect (similar to the leaves of a book). *Stricto sensu,* the word *liber* should thus refer only to the secondary stratified phloem which is found in numerous Dicotyledons and Gymnosperms. The number of fibre bands is not constant from season to season and the zonation cannot be used to determine the age of the organ, as for the zonation of the xylem.

In fact, the phloem thickening is limited by the periodic formation of periderm which exfoliates its earlier layers.

The growth in thickness increases more and more the periphery of the phloem formation derived from the generating zone. Usually, only the phloem rays and not the vertical cells support the tangential growth. Sometimes, the cells of the horizontal parenchyma simply widen, but most of the time, they divide along the longitudinal radial orientation which in turn increases the number of cells in each ray, tangentially calculated.

The secondary phloem of Gymnosperms and Dicotyledons can be dinstinguished according to several characteristics.

6.7.1 Secondary Phloem of Gymnosperms

The secondary phloem of Gymnosperms can be distinguished first according to the characteristics of the sieve cells. They are elongated and slender, comparable to the fusiform initials of the cambial zone. This spindlelike form implies that the terminal wall are *very oblique* or even than they cannot be distinguished from the radial lateral walls. In this latter case, the sieve areas appear all along the radial faces. These sieve areas are more or less circular and are separated by areas without pores (compound sieve plates, Fig. 6.23). The pores seem to be very slender but they can be distinguished because their lumen bears an accentuated amount of chromophil content.

The vertical parenchyma is more or less developed depending on the groups and forms strands of cells which are shorter than the sieve cells. It often forms tangential bands which alternate with sieve cells. These bands are uncertain in

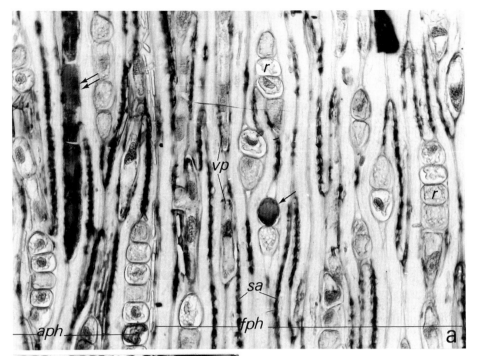

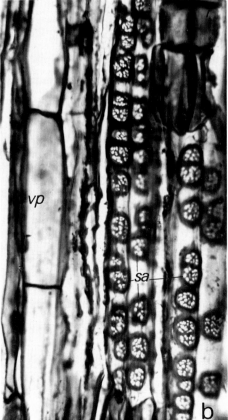

Fig. 6.23 a, b. *Pinus silvestris,* stem phloem.
a Tangential section; *r* uniseriate rays of
small height; the sieve areas *sa,* are seen in
profile; *vp* vertical parenchyma cells. Tanniniferous cells are found in the rays *(arrow)* and in the vertical system *(double arrow);* they are more abundant in the ancient phloem *aph* than in the functional
phloem *fph.* **b** Radial section showing the
sieve areas *sa* in front view; *vp* series of
vertical parenchyma cells, shorter than the
sieve cells

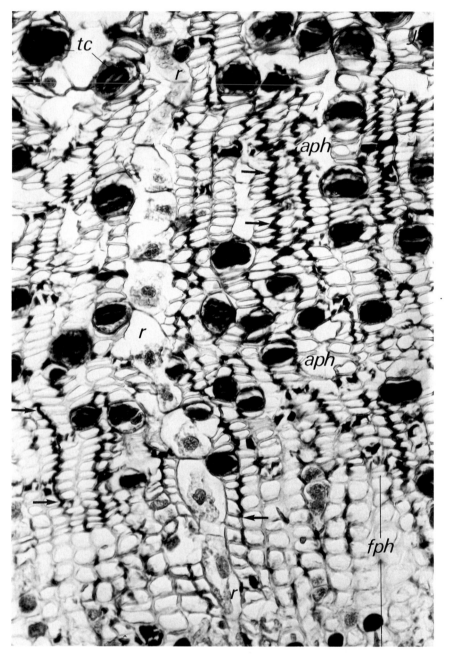

Fig. 6.24. *Pinus silvestris,* stem phloem, transversal section. *fph* Functional phloem, isodiametric cells, quadrangular in section; *aph* ancient phloem, more or less flattened sieve cells; *arrows* indicate the sieve cell walls, pressed and more contrasted in the ancient phloem; *tc* tanniniferous cells, irregularly scattered; *r* uniseriate phloem ray

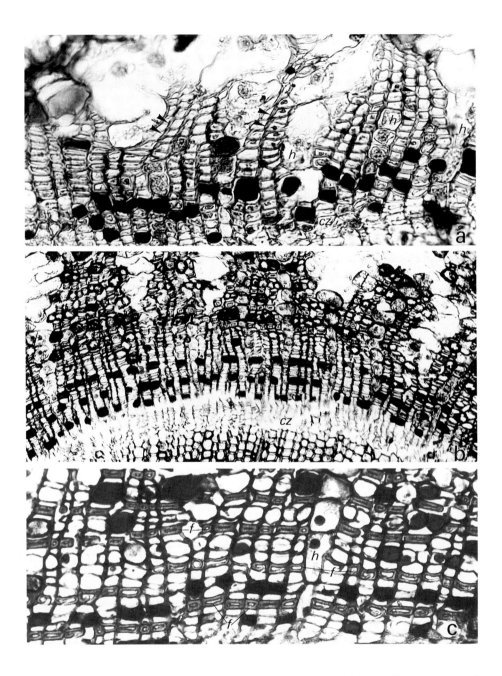

Fig. 6.25 a–c. Secondary phloem of Gymnosperms. **a** Young stem of *Pinus silvestris*, transversal section. Non-ordered scattering of tanniniferous cells into the vertical phloem parenchyma *(black); cz* cambial zone; *h* horizontal parenchyma (phloem rays); the *arrowheads* indicate sieve area on radial sieve cell wall. **b** Young stem of *Cupressus sempervirens*, transversal section. Ordered arrangement of tanniniferous cells, in concentric zones, in the vertical phloem parenchyma; *cz* cambial zone. **c** Id, more or less regular alternation of phloem fibre layers *f* and other cells of the vertical system; *h* horizontal parenchyma

Abietineae but clearer in the other conifers (Taxaceae, Taxodiaceae, Cupressaceae) (Figs. 6.24, 6.25). The cells of the vertical parenchyma store various storage products (starch) as well as secretory products (resinous, tanniniferous inclusions and oxalate crystals). Some cells, near the sieve cells, retain a dense cytoplasm and their vacuoles include chromophil substances ("albuminous cells" or STRASBURGER cells): they replace the *companion cells,* which do not exist in Gymnosperms. Their essential cytological characteristics were described previously (see p. 333; Fig. 6.14a). They can be differentiated from cells of the vertical parenchyma or mainly of the horizontal parenchyma (Fig. 6.14). They die at the same time as the sieve cells with which they are associated (Fig. 6.14b).

In Abietineae, the secondary phloem does not include fibres, but the sieve cells seem to differentiate a secondary pluristratified wall, which is exceptional. The other conifers have fibre beds in their secondary phloem. The alternation fibres – sieve cells – vertical parenchyma – sieve cells – fibres is sometimes regular (ABBE and CRAFTS 1939) (Fig. 6.25b, c). However, precise research is still necessary for the interpretation of the sieve cell walls, into terms of primary and (or) secondary parietal layers.

The horizontal parenchyma generally forms rays of one layer of cells, whith a height of several cells only (Fig. 6.23a). It stores the same substances as the vertical parenchyma. In particular, some cells differentiate into "Strasburger cells". According to the species and the age of the organs, both systems of parenchyma or only one include such cells (STRASBURGER 1891).

The secondary phloem of Gymnosperms may contain resiniferous secretory canals. According to some histologists, these canals seem to be due to reaction to traumatic excitations but this fact is difficult to prove.

6.7.2 Secondary Phloem of Dicotyledons

6.7.2.1 Various Cell Types and their Arrangement

The secondary phloem in Dicotyledons shows a more complex and a wider diversity of patterns than the Gymnosperm phloem. Among the "vertical" elements, sieve cells, companion cells, cells of the vertical parenchyma and fibres can be found in most species.

             ▷

Fig. 6.26a–c. Secondary phloem (also called "liber") of *Castanea vulgaris.* **a** Radial section; *ray* phloem rays (horizontal parenchyma) in front view; several layers of fibres *f* alternate with vertical parenchyma layers and are coated by a layer of cells, each of them containing a calcium oxalate crystal *(arrows); cb* cambium; *phl* functional phloem. **b** Magnified detail showing the layer of isodiametric cells, each one containing a crystal of calcium oxalate in the shape of a tablet *(oxt)* and bordering a set of fibres, on each side; *vp* vertical parenchyma, compound of elongated pitted cells and isodiametric cells, coming from transversal cleavages of fusiform initials and containing a macled crystal in the shape of a sea urchin of calcium oxalate *(oxm); ray* phloem ray, the cells of which are strongly tanniniferous, except at the contact of fibre layers. **c** Tangential section through the parenchyma layers; *vp* vertical parenchyma, constituted of fairly elongated cells and isodiametric cell files with a macled crystal of calcium oxalate; *ray* uniseriate horizontal parenchyma, here seen in transversal section

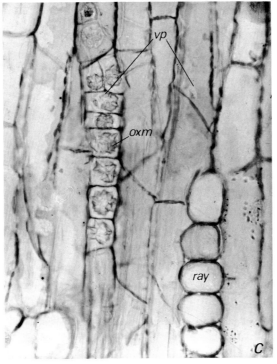

The cells of the vertical parenchyma can be of various kinds in the same phloem (Fig. 6.26). They are then distinguishable by their length, their thicker walls in some cases with variable pits, and by the products that they elaborate. Thus, some short cells of *Castanea* phloem produce calcium oxalate crystals in the shape of a sea urchin (Fig. 6.26 b, c), whereas others contain one single crystal in the shape of a tablet (Fig. 6.26 b). On the periphery of the fibre bundles, cells can be found having, in transversal section, three thickened and sclerified sides, whereas the fourth, close to the parenchyma, remains cellulosic. These intermediary cells often include oxalate crystals (Fig. 6.26 a, b).

The vertical parenchyma becomes dominant in the storage organs: tuberized roots, rhizomes, etc. where it is often the main storage tissue (carrot, chicory, dandelion).

The horizontal parenchyma forms rays of various shapes and sizes according to the species. Some of them may be transversally uniseriate (*Castanea*, Fig. 6.26 c) or multiseriate (*Robinia*, Fig. 6.27) and may be constituted, in hight, of just a few cells or a number of tiers (Fig. 6.27). It may also be lacking, when the cambium, remaining discontinued, produces only isolated conducting strands *(Cucurbita)*.

Besides these widely existing cell types, the secondary phloem can include sclereids (Fig. 6.28) and, most of all, secreting cells which are often characteristic of one family or a group of families. This is the case, for instance, for the *secretory*

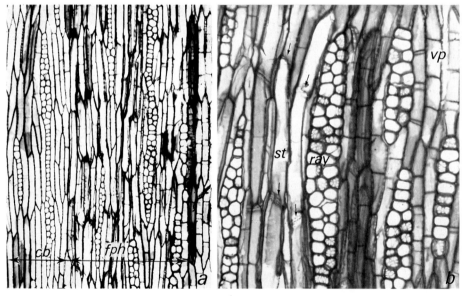

Fig. 6.27 a, b. *Robinia pseudoacacia,* secondary phloem, subtangential sections passing outside of the fibre layers. **a** Toporaphical view showing the differentiation of the functional phloem *fph,* starting from the cambium *cb,* extended into the tangential section. On the *right,* three rows of cells are already formed in the ancient phloem, sieve tubes and companion cells are crushed and form strongly stained vestiges. **b** Detail of a functional phloem area; *st* sieve tubes, four sieves are indicated by *arrows; ray* phloem ray (horizontal phloem parenchyma); *vp* vertical parenchyma with redivided cells, issued from cambial derivatives, and relatively short

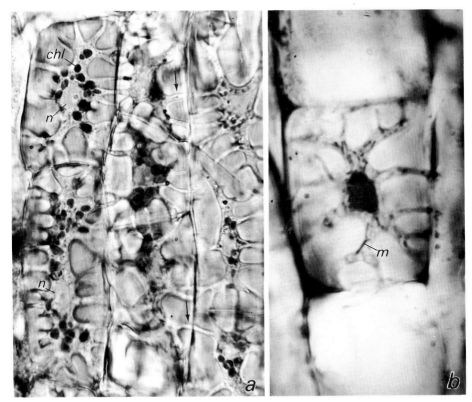

Fig. 6.28 a, b. Sclereids in the ancient phloem of *Robinia pseudoacacia*. **a** Cells, often adjacent to fibres, with considerably thickened and lignified cell walls, but connected together through long tubular secondary pits *(arrows)*. They include a nucleus *n* and numerous chloroplasts, and are very active. **b** Sclereid showing the abundance of mitochondria *m* in the cytoplasm, reduced by the cell wall thickening. Fixation: REGAUD – hematoxylin

canals of the phloem of Umbellifers and its close families, for true *laticifers* of Euphorbiaceae, Moraceae, etc. which furrow the phloem like the more external parenchyma, and for the pseudolaticifers of Chicoraceae which are essentially part of the phloem (see Chap. 9).

Such a diversity of cell types is the cause of the great complexity of the secondary phloem, but the most visible diversity is linked to the fibres. The phloem of many trees (*Castanea, Tilia, Magnolia,* etc.) and of liana *(Vitis)* presents regular alternations of cells with pectocellulosic membranes and fibres (Fig. 6.29). Here again, we find this foliated aspect of the *"liber"* of many Gymnosperms. However, in other species (*Laurus, Nicotiana,* etc.) the fibres are scattered among other non-sclerified cells of the vertical system (Fig. 6.30). Two extreme cases can be found: one where the fibres are so abundant that it is the cellulosic elements which are dispersed in slerified tissue *(Carya),* and the other where no fibres are present (*Aristolochia,* functional phloem of *Prunus*) (Fig. 6-31). In *Prunus avium,* fibres differentiate only in the matured phloem, once it has ceased to be functional. In ad-

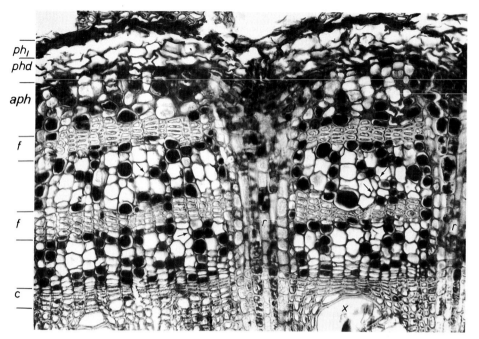

Fig. 6.29

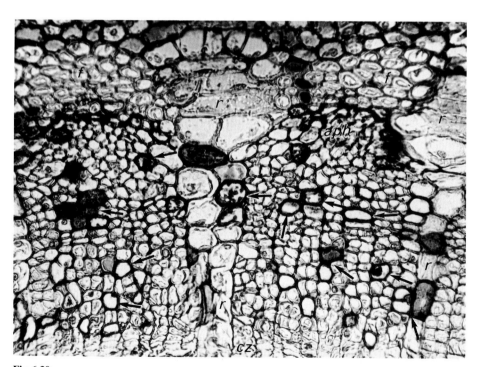

Fig. 6.30

356

◁ **Fig. 6.29.** *Vitis vinifera* phloem, composed of layers of thin cell-walled cells, alternating with fibre layers *f;* this structure is typical of the true "liber". The cells with thin cell walls are of three kinds: sieve cells (large diameter, content not preserved), companion cells (small diameter, *arrows*) and vertical parenchyma cells (tanniniferous, opaque content with chromic fixative). *c* Cambium; *ph₁* crushed primary phloem, capped with phloem fibres; *aph* ancient phloem; *phd* phelloderm; *r* horizontal parenchyma rays, externally enlarged because of diametral growth of the stem; *x* xylem. Fixation: Navaschin – hematoxylin

Fig. 6.30. *Laurus nobilis,* stem phloem, transversal section. Fibres, isolated or arranged in groups of two or three, are scattered without obvious ordering into the functional phloem (most of them, but not all, are indicated by *arrows*). More differentiated fibre bundles *f* capping the ancient phloem *aph,* may be of metaphloem type; *r* phloem rays, becoming multiseriate when more distal from the cambial zone *cz,* and widening in the ancient phloem, because of the transversal growth of the stem

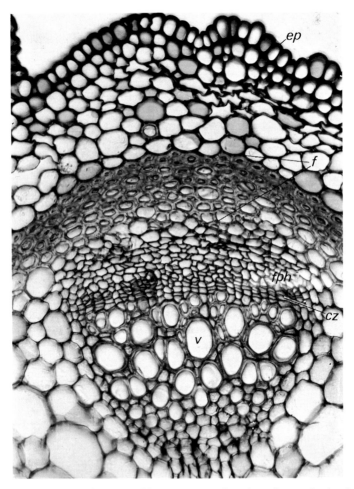

Fig. 6.31. *Aristolochia pistolochia,* stem phloem, transversal section of a conducting bundle. The functional phloem *fph* is devoid of fibres. The latter form a periphloem covering *f* particularly over the ancient, more or less crushed, phloem; *ep* epidermis, doubled by a collenchymatous subepidermal layer; *v* vascular bundle; *cz* cambial zone

357

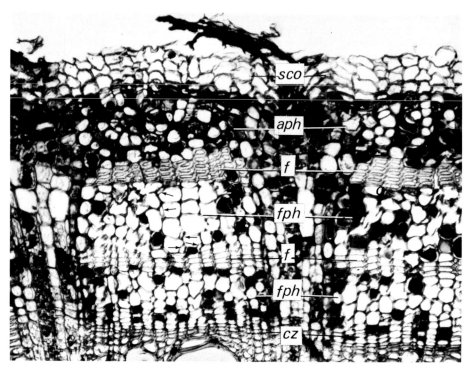

Fig. 6.32. *Robinia pseudoacacia,* 2-years old stem, stratified phloem (liber). *f* Fibre layers, alternating with other cells of functional phloem *fph; aph* ancient phloem, the sieve cells and companion cells of which are dead; *sco* secondary cork; *cz* cambial zone; *arrows* indicate companion cells

dition to their supporting role, fibres often present a significant physiological activity. For instance, in *Vitis,* there are redivided fibres which store starch.

Sieve tubes may also be distributed in tangential series *(Robinia, Aristolochia)* (Fig. 6.32). Bands of sieve cells alternate then with bands of parenchyma which separate them from each other, or from the fibres if there are any. The "liber" is then clearly stratified. In other species, sieve tubes form radial series *(Prunus).*

Most of the woody Dicotyledons of so-called little-developed species have non-stratified phloem.

6.7.2.2 The Sieve Cell in the Series of Dicotyledons

In arborescent species considered as primitive, end wall of sieve cells are very much inclined which approaches the fusiform cells in Gymnosperms. This is also the case in some *Rosaceae* (subfamily of Pireae). End wall bear in this case compound sieve areas with numerous pit plates, not much different from those existing on the lateral walls of the same cells. In other lignified Dicotyledons (*Betula, Quercus, Castanea, Populus,* etc.) end wall are still very much inclined bearing compound sieve plates but which are more differentiated than the lateral sieve areas (Fig. 6.33 b).

Fig. 6.33 a–c. *Castanea vulgaris,* secondary phloem; terminal compound sieve plates, markedly oblique. **a, b** In radial section, they are obviously in front view; **c** in tangential section; *sa* sieve areas; *b* bars of the compound sieve plates; *ls* lateral sieve plates. Each pore is lined with a callose sheath which narrows its lumen

In the cases where the ends of the cells are wedge-shaped, the sieve walls are parallel to the *radial* faces and therefore seen almost in front views in the radial sections and in profile views in the tangential (Fig. 6.33 b, c).

These walls are much less inclined in the phloem of other trees *(Fagus, Acer)* and become transverse in the species considered as developed for their sieve cells, such as *Fraxinus, Ulmus, Robinia* (Fig. 6.34).

Whereas the inclination diminishes, the number of sieve plates also decreases and species with transverse end walls usually have *single sieve areas.* Moreover, sieve cells of these species are shorter. The species which include such cells often show a "storied" secondary phloem, meaning that end walls of long cells are approximately aligned in the same plane perpendicular to the elongation orientation *(Robinia)* (Fig. 6.27 a).

In herbaceous Dicotyledons, there are relatively short sieve cells in which the sieve fields are transversal and simple *(Cucurbita)* (Fig. 6.1).

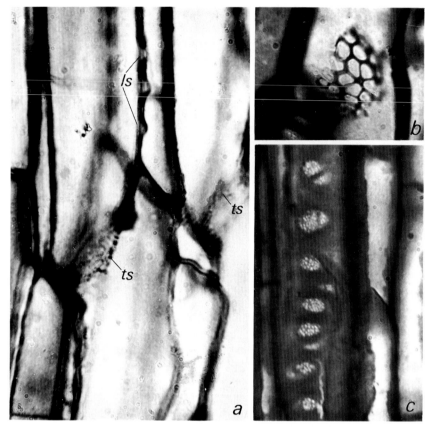

Fig. 6.34 a–c. *Robinia pseudoacacia,* secondary phloem. Simple terminal sieve plates with weak obliquity. **a** Tangential section; *ls* lateral sieve plates; *ts* terminal sieve plates. *b* Part of a terminal sieve plate in front view, note the large diameter of the pores, considered as a character of an advanced plant. **c** Lateral sieve areas *ls,* the pores of which are much more exiguous than those of terminal sieve plates

6.8 The Phloem in the Series of Vascular Plants

The preceding has shown that the structure of phloem varies much in the series of superior plants, thus providing very important systematical and phylogenetic characteristics. However, because of the difficulty in studying the phloem and its poor conservation, compared to that of vascular tissues, in fossil remnants of past flora, this tissue has been less used than the xylem tissue in research in the field of compared anatomy of continental plants.

Let us first consider the sieve cell itself. The main variations concern the diameter of junctions between sieve cells, the inclination of end walls and their simple or compound sieve plates.

The very thin pits of Gymnosperms sieves have been considered as a primitive characteristic as well as the strong inclination of sieve walls. These characteristics

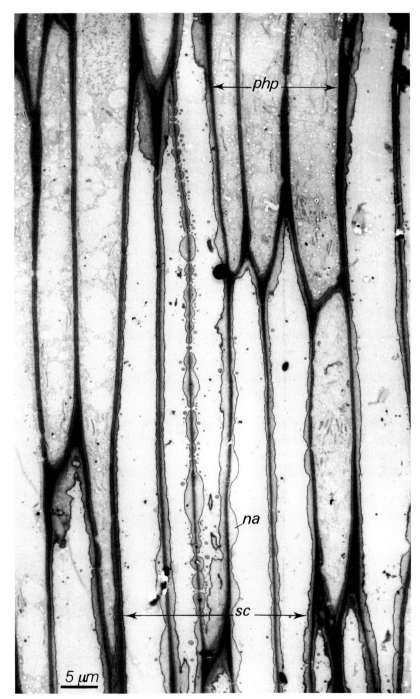

Plate 6.24. *Polypodium vulgare,* metaphloem, longitudinal section.

sc Long and fusiform sieve cells; the section of the median cell shows a longitudinal series of pores; *na* "nacreous" cell wall; *php* phloem parenchyma; × 2000. Fixation: glutaraldehyde – O$_S$O$_4$; PATAg technique, according to Thiéry (1967); (after Liberman–Maxe 1983)

361

are also encountered in Dicotyledons considered as not very advanced (lignified Ranales, some Apetals).

But one can also speculate as to whether this is a feature of arborescent plants; if it were not, Cupulifers, among others, should be considered as primitive and this fact has been denied.

Moreover, in herbaceous plants acknowledged as being primitive, such as *Botrychium*, a type of poorly developed ferns, sieve areas seem to be simple. Conversely, ferns such as *Pteridium* and *Polypodium vulgare* have long sieve cells forming sieve areas uniformly distributed all along the wall faces (Plate 6.24, *sc*).

However, it is suggestive to think that some arborescent species considered as very developed *(Fraxinus, Ulmus, Robinia)* have sieve cells with clearly transversal end walls and simple sieve areas, and that the same occurs in gamopetal Dicotyledons such as *Cucurbita*.

The absence of companion cells in Pteridophytes and Gymnosperms is a more obvious systematical feature. Companion cells occur only in Angiosperms, Dicotyledons and Monocotyledons. The Strasburger cells of Gymnosperms could play the role of less specialized replacing cells for the companion cells of Angiosperms.

The vertical parenchyma is commonly found in all types of phloem tissues, but it would be interesting to elucidate its development more accurately in the various groups of Pteridophytes and Gymnosperms. This study has not been carried out in the phloem tissue as accurately as in wood for which it has yielded important phylogenetic data.

The horizontal parenchyma, specific for secondary phloem, is generally arranged in narrow rays, transversally uniseriate, in Gymnosperms whereas, in Dicotyledons, the width and hight of rays are variable and are often of importance.

Phloem fibres do not show features quite particular to certain groups of vascular plants. Their wide distribution suggests that the phloem, through its physiological functioning, favours the differentiation of sclerified tissues in neighbouring areas. It is significant, first considering the so-called pericyclic fibres for which the term "periphloem" seems more suitable, to relate to these structures the fact that, when isolated "medullary" fibres are present, they generally occur on the periphery of the internal phloem, itself designated as "perimedullary".

6.9 Histology and Cytology of the Phloem and Mechanisms of Translocation

Despite the relative advances made in the study of the histology of phloem and the infrastructure of sieve cells and companion cells, the mechanisms of phloem circulation, the *translocation,* have not been clearly established. The existence of overpressure in sieve tubes has long been acknowledged suggesting that the translocation could be achieved by means of higher pressures induced in those areas where metabolites rush along and by means of depressions caused by the flowing back of metabolites towards the sites where they are used (sites of growth, storage, etc.). In this so-called *Flüssigkeitsströmung* theory, mass flow or *ecoulement de masse* (MÜNCH 1930; CRAFTS 1938) the cytoplasm of sieve cells would be inactive.

This way of thinking implies that, from one sieve cell to the next, sieve areas allow the assimilates to circulate freely. As it is difficult to avoid artefacts on the sample fixation or preparation, which often shifts the content of sieve cells against the sieve plates, with the risk of occluding pores, this fact prevents the accurate study of the pore content in the in situ functional phloem. This question remains without a definitive answer. However, the pore canal, obviously lined with plasmalemma (Plate 6.19 A), generally encloses mictoplasm (Plates 6.1 and 6.19 A) in some cases surrounded by a thin layer of non-diluted cytoplasm applied against the plasmalemma, and remains "open" if it is not occluded by callose.

Of course, this ultrastructure is favourable to the theory of mass flow induced by hydrostatic pressures (*Druck strom* for German authors, *pressure flow* for English authors) but this theory hardly matches other cytological features.

Thus, functional sieve cells, although devoid of true vacuoles as the tonoplast has disappeared, are still plasmolyzable, then deplasmolyzable (KOLLMANN 1960). They maintain structures of endoplasmic reticulum type in the thin layer of non-diluted cytoplasm which persists against the plasmalemma, along with the usual permeability characteristics of common cells. Moreover, contrarily to that believed for a time, they preserve mitochondria (BUVAT 1963a, b) which are apparently functional (CATESSON and LIBERMAN MAXE 1974). But when these protoplasmic remnants degenerate in turn, sieve tubes *cease to function*. The passivity of the cytoplasm at the onset of translocation is then very doubtful.

Moreover, a saccharose transfer, among other metabolites, was demonstrated from the photosynthesizing cells of the leaf mesophyll for example, towards the phloem (GIAQUINTA 1983). This migration occurs partly through symplasmatic routes, i.e. form cytoplasm to cytoplasm, probably through plasmodesmata, and partly through apoplasmatic routes, i.e. by diffusion into the walls of parenchyma cells (GIAQUINTA 1983). According to these routes, the saccharose reaches the sieve tube-companion cell complex and is released in sieve tubes where its concentration is higher than in the neighbouring parenchyma. Depending on species, the transfer could be mainly achieved by the apoplasmatic route, i.e. through differentiated cell walls, such as nacreous walls, or also by the symplasmatic route. In any case, it seems that the plasmalemma and sometimes the plasmodesmata are bound to participate in the release of sucrose in sieve cells, release which occurs against the concentration gradient.

The transfer of saccharose from companion cells (or from transfer cells) into sieve cells, where it is commonly found to be more concentrated requires some energy. This energy is provided through ATP hydrolysis at the level of the plasmalemma of companion cells (or transfer cells) and sieve cells, in which the nucleoside-phosphatase activity is particularly pronounced (cf. p.332). With reference to the present hypotheses, the energy delivered by ATP hydrolysis could activate a "proton pump" which could induce a dissymmetric distribution of protons and hydroxyl ions on each side of the plasmalemma. An active pumping maintained by K^+ ions entering sieve cells, together with proton exiting, has been considered as related to the activity of plasmalemma ATPases of phloem (MALEK and BAKER 1977).

A proton gradient of about 3 pH units, between the cavity of the sieve tube (alkaline side) and the wall (acid side), would determine the energy required for saccharose entering sieve cells. If this is the case, the saccharose storage induces an increased osmotic pressure in sieve cells and consequently produces the well-known over pressure accountable for the "mass flow" mechanisms, assessed by a number of physiologists.

This type of molecular mechanisms can be compared, on the one hand, to the finding of a microperistaltism in the fibrillar content of sieve cells of *Heracleum mantegazzianum* (LEE et al. 1970; FENSOM 1972) and, on the other hand, to the macromolecular configuration in double helices with variable pitch, of P-proteins, described by ARSANTO (1977, 1979, 1982 (see pp. 303-309).

Without excluding the contribution of "pressure flow" processes in translocation, one should keep in mind the action of molecular phenomena, occurring either in the assimilate concentration or in their translocation. The residual cytoplasmic structures of sieve cells surely participate in these processes, as probably also the companion cells.

To conclude, one can quote the opinion of GIAQUINTA (1983) on the state of knowledge related to the loading of the phloem and to the translocation: *"clearly, we are in the early stages of discovering the rules of assimulate transport and partitioning."*

For further details which are beyond the scope of this book, the author advises the reader to refer to the work of EVERT (1984); Comparative structure of phloem (in Academic Press, Contemporary problems in plant anatomy, pp. 145-234).

References

Abbe LB, Crafts AS (1939) phloem of white pine and other coniferous species. Bot Gaz 100: 695-722

Aldaba VC (1927) The structure and development of the cell wall in plants. I. Bast fibers of *Boehmeria* and *Linum*. Am J Bot 14: 16-24

Arsanto JP (1970) Infrastructures et différenciation du protophloème dans les jeunes racines du Sarrasin (*Polygonum fagopyrum*, Polygonacées). C R Acad Sci Paris 270 D: 3071-3074

Arsanto JP (1977) Sur l'ultrastructure, la biogenèse et les changements conformationnels des constituants protéiques-P du phloème de trois Dicotylédones. C R Acad Sci Paris 285 D: 93-96

Arsanto JP (1979) Ontogenèse du phloème, en particulier des protéines-P, chez quelques Dicotylédones. Thèse Dr es-Sci Univ Aix-Marseille II

Arsanto JP (1982) Observations on P-proteins in dicotyledons. Substructural and developmental features. Am J Bot 69: 1200-1212

Arsanto JP, Coulon J (1974) Détection radio-autographique et cytochimique des sites d'élaboration ou de transit des précurseurs polysaccharidiques pariétaux dans les cellules criblées en cours de différenciation du métaphloème caulinaire de deux Cucurbitacées voisines (*Cucurbita pepo* L. et *Ecballium elaterium* R.). CR Acad Sci Paris 278 D: 2775-2778

Arsanto JP, Coulon J (1975) Application des méthodes cytochimique et radio-autographique de détection ultrastructurale des polysaccharides à l'étude de la différenciation des plateaux criblés du métaphloème caulinaire de deux Cucurbitacées voisines (*Ecballium elaterium* R. et *Cucurbita pepo* L.). CR Acad Sci Paris 280 D: 601-604

Bailey IW (1923) Slime bodies of *Robinia pseudoacacia*. Phytopathology 13: 332-333

Bailey IW, Swamy BGL (1949) The morphology and relationships of *Austrobaileya*. J Arnold Arbor Harv Univ 30: 211-266

Behnke HD (1965) Über das Phloem der *Dioscoreaceen* unter besonderer Berücksichtigung ihrer Phloembecken. II. Elektronenoptische Untersuchungen zur Feinstruktur des Phloembeckens. Z Pflanzenphysiol 53: 214-244

Behnke HD (1968) Zum Aufbau gitterartiger Membranstrukturen im Siebelementplasma von *Dioscorea*. Protoplasma 66: 287-310

Behnke HD (1971) The contents of the sieve-plate pores in *Aristolochia*. J Ultrastruct Res 36: 493-498

Behnke HD (1972) Sieve-tube plastids in relation to angiosperm systematics. An attempt towards a classification by ultrastructural analysis. Bot Rev 38: 155-197 (références des multiples publications antérieures)

Behnke HD (1973) Plastids in sieve elements and their companion cells. Investigations on monocotyledons, with special reference to *Smilax* and *Tradescantia*. Planta (Berl) 110: 321-328

Beyenbach J, Weber C, Kleinig H (1974) Sieve tube proteins from *Cucurbita maxima*. Planta 119: 113-124

Bouck GB, Cronshaw J (1965) The fine structure of differentiating sieve tube elements. J Cell Biol 25: 79-95

Burr FA, Evert RF (1973) Some aspects of sieve element structure and development in *Selaginella Kraussiana*. Protoplasma 78: 81-97

Buvat R (1963a) Infrastructure et différenciation des cellules criblées de *Cucurbita pepo*. Evolution du tonoplaste et signification du contenu cellulaire final. CR Acad Sci Paris 256: 5193-5195

Buvat R (1963b) Les infrastructures et la différenciation des cellules criblées de *Cucurbita pepo* L. Port Acta Biol sér 7 A: 249-299

Buvat R (1963c) Infrastructure et differenciation des cellules criblées de *Cucurbita pepo*. Relations entre la membrane pectocellulosique et les membranes plasmiques du cytoplasme. C R Acad Sci Paris 257: 221-224

Buvat R (1981) Vésicules "alvéolées" et vésicules "épineuses" dans les racines de l'Orge. CR Acad Sci Paris sér III 292: 825-832

Buvat R, Robert G (1979) Activités golgiennes et origine des vacuoles dans les cellules criblées du protophloème de la racine de l'Orge *(Hordeum sativum)*. Ann Sci Nat Bot Paris 13e sér 1: 51-66

Catesson AM (1966) Présence de phytoferritine dans le cambium et les tissus conducteurs de la tige de Sycomore, *Acer pseudoplatanus*. CR Acad Sci Paris 262 D: 1070-1073

Catesson AM (1973) Observations cytochimiques sur les tubes cirblés de quelques Angiospermes. J Microsc 16: 95-104

Catesson AM (1980) Localization of phloem oxidases. Ber Dtsch Bot Ges 93: 141-152

Catesson AM, Czaninski Y (1967) Mise en évidence d'une activité phosphatasique acide dans le reticulum endoplasmique des tissus conducteurs de Robinier et de Sycomore. J Microscopie 6: 509-514

Catesson AM, Liberman-Maxe M (1974) Les mitochondries des cellules criblées: réactions avec la 3-3'-diaminobenzidine. CR Acad Sci Paris 278 D: 2771-2773

Crafts AS (1938) Translocation in plants. Plant Physiol 13: 791-814

Cronshaw J (1980) Histochemical localization of enzymes in the phloem. Ber Dtsch Bot Ges 93: 123-139

Cronshaw J, Esau K (1967) Tubular and fibrillar components of mature and differentiating sieve elements. J Cell Biol 34: 801-816

Cronshaw J, Esau K (1968) P-protein in the phloem of *Cucurbita*. I. The development of P-protein bodies. J Cell Biol 38: 25-39

Currier HB (1957) Callose substance in plant cells. Am J Bot 44: 478-488

Czaninski Y, Catesson AM (1970) Activités peroxydasiques d'origines diverses dans les cellules d'*Acer pseudoplatanus* (Tissus conducteurs et cellules en culture). J Microsc 9: 1089-1102

Deshpande BP (1974) On the occurrence of spiny vesicles in the phloem of *Salix*. Ann Bot 38: 865-868

Dunoyer de Segonzac G (1958) L'ontogénie du phloème chez *Vanilla planifolia* Andr. Rev Cytol Biol Vég 19: 153-184

Engleman EM (1965) Sieve element of *Impatiens sultanii*. II. Developmental aspects. Ann Bot 29: 103-118

Esau K (1943) Vascular differentiation in the vegetative shoot of *Linum*. III. The origin of the bast fibers. Am J Bot 30: 579-586

Esau K (1947) A study of some sieve-tube inclusions. Am J Bot 34: 224-233

Esau K (1961) Plants, viruses and insects. Harvard University Press, Cambridge Mass

Esau K (1965) Plant anatomy. Wiley New York

Esau K, Cheadle VI (1958) Wall thickening in sieve elements Proc Natl Acad Sci USA 44: 546–553

Esau K, Cheadle VI (1965) Cytologic studies on phloem. University of California Press, Berkeley and Los Angeles 36: 253–344

Esau K, Cronshaw J (1967) Tubular components in cells of healthy and tobacco mosaic virus-infected *Nicotiana*. Virology 33: 26–35

Esau K, Cronshaw J (1968a) Plastids and mitochondria in the phloem of *Cucurbita*. Can J Bot 46: 877–880

Esau K, Cronshaw J (1968b) Endoplasmic reticulum in the sieve element of *Cucurbita*. J Ultrastruct Res 23: 1–14

Esau K, Gill RH (1971) Aggregation of endoplasmic reticulum and its relation to the nucleus in a differentiating sieve element. J Ultrastruct Res 34: 144–158

Eschrich W (1956) Kallose. Protoplasma 47: 487–530

Eschrich W (1963) Beziehungen zwischen dem Auftreten von Callose und der Feinstruktur des primären Phloems bei *Cucurbita ficifolia*. Planta (Berl) 59: 243–261

Evert RF (1984) Comparative structure of phloem. In: Contemporary problems in plant anatomy, pp 145–234. Academic Press

Evert RF, Deshpande BP (1969) Electron microscope investigation of sieve-element ontogeny and structure in *Ulmus americana*. Protoplasma 68: 403–432

Evert RF, Eichhorn SE (1974) Sieve element ultrastructure in *Platycerium bifurcatum* and some other polypodiaceous ferns: the refractive spherules. Planta 119: 319–334

Fensom DS (1972) A theory of translocation in phloem of *Heracleum*. Can J Bot 50: 479–497

Fisher DG, Evert RF (1979). Endoplasmic reticulum-dictyosome involvement in the origin of refractive spherules in sieve elements of *Davallia fijiensis* Hook. Ann Bot 43: 255–258

Fisher DG, Evert RF (1982) Studies on the leaf of *Amaranthus retroflexus* (Amaranthaceae) quantitative aspects and solute concentration in the phloem. Am J Bot 69: 1375–1388

Friend DS, Farquhar MG (1967) Functions of coated vesicles during protein absorption in the rat *vas deferens*. J Cell Biol 35: 357–371

Giaquinta RT (1983) Phloem loading of sucrose. Annu Rev Plant Physiol 34: 347–387

Gomori G (1952) Microscopic histochemisty; principles and practices. University of Chicago Press: Chicago

Johnson RPC (1973) Filaments but no transcellular strands in sieve pores in freeze-etched, translocating phloem. Nature 244: 464–466

Jupin H, Catesson AM, Giraud G, Hauswirth N (1975) Chloroplastes à empilements granaires anormaux appauvris en photosystème I dans le phloeme de *Robinia pseudoacacia* et de *Acer pseudoplatanus*. Z Pflanzenphysiol 75: 95–106

Kessler G (1958) Zur Charakterisierung der Siebröhrenkallose. Ber Schweiz Bot Ges 68: 5–43

Kollmann R (1960) Untersuchungen über das Protoplasma der Siebröhren von *Passiflora coerulea*. I. Lichtoptische Untersuchungen. Planta (Berl) 54: 611–640

Kollmann R (1964) On the fine structure of the sieve element protoplast. Phytomorphology 14: 247–264

Kollmann R, Schumacher W (1962) Über die Feinstruktur des Phloems von *Metasequoia glytostroboides* und seine Jahreszeitlichen Veränderungen. III. Die Reaktivierung der Phloemzellen im Frühjahr. Planta (Berl) 159: 195–221

Kollmann R, Schumacher W (1963) Über die Feinstruktur des Phloems von *Metasequoia glytostroboides* und seine Jahreszeitlichen Veränderungen. IV. Weitere Beobachtungen zum Feinbau der Plasmabrücken in den Siebzellen. Planta (Berl) 60: 360–389

Kollmann R, Dörr I, Kleinig H (1970) Protein filaments; structural components of the phloem exsudate. I. Observations with *Cucurbita* and *Nicotiana*. Planta 95: 86–94

Kundu BC (1942) The anatomy of two Indian fibre plants, *Cannabis* and *Corchorus*, with special reference to the fibre distribution and development. Indian Bot Soc J 21: 93–128

Kursanov AL, Kulikova AL, Turkina MW (1983) Actinlike protein from the phloem of *Heracleum sosnowskyi*. Physiol Veg 21: 353–359

Laflèche D (1966) Ultrastructure et cytochimie des inclusions flagellées de *Phaseolus vulgaris*. J Microsc 5: 493–510

Lawton DM, Johnson RPC (1976) A superhelical model for the ultrastructure of "P-protein tubules" in sieve elements of *Nymphoides peltata*. Cytobiologie 14: 1–14

Lecomte H (1889) Contribution à l'étude du liber des Angiospermes. Ann Sci Nat Bot sér 7 10: 193–324

Lee DR, Fensom DS, Costerton JW (1970) Particle movement in intact phloem in *Heracleum*. Can Natl Film Library, Ottawa, Canada

Lee DR, Arnold DC, Fensom DS (1971) Some microscopical observations of functioning sieve tubes of *Heracleum,* using Nomarski optics. J Exp Bot 22: 25–38

Liberman-Maxe M (1968) Différenciation des pores dans les cellule criblées de *Polypodium vulgare* (Polypodiacée). CR Acad Sci Paris 266 D: 767–769

Liberman-Maxe M (1971) Étude cytologique de la différenciation des cellules criblées de *Polypodium vulgare* (Polypodiacée). J Microsc 12: 271–288

Liberman-Maxe M (1974) Localisation ultrastructurale d'activités peroxydasiques dans la stèle de *Polypodium vulgare* (Polypodiacée). J Microsc 19: 169–182

Liberman-Maxe M (1978) La paroi des cellules criblées dans le phloème d'une Fougère, le Polypode. Biol Cell 31: 201–210

Liberman-Maxe M (1983) Étude ultrastructurale et cytochimique de la différenciation des tissus de la stèle d'une Fougère, le *Polypodium vulgare* L. Thèse Doct es-Sci Nat, Université P et M Curie, Paris

Malek F, Baker DA (1977) Proton co-transport of sugars in phloem loading. Planta 145: 297–299

Magenot G (1929) Sur les phénomènes dits d'agrégation et la disposition des vacuoles dans les cellules conductrices. CR Acad Sci Paris 188: 1431–1434

Mangin L (1890) Sur la callose, nouvelle substance fondamentale existant dans la membrane. CR Acad Sci Paris 110: 644–647

Markham R, Frey S, Hills GJ (1963) Methods for the enhancement of image and accentuation of structure in electron microscopy. Virology 20: 88–102

Maxe M (1964) Aspects infrastructuraux des cellules criblées de *Polypodium vulgare* (Polypodiacée). CR Acad Sci Paris 258: 5701–5704

Maxe M (1966) tude de la dégénérescence nucléaire dans les cellules criblées de *Polypodium vulgare* (Polypodiacée) CR Acad Sci Paris 262 D: 2211–2214

Milburn JA (1971) An analysis of the response in phloem exudation on application of massage to *Ricinus*. Planta 100: 143–154

Münch E (1930) Die Stoffbewegungen in der Pflanze. Fischer, Jena

Murmanis L, Evert RF (1966) Some aspects of sieve cell ultrastructure in *Pinus strobus*. Am J Bot 53: 1065–1078

Nägeli CW (1858) Das Wachstum des Stammes und der Wurzel bei den Gefäßpflanzen und die Anordnung der Gefäßstränge im Stengel. Beitr Z Wiss Bot Heft 1: 1–156

Newcomb EH (1967) A spiny vesicle in slim-producing cells of the bean root. J Cell Biol 35: C17–C22

Northcote DH, Wooding FBP (1966) Development of sieve tubes in *Acer pseudoplatanus*. Proc R Soc Lond B Biol Sci 163: 524–537

Parameswaran N (1971) Zur Feinstruktur der Assimilatleibahnen in der Nadel von *Pinus silvestris*. Cytobiology 3: 70–88

Parthasarathy MV (1974a, b) Ultrastructure of phloem in palms. I. Immature sieve elements and parenchymatic elements. Protoplasma 79: 59–91. II. Structure changes and fate of the organelles in differentiating sieve elements. Ibid pp 93–125

Parthasarathy MV, Mühlethaler K (1969) Ultrastructure of protein tubules in differentiating sieve elements. Cytobiologie 1: 17–36

Roth TF, Porter KR (1964) Yolk protein uptake in the oocyte of the mosquito *Aedes aegypti* L. J Cell Biol 20: 313–332

Salmon J (1946) Recherches cytologiques sur la différenciation des tubes criblés chez les Angiospermes. Rev Cytol Cytophysiol Vég 9: 55–168

Sauter JJ (1974) Structure and physiology of Strasburger cells. Ber Dtsch Bot Ges 87: 327–336

Sauter JJ (1980) The Strasburger cells. Equivalents of companion cells. Ber Dtsch Bot Ges 93: 29–42

Sauter JJ, Braun HJ (1972) Cytochemische Untersuchung der Atmungsaktivität in den Strasburger Zellen von *Larix* und ihre Bedeutung für den Assimilattransport. Z Pflanzenphysiol 66: 440–458

Sauter JJ, Dörr I, Kollmann R (1976) The ultrastructure of Strasburger cells (= albuminous cells) in the secondary phloem of *Pinus nigra* var *austriaca* (Hoess) Badoux. Protoplasma 88: 31–49

Schoch-Bodmer H, Huber P (1951) Das Spitzenwachstum der Bastfasern bei *Linum usitatissimum* und *Linum perenne*. Schweiz Bot Ges Ber 61: 377–404

Srivastava LM, O'Brien TP (1966) On the ultrastructure of cambium and its vascular derivatives. II. Secondary phloem of *Pinus strobus* L. Protoplasma 61: 277–293

Strasburger E (1891) Über den Bau und die Verrichtungen der Leitungsbahnen in den Pflanzen. Histologische Beiträge, Band 3. Fischer, Jena

Thiery JP (1967) Mise en évidence des polysaccharides sur coupes fines en microscopie électronique. J Microsc 6: 987–1018

Van Thieghem P (1882) Sur quelques points de l'anatomie des Cucurbitacées. Bull Soc Bot Fr 29: 277–283

Ullrich W (1962) Beobachtungen über Kalloseablagerungen in transportierenden und nicht transportierenden Siebröhren. Planta (Berl) 59: 239–242

Vian B, Roland JC (1972) Différenciation des cytomembranes et renouvellement du plasmalemme dans les phénomènes de sécrétions végétales. J Microsc 13: 119–136

Weber C, Franke WW, Kartenbeck J (1974) Structure and biochemistry of phloem isolated from *Cucurbita maxima*. Exp Cell Res 87: 79–106

Wilhelm K (1880) Beiträge zur Kenntnis des Siebröhrenapparates Dicotylerpflanzen. Engelmann Leipzig

Wooding FBP (1966) The development of sieve elements of *Pinus pinea*. Planta 69: 230–243

Wooding FBP (1968) Fine structure of callus phloem in *Pinus pinea*. Planta 83: 99–110

Wooding FBP (1969) P-protein and microtubular system in *Nicotiana* callus phloem. Planta 85: 284–298

Zee SY, Chambers TC (1968) Fine structure of the primary root phloem of *Pisum*. Aust J Bot 16: 37–47

CHAPTER 7

Xylem (Vascular Tissue)

The term xylem was introduced by NAEGELI (1858) and refers to the tissues, generally lignified, which together with phloem are the conducting tissues of superior plants. These tissues conduct the water and the dissolved mineral substances taken from the ground through roots. Thus, xylem is complementary to phloem which conducts the products elaborated in those organs where the syntheses occur and in those where organic substances are stored. However, this specialization of conduction does not exclude the fact that the xylem may also store elaborated organic substances such as starch.

Structurally, like the phloem, the xylem is a complex tissue, including a great variety of cells but, whereas the sieve cell was characteristic of phloem, the xylem is characterized by cells which, after differentiation, generate *vessels* or *tracheids*. These cells symmetrically correspond to the phloem *sieve cells*.

The first elements of the xylem develop from the derivatives of the apical meristems, typically from the procambium, like the phloem. Depending on the plants and the organs of the same species, the primary xylem tissue may develop progressively and may or may not be followed by the formation of xylem tissue through the activity of *the vascular cambium*. In that case, in addition to the primary xylem, a *secondary xylem*, the wood, occurs which has a different structure. However, in many cases, the structural change is progressive (ESAU 1965).

7.1 Characteristic Elements: Tracheids and Vessel Members

Tracheids and vessel members are originally arranged in longitudinal series, either in the procambium (primary tracheids or vessels) or in the internal productions of the vascular cambium (secondary tracheids or vessels). The tracheids are imperforate cells with continuous walls which keep the cells separated, whereas *vessel members* are perforated and in union with other vessel members.

Depending on the site and the earliness of the structure arising, these cells may or may not elongate, but in almost all cases their diameter increases more or less. With regards to tracheids, and sometimes vessel members, when the end alls are originally transverse (primary tracheids), the cells tips may extend and become very oblique and fusiform.

7.1.1 General Structures of the Cell Walls of Differentiated Tracheids and Vessel Members

Differentiated tracheids and vessel members are characterized by their lignin impregnated *secondary walls*. The lignified deposits are discontinuous with ornamentations which distinguish the various types. The term "tracheid" refers to the "tracheal" aspect given to these cells by the lignified thickenings, most often arranged in transverse strips.

After their differentiation, these cells lose their protoplasm: they are dead cells. Those which develop vessels are perforated and in union with other vessel mem-

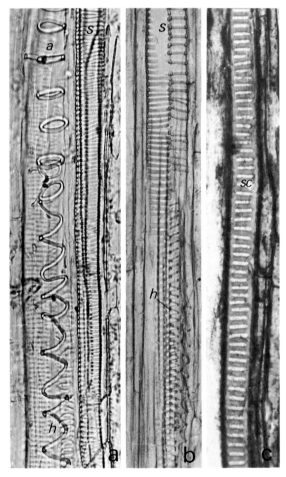

Fig. 7.1 a–c. Types of ornamentation in tracheids and vessels (I). **a** *Ranunculus acris* tracheids; *a* annular ornamentation changing to the helical type *h*. **b** *Hordeum vulgare:* constriction of the helix pitch, passing from the loose helical type *h* to the tight pitch type, often referred to as striped ornamentation *s*. **c** *Vitis vinifera,* narrow vessel, ornamentation in the form of transversal strips, with angular thickenings: scalariform type *sc*

Fig. 7.2 a–c. Ornamentation types of tracheids and vessels (II). **a** *Ranunculus acris;* *a* annular tracheids; *s* striped tracheids, resulting from the pitch tightening of helical ornamentation, nearby a xylem pole. **b** Id, distension and flattening of an annular tracheid *a,* no longer functional, near helical tracheids *h,* passing to the striped type *s.* **c** *Onobrychis sativa,* pitted vessels; *p* bordered pits with aperture in the shape of "buttonhole", in opposite distribution and longitudinal files; *per* simple terminal perforation between two vessel elements

bers through the end walls. So, tubular elements, i.e. vessels, are derived from a various number of cells. Conversely, tracheids are imperforated and separated.

Although perforated walls are mainly end walls, pit fields (see p. 383) occur on the lateral walls between two contiguous cells. Thus, vessel members communicate. The perforated plate may be simple (a single perforation, Figs. 7.2, 7.26) or multiple. In this case, the openings are arranged in parallel splits (scalariform perforation plate, Fig. 7.28 a, b), in a group of circular pores (foraminate plate, Fig. 7.27) or in a reticulate manner (reticulum perforation plate, a rare case).

On arising in a young organ, the primary tracheids are more slender and less lignified than those which differentiate further. In plants with true vessel members, tracheids constitute cellular series designated as *imperfect vessels* compared to *per-*

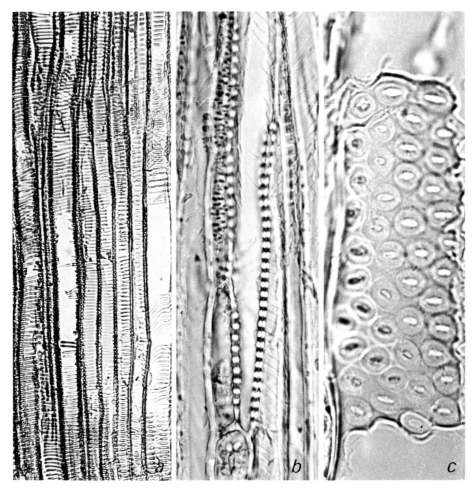

Fig. 7.3 a–c. Ornamentation types of vessels (III). **a,b** *Magnolia grandiflora*, scalariform bordered pits, in front view (**a**) and in profile (**b**). **c** *Populus alba*, bordered pits between two vessels, in irregularly alternating distribution, in front view

fect vessels. The "imperfect vessels" are only files of imperforate tracheids, whereas "perfect vessels" are derived from cells with perforations in the end walls. The simple term vessels refers only to the latter.

Secondary lignified rings (Figs. 7.1 b; 7.2 b) thicken the narrow tracheids, the first to occur in the procambium, i.e. the annular tracheids or erroneously called annular vessels. Then, tracheids differentiate secondary thickenings in spirals or helices (*helical tracheids* or helical vessels; Figs. 7.1 b; 7.2 b). In other elements, the close tightening of whorls gives a *striped* aspect (Figs. 7.1 b; 7.2 a, b). Anastomoses between parietal thickenings result in a *reticulated* ornamentation, leaving non-lignified oblong areas, especially in late elements. The reduction of the surfaces which are not thickened results in *secondary pits* in cell walls, which, in such cases are *pitted tracheids* or *pitted vessel elements* (Fig. 7.2 c).

In Pteridophytes and a few other plants, most tracheids, prismatic in shape, are lignified at the angles also; they are called "scalariform" with reference to the ladder aspect of the lateral walls (Figs. 7.3 a; 7.7 a).

Most cells which have thus developed through primary differentiation persist as tracheids. However, whereas, during the ontogeny of xylem, annular, helical, striped and scalariform, reticulate, then pitted elements (if the series is complete) occurs successively, the first cells which perforate, changing tracheids into true vessels, are more and more precocious in the previous series, from the plants considered as primitive in this respect to the plants considered as more advanced.

Helical, striped, reticulate elements are most often tracheids but in certain, very superior species, tracheids may become perforated to produce vessels (BAILEY 1953; CHEADLE 1953; DUERDEN 1934, 1940; Sect. 7.6). Pitted cells remain in the state of tracheids in the primitive groups (almost all Pteridophytes and Gymnosperms) but many, if not all, depending on species, become perforated, resulting in true vessels in Gnetales and Angiosperms.

In the xylem derived from the vascular cambium, the conducting elements of secondary differentiation immediately assume the more advanced pattern of a given species. In Gymnosperms, these elements are still exclusively tracheids, whereas the secondary xylem of Angiosperms has true vessels with or without tracheids.

In addition to the wall perforation, tracheids are also different from vascular cells with respect to the growth features. During ontogenic differentiation, tracheids often become more elongated than vessel elements in the same species. Conversely, vessel elements increase much more in diameter. Thus, their section is circular, whereas that of tracheids is often prismatic (scalariform tracheids of Pteridophytes, pitted tracheids of Phanerogams).

7.1.2 Xylem Pits

The pits of the tracheids and the vessel elements are generally more differentiated than the simple secondary pits of the other cells which have a secondary wall. This bulging wall reduces the aperture of the cellular lumen and delimits a kind of intraparietal chamber or areola (Figs. 7.4–7.7).

The most differentiated bordered pits, and the most easily observed, occur on the tracheids of Gymnosperms (Fig. 7.4). Within the areola, the primary wall along with the middle lamella form a very thin and circular velum, with a small lenticular thickening in the centre, called the *torus*. It is constituted of primary parietal substance (Fig. 7.4 f). Above the annular space, where only the middle lamella is found, the secondary wall extends into a roughly conic lamina, forming a bulging diaphragm which delimits a circular pit, slightly smaller than the torus (Fig. 7.4 f; 7.7 c, d). On the primary wall, an annular thickening underlines the rim around which the secondary wall forms the areola border. In front view, the pit field has the feature of two concentric circles (Fig. 7.7 c). In side view, it is symmetrical when it is located between two tracheids, which is the more frequent case. This description concerns in particular the areola of Abietaceae (or Pinaceae, depending on the author) and particularly the genus *Pinus* which has been the most studied. The images obtained with electron microscopy, either with thin sections or with replicas, showed that the thin median lamella of the areola of *Pinus silvestris* is a lace of thin cellulosic fibrillae mainly oriented radially around the torus. The latter results from an intrication of fibrillae thicker than in the peripheral ring and more or less hidden by a lining of amorphous parietal substances, sometimes producing warty layers on its surface (*Pinus silvestris,* Fig. 7.4 e). In contrast, the fibrillar ring is typically devoid of this lining although some granules occur in some cases, so that the spaces between the fibrillae correspond to free apertures. The density of peripheral fibrillae varies in the same specimen. Circular fibrillae complete the lamella structure. These fibrillae, closely arranged in the external border of the velum, form the bordering ring well visible with light microscopy (Fig. 7.4 a, c; LIESE 1965; LIESE and HARTMANN-FAHNENBROCK 1953; BAUCH et al. 1972).

In Angiosperms, bordered pits occur also on pitted vessel elements, but they are simple and *the torus is usually omitted.* The lips are usually generated only by the latest wall thickenings and are not very swollen. In many cases, these pits are oval or oblong (scalariform), resembling a buttonhole in front view (Figs. 7.2 c; 7.6 7.7 b).

The study of the comparative anatomy of xylem in which BAILEY and co-workers participated greatly, shows that there are many intermediates between the pits directly derived from scalariform thickenings and the circular or oval bordered pits of the highest structures. In fact, numerous species, living or fossil, of Pteridophytes (advanced) or of Phanerogams (primitive) contain scalariform tracheids, but the lignified bars become thickened and bordered with lips which partially cover the non-thickened areas. In such cases, elongated bordered pits occur similar to buttonholes, called *scalariform pits* (Fig. 7.7 a).

Such pitting occurs also in the end walls of the vessel elements of the primitive Angiosperms (Fig. 7.3 a, b). But, at least in some pits, the middle lamella disappears and pits become scalariform perforations (Magnoliales; Fig. 7.28 a, b).

In Gymnosperms and Angiosperms, the wall of tracheids may become thickened giving them an intermediate aspect between normal tracheids and fibres ("fibre tracheids"). In Angiosperms, more thick-walled fibres differentiate, the "libriform fibres", all of them present particular pits, the very thick secondary wall resulting in a canal which widens towards the middle lamella, thus becoming a

---

▷

Fig. 7.4 a–f. *Pinus silvestris* xylem (wood). **a–c** Bordered pits in front view (radial sections). *cr* Crassulae; *la* perforated lamella of the primary cell wall-middle lamella complex, also called "margo", underlying the torus *t; so* separation outline of primary and secondary cell walls, marking the peripheral limit of the pit chamber. **d** Tangential half-bordered pits *thp* (radial section) at the limit between the cambium *cb* and the last tracheids of the late fall wood *fw* of the preceding season. These pits are not bordered on the cambium side, they might play an important function in spring, during the cambium rehydration, at the time of its resumption of activity. **e** *Pinus silvestris:* membrane replica of a bordered pit in late secondary xylem; fibrils of the margo are particularly abundant, especially tangential fibrils. These fibrils are less numerous in the spring xylem and there is more space between radial fibrils. The torus is thickened by irregular concretions, characteristic of *Pinus silvestris;* × 5100. (Electron micrograph of a replica; courtesy of BAUCH, LIESE and SCHULTZE, 1972.) **f** *Thuja occidentalis,* ultrathin transversal section of a bordered pit, showing the very slight thickening, forming a discrete torus *t; a* section of the marginal ring, circular extension of the middle lamella – primary cell wall complex, visible in the light microscope (see **a** and **c**); × 5300. (Electron micrograph; courtesy of LIESE and BAUCH, 1967.)

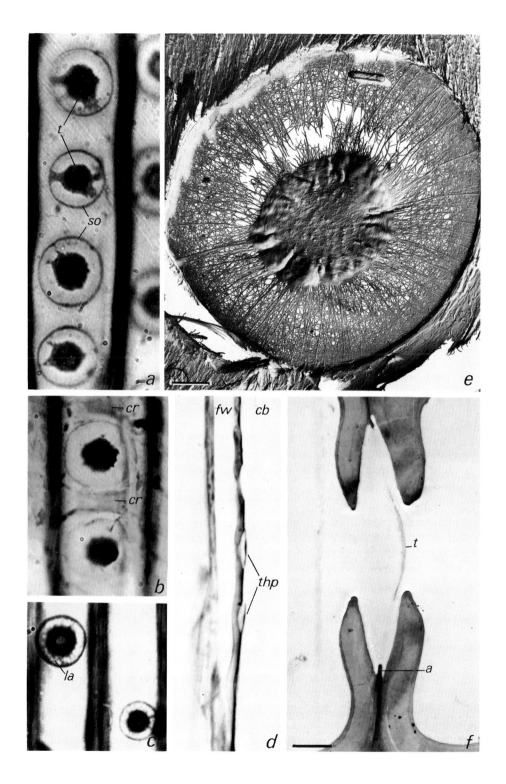

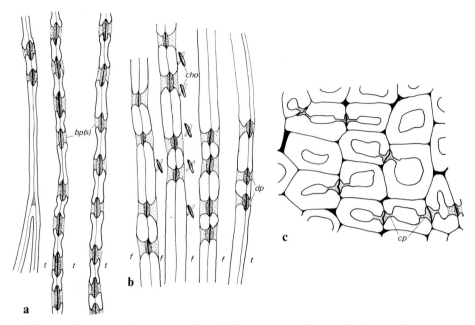

Fig. 7.5 a–c. *Castanea vulgaris,* types of wood pits. **a** Symmetrical bordered pits, *bp(s),* between tracheids (tangential section). **b** Non-bordered pits between libriform fibers *f:* evolutionary character; pits between radial cell wall have chink-like openings stretched in splits *cho;* the openings on the face of the other cell wall, *intercrossed* with the preceding ones, are indicated by *dots; dp* dissymetrical pits between a fibre *f* and a tracheid *t,* only bordered on the tracheid side. **c** Transversal section of fall wood fibres, showing canalicular pits *cp,* widened in the shape of a funnel at the level of middle lamella

small chamber, and towards the cellular cavity into an oval space (Fig. 7.5 c). Usually, the orientation of these openings varies on both sides of the pits (Fig. 7.5 b).

In vessel elements, pits may occur either between two vessels, or between a vessel element and vertical parenchyma cells or horizontal parenchyma cells. In the first case, pits are more or less symmetrical with respect to the middle lamella. In other cases, they are most often bordered on the element side, simple on the side of the living lignified cells between which, moreover, simple pit pairs occur (Fig. 7.6).

Tracheid pittings have the same peculiarities (Fig. 7.5 a). Between the tracheids of Gymnosperms and the wood rays, pit pairs are moderately bordered on the tracheary side and simple on the other.

Theoretically, the sap moves freely from vessel element to vessel element.

In tracheids, the sap passes through the pits, i.e. through the thin middle lamella which may or may not bear a torus. BAILEY (1913), FRENZEL (1929) and other authors demonstrated that this lamella is perforated in the Gymnosperms studied by these authors, and that it is reduced to a fine lace (Fig. 7.4 e). It becomes very thin through partial hydrolysis processes described later, but persists in Angiosperms.

376

In fact, the fine structure of the middle lamella and the primary wall of the bordered pits varies much depending on families, genera and even species. Using thin sections and replicas with electron microscopy, LIESE and co-workers (LIESE 1965; BAUCH et al. 1972) observed that in some Gymnosperms, the lamella of areolae has no torus or a poorly differentiated torus (*Thuja, Thujopsis;* Fig. 7.4 f) and in others, the torus is differently thickened by a ground substance lining which involves the cellulosic fibres and may produce warty deposits (*Pinus silvestris;* Fig. 7.4 e). Moreover, these authors reported that in the genera *Gnetum, Welwitschia* and *Cycas* no torus occurs in the areolae, whereas it occurs in the *Ephedra* genus.

Conversely, structures similar to a torus have been observed between the fibres of *Prunus, Ulmus* and *Pyrus* (PARAMESWARAN and LIESE 1981).

In fact, vessel elements have a limited length and do not form continuous canals from one extremity of the plant to the other. The sap passes from one vessel to another. This movement occurs laterally, between contiguous vessels through the pits as well as between tracheids. Finally, one should keep in mind that in association with the vessel elements, living cells contribute significantly to the circulation of the sap, ensuring in particular a role of deviation.

In the functional wood of Gymnosperms, the torus underlined by the median lamella is located in the middle of the areola chamber (Fig. 7.4 c). In non-function-

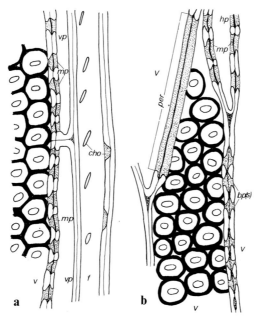

Fig. 7.6 a, b. *Aesculus hippocastanum,* types of wood-bordered pits. **a** Mixed pits, *mp,* dissymmetrical, between vessels *v* and vessel-associated cells of the vertical parenchyma *vp,* bordered on the vessel side, simple on the parenchyma side; *cho* chink-like opening of pits between fibres (*f*) (radial section). **b** Bordered symmetrical pits *bp(s),* between vessels *v;* mixed pits, dissymmetrical, *mp,* between vessel and horizontal parenchyma *hp; per* simple oblique terminal perforation between two vessels (tangential section). The two figures show the features of bordered pits between vessels in front view

377

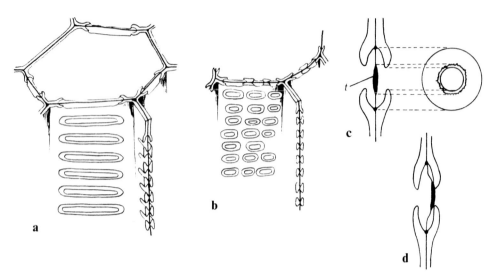

Fig. 7.7 a–d. Schemes of some bordered pit types. **a** Scalariform pitting of Pteridophytes. **b** Angiosperm type of pitting *(Aesculus hippocastanum).* **c** Gymnosperm type of pitting (*Pinus silvestris,* areolate pit) profile and front view; *t* torus. **d** Areolate pit of *Pinus silvestris,* the torus of which is shifted against the secondary wall border of one of the tracheids

al ancient wood (core wood) in most areolae, the torus is applied against one of the pores of the pitting (Fig. 7.7 d) which is thus occluded. This lateral shifting has been considered to be induced by the dissymmetric drying of tracheids, probably when the content of one of them becomes gaseous (HARRIS 1954). This arrangement is called "aspirated tracheids".

7.1.3 Cell Wall Differentiation of Tracheids and Vessel Elements

The cells which generate the tracheids or the vessel elements are originally arranged in longitudinal rows, in the procambium or in the vascular derivatives of the cambium. In the latter case, longitudinal arrangement may be less obvious for the future tracheids. Most often, cambial initials are fusiform and sometimes very elongated and the linear disposition is less obvious (Gymnosperms).

▷

Plate 7.1 A, B. *Cucurbita maxima,* stem xylem.

A Young vessel element in the process of differentiation, presenting ordinary cytoplasmic organelles; *d* dictyosomes; *m* mitochondria; *n* nucleus, already swollen and slightly lobed, with voluminous nucleolus *nu* and chomocentres *ch; p* plastids, poorly differentiated; *lw* lateral cell walls, still primary; *er* endoplasmic reticulum; × 11000. Fixation: O_sO_4; contrast: $KMnO_4$.

B Cell wall of a transfer cell *tc,* contiguous to a vessel *v,* in the nodal xylem of a wheat stem. PATAg technique for polysaccharide detection: the primary cell wall, *pw,* which remains pectocellulosic, is more reactive than the lignified secondary cell wall of the vessel *sw;* × ≈ 6000.

(Courtesy of CZANINSKI 1977 b, relettered)

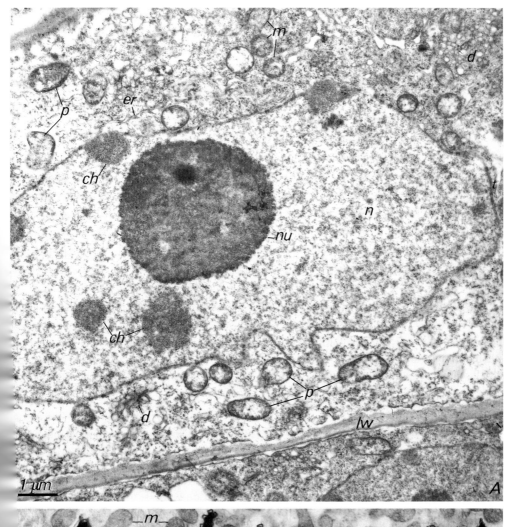

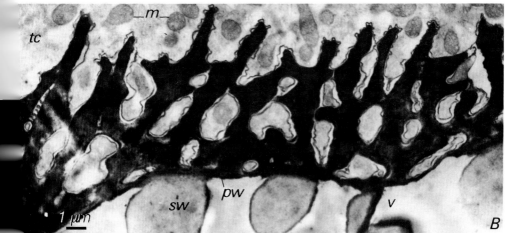

7.1.3.1 Infrastructural Cytology of Secondary Cell Wall Elaborations: Transversal Growth, Dictyosomes, Endoplasmic Reticulum, Microtubules, Intercellular Connections

The most significant growth of the future conducting cells occurs *transversally* rather than *in length*. Above all, the future vessel elements are characterized early by their diameter (see Figs. 6.18, 6.19). The surface of the transverse walls (or end walls) may be multiplied by 100, mainly in spring.

Secondary wall structures develop at the end of the transverse growth, sparing some spaces of primary wall on which pits and perforations differentiate.

A high metabolic activity is present in the future conducting cells where the secondary wall structures arise. They have been extensively studied with electron microscopy (NEWCOMB 1963; BUVAT 1964 a, b, c; WARDROP 1964; WOODING and NORTHCOTE 1964; ESAU et al. 1966a; see other references in CZANINSKI 1968b). When they have significantly ceased to grow, they contain one or two large vacuoles, the tonoplast of which is well preserved in experimental preparations, a cytoplasm rich in ribosomes but reduced to a thin pelicle applied against the walls (Plate 7.1 A).

The secondary thickenings involve particularly but not exclusively the Golgi apparatus *(dictyosomes)* and the endoplasmic reticulum. From the onset of cell wall differentiation, dictyosomes are very abundant but they increase in number up to the end of the differentiations. They generate numerous vesicles which are shown to be transparent with the usual electronic techniques; the vesicles move along towards the plasmalemma, mainly facing the developing structures, and associate with it through anastomoses, thus releasing their content in the periplasmic space (Plates 7.2 and 7.4 A). This activity continues along with the development of the walls and it now seems possible to acknowledge the fact that the Golgi derivatives yield some precursors, mainly glucids, for this development. Among several works, those of PICKETT-HEAPS (1966) demonstrated the incorporation of tritium-labelled glucose in Golgi vesicles located in the neighbourhood of the lignified thickenings and in the walls of the xylem of wheat in the course of differentiation. The labelling is even clearer with the PATAg technique according to THIERY (1967).

The *endoplasmic reticulum* is well developed in the future conducting cells. It occurs first in the shape of elongated flat saccules more or less parallel to the walls and does not take long to become enlarged, producing numerous vesicles. These productions and derivatives are frequently observed in contact with the plasmalemma or close to it. Like dictyosomes, they seem to bring preparietal substances to the surface of the cytoplasm. This fact was also supported by the results of PICKETT-HEAPS (1966) which showed a strong incorporation of labelled glucose in the reticulum of the conducting cells of the xylem of wheat in the course of differentiation. Moreover, in some cases, an acid phosphatase activity takes place in these structures, mainly in the neighbourhood of the new secondary thickenings (CZANINSKI 1968b), an unusual fact in the endoplasmic reticulum. The biosynthesis of the lignin monomeres begins also in the endoplasmic reticulum through fixation of -OH and $-OCH_3$ radicals on the cinnamic acid (ALIBERT et al. 1977).

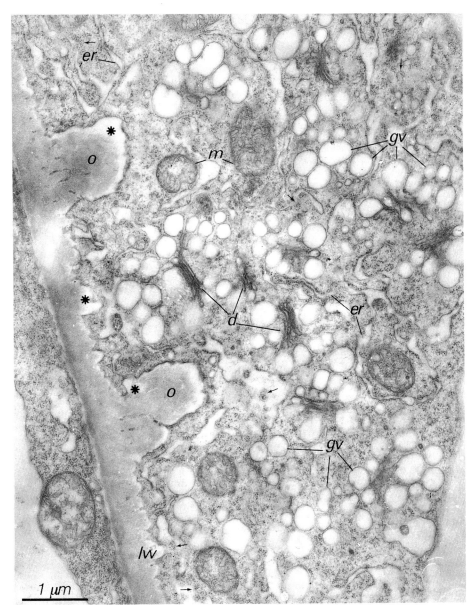

Plate 7.2. *Cucurbita maxima*, stem xylem.

Activities of the Golgi apparatus during the differentiation of a vessel element; *d* numerous dicty-osomes (three of which are labelled) surrounded by a profusion of Golgi vesicles *gv; o* secondary cell wall ornamentations, in formation, at the periphery of which, mainly but not exclusively, the plasmalemma incorporates Golgi vesicles by anastomosis *(asterisks); m* mitochondria; *er* rough endoplasmic reticulum, with numerous ribosomes; *lw* lateral cell wall of the vessel element; *arrows* indicate some of the polysomes scattered in the cytoplasm; × 18000.

Fixation: O_SO_4; contrast: $KMnO_4$

Microtubules parallel to the latest microfibrils and in the vicinity of the secondary wall structures were reported by several authors (HEPLER and NEWCOMB 1964; WOODING and NORTHCOTE 1964; CRONSHAW 1965a; CRONSHAW and BOUCK 1965; ESAU et al. 1966b; SRIVASTAVA and O'BRIEN 1966; CZANINSKI 1968b etc). It is admitted that usually microtubules determine the orientation of the cellulose microfibrils. Before the occurrence of secondary wall deposits, microtubules are grouped into bundles against the primary wall and the earliest wall thickenings are parallel to these bundles (BROWER and HEPLER 1976; FALCONER and SEAGULL 1985). This fact suggests that the side by side association of microtubules concentrate the enzymatic complexes and the precursors of the secondary wall differentiations in determining their localization. Experiments carried out with in vitro cultures of *Zinnia* mesophyll cells confirmed this hypothesis (FALCONER and SEAGULL 1986). Some of these cells develop into elongated or isodiametric tracheids. In the elongated cells, the microtubule bundles are mostly transverse and develop a secondary wall deposit of ray type (band pattern). In isodiametric cells, microtubule bundles are not significantly oriented and differentiate secondary deposits in networks (web pattern). After experimental depolymerization of the microtubules, the elongated cells become more or less isodiametric, producing secondary structures of a web type. After the removal of the depolymerization agent, microtubules polymerize again and become oriented in the elongated parts of the cells which, then, generate secondary layers of a ray type.

The mechanisms inducing the associations of tubules remain unknown. They could be induced by specific proteins, which have not yet been clearly identified. According to FUKUDA and KOMAMINE (1983), the differentiation of tracheids in the in vitro cultures of *Zinnia* is inhibited by inhibitors of protein synthesis such as cycloheximide. In addition, two "new" polypeptides are synthesized during the differentiation of these tracheids, whereas *two* other polypeptids are no longer produced.

Intercellular Connections. CZANINSKI's works (1968b) on the *Robinia* xylem showed that there are no plasmodesmata on the walls of the vessel elements even in the non-thickened areas with the primary wall only. Therefore, the connections between vessels and tracheids and between these elements and the other xylem cells are ensured without plasmodesmata.

The development of secondary walls, or "ornamentations" of tracheids and vessel elements, is certainly rapid. It is often difficult to observe the successive stages of the wall thickening but in many cases, the lignification progression may be recognized. For example, in the xylem of the stem of *Dianthus caryophyllus,* the lignification of the secondary walls begins against the primary wall and extends to the latest pectocellulosic deposits which increase on the plasmalemma side (CATESSON 1983). Before the end of the development, the whole protoplasm starts degenerating (see p.388).

Plate 7.3. *Cucurbita maxima,* stem xylem; ultrastructural aspects of terminal cell wall lysis, resulting in the perforation between two vessel elements.

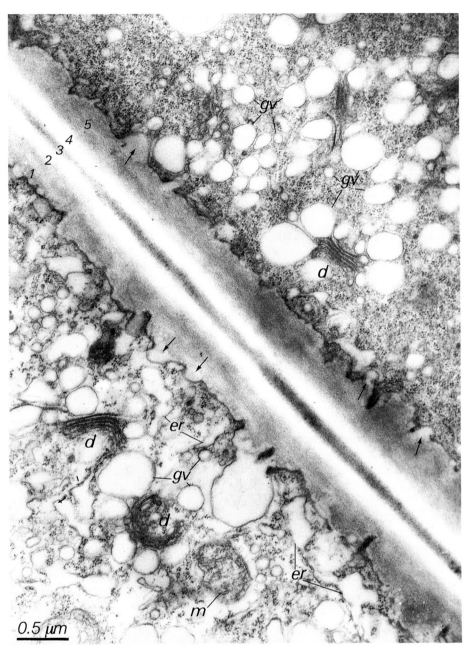

1–5 Development of five strata in the hypertrophied cell wall; **1,5** pectocellulosic primary wall, thickened and swollen; **2,4** clear layer related to lytic processes; **3** middle lamella. *d* Numerous dictyosomes, surrounded by a profusion of Golgi vesicles *gv; m* mitochondria; *er* distended endoplasmic reticulum, occasionally in contact with the plasmalemma which is very uneven; *arrows* indicate anastomosis figures between plasmalemma, Golgi vesicles or endoplasmic expansions, abundant in the neighbourhood of these figures; × 28000.

Fixation: OsO_4, injected into the medullary cavity; contrast: $KMnO_4$

7.1.3.2 Cytology of Perforation Processes Between Vessel Elements

Perforations between vessel elements occur along with the degeneration of the protoplasm. In light microscopy, the perforations were observed particularly by ESAU and HEWITT (1940) and interpreted as mainly induced by a gelation and hypertrophy of the middle lamella followed by the primary wall lysis. In fact, the primary wall itself may become thicker as shown on various figures obtained with electron microscopy displaying a set of five layers:

1. A more or less hypertrophied primary wall;
2. A clear layer obtained with the routine electron technique;
3. More or less modified middle lamella;
4. A clear layer as in point 2;
5. Primary wall of the next cell (Plate 7.3).

The primary wall does not become exclusively hypertrophied through lytic processes. At the onset of the differentiation of the vessel elements, the transverse enlargement, i.e. of the end wall, involves an *active* elaboration of the primary wall substances, particularly observed in the pink *Dianthus caryophyllus,* which goes on even after the enlargement has ceased (BENAYOUN et al. 1981).

Dictyosomes play an important part in the process. Lytic phenomena occur only later, first in the middle lamella, then in the other layers (e.g. in the poplar, according to BENAYOUN et al.). Similar processes were described in the bean by ESAU and CHARVAT (1978).

The ultrastructural cytochemical studies carried out by BENAYOUN showed that the portion of wall which will be further removed does not become lignified and seems to contain little cellulose or none, depending on species. The extraction techniques demonstrate that it is mainly composed of hemicelluloses and pectins.

Common organelles are still present in the cytoplasm which participates actively in the end wall lysis. An acid phosphatase activity takes place in the endoplasmic reticulum near the future perforation (CZANINSKI 1968b), the plasmalemma becomes significantly rough and the dictyosomes remain very active here (Plates 7.3 and 7.4 A). These characteristics seem to demonstrate the exchanges between the cytoplasm and the wall and the probably numerous lytic mechanisms.

Usually, the partial desintegration of the end wall is concomitant with the death of the vessel elements or it occurs just before. The perforation is bordered

---

Plate 7.4 A, B. *Cucurbita maxima,* stem xylem.

A Terminal cell wall *tw* between two vessel elements, during hydrolysis preceding the perforation. Golgi vesicles *gv,* issued from dictyosomes *d,* come in contact with the plasmalemma and are discharged by anastomosis into the periplasmic space *(arrows),* where they seem to contribute to the primary wall hypertrophy; × 20000.

B Young vessel element in the course of differentiation. *gv* Numerous Golgi vesicles surrounding dictyosomes; *er* endoplasmic reticulum, strongly dilated and extending branches in contact with the plasmalemma *(arrows); o* secondary cell wall ornamentations; *lw* lateral cell wall. The underlying cell is probably a future, still undifferentiated, vessel element; × 15200.

Fixations: OsO_4; contrast: $KMnO_4$

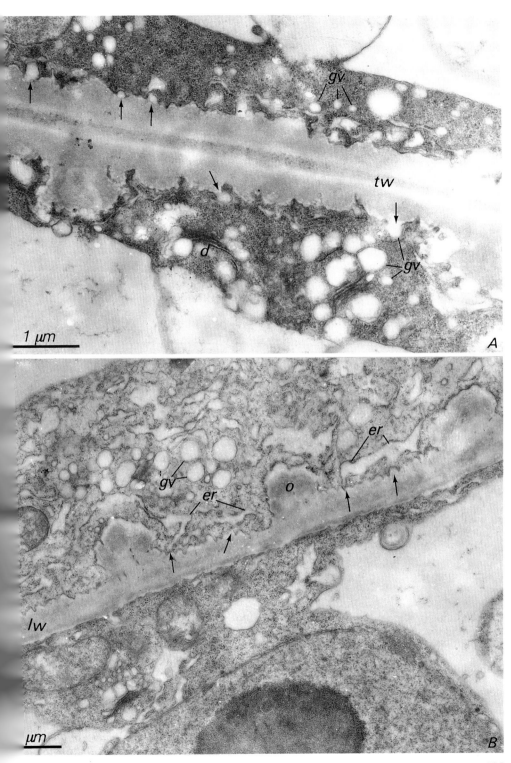

by a wall rim thickened by layers of lignified secondary wall (Figs. 7.6b, 7.26 a, b; Plate 7.5 A).

7.1.3.3 Partial Hydrolysis in Bordered Pits Between Vessels or Between Tracheids

Hydrolytic processes different from those inducing the perforations occur during the differentiation in the pittings between vessels, between vessels and tracheids, or between tracheids (O'BRIEN and THIMANN 1967; O'BRIEN 1970; CZANINSKI 1972a; LIBERMAN-MAXE 1982). According to CZANINSKI, this limited hydrolysis affects the polysaccharide compounds positive to the test of THIÉRY (1967) and to ruthenium red in the primary wall of the pits. These processes occur with a determined chronology and along with the differentiation of the conducting cells. Pectic substances only are spared, which can be stained with ferric hydroxylamine, mainly methylated pectins which are most resistant together with proteins which are contrasted with uranyl acetate. The hydrolysis is more significant in the centre of the pit lamella, whereas the peripheral areas, located between the secondary walls, lignified but positive to PATAg test, extent moderately on the border of the pits (CZANINSKI 1972a, 1979; CATESSON 1983; Plate 7.5 B–D).

These processes were clearly acknowledged in the fern *(Polypodium vulgare)* metaxylem by LIBERMAN-MAXE (1982).

7.1.3.4 Infrastructural Cytochemistry of the Lignification

Infrastructural cytochemical studies have been carried out on the lignification of the walls of the tracheary elements of the primary xylem and the wood. In the wheat xylem (metaxylem) CZANINSKI (1978) noted a significant peroxidase activity in the middle lamellae and the primary walls still pectocellulosic or in the process

---

▷

Plate 7.5.

A *Cucurbita maxima,* stem xylem. Part of transversal cell wall *tw,* in the course of lysis, between two successive vessel elements, fixed just before its breaking down. Around the resulting perforation *per,* only a ring *r* subsists made of lignified secondary wall, similar to lateral cell wall ornamentations *o; d* dictyosomes; *m* mitochondria; *l* leukoplast; $\times 8500$; Fixation: O_sO_4; contrast: KMnO$_4$.

B–D *Dianthus caryophyllus,* stem internode xylem. Three successive states of primary cell wall partial lysis in pits between contiguous vessels; *yv* youngest vessel; *av* more ancient vessel, each one on one side of each pit. In **B** the lysis hardly begins in the more ancient vessel only (places marked by *arrows*); In **C** the lysis is more advanced in the ancient vessel, but has not started in the younger one, where a dictyosome *d* subsists, surrounded by Golgi vesicles already devoid of polysaccharides; In **D** the lysis has reached the whole thickness of the pit wall. This wall is no longer or only slightly reactive to the PATAg test for acid polysaccharides, with the exception of the margin which borders it and links it to the primary cell wall which remains reactive between the ornamentations *(arrowheads);* still active dictyosomes *d* persist on the younger vessel side; $\times 20000$; PATAg technique, according to THIÉRY (1967) for acid polysaccharide detection.

(Courtesy of A.M.CATESSON 1983)

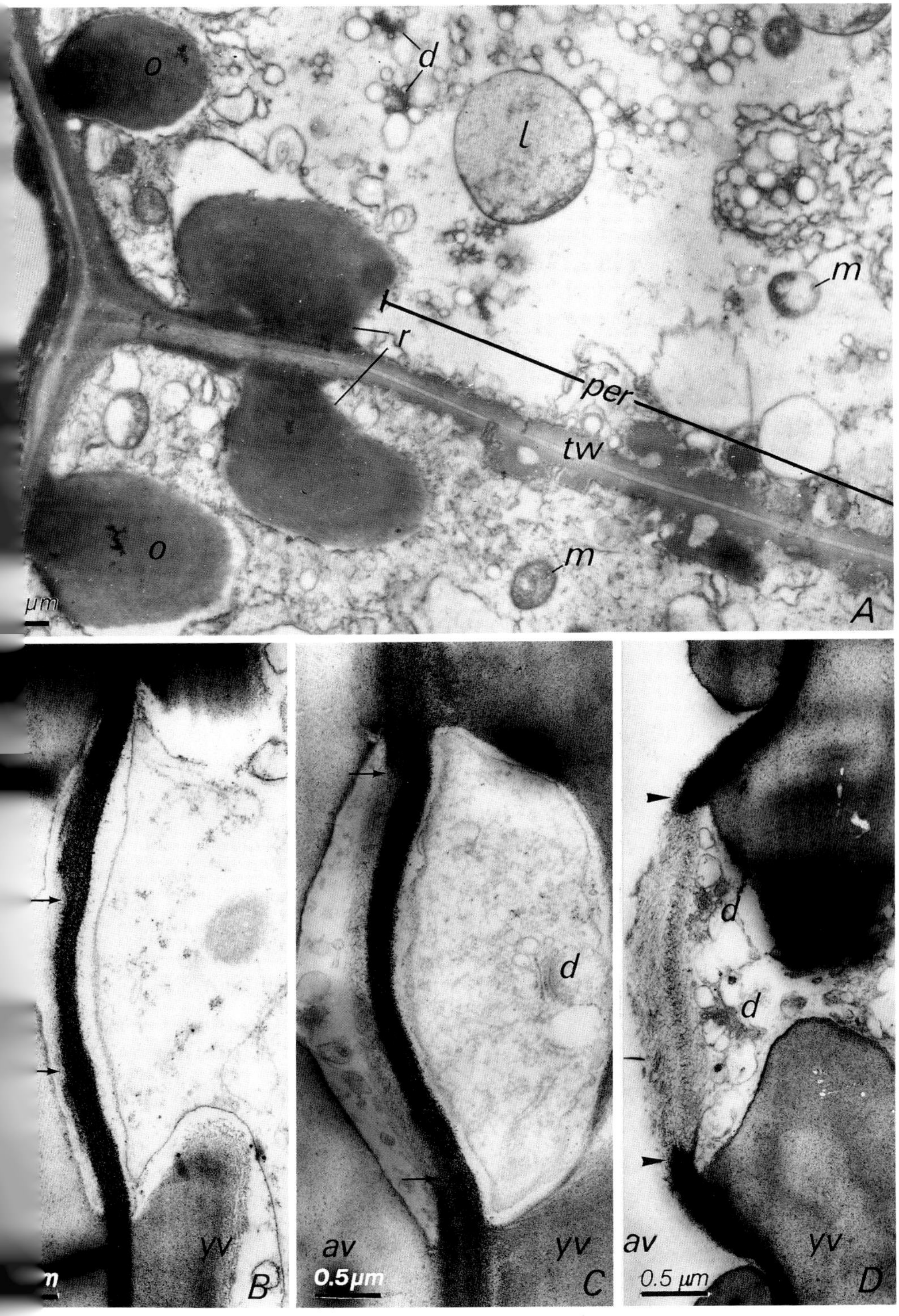
o
d
l
m
r
per
tw
o
m
μm
A
yv
B
av
0.5μm
yv
C
d
d
d
av
yv
0.5 μm
D

of lignification. A discrete and transient activity was observed during the formation of the secondary thickenings of the longitudinal walls and in the transverse walls before they become swollen. In any case, the peroxidase activity ceases when the lignification progresses. The lignin synthesis, from by-products of *cinnamic acids,* is induced by an oxidative polymerization which probably requires the contribution of a peroxidase. Similar results were obtained on *Coleus* vessels in the process of diffentiation by HEPLER et al. (1972).

In fact, the protoxylem primary walls do not lignify. In the metaxylem, a lignification gradient affects the primary walls; the impregnation is mainly significant between the secondary ornamentations of contiguous cells, in Angiosperms and Gymnosperms. In Pteridophytes (LIBERMAN-MAXE 1982), the primary wall of *Polypodium* metaxylem are poorly lignified (in scalariform tracheids), but they are not in the protoxylem.

In the differentiated secondary xylem, no peroxidase activity is noted, although it may last in the primary wall for some time. Middle lamellae and primary walls are strongly lignified when they are located between secondary thickenings. In pits, there are three cases: (1) between two vessels, the wall film *is not lignified;* (2) it is the same between a vessel and a parenchyma cell; (3) between two parenchyma cells, the primary wall and middle lamella are lignified (CZANINSKI 1979).

The lignification of the secondary walls reduces or suppresses the positivity of the polysaccharides to vic-glycol radicals in the test of THIÉRY, a fact which does not occur in primary walls. One can suppose that the lignin is mainly fixed on the carboxyl groupings of pectic acids, in abundance in the primary wall and on the vic-glycol radicals in secondary walls. Moreover, the secondary walls differ from the primary walls according to the amount of their various components; in particular, the primary walls seem to contain more coniferilic aldehydes (positive to chlorhydric phoroglucin; CZANINSKI 1979). The lignin in associating with polysaccharides may reduce the accessibility of the vic-glycol radicals. In fact, extractions performed with methylamine (CZANINSKI and MONTIÈS 1982), which removes part of the wall components (32% hemicelluloses, 23% cellulose, 5% lignin), or the treatment with chlorine-ethanolamine-silver nitrate water (technique of COPPICK and FOWLER 1939), used for the detection of lignin, alter the intermolecular relations between lignin and polysaccharides and enhance the test of THIÉRY in lignified secondary walls CATESSON and CZANINSKI 1979; CZANINSKI et al. 1982; LIBERMAN-MAXE 1982).

Table 7.1 summarizes the contrasts obtained in the detection of the wall components of the secondary xylem (CZANINSKI 1979).

7.1.4 Nucleus and Cytoplasm Degeneration

During transverse growth, i.e. the first differentiation phase of the conducting cells, the vacuolar apparatus which induces this growth stores most of the water absorbed by the cells and shifts the cytoplasm against the walls, where it forms a thin peripheral lining, swollen only around the nucleus.

The whole protoplasm is soon saturated by the significant volume of water and most of its components are hypertrophied. Firstly, the volume of the nucleus is

Table 7.1. Contrasts obtained using different techniques to distinguish the parietal components (+ + very contrasted, + contrasted, ± poorly contrasted, − no contrast)

		Primary walls				Secondary walls	Protecting layer
	Techniques[a]	In pits			Between the secondary lignified structures[d]	In vessels and parenchyma	In parenchyma cells associated with vessel elements (c. par. vss)
		Between two vessels	Between a parenchyma cell and a vessel	Between two parenchyma cells			
Polysaccharides	PATAg	−	+ +	+ +	+ +(vss) ou +(c.par) ou ±(c.par vss)	− ou ±[b]	+ ou + +
	Ferric hydroxylamine	+ ou + +[b]	+	+	+ +	− ou ±	+ ou ±
	Ruthenium red	− ou +[c]	+ ou + +[b]	± ou +[b]	±	− ou ±	+
Lignin	K.Mn O₄	− ou +[c]	−	+	+ +	+	−
	Chloride ethanolamine water A g NO₃	− ou ±[c]	−	+ +	+ +	+	−

[a] Chemical radicals PATAg: polysaccharides with vic – glycol radicals. Ferric hydroxylamine: methylated acid polysaccharides. Ruthenium red: acid polysaccharides.
[b] Contrasts obtained in plastids belonging to two different families.
[c] Contrasts obtained *at the level* of *the torus only* (Ulmus).
[d] (vss: between vessels; c. par.: between parenchyma cells; c. par. vss: between parenchyma cells and vessels).

(Courtesy of Czaninski 1979).

strongly increased, which involves the enlargement of the nuclear envelope surface (Plate 7.6 A). Then, the nucleus is distorted becoming deeply sinuous and lobed (Plate 7.6 A).

In the metaxylem of *Zea mays* roots, nuclei undergo cycles of endoreduplication with a DNA content of 4C, 8C, 16C and 32C. The synthesis phase lasts about 8 to 10 h, the interphase about 8 to 12 h (BARLOW 1985).

The nucleolus increases simultaneously, reflecting a high metabolic activity (Plate 7.6 A), which induces the increase of the messenger RNA and multiplies by six the amount of ribosomal RNA in the root metaxylem of *Allium cepa* (AVANZI et al. 1973). Later, the nucleolar substance fades and is diluted in the nucleoplasm. The nucleolar evolution, in the striped vessels of the *Cucurbita* metaxylem is slow compared to that of the cytoplasm. Chromocentres become lacunar and less osmiophilic. The nuclear vesicles, more and more deeply lobed, degenerate probably through dechromatinization but persist until the cytoplasm remnants have disappeared; the end of their involution has not yet been observed (Plate 7.7 B; BUVAT 1964a).

Similar processes were reported by CZANINSKI (1968b) in the *Robinia* secondary xylem. Following the swelling of the nucleus, the technique of FEULGEN shows a chromatin degeneration. The nucleus which has become lobed does not change until the end of the vessel element differentiation. Other examples have been described by CRONSHAW and BOUCK (1965), ESAU et al. (1966a), etc.

In the very young conducting elements, the *hyaloplasm* is strongly *basophilic;* it contains a high density of ribosomes. In places where ribosomes are not too close, it is possible to see that they assume a pattern of polysomes, which is a sign of activity. Moreover, the profiles of the endoplasmic reticulum are mainly of granular type at the onset of differentiation (Plates 7.2 and 7.8 A).

The endoplasmic reticulum is exceptionally developed since the beginning of the wall thickenings and the hydration induces a significant dilatation of the saccules, the production of fingerlike extensions and, on this sections, the appearance of vesiculations. Even when they are dilated, profiles are partially rough and partially smooth (Plate 7.2).

The dilated profiles of the reticulum stretch out, assuming an irregular and sinuous outline, whereas the hyaloplasm becomes clearer and loses ribosomes. In many cases, reticulum derivatives come into contact with the plasmalemma (Plate 7.4 B), suggesting metabolic exchanges between their content and the periplasmic space which develops along with the wall productions.

These observations can be compared with the above mentioned results of PICKET-HEAPS (1966) concerning the incorporation of tritiated glucose (see Sect. 7.1.3.1).

Dictyosomes are among the most active cytoplasmic ultrastructures during the differentiation of tracheids and vessels, according to their morphology. From the onset, they occur in great number, surrounded by many vesicles with a transparent content, about 0,1 to 0.4 μm in diameter in the future vessel elements of the *Cucurbita* metaxylem (Plates 7.2 and 7.3). These vesicles sometimes give a vesicular aspect to the cytoplasm. Their almost regularly circular outline allows their distinction from the distorted profiles of the endoplasmic reticulum. They come into contact with the plasmalemma and it is possible to obtain figures of fusion of the Golgi vesicles with the plasmalemma, like snapshots, with a fixative; the vesicle content is released through exocytosis, in the similarly transparent periplasmic space (Plates 7.2, 7.3, 7.4 A). Here again, the works of PICKETT-HEAPS and those of VIAN and ROLAND (1972) on different cellular types, suggest that the polysaccharide precursors of walls are thus brought where secondary thickenings occur. The same thing was observed in the *Robinia* secondary xylem (CZANINSKI 1968b with other references).

In great number from the beginning to the end of the differentiation, dictyosomes are among the last components to disappear when the protoplasm dies (Plate 7.7 C).

Plate 7.6 A, B. *Cucurbita maxima,* stem xylem.

A Vessel element in the course of differentiation, swollen and deeply lobed nucleus; the ultrastructure is still preserved, without degeneration signs; *ch* chromocentres; *chnu* nucleolar chromocentre; *m* mitochondria; *nu* nucleolus; *p* little differentiated plastids; *t* tonoplast; ×9600.
B "Parietal" cytoplasm of a young vessel element at the time of the tonoplast *t* detachment; *d* dictyosomes; *m* mitochondria; *o* secondary wall thickening (ornamentation); ×10000.

Fixations: OsO_4; contrast: $KMnO_4$

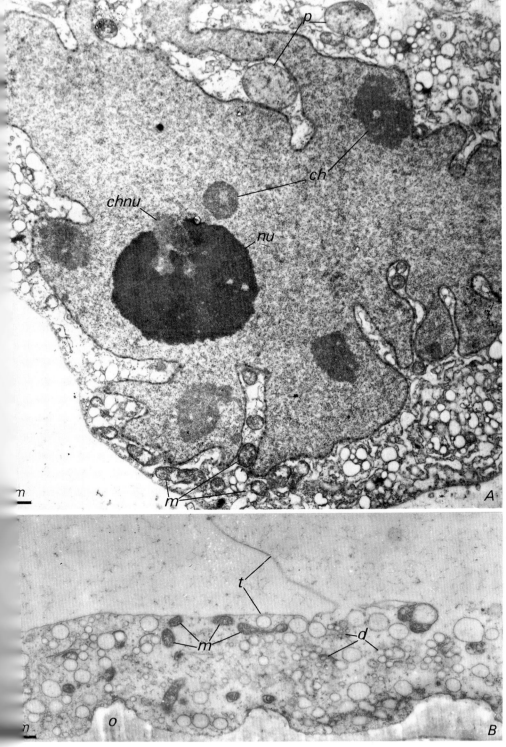

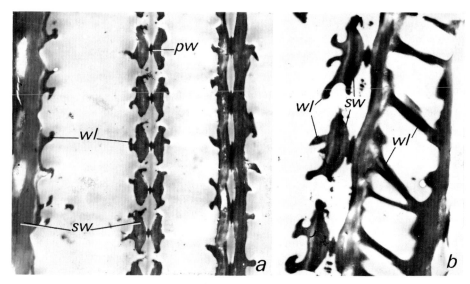

Fig. 7.8 a, b. Parts of differentiated vessels, no longer containing cytoplasm. The lignified primary wall *pw* is visible between the pits; *sw* lignified secondary wall; *wl* warty layer, forming outgrowths in the vessel cavity, and sometimes with striated or helical arrangements (**b**); (after CZA-NINSKI 1968 b)

The *plasmalemma,* involved in the active exchanges between the cytoplasm and the walls, becomes very uneven and distorted along with the first wall elaborations, either in the lateral walls or in the end walls, before the latter is destroyed. An electron-transparent periplasmic space develops together with the parietal productions, it is reduced with the completion of these productions and the plasmalemma surface becomes less uneven. At this time, the cytoplasm is very poor and its features indicate the beginning of the degeneration of the plasmalemma which is no longer visible when the cytoplasm itself has almost disappeared (Fig. 7.8).

The *tonoplast* is also affected by the hydration and involution of the cytoplasm from which it is detached quite late, freely moving in the cellular cavity, then unre-

▷

Plate 7.7 A–C. *Cucurbita maxima,* stem xylem; vessel element nearing the completion of differentiation.

A Degeneration of part of mitochondria *m,* swollen and vesiculated, whereas others keep an apparently normal ultrastructure *(arrows). vac* Vessel-associated cell, containing chlorophyllous plastids *p; d* numerous dictyosomes, surrounded by Golgi vesicles; *o* secondary wall ornamentation; *t* waving tonoplast, detached from impoverished cytoplasm; × 6400.
B Nucleus at the end of degeneration, reduced to a vesicle with clear content, into which some heterochromatic masses subsist; *m* persistent mitochondria; × 4200.
C Vessel element *v* at the end of differentiation; secondary cell wall completed; the tonoplast has disappeared; *m* and *d* persistent mitochondria and dictyosomes; *mic* mictoplasm; *vac* vessel-associated cell; *vxp* vertical xylem parenchyma; × 6800.

Fixations: O_SO_4; contrast: $KMnO_4$

392

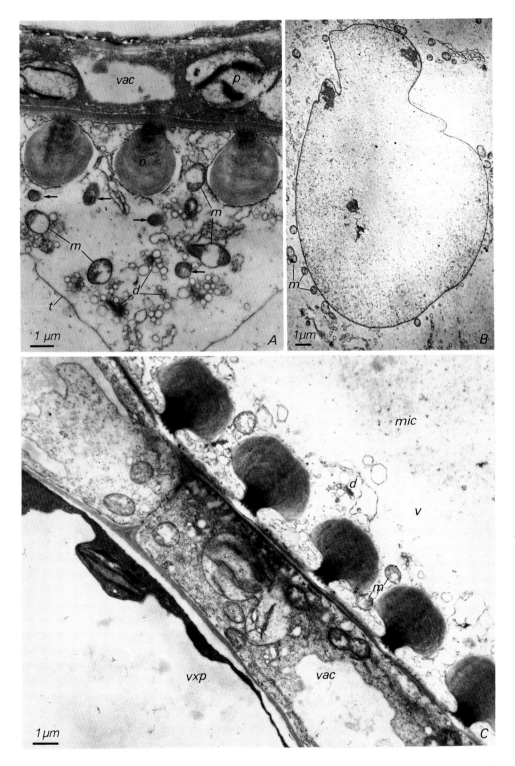

393

cognizable, the cavity being invaded by the cytoplasm dilution for some time (Plates 7.6, 7.7 A).

Plastids remain poorly differentiated during the maturation of tracheids and vessels. In several examples: *Cucurbita* xylem (BUVAT 1964a), *Robinia* secondary xylem (CZANINSKI 1968b), they contain no starch. In other cases, plastids are slightly amyliferous, particularly in the *Cucurbita* secondary vessels (BUVAT 1964a; ESAU et al. 1966a) and the *Pinus radiata* tracheids (CRONSHAW and WAR-DROP 1964). Furthermore, plastids do not grow much and remain poor in lamellae; their stroma clears up until they disappear concomitantly with the cytoplasm degeneration (Plate 7.8 A, B).

Mitochondria have a common infrastructure with several cristae in the very young conducting cells (Plate 7.2). But, very early, some cristae are affected by the hydration, thus becoming swollen and vesiculous (BUVAT 1964a; CZANINSKI 1968b; Plates 7.7 A and 7.8 B). These organelles together with the dictyosomes are among the last recognizable structures throughout the cytoplasm degeneration (Plate 7.7 B, C).

7.1.5 Tertiary Thickenings and "Warty Layer"

In the differentiated vessel elements or tracheids, irregular or spiral deposits sometimes line the S3 layer of the secondary wall (Fig. 7.8 a, b). This lining, usually amorphous, forms warty structures within the cellular cavity, called the "warty layer" (WARDROP and DAVIES 1962; LIESE 1963; LIESE and LEDBETTER 1963). This lining is independent of the secondary wall but it is not excluded that it comes in addition on this wall surface, to a latest deposit produced by the degenerating cytoplasm, forming localized thickenings (CRONSHAW 1965b). According to the authors, this lining may be partially composed of the remnants of the protoplasm necrotized along the differentiation of the conducting elements (LIESE 1963).

Plate 7.8 A, B. *Cucurbita maxima,* stem xylem.

A Vessel element in the course of differentiation; *p* ultrastructure of only slightly differentiated plastids; *d* dictyosomes; *m* mitochondrion; *o* secondary cell wall ornamentation; *arrows* indicate polyribosomes; × 27000.
B Vessel element, advanced state of differentiation; cytoplasm and nucleus *(n)* are degenerating; *er* hypertrophied and degenerating endoplasmic reticulum; *p* amyloplasts, the content of which is strongly cleared up; *s* starch grains; *m* mitochondria; *t* tonoplast, still in situ; × 11000.

Fixations: OsO_4; contrast: $KMnO_4$

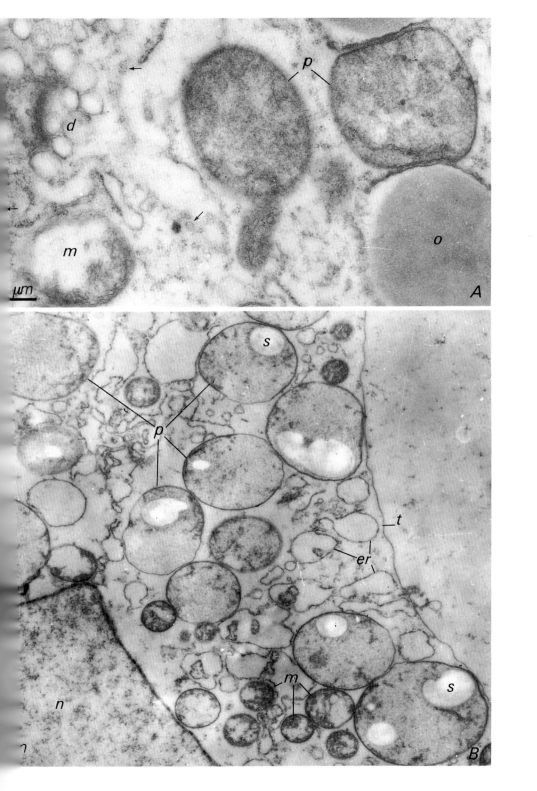

7.2 Xylem Fibres

The cells of the vascular tissue fibres are less different from the other cells of this tissue than the cells of the phloem fibres compared to the other phloem cells. They evolve from the same cells as tracheids and the differentiation occurs in the same way in both types of cells. In fact, phylogenetically, they seem to have evolved from tracheids. However, along their differentiation, they become longer than the tracheids of the same xylem. This is particularly true in the secondary xylem in which the fibre cells are the longer cells with respect to the fusiform cambial initials.

7.2.1 General Features and Various Categories

Xylem fibres occur mainly in the wood of Angiosperms and, in many cases, next to the vessels there is some intergrading betwen tracheids and fibres. Fibres can also occur in metaxylem (primary formation).

Thus, fibres differentiate a lignified secondary pitted wall. In the walls of the less evolved fibres, the pits are bordered as in tracheids. In species where the differentiation of fibres is more significant, the lips of the pit are less pronounced and the fibre pits develop into non-bordered pits (Fig. 7.5 b). The "libriform" fibres (see below) of about 60 species of ligneous plants, most of them bushy, mainly originating from the Negev and belonging to 24 families, were observed by FAHN and LESHEM (1962). With two exceptions, pits were simple in all of them. It is interesting to note that they were mainly families in which the herbaceous species, considered as more advanced than the arborescent ones, are prevalent in our regions: Caryophyllaceae, Chenopodiaceae, Compositae, Labiae, etc.

Correlatively, the wall becomes thicker and thicker. In this case, pits are canaliculi which become flattened and widened out in the cellular cavity through a slit-like aperture. The orientation of the splits on each extremity of the same canal often intersect (Fig. 7.5 b, c).

According to the degree of development of the secondary wall, *fibre tracheids* with a still not very thickened wall are distinguished from *libriform fibres* in which

---

▷

Plate 7.9 a–d. *Robinia pseudoacacia,* xylem; cytology of libriform fibres differentiation.

a Young fibre, fixed in *May;* cell wall still thin and non-lignified; *d* dictyosome; *m* little structured mitochondrion; *er* short profiles of endoplasmic reticulum; *v* vacuole; × 6700.

b Fibre during lignification (fixation in *May*); secondary cell wall *sw* already thickened; *mt* microtubules; *pl* poorly differentiated plastids; *m* mitochondrion, denser as in **a**; *d* active dictyosomes, forming numerous vesicles; × 9000.

c Differentiated fibres from the preceding year (fixation in *April*); maximal development of starch grains *s,* huddled together; cytoplasm reduced to a parietal coat; *lsw* lignified secondary cell wall; × 2900.

d Persistence of the nucleus *n* in a differentiated fibre, the cytoplasm of which has almost disappeared; *lsw* lignified secondary cell wall; × 2750.

Fixation: glutaraldehyde – OsO_4; contrast: $KMnO_4$; (courtesy of Y. CZANINSKI 1967)

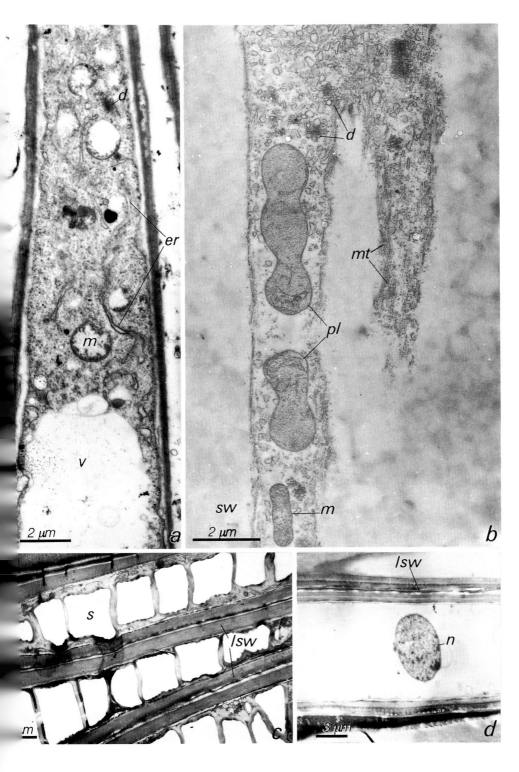

the walls are very thick as in the liber fibres, hence their name; but the developmental variation is continuous between tracheids, fibre tracheids and libriform fibres. To a certain point, flattened non-bordered pits in splits occur in the latter. In both categories, fibres may *divide again* belatedly. Transverse walls arise then, following the lateral wall lignification. They are generated through apparently regular mitoses, in which karyokinesis is followed by an ordinary cytodieresis (VESTAL and VESTAL 1940). *Septate fibres* have a longer life than the others and often store material (starch, lipids) or various secretions (resins, calcium oxalate). They are a compound cellular type showing both features of *fibres* and storage *lignous parenchyma* cells. Moreover, like tracheids, all these cells have the same histogenic origin.

Septate fibres occur in many Dicotyledons. In some of them, walled phloem fibres also occur *(Vitis)*.

In several woods, called "reaction wood" *(Quercus, Robinia)*, the fibres present hygroscopic wall layers which can swell up to suppress the cellular cavity. These fibres are of "mucilagenous" or "gelatinous" type, although, according to BAILEY and KERR (1937) they do not contain much mucilages or pectin compounds. The water could be absorbed by hydroscopic layers, mainly the most internal one, which may or may not be poorly lignified. These layers develop in the upper parts of the bowed branches of Angiosperms ("tension wood"; SINNOTT 1952).

In the wood of the horizontal or distorted branches of Gymnosperms ("compression wood") the gelatinous fibres are replaced by tracheids, the secondary wall S_2 of which is particularly developed and lignified.

In a same wood, fibres are usually longer than the nearby tracheids. The length ratio between the cambial cells and their derivatives, the fibre tracheids studied in several species of trees, is about 1.5 to 5. Approximately, tracheids may be half the length of fibres (BAILEY 1936).

Usually, fibres are all the more differentiated as vessel elements are more developed. Wood with tracheids only, without developed vessels or with primitive vessels *(Gnetales)*, have no fibres *(Pinus)* or the fibres do not develop in the libriform type *(Ephedra)*.

7.2.2 Cytological Data, Seasonal Variations and Life Duration

The *Robinia* xylem (CZANINSKI 1964 a, b, 1967) contains two categories of libriform fibres which start differentiating from May, forming a continuous sheath in the 3-year-old branches at the end of summer, before the last formations of vascular parenchyma. Some of them die early after they have lignified, whereas the others become loaded with starch and live until the next spring.

▷

Plate 7.10a, b. *Robinia pseudoacacia,* xylem; cytophysiology of libriform fibres; hydrolysis of starch, in *April,* in differentiated fibres of the preceding year.

a Beginning of amylolysis; splitting of starch grains *s; n* nucleus; ×3200.
b More advanced amylolysis; accompanied by the cytoplasm degeneration; some vestiges of membranes are still visible; *lsw* thickened and lignified secondary cell walls; ×3000.

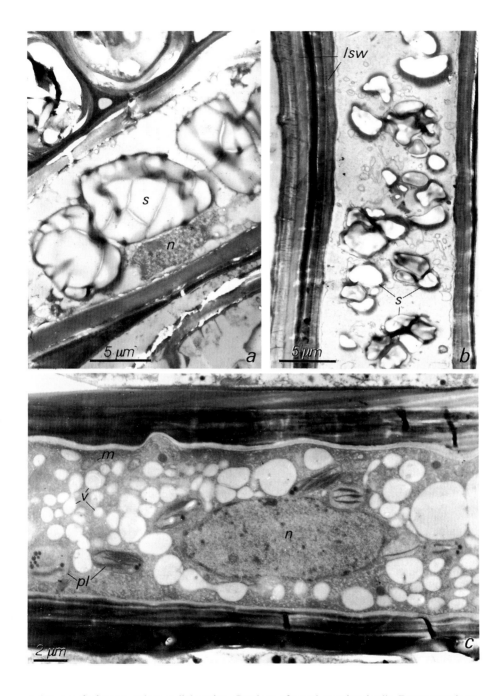

c *Acer pseudoplatanus,* xylem, radial section. Cytology of vessel-associated cells. Dense cytoplasm, containing numerous small vacuoles *v; pl* relatively little differentiated chloroplasts; *m* mitochondria; *n* nucleus; × 4500.

Fixations: glutaraldehyde – O$_S$O$_4$; contrast: KMnO$_4$; courtesy of Y. CZANINSKI 1967 **(a, b)** and 1977 a **(c)**

Young fibres which are not yet lignified are widely vacuolated. The cytoplasm, applied against the wall, is poor in ribosomes and the endoplasmic reticulum is mainly agranular. Mitochondria are poorly structured and the plastids are not differentiated. Dictyosomes are not yet numerous and generate vesicles of small diamter only (Plate 7.9 a, b).

During the deposition of the secondary wall and the lignification of fibres, the number of dictyosomes increases, surrounded by many clear vesicles which probably carry precursors to the wall. Mitochondria become more structured and numerous dense but poorly lamellated plastids occur in the cytoplasm (Plate 7.9 b).

These same features are observed in the fibres which degenerate shortly after they have differentiated and in those which will remain alive. In the first ones, the cytoplasm content is reduced to some membrane remnants; only the nucleus survives longer, contrarily to that which occurs in the differentiation of tracheids and vessel elements, where the last recognizable organelles are mitochondria and dictyosomes (Plate 7.9 d).

The fibres which remain alive until the next spring, store huge starch grains during the summer. The swollen amyloplasts fill the cellular cavity almost completely (Plate 7.9 c). The endoplasmic reticulum is poorly visible, a few dictyosomes and mitochondria occur in the impoverished and vesiculous cytoplasm.

In the next spring, the amylaceous storage of the fibres undergoes changes. Contrarily to that which occurs in vascular parenchyma (see later), plastids are not restructured after amylolysis, they degenerate along with the cytoplasm. Starch grains are split when they are released in the cytoplasm and this can be compared to the corrosion of similar grains contained in seed albumens in which amyloplasts also degenerate, e. g. in wheat (Plate 7.10 a, b). As in the other fibres, the nucleus persists for a time in the cytoplasm remnants, reduced to a few unidentified membranes, then degenerates through dechromatinization. In fibres without cytoplasm, CZANINSKI (1967) reported the existence of protuberances against the secondary wall, but independent of it, which are comparable to the structures described in vessel elements and tracheids under the term "tertiary layer" or "warty layer" (LIESE and LEDBETTER 1963).

Previously, CZANINSKI (1964 a, b) had described the seasonal variations of the chondriome and plastids. The author showed that their development is comparable to that of the vascular parenchyma cells: The formation of threadlike chondriocontes in spring, during amylogenesis, their fragmentation in fall, the almost selective occurrence of short mitochondria in winter, which become associated again in chondriocontes in April and during the spring amylolysis (Fig. 7.9).

Thus, in the *Robinia,* the libriform fibres may live for about a year, but the duration of life of the wood fibres is very variable, according to the species, and possibly depending on environmental conditions. According to FAHN and LESHEM (1962), the libriform fibres live longer in bushes or small tress growing under unfavourable conditions. In this case, the life duration of fibres is similar to that of the parenchyma and is a factor of adaptation. In the wood of *Tamarix aphylla,* using the existence of a nucleus as a criterion and the reaction to triphenyl tetrazolium chloride, FAHN and ARNON (1962) singled out living fibres in the last 16 to 21 annual growth rings.

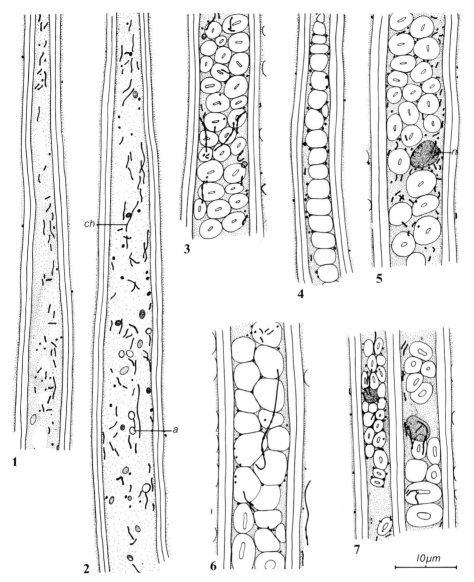

Fig. 7.9. Seasonal variations of chondriosomes and plastids in the libriform fibres of *Robinia pseudoacacia* xylem. **1** Recent fibre, observed in *May,* devoid of starch; **2** fibre fixed in *June;* onset of amylogenesis; *a* amyloplast; *ch* chondrioconte; **3** in *September;* end of amylogenesis, "moniliform" chondriocontes; **4** in *October;* chondriosomes already divided into short elements; **5** in *November;* chondriosomes in the shape of granular mitochondria and short bacilliform organelles; *n* nucleus; **6** in *April* of the next year; partial reassociation of chondriosomes in long filaments; **7** in *May;* beginning of amylolysis in differentiated fibres of the preceding year; chondriosomes associated in moniliform chains of filaments of various length; (after Y. CZANINSKI 1964b)

7.3 Xylem Parenchyma

All xylem tissues contain parenchymatous living cells. As in phloem, there are two categories: usually in primary xylem tissues and in the most differentiated woods, elongated cells parallel to conducting elements and to fibres occur. They form the *vertical xylem parenchyma* which itself may include different types of cells.

In the secondary xylem, or wood, radial rows of cells occur typically, in addition, producing the *wood rays* or *horizontal xylem parenchyma*.

7.3.1 Vertical Xylem Parenchyma

In this tissue, the cells may be as long as the initial cells of the procambium (primary xylem) or the cambium (*fusiform* initials, secondary xylem), but, quite often, transverse divisions occur along with differentiation, resulting in rows of cells shorter than the initials.

Frequently, mainly in the primary xylem, the walls of xylem parenchyma cells remain thin and cellulosic, but, in the wood, they usually become lignified. They may elaborate secondary layers but this fact is not general. In such cases, secondary pittings occur naturally on the walls. The walls which separate the parenchymatous cells from the tracheids and vessel elements have bordered pits which may be symmetrical or mixed, i.e. simple on the side of the parenchyma cell (Fig. 7.6). Between parenchymatous cells, pits are simple.

Vertical parenchyma cells have different shapes. Some are elongated in the orientation of the organ, parallel to tracheids and vessel elements; they are about 100 μm in length and their diameter ist often greater than 10 μm (in particular in the *Robinia* wood). They constitute the *vertical storage parenchyma*.

Other cells encircle the vessels with a more or less continuous sheath. They appear to be stretched and flattened on the vessel surface; they are about 25 to 30 μm in length in the *Robinia* (CZANINSKI 1966); they are the *vessel-associated cells* (Fig. 7.10).

In addition to their morphology and distribution, these two categories of cells vary in their functions; the cytological features and seasonal variations summarized in Section 7.3.3 reflect these physiological differences.

Locally, xylem parenchyma tissues have another type of cell on the walls of which protuberances develop, enlarging the plasmalemma surface. Usually the protuberances are not lignified. Such cells occur in the primary xylem of herbaceous plants at the level of the stem nodes and in the foliar ribs in those places where exchanges are most intense between cells and neighbouring tissues.

⊳

Fig. 7.10a, b. *Robinia pseudoacacia,* topography and cytology of xylem parenchyma cells. **a** Longitudinal radial section, showing the localization of storage parenchyma cells, pertaining either to the vertical system *(cprv)*, or to the horizontal system *(cprh)*, and the localization of vessel-associated cells *(cav)*, which can equally pertain to the two systems *(cav* and *cavh)*; *va* vessel of relatively small diameter. **b** Enlargement of the framed area of **a**. The chondriome of vessel-associated cells

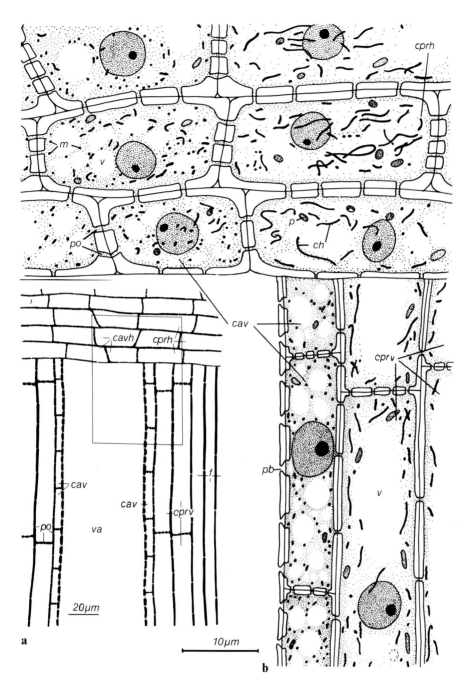

is strongly divided, contrarily to that of storage parenchyma cells, which comprises numerous filamentous chondriocontes *ch;* plastids are more scarce and exiguous in the vessel-associated cells
than in other cells, *cprv* and *cprh,* the vacuoles *v* of which are equally more voluminous. *m* Mitochondria; *pb* bordered pit; *po* simple pits (original lettering of the author). Fixation: Regaud,
staining: hematoxylin; (after Y. Czaninski 1968a)

These *"transfer cells"* occur also in other organs or tissues (nectaries, secretory tissues, placentas, hydatodes, etc; see PERRIN 1971; GUNNING et al. 1970; PATE and GUNNING 1972). In the primary xylem, they may correspond to the cells associated to vessels which contribute to the lateral transfer of substances in the secondary xylem between vessels and parenchyma tissues (CZANINSKI 1977b; Plate 7.1 B).

7.3.2 Horizontal Xylem Parenchyma (Xylem Rays)

The shape of the horizontal parenchyma cells varies, depending on species and on their distribution in wood rays. Usually, the median cells are radially elongated, and are called *procumbent cells*. On each end of the rays, at the lower and upper level, cells may be higher (in the longitudinal orientation of the organ) than wide (in the radial orientation; Fig. 7.11 a, b), they are called the *upright cells*. They all become lignified in most woods after having developed secondary walls with pits

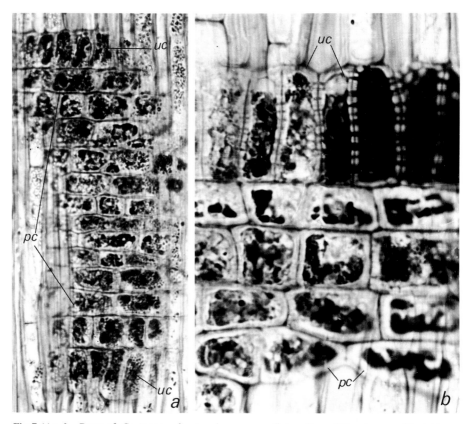

Fig. 7.11 a, b. Parts of *Castanea vulgaris* xylem rays, radial section of the stem. *uc* Upright ray cells, elongated parallel to the stem axis; *pc* procumbent cells, radially elongated; **b** shows simple pits between upright cells. Fixation: REGAUD, staining: hematoxylin

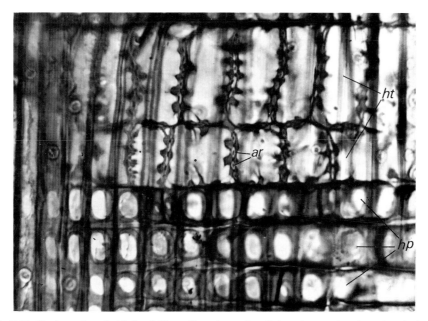

Fig. 7.12. Horizontal tracheids *ht* of *Pinus silvestris* wood, radial section. *ar* Areolate pits; *hp* horizontal parenchyma with particularly large pits between horizontal cells and underlying vertical tracheids

similar to those of the vertical parenchyma cells. In some cases, the cells of ligneous rays evolve into *horizontal tracheids* with areola pits (*Pinus silvestris;* Fig. 7.12) and sometimes even in vessels (Annonaceae, Botosso and Gomes 1982).

Finally, horizontal parenchyma cells may be associated with vessels; in this case they acquire particular features, including pits (see Sect. 7.3.3). In both categories of xylem parenchyma, cells live as long as the wood is functional and sometimes several years after. They seem essential for the functioning of the vessels per se. Furthermore, except for part of the fibres, with regards to living cells, the wood contains only those of these parenchyma tissues. The cytology of these elements gives evidence of the important contribution of the xylem to the transport and storage of *organic* substances, a contribution which comes in addition to its function of the circulatory system of the mineral sap.

7.3.3 Cytology and Seasonal Behaviour of Xylem Parenchyma

The vertical xylem parenchyma cells differentiate from the xylem derivatives of the long cambial initials, those of the woody rays are derived from the short initials.

7.3.3.1 Seasonal Parenchymatous Productions of the Cambium

At the end of winter, the cytoplasm of the cambium cells includes many small vacuoles, the chondriome is reduced to mitochondria, and plastids, undifferentiated, are hardly identified with light microscopy (BUVAT 1956); the nucleus is at the centre; in *Robinia,* it has one or two nucleoli about 3 μm in diameter (see Fig. 2.50 A). As soon as the generating zone is reactivated, the xylem derivatives widen and become more vacuolated, the radial size of the vertical cells increases from 5 to 10–12 μm (Fig. 7.13/1), numerous mitochondria become united into chondriocontes, and tiny plastids develop; the nucleus is shifted against the wall and the nucleolar substance is reduced. Usually, vertical cells are the seat of an intrusive apical growth; the apical cytoplasm is denser with spherical small vacuoles (Fig. 7.13/2). Before the end of differentiation, some vertical cells divide transversally so that a single cambial derivative generates a series of two to six cells (CZANINSKI 1965).

There are no processes of intrusive growth or further division in horizontal parenchyma cells; they become elongated radially, their plastids differentiate earlier and produce small starch grains (Fig. 7.13/5,6).

In the same way, during summer, parenchymatous productions go on developing. In September, the fusiform initials of *Robinia* give rise to two kinds of particular cells of late xylem parenchyma. Some of them, issued from further divisions, are short (12 to 16 μm), less vacuolated and each of them elaborates a crystal of calcium oxalate in tablet. They are poorly amyliferous in winter. Others, as thick as the cambial initials, divide into elements of 30 to 50 μm; their vacuoles are rather reduced and store tannins, plastids become slowly amyliferous (CZANINSKI 1965; Fig. 7.13/3,4). In addition to these specialized cells, the common xylem parenchyma cells differentiate in the same way as the cells of libriform fibres, they become highly amyliferous in summer, with the exception of the vessel-associated

---

▷

Fig. 7.13. *Robinia pseudoacacia;* cambial productions of xylem parenchyma and onset of their differentiation *(April–September).* **1** Arising of the first long cambial productions towards the xylem *(April)* and onset of vertical parenchyma differentiation; mitochondria become associated in little chains; beginning of plastid development. **2** Id *(April),* cellular lengthening by intrusive growth; apical cytoplasmic condensations. **3** Transversal cleavages giving rise to future crystalliferous cells *(September).* **4** Future tanniniferous cells *(September).* **5** Production of procumbent and upright cells derived from cambial short initials forming the horizontal parenchyma completed by radial elongation; linear associations of chondriosomes *(April).* **6** Horizontal xylem parenchyma cells at the beginning of differentiation *(April);* development of plastids, some of them already amyliferous. *a* Starch; *ca* bordering cell walls of differentiated cells of the preceding year; *cc* cambial cells; *ch* chain-shaped chondriosome; *p* plastids; *pa* amyloplasts (original lettering of the author). Fixation: REGAUD, staining: hematoxylin (after Y. CZANINSKI 1965)

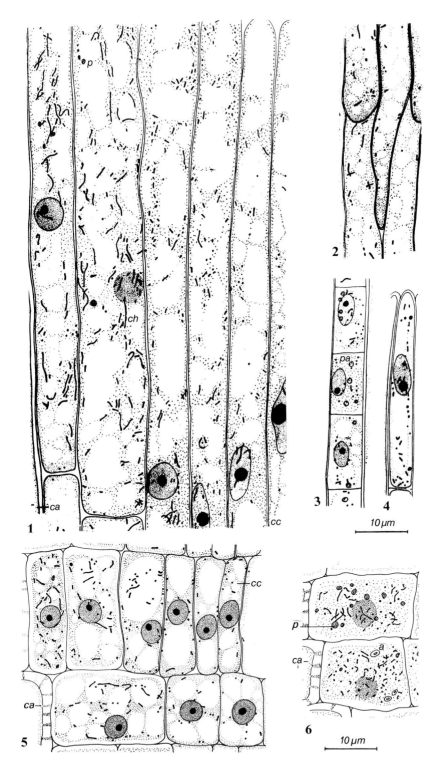

10 µm

10 µm

407

cells (see Sect. 7.3.1). The seasonal variations of these various types of cells are discussed later (see Sect. 7.3.3.2).

7.3.3.2 Ultrastructural Cytology of the Differentiation of Xylem Parenchyma

The works of Y. CZANINSKI (1970) have compleded the ultrastructural data on the xylem parenchyma differentiation in the *Robinia*. Following the growth phase and vacuolization mentioned above, the plasmalemma is activated and distorted, producing numerous vesicles prior to the secondary wall thickening, then a secondary growth process takes place, during which thickenings are quickly lignified; this phase coincides with a characteristic Golgi activity (Plate 7.12 A, B) and with the presence of an unusual acid phosphatase activity in the endoplasmic reticulum (CATESSON and CZANINSKI 1967).

The rough profiles of the plasmalemma (Plate 7.11 B, C) suggest an increase in the exchanges between the cytoplasm and the primary wall, but the nature of these exchanges has not been elucidated.

The period of Golgi activity (Plate 7,12 A, B) lasts as long as the elaboration of the secondary thickenings. The lignification of these increasing depositions progresses from the vessels, first reaching the cells to which they are associated, then the storage cells (CZANINSKI 1970). Along with the polysaccharide deposits on the secondary wall, numerous microtubules are concentrated in front of the thickenings; they are parallel to the latest fibrils deposited (Plate 7.11 A).

The further division of the vertical parenchyma cells gives rise to cells which contain a large vacuole, and develop into storage parenchyma cells and smaller cells with multiple vacuoles. The latter are the future "vessel-associated cells". The horizontal parenchyma cells include several vacuoles of variable sizes.

The chondriome, partially in the shape of filaments in the storage parenchyma (Plate 7.13 A), remains in the shape of mitochondria in the vessel-associated cells (Fig. 7.10 b; Plates 7.14 A and 7.10 C), presenting in any case a common infrastructure. Plastids differentiate slightly as soon as the walls are thickened. In the vertical storage parenchyma, they produce a few lamellae and become chlorophyllous in young stems (1 to 3 years); only later they synthesize starch grains. Conversely, amylogenesis occurs earlier in xylem rays. In vessel-associated cells, plastids soon assume a particular characteristic: confluences occur between thylakoids, thus encircling and imprisoning parts of the stroma (CZANINSKI 1968 a; Plate 7.13 B).

The amount of ribosomes decreases during the differentiation of the xylem parenchyma cells but remains higher in the vessel-associated cells.

Finally, lipidophilous granules, which do not seem to occur in young vessels, are constant in parenchyma cells in the process of differentiation. Thus, with the end of their development, considering their cytological features, it is possible to distinguish *vertical storage xylem parenchyma cells, horizontal storage xylem parenchyma cells* and *vessel-associated cells* which are derived from one or the other of the previous types.

In fall, the first two categories of xylem parenchyma accumulate large quantities of starch except for the cells of late vertical parenchyma, which accumulate oxalate or tannins and synthesize only little starch. In cells with oxalate CZANIN-

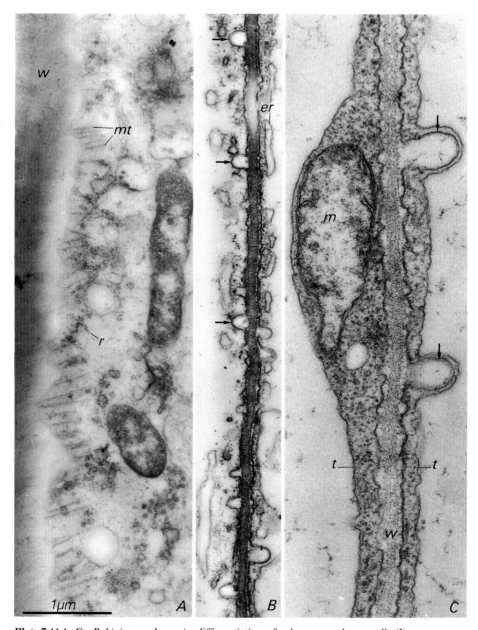

Plate 7.11 A–C. *Robinia pseudoacacia,* differentiation of xylem parenchyma cells (I).

A Abundance of microtubules, *mt,* at the time of dictyosome activity, contributing to the construction of the secondary cell wall, *w; r* ribosomes (associated in polyribosomes); ×23000.

B,C. Plasmalemma activity, demonstrated by invaginations *(arrows)* just before the onset of secondary wall elaboration; in fact, the cell wall *w* is still thin; *er* endoplasmic reticulum; *t* tonoplasts; ×23000.

Fixations: glutaraldehyde – O$_S$O$_4$; contrast: KMnO$_4$; (courtesy of Y. CZANINSKY 1970, relettered)

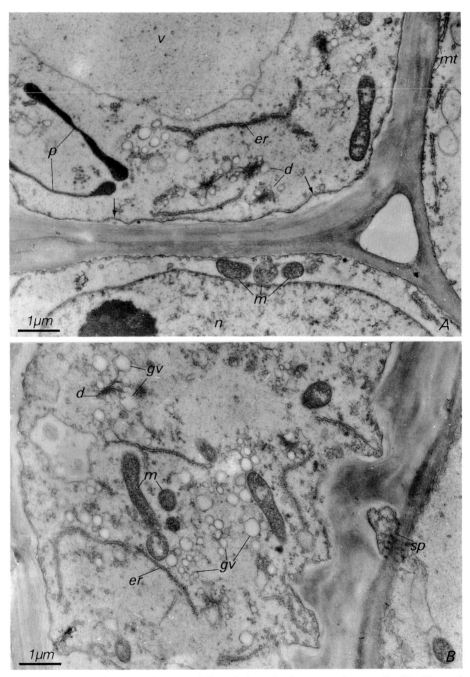

Plate 7.12 A, B. *Robinia pseudoacacia*, differentiation of xylem parenchyma cells (II). Phase of Golgi apparatus activity.

A Horizontal parenchyma cells at the beginning of the secondary cell wall layer deposit. The plasmalemma does not show invaginations, but only some bulges over clear periplasmic spaces *(arrows)*; × 11 200.

410

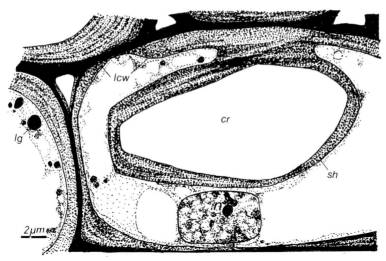

Fig. 7.14. *Robinia pseudoacacia,* xylem, oxalate cell of late xylem parenchyma. *cr* Site of the calcium oxalate, removed by the preparation of thin sections. The crystal is enclosed into a sheath *sh,* formed by an outgrowth of the lignified secondary cell wall *lcw,* linked to this wall by a kind of pedicel; so, the crystal is situated out of the cytoplasm; *lg* lipid granule; *n* nucleus (drawn after an electron micrograph of Y. CZANINSKI 1968 c)

SKI (1968 c) elucidated the unusual position of the crystal in tablet. This crystal is enclosed in an outgrowth of the wall to which it is linked by a wall-generated peduncle: in this case, the location is not intravacuolar as it is usually the case with oxalate crystals (Fig. 7.14).

In the *vessel-associated cells,* the cytoplasm remains rich in ribosomes (basophil) and maintains a divided vacuome (Plate 7.10 C). Plastids, usually poorly differentiated, have some thylacoids like in sycamore (CZANINSKI 1972b) and, in the *Robinia,* plastids are characterized by stroma islets encircled by anastomosed thylacoids as mentioned above. *They contain a small amount of starch only,* very short mitochondria, rare dictyosomes, a normal endoplasmic reticulum and the plasmalemma does not reflect any particular features.

Moreover, is should be noted that in the pits between the associated cells and the vessels, in which the wall (primary wall) remains cellulosic, *there are no plasmodesmata, whereas they usually occur in pits (simple) between parenchyma cells in which the primary wall becomes lignified (*CZANINSKI 1977 a).

Two other characteristics distinguish the vessel-associated cells from the other components of xylem parenchyma.

⊲————————————————————————————————————

B Vertical parenchyma cell, section tangential to one of the faces, showing the cytoplasmic film in front view containing numerous Golgi vesicles *gv. d* Dictyosomes producing, in the two figures, numerous transparent vesicles; *m* mitochondria; *mt* microtubules; *n* nucleus; *p* plastids; *sp* secondary pit, with plasmodesmata crossing out the primary cell wall; *er* endoplasmic reticulum; *v* vacuole; × 11 000.

Fixations: glutaraldehyde – OsO_4; contrast: $KMnO_4$; (courtesy of Y. CZANINSKI 1970, relettered)

411

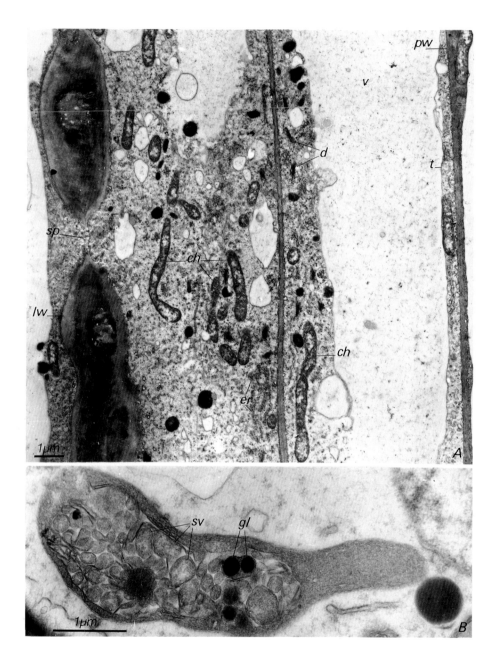

Plate 7.13 A, B. *Robinia pseudoacacia,* differentiation of xylem vertical parenchyma cells.

A Very young xylem derivatives; primary cell walls *pw* still devoid of secondary thickenings and not lignified on the cambial side, contrarily to the more ancient cells, the cell walls of which are thick and lignified *(lw); v* large axial vacuole (on the *left,* the cell is tangentially cut); *ch* filamentous chondriosomes; *d* little active dictyosomes; *sp* secondary simple pit (not bordered) with plasmodesmata; *er* endoplasmic reticulum; *t* tonoplast; ×8200 (courtesy of Y.CZANINSKI 1970).

B Plastid with anastomosed lamellae, which encircle "stroma vesicles" *sv,* observed in vessel-associated cells; *gl* lipid globules; ×19700.

Fixations: glutaraldehyde – O_SO_4; contrast: $KMnO_4$ (courtesy of Y.CZANINSKI 1968 a)

412

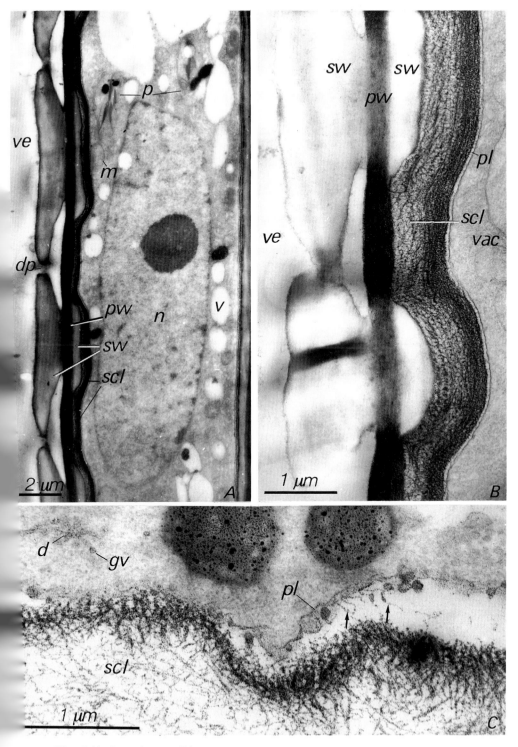

Plate 7.14. Legend see p. 414

1. A few weeks after they have differentiated, these cells generate an additional parietal layer exclusively applied on the wall that they have in common with the vessel (Plate 7.14 A, B). This structure, first mentioned by SCHMID (1965) under the term *"protection layer"* was also described by O'BRIEN (1970), who called it *"protective wall layer"*. The modalities of its formation and its cytochemical characteristics were elucidated by CZANINSKI (1973) in the *Robinia* and the sycamore. The first deposits occur on the secondary lignified thickenings, then they are extended to the surface of the primary cellulosic wall of the pits, a wall which is common to the associated cell and the vessel. This structure is elaborated, whereas the plasmalemma and the dictyosomes do not manifest the particular activity which goes along with the differentiation of the associated cell. In fact, this layer is independent of the usual wall, and hypothetically, it could be induced by the dead cells, i.e. the vessel element. However, during the formation of this layer, the PATAg test, i.e. THIERY's technique (1967) for the detection of polysacchardies with vic-glycol groupings, shows that the dictyosomes produce small diameter vesicles containing such polysaccharides. These vesicles are probably released, through exocytosis, in the periplasmic space where the same technique permits the detection of thin fibrils which elaborate the "protective wall layer" (Plate 7.14 C). Besides, this parietal structure is composed of cellulose and pectin and does not become lignified. Its role has not yet been established. SCHMID (1965) and O'BRIEN (1970) suggested its protective role towards the vessel-associated cells against the hydrolytic processes such as those which degrade the primary wall or perforate it, resulting in the pits between vessels. However, it is difficult to imagine how, from the inner part of cells, this layer could impair the activity of enzymes carried by the vessel. Considering that the cells associated to vessels are the main agents of the production of tyloses (see Sect. 3.4), CZANINSKI suggested that this structure has the role

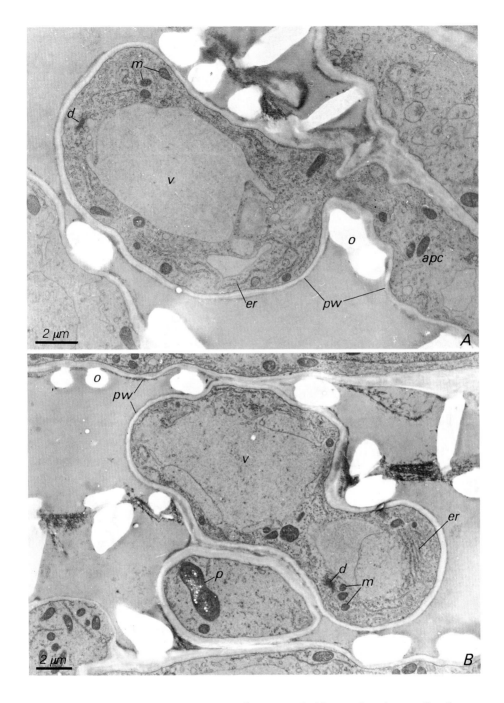

B Penetration of a tylose from a vessel to an adjacent one. In this case, the primary cell wall, common to the two vessels, is broken between the lignified ornamentations *o; p* plastids; other indications as in **A;** ×4500.

Fixations: glutaraldehyde – O_SO_4; contrast: $KMnO_4$; (courtesy of Y. CZANINSKI 1974)

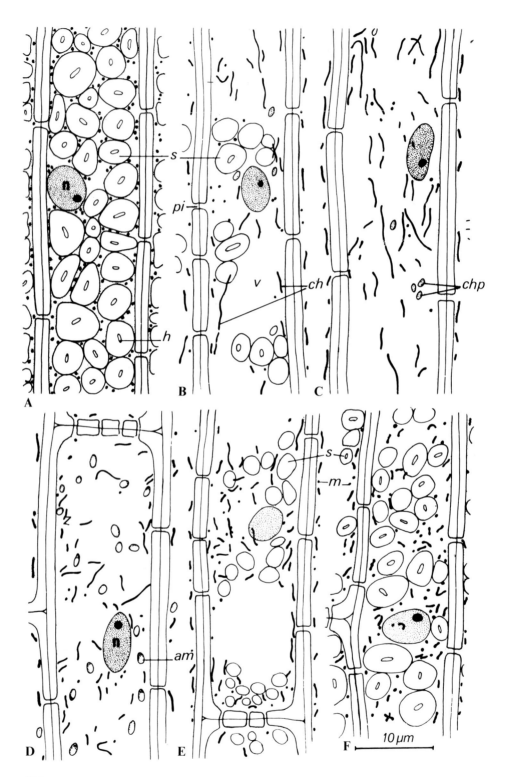

416

of parietal storage, and proposed the term *"secondary cellulosic layer"* which could be used in the development of these tyloses.

2. Contrarily to storage cells, the vessel-associated cells *do not undergo seasonal cytological variations.*

However, their metabolism may be affected by influences different from the seasonal ones, for example a secretory activity of gums and mucilages in response to parasitic aggressions (xylem of *Dianthus caryophyllus,* CATESSON and MOREAU 1985; CZANINSKI et al. 1973).

Seasonal variations are mainly concerned with the plastids and chondriosomes of vertical parenchyma cells and horizontal storage parenchyma cells. From May and June, the cells differentiate amyloplasts (Fig. 7.15 D) which can become quite numerous and voluminous in winter and almost fill the whole cellular volume completely (Fig. 7.15 A). Chondriosomes which are long and distorted in spring (Fig. 7.15 C) are divided into short mitochondria from the end of summer and in fall (Fig. 7.15 E, F). In the spring of the following year, along with the hydration of tissues, the starch is hydrolyzed as in April and May; plastids are somehow regenerated and mitochondria become associated in small chains (Fig. 7.15 B). One-year-old cells are then similar to the new cells derived from the cambium and enter into another seasonal cycle. Such cycles are repeated for several years as long as the xylem remains alive by its parenchymas (CZANINSKI 1968a).

The small vacuoles of the vessel-associated cells are generally positive to the peroxidase tests, unlike other parenchyma cells (CZANINSKI and CATESSON 1969).

7.3.4 Tyloses

When the vessels of the aged wood no longer function, or when their functioning is impaired by a traumatism or a parasitic infection, they are usually occluded by cellular protrusions issued from the xylem parenchyma, the tyloses. The parenchyma cells sheathing the vessels and particularly the vessel-associated cells produce protoplasmic outgrowths which penetrate into the vascular cavity through the pits (Fig. 7.16 a, b; Plate 7.15). One does not know what happens to the thin primary wall which subsists in the pits; it might either increase to envelop the cytoplasmic protrusion or it might be lysed and in this case, it is the invading cell which ensures the expansion of the wall. In any case, a wall covers the surface of tyloses; thin and pectocellulosic at first, it may become thickened and lignified.

◁

Fig. 7.15 A–F. *Robinia pseudoacacia,* xylem parenchyma; seasonal variations of plastids and chondriosomes. **A** Cell fixed in *January,* huge starch grains *s,* chondriome granular. **B** *May,* advanced hydrolysis of starch grains, chondriosomes *ch* associated in filaments (chondriocontes). **C** Recent cell, formed in *April;* long chondriocontes, small chloroplasts *chp.* **D** June, beginning of amylogenesis, filamentlike chondriosomes. **E** *September,* growth of starch grains, shorter chondriosomes. **F** *October,* completion of amylogenesis and breaking up of chondriosomes, in the shape of granular or short mitochondria. This development will be accentuated in November, resulting in the structure in **A**; *am* amyloplast; *h* hilum; *m* mitochondria; *pi* pit. Fixation: REGAUD – staining: hematoxylin; (after Y. CZANINSKI 1968a)

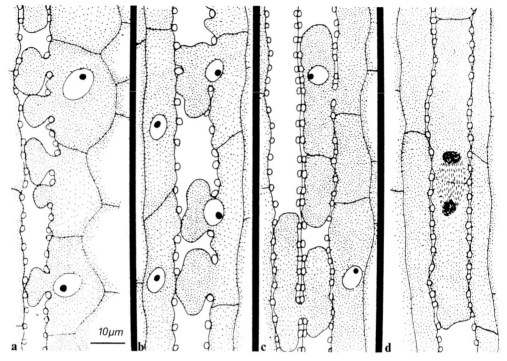

Fig. 7.16 a–d. *Daucus carota,* xylem. Development of tyloses from cells adjacent to vessels. **a** Protoplasmic protrusions penetrating into a vessel cavity through bordered pits; **b,c** passage of the nucleus into the outgrowth; **d** partitioning mitosis of a tylose; (after Y. CZANINSKI 1974)

In some cases, nuclei go into the tyloses which can then divide (Fig. 7.16 c, d) and sometimes develop into sclereids. In case of injury (section of the organ) tyloses seem to have a protecting role, they occlude the aperture of the vessels, resulting in a barrier against parasitic or saprophytic infections (BECKMAN 1971). At the level of the vessels, they contribute to the arising of scar tissues which can develop only from living cells.

The development of tyloses requires the resumption of significant growth of the primary wall of the parenchymatous cells adjacent to the vessels, and this fact supposes that the wall regains a plasticity comparable to that of the young cells of elongating organs. According to BECKMAN (1971), injuries and parasitic infections result in the acidification of the walls, comparable to that which occurs in young organs with a more or less circadian rhythm. The acidification "plastifies" the wall by weakening the hydrogen bridges which contribute to the stabilization of the structure of pectic compounds (including calcium pectates) and hemicelluloses of the wall matrix.

Very few studies have been devoted to the cytology of tyloses. In the rhizogenic rays of tubercles of *Daucus carota,* CZANINSKI (1974) showed that the tylose cytoplasm contains all the usual organelles and is not different from the cytoplasm of the cells which produce them. It is relatively rich in ribosomes, at least during

418

growth; the endoplasmic reticulum is well individualized, but the dictyosomes do not manifest any significant activity (Plate 7.15 A, B).

Although difficult to carry out, cytological studies on the tyloses of the secondary xylem of trees would be useful.

The main types of cells which form the vascular tissues have been described previously, we shall now discuss their distribution in the different categories of xylem tissues.

7.4 Primary Xylem

The primary xylem is derived from the cells of the procambial strands. When phloem and xylem are originally superimposed, the xylem results from the direct differentiation of the cells located on the opposite side to that which generates the primary phloem after they have ceased to proliferate. When the primary differentiation is completed, between the primary phloem and xylem, a variable number of cells of the procambial strand remains, which form a generating zone in the case of a plant with secondary tissues, or which are inactive in the other case.

In the primary xylem, two parts are usually distinguished, the *protoxylem* in which the vascular differentiation is completed before the end of organ growth, and the *metaxylem* in which differentiation may appear while the organ is elongating, but which matures only after this elongation is finished.

7.4.1 Protoxylem

The "vessel elements" of the protoxylem are not of a true vessel element type. In fact, they are tracheids with a small diameter. As the term of their differentiation, i.e. the death of the protoxylem constitutive cells, is completed before the end of organ growth, they can not longer follow this growth process and become stretched, torn or flattened. Thus, they do not live a long time and their remnants result in the formation of *"xylem poles"*.

Moreover, there are rather *few* conducting elements in the protoxylem, which mainly has thin-walled and cellulosic *parenchymatous cells*. After tracheids have been occluded, these cells either remain cellulosic or become lignified with or without the arising of secondary wall layers.

The tracheids of the protoxylem have about the *same structure* in the whole series of vascular plants, i.e. mainly *annular* or *helical* tracheids. One should note that these ornamentations only allow the elongation of the cells which are differentiating along with the organ elongation. In fact, if the elongation of the young organs is inhibited, perforated vessel elements are differentiated closer to the apical meristem (KOERNIKE 1905; GOODWIN 1942); however, this only reflects an accelerated general differentiation of the tissues in which growth has been interrupted.

Moreover, the study of the differentiation process of the protoxylem tracheids shows that the secondary wall layer is never continuous and that the formation of rings and helices *does not result from the elongation* which would stretch and dilacerate a uniform deposit, i.e. from the onset and during the whole process, the wall differentiations foreshadow the further formations.

7.4.2 Metaxylem

The maturation of the tracheids and vessel elements of the metaxylem, *after the organ elongation* is completed, characterizes this tissue. This temporal feature more than the vascular ornamentation, which depends on the modalities of organ growth, is a main distinction from the protoxylem.

However, the metaxylem tracheids or vessel elements are usually wider than the conducting cells of the protoxylem. The metaxylem may have striped and pitted thickenings, including all the intermediate ornamentations. True vessels may occur, but after a phase of tracheid differentiation only, which may or may not be early, depending on the developmental process of this xylem and on the group of plants considered.

In fact, the development of metaxylem is variable, depending on the organs of the same plant and whether or not secondary tissues occur in this plant. Thus, in the shoot of Dicotyledons, the metaxylem may be poorly developed as the secondary xylem arises early from the cambium. Conversely, in the stem of perennating Monocotyledons without secondary growth, it forms the whole functional xylem. In this case, numerous true vessels occur together with "imperfect vessels". Moreover, the metaxylem of Dicotyledons is more significant in foliar ribs and roots than in stems.

With a few exceptions, tracheids only occur in the metaxylem of Pteridophytes and Gymnosperms (see p.373).

In addition to tracheary elements, lignified or not, cells of *vascular parenchyma* and, frequently, *fibres* are found in metaxylem. Secondary-walled cells are therefore more significant in the metaxylem than in the protoxylem.

More complex than the protoxylem, the metaxylem is also more varied, depending on the systematic groups of vascular plants. Thus, the scalariform tracheids of the metaxylem of the present Pteridophytes characterize this wide set of plants.

The areolate tracheids which are the main part of the secondary xylem of Gymnosperms are already present in the metaxylem of these plants.

The *peripheral vessel elements* of the Equisetales and Monocotyledons, forming true vessels in the latter, are part of the metaxylem. They also occur in some Dicotyledons *(Dianthus)*.

When this tissue is very developed and when its maturation is going on for a long time in an organ, the oldest parts become inactive like the protoxylem, but usually vessel elements are not occluded. They are often invaded by tyloses.

7.5 Secondary Xylem

7.5.1 Distinctive Characters

Theoretically, the secondary xylem differs from the primary xylem because it is formed of derivatives of the vascular cambium. The cambium develops between the primary conducting tissues, through oriented mitoses in cells which, in typical cases, *were inactive or non-histogenic for some time.* At first, these cells are part of the procambial strands, then casually, of the neighbouring parenchyma. The transition between metaxylem and secondary xylem depends on the duration of the non-histogenic phase of the procambial cells, on the modalities of their further histogenic activity and on the development of metaxylem. Therefore, with regards to the anatomy, the classification into primary and secondary xylem is not very clear, due to the progressive pericline orientation of the mitoses in the procambium and to the arising of fusiform and ray initials which do not occur simultaneously and vary, depending on species (LARSON 1982).

However, when the cambium is mature, it usually consists of both types of initials, i.e., fusiform and ray initials, and the secondary xylem issued from it is therefore composed of two systems, the "axial" and the "ray" systems like in the secondary phloem. This architecture is the main characteristic of the secondary xylem. In contrast, in some cases, differentiations of the axial system constitute a transitional tissue between the primary and secondary xylem; the radial arrangement of cells of the secondary xylem, as seen in transections and often considered as a criterion for distinguishing their origin, is unreliable, for features of the secondary xylem may intergrade with the primary features (ESAU 1943).

According to BAILEY (1944b), the length of tracheary cells is the best criterion to distinguish the primary from the secondary xylem. When the differentiation is completed, the secondary elements are generally much shorter than those of the metaxylem which are shorter than those of the protoxylem. This is true in Dicotyledons as well as in Gymnosperms, perhaps with the exception of the most primitive and Prephanerogams. For example, in a species considered as advanced, with "storied cambium", the *Robinia,* vessel elements of the late metaxylem (pitted vessels) are 240 µm in length, whereas in the secondary xylem, the first elements are 170 µm in length (BAILEY 1944b; see Sect. 7.6).

7.5.2 General Features

The wood of Gymnosperms and arborescent Dicotyledons is the typical secondary xylem. In this case, it constitutes the most significant tissue of the shoots (trunks and branches) and the roots, but only the latest parts, i.e. closest to the cambium, are functional. The ancient parts, which form the wood core, have no living cells and no conducting function. The functional wood is part of the peripheral wood of the trunk or the branch.

There are two interpenetrating systems in the wood, the "wood rays" or *horizontal system* and the *axial or vertical system*. The nature and function of these systems will be disussed later. *The living cells* of the vertical system (lignified vertical

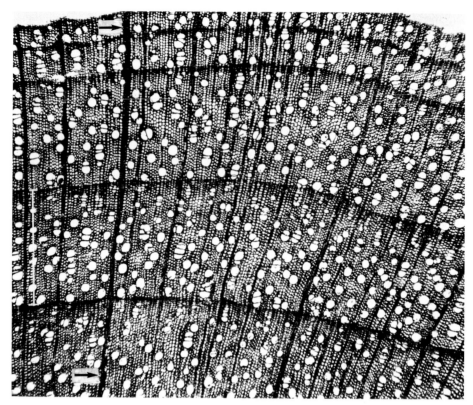

Fig. 7.17. *Acer campestre,* secondary xylem; concentric rings of annual growth, marked by the discontinuity between the autumnal late wood *(dark striae)* and the spring early wood (appearance of large vessels). *Arrows* indicate one of the horizontal parenchyma rays; the *white line* indicates one of the annual growth rings

parenchyma) or at least tracheids are connected, through pitted walls, with the living cells of the horizontal system (horizontal xylem parenchyma). Specific pits between the latter and the vessel elements stress the importance of these connections. This system of cells, living at least in part, is essential for the functioning of the wood.

The cambium activity is affected by seasonal variations not only in the plants living in moderate or cold regions, in which the proliferation ceases in winter, but very often also in tropical or subtropical plants, in which growth depends sometimes on particular environmental conditions (BAILEY 1944b), resulting in an annual zonation in the wood (Fig. 7.17).

Under moderate and cold climatic conditions (trees become rare in the latter), tow types of wood occur in plants: the spring wood with the conducting elements of great diameter and relatively thinner walls; the summer wood (or autumn wood) with slender cells and thicker walls: the latter is denser (Fig. 7.18). Spring wood passes more or less gradually to summer wood and autumn wood, the activity of the cambium being variable, but continuous. In contrast, the seasonal cam-

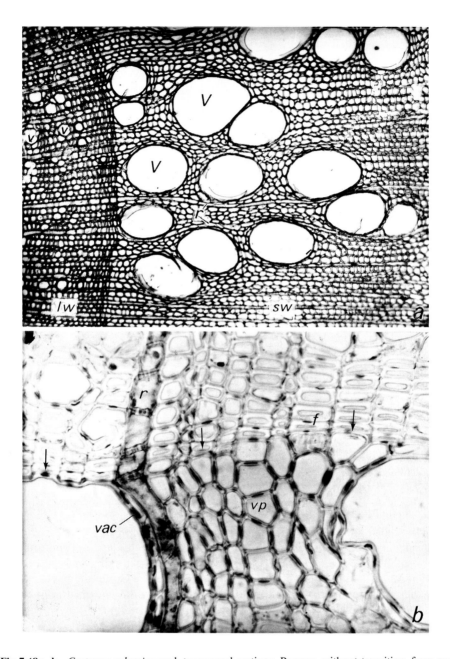

Fig. 7.18a, b. *Castanea vulgaris* wood, transversal sections. Passage, without transition, from autumn late wood *lw* to spring wood *sw* of following year. The autumn wood contains only a few vessels *v* of little diameter, it is essentially composed of fibres (*f, in* **b**). In contrast, in the spring wood, vessels *V* have large diameters and are surrounded by cells, most of which pertain to the vertical parenchyma and bear numerous pits, as shown in **b**. *vac* Vessel-associated cells; *vp* vertical parenchyma in the spring wood (**b**); *r* xylem rays (horizontal parenchyma); *arrows* (in **b**) show the limit between the autumn and spring wood, determining the seasonal zonation of the xylem

bial activity is interrupted at the onset of the cold season and resumed the next spring, directly generating the wood with large elements, so that the passage of the late wood to the further spring wood is discontinued. The contact of these types of wood is the most striking division line between the wood of the late season and the wood of the following season and the number of contacts indicates the age of the sectioned organ. In fact, in most shoots, the initiation of wood occurs with the first year of growth. Annual rings showing the discontinued growth of wood are found in deciduous as well as in evergreen trees.

In herbaceous plants with secondary growth (Dicotyledons) the wood structure is simple. Moreover, the amount of the different types of cells is variable, depending on organs. Thus, in the organs which become tuberized and store food, the vascular parenchyma, which remains cellulosic, becomes prevalent compared to the vessel elements and fibres (*Raphanus sativus, Daucus carota, Helianthus tuberosus*, etc.).

7.5.3 Gymnosperm Wood

The wood of Gymnosperms *(sensu stricto)* is chiefly characterized by the absence of vessels and the relative uniformity of the axial system, mainly composed of *tracheids*. These cells are almost all the same. They are prismatic, with four to six lateral faces, and very long. They are only wider in the spring wood and their wall is thicker in the late wood in which they gradually change into "fibre tracheids" (Fig. 7.19a). Conversely, there are no differentiated libriform fibres in the wood of Gymnosperms.

7.5.3.1 Tracheids

Tracheids may reach 4000 µm in length *(Pinus stobus)* or even 8000 µm in old trees (BAILEY 1920). They are derived from very long fusiform cambial initials and increase slightly in length through intrusive apical growth, their diameter is about 25 to 50 µm. In tangential sections, the extremities of tracheids, rather difficult to recognize, are not arranged in horizontal lines as it is the case in the wood of some advanced Dicotyledons. They are more or less intricated through intrusive apical growth, thus hiding the distribution in longitudinal series. Originally bevel-edged, the extremities are more or less distorted during growth. They are sometimes forked. However, usually the flat part is approximately seen in front on radial sections (Fig. 7.20b).

The areolae of tracheids are typically circular and located on the *radial* faces (Fig. 7.19 b, c). However, in the late wood, thick-walled slender tracheids have are-

---

                                                                ▷

Fig. 7.19 a–c. *Pinus silvestris* wood, transversal sections (**a, b**) and tangential section (**c**). **a** Stem area showing three spring resumptions of xylem formations; autumnal dormancy phases of cambial activity are indicated by *arrows*. In each annual ring of this homoxylate wood, vertical tracheids are more and more slender and have more and more thickened cell walls, from spring to

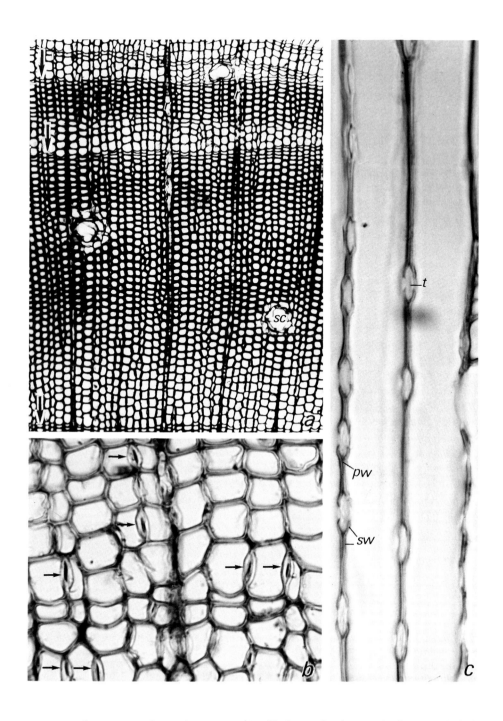

autumn. *sc* Secretory canals; *r* xylem rays. **b** Magnified part of spring wood; pits are exclusively localized on radial cell walls. **c** Longitudinal tangential section of spring wood; pits in profile view. *pw* Primary cell wall (and middle lamella); *sw* secondary cell wall; *t* torus, here situated in the middle of the pit chamber ola on both radial and *tangential* faces (Fig. 7.4 d). Above and below the areolae, transverse thickenings are often observed, deriving from the middle lamella and the primary wall, the "crassulae" (Fig. 7.4 b). The abundance of distribution of areolae are useful characters in systematics and phylogeny.

7.5.3.2 Vertical Parenchyma and Resin Ducts

The vertical parenchyma is more or less developed in Gymnosperms and is lacking in some genera *(Taxus, Araucaria, Torreya)*. It is frequently selectively localized around the secretory canals (*Pinus;* Fig. 7.19 a). In other cases, it consists of rows of cells scattered among tracheids, issued from fusiform initials which divide again transversally *(Thuja)*.

In spite of its practical interest, the problem of resin ducts in the wood of Gymnosperms has not been elucidated. In fact, several authors observed a close relationship between the development of resin ducts and the injuries which affect the trees naturally or experimentally. These observations suggest that resin ducts would be a response to a traumatism caused by injury, pressure, frost or wind.

Depending on species, resin ducts occur only in the axial system or in both systems (*Pinus silvestris;* Fig. 7.20 d). They are generally of the schizogenous type (see below).

The responses to traumatisms vary with the groups of conifers. In *Abies* or in close genera (Abietaceae), there are only short resin ducts which do not extend far from the traumatism; in *Pinus* and in close genera *(Pinaceae)* resin ducts are longer, extending far from the induction point. In the first case *(Abies)* the walls of the secretory cells lignify relatively early and resin is only produced when the canal is in formation for a year. In contrast, in *Pinus,* the secretory cells remain cellulosic and resin is produced in abundance for several years.

Compared to the frequency of traumatisms, the duration of life and the extension of the resin ducts could explain why they occur almost constantly in the wood of Pinaceae.

Fig. 7.20 a–d. *Pinus silvestris* wood, radial sections (**a, b**) and tangential sections (**c, d**). **a** Autumnal wood, made of fibre - tracheids *ft* and spring wood, where vertical tracheids *vt* are wider and with less thickened cell walls. *ar* Areolate pits; *p* large particular pits, between horizontal parenchyma cells and vertical tracheids; *ht* horizontal tracheids. **b** Spring vertical tracheids *vt,* in contact with the last fibre - tracheids of the preceding autumn, *ft,* which, in front view, show their flattened extremities *(arrow); r* xylem rays, composed of three series of horizontal parenchyma cells and, at each extremity, of a series of horizontal tracheids. **c** Aspect of the wood in tangential section. *r* Uniseriate rays, moderately high; this orientation shows the tapered extremity of tracheids, here in profile view *(arrows).* **d** Same tangential orientation as in **c**; one of the rays is thickened and contains a horizontal secretory canal *sc; arrows* indicate the particular pits between horizontal parenchyma cells and vertical tracheids

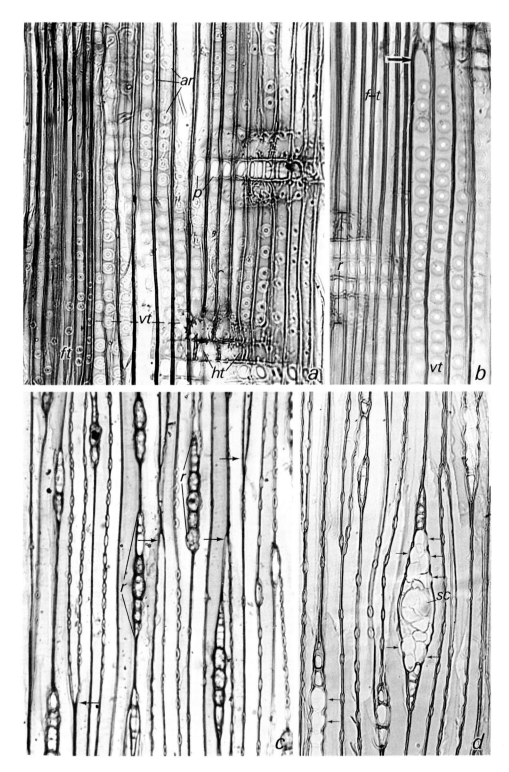

7.5.3.3 Horizontal Parenchyma

In Gymnosperms, wood rays are usually narrow. They consist of a single tangential layer (uniseriate rays) of a few cell tiers in height (1 to 20, more in rare cases; Fig. 7.20). Such rays enlarged by a resin duct are an exception (Fig. 7.20 d).

In some cases, horizontal parenchyma cells have thin walls. In Taxaceae, Taxodiaceae and Podocarpaceae, BAILEY and FAULL (1934) considered that cells have only a primary wall, whereas a secondary layer occurs in Abietaceae and Pinaceae.

The wood rays of Gymnosperms sometimes include "horizontal" tracheids, less elongated than the vertical ones, with areolae and therefore secondary lignified walls, ornamented with indented extensions in the cellular cavity (*Pinus;* Fig. 7.12).

When horizontal parenchyma cells have secondary walls, mixed pittings occur on the radial walls. In fact, these walls lean against perpendicular vertical tracheids. Pits are bordered on the side of tracheids, simple on the side of the horizontal parenchyma cell. The common part between a vertical tracheid and a horizontal cell usually assumes a rectangular or square shape (Figs. 7.12; 7.20 a, b), the distribution, number and shape of the mixed pittings in this quadrangular area are used as criteria for the subdivision of conifers and for phylogenetic study (JACQUIOT 1955).

The intricacy and frequency of the horizontal parenchyma rays are so important that *along its radial faces, each vertical tracheid abuts on one or several rays;* thus it is in close contact with *living cells* through the particular, mixed pits. This fact may explain why the wood of many Gymnosperms functions in the absence of vertical parenchyma (Fig. 7.20 d).

7.5.4 Ligneous Dicotyledon Wood

Ordinarily the wood of arborescent Dicotyledons is more complex than that of Gymnosperms. The chief distinction is the presence of *true vessels* which are strongly individualized among the other cells (Figs. 7.17, 7.18). However, in some very primitive Dicotyledons, there are no true vessels, but only tracheids. An accurate observation only allows one to distinguish their wood from Gymnosperm wood. This is the case of the Winteraceae, among the Ranales (BAILEY 1944a).

In contrast, in the more complex woods, vessels, tracheids, fibre tracheids, libriform fibres, vertical and horizontal parenchyma are found all together, e.g. in *Quercus* (Fig. 7.21). Moreover, the most complex woods do not necessarily belong to the most advanced species.

An extensive terminology has been established to distinguish the various types of wood according to the following features:

1. Distribution of vessels and fibres and size of the vessels in each annual ring;
2. Distribution of vertical parenchyma;
3. Composition of the rays (horizontal parenchyma);
4. Orientation of the extremities of cells in the vertical system.

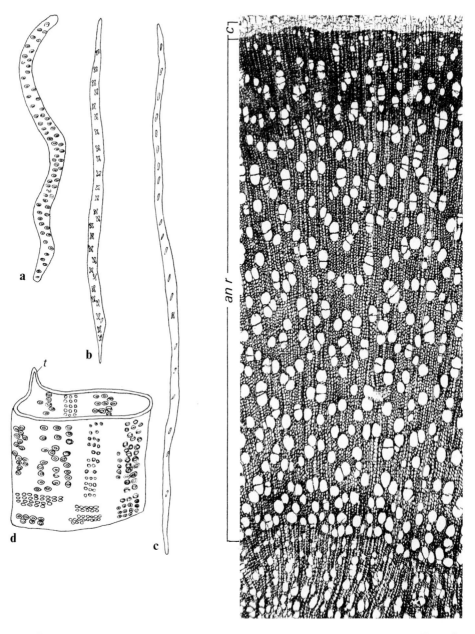

Fig. 7.21 a–d. Isolated elements of *Quercus* wood. **a** Tracheid; **b** fibre – tracheid; **c** libriform fibre; *d* vessel element of large diameter; very slight longitudinal growth, simple and transversal perforation, intrusive growth vestige, indicated by the "tail" *t*

Fig. 7.22. *Populus alba*, secondary xylem; an example of diffuse porous wood; vessels have the same diameter in thickness of an annual growth ring *anr; c* cambium

429

The main variations in structure, depending on these four characteristics, are summarized in the following sections.

7.5.4.1 Distribution and Size of Vessels

In various woods, the vessels have about the same size and are regularly scattered within in annual production (diffuse distribution; Figs. 7.17, 7.22; e.g. *Acer, Betula,* some primitive species such as *Liriodendron tulipifera*). In others, there are very large vessels in the spring wood and smaller ones or none in the late wood (annular distribution). This is the case for most deciduous oaks and the genera *Fraxinus, Robinia, Castanea* (Figs. 7.18, 7.25).

In fact, there are many intermediate types between the diffuse and annular structures. These features are specific but affected by the climatic conditions. The annular distribution is more frequent in the cold temperate zone (ZIMMERMANN 1983).

Usually this distribution is considered as more advanced. The physiological activity of this type of wood seems also different from that of the wood with a diffuse distribution. The vessels are longer and, according to HUBER (1935), the sap circulates ten times quicker than in the vessels of the wood of the diffuse type. Conversely, the sap circulates only during spring.

Late vessels of small size are less efficient but remain active for a longer time, when the sap circulation is less active (ZIMMERMANN 1983).

The spring wood of annular type differentiates usually earlier than the corresponding xylem system in wood with diffuse distribution, but this fact is not general.

The distribution of fibres, particularly of the most differentiated fibres (libriform) is apparently inverse to that of large vessels. Thus, they form the main part of the late wood in the case of a ringlike arrangement but in the diffuse condition, a few layers exclusively made of fibres indicate the winter discontinuity (Figs. 7.17, 7.22).

During differentiation, the cells derived from the fusiform initials, which generate the vessels, do not usually grow in length, except in some cases, in which a slight intrusive apical growth produces a short tapered end which will not widen any further. Conversely, the cell increases in diameter significantly. These processes result in the formation of cyclindrical vessel elements, wider than high in some cases, with a small protuberance, the "tail" (Fig. 7.21 d).

The more advanced the wood structure, the shorter the "fusiform" cambial initials and the shorter the vascular element with regards to its diameter.

The transverse growth of the vascular elements shifts the other wood cells and significantly modifies the "radial" arrangement of this secondary tissue. Very often, the neighbouring cells do not increase along with the vascular growth: they are stretched and sometimes separated from one another partially or completely. When they are partially separated, the stretched cells are in contact through the pit-richest areas. Between these areas, detached parts constitute some archs and the whole structure resembles a lace against the vessels, visible on longitudinal sections (Fig. 7.23 c).

430

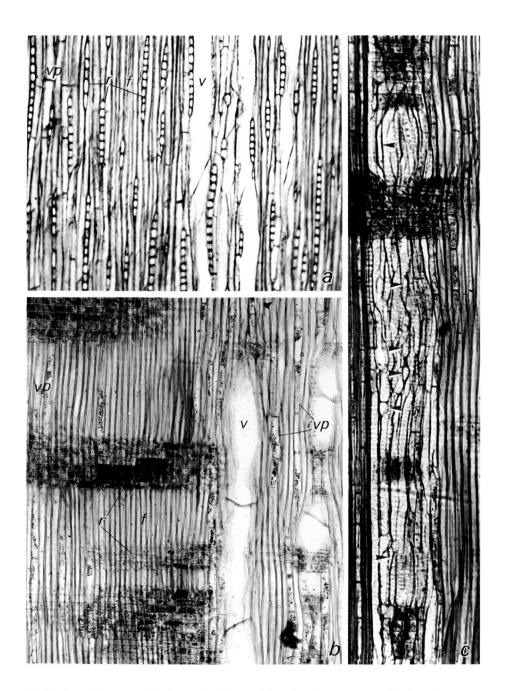

Fig. 7.23 a–c. *Castanea vulgaris* wood. **a** Tangential section in summer wood, relatively narrow vessels *v; r* uniseriate rays of moderate height; *f* fibres abundant in the late wood; *vp* vertical parenchyma, less abundant, contrarily to the fibres *f*. **b** Radial section through several rays *r*, in front view. Other indications as in **a. c** Radial section, hardly crossing associated cells of a spring vessel, which become distended by the vessel lateral growth and form, here and there, a perforated pellicle *(arrowheads)*

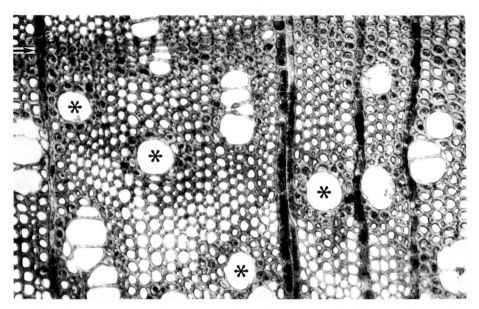

Fig. 7.24. *Acer campestre,* secondary xylem; preferential paratracheal localization of the vertical xylem parenchyma, particularly around vessels indicated with an *asterisk.* Transversal sections treated with the iodic-iodine reactive, which stains the starch in vertical parenchyma cells and fibres in autumnal wood, and in horizontal parenchyma cells where it is abundant *(arrow)*

7.5.4.2 Distribution of the Vertical Parenchyma

The amount of axial parenchyma is usually more significant than in the wood of Gymnosperms. Only a few species of Dicotyledons "without vessels" such as the *Drimys winteri* have a small amount or none (BAILEY 1944a). These living cells are variably distributed among the other cells of the axial system, depending on species.

In some woods, the rows of parenchymatous cells are dispersed among the other elements (diffuse parenchyma, e.g. *Quercus*) or appear in parallel bands giving each growth ring of wood a stratified aspect (some *Acer*). But, in the second

▷

Fig. 7.25 a–c. *Robinia pseudoacacia,* secondary xylem. **a** Transversal section; *sw* spring wood; differentiation, from the resumption of cambial activity, of wide spring vessels *sv,* mostly surrounded by parenchyma cells (large dimensions in sections); *lw* summer and autumnal late wood, narrow late vessels *lv,* progressive abundance of fibres *f* (narrow diameter) in the vertical system; the *arrow* indicates the discontinuity between autumnal and spring wood; *r* xylem rays (horizontal parenchyma); *vp* vertical parenchyma. **b,c** Two tangential sections showing more or less clear vestiges of the cambial storied structure, in non-redivided cells of vertical parenchyma *vp* and in redivided cells *rc* of this parenchyma, when initial extremities remain visible (particularly between *arrows*). This arrangement is no longer apparent in zones of fibres *f* where it has been obliterated by the intrusive growth. *vac* Vessel-associated cells; *r* xylem rays, either uniseriate or multiseriate, and of variable height; *v* vessel, constituted of short elements, with obviously transversal perforations

432

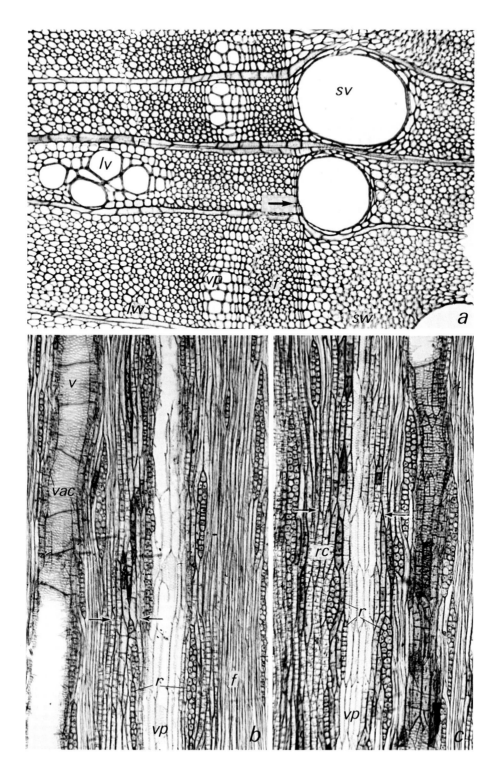

433

case, there are two types of distribution with regards to the relationship between vessels and parenchymatous cells. In the *apotracheal type,* although parenchymatous cells may touch the vessels, the position of the parenchyma is independent of that of the vessels. In contrast, in the *paratracheal type,* the two elements are closely associated with one another in the species in which ligneous parenchyma cells occur only around the vessels (Fig. 7.24). Of course, the diffuse distribution is also apotracheal. The phylogenetic sequence of these structures seems to be the following: diffuse apotracheal, stratified apotracheal, paratracheal, but this has not been checked (MONEY et al. 1950). Moreover, through their arrangements as well as through their cytological features, the septate fibres, when they occur, are comparable to the axial parenchyma cells. In addition, non-zonate paratracheal distributions of parenchymatous cells form a more or less cylindrical sheath around vessels (some leguminous).

7.5.4.3 Structure of the Xylem Rays

Dicotyledonous wood contains either uniseriate rays (a single layer of cells; Fig. 7.23) or multiseriate (several layers of cells, Fig. 7.25 b, c) or both types. The height of rays is very variable and sometimes may exceed that of the rays of Gymnosperms, particularly in multiseriate rays.

The Dicotyledons typically contain only horizontal parenchyma cells in the rays. But in some exceptional cases, horizontal cells generate tracheids and even vessels (Annonaceae, BOTOSSO and GOMES 1982). The rays may consist of only procumbent or only upright cells (*homogeneous* or *homocellular* rays) or both types of cells (*heterogeneous* or *heterocellular* rays).

The most primitive woods have both heterocellular uniseriate rays and multiseriate rays, usually very high. In more advanced woods, one can observe:

1. Either high multiseriate rays and shorter uniseriate rays;
2. Either a single type of rays, uniseriate or multiseriate;
3. Exceptionally, rays disappear (BARGHOORN 1941 b).

The development of *homocellular* rays may be induced by the Evolution from the primitive heterocellular types.

In some cases, this evolution may partially occur during the ontogenesis of the same trunk. By means of successive tangential sections, the structural evolution of the same ray may be revealed, displaying the phases from the earliest wood to the latest (BARGHOORN 1940, 1941 a).

Such modifications are obviously induced by changes in the structure of the *cambium.* This meristem has a remarkable plasticity. Short initials may disappear and be replaced by long ones or conversely. BARGHOORN, BANNAN (1951, 1953, 1960, 1963, 1968) and other histologists have particularly demonstrated this pasticity in the cambium of Gymnosperms, but that of Dicotyledons seems to have the same peculiarity. Thus, ray initials may arise through transverse divisions, either from the extremities of fusiform initials only or from their whole length.

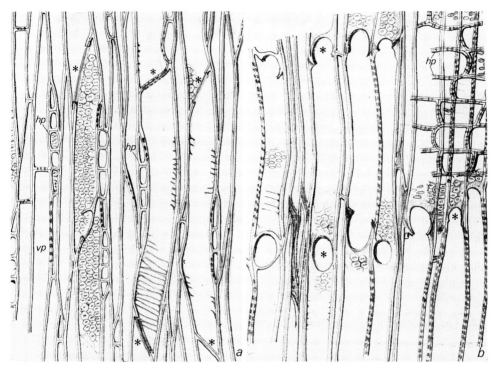

Fig. 7.26 a, b. *Aesculus hippocastanum,* secondary xylem, an example of wood with roughly storied structure of vessel elements. **a** Tangential section; **b** radial section; *asterisks* show the simple and oblique perforations of vessel elements, which are arranged in approximately horizontal lines (perpendicular to the stem axis). These perforations are seen in profile in **a**, in front view in **b**; *hp* horizontal parenchyma; *vp* vertical parenchyma

7.5.4.4 Disposition of Terminal Walls of Vertical Parenchyma Cells

Like in the secondary phloem, in the most advanced wood structures, the cells of the vertical system are relatively short. In the process of differentiation, there is a slight "intrusive growth" or none at all and the terminal walls keep about the same characteristics as in the cambium. When fusiform initials are arranged in parallel rows, on longitudinal views (cf. Fig. 2.51 c), the terminal walls being approximately in a transverse plane, the wood follows more or less this *"storied distribution"*, e. g. in the *Robinia* (Fig. 7.25 b, c) and less clearly in the *Aesculus* (Fig. 7.26).

In primitive species with regards to the structure of their wood, either non-storied long initials occur in the cambium, or the storied distribution is more or less impaired by the variable apical growth which takes place during the differentiation of some cells of the axial system (tracheids, vessels and fibres). In such cases, the wood includes longer elements, even those arranged in vessels, and there is no more storied distribution than in the wood of Gymnosperms (Fig. 7.23).

7.5.5 Histology and Mechanical Properties of Woods

The various distributions of the cellular types of wood play a part in the mechanical characteristics as well as the ratio of its components. The areas with large vessel members of the annular system are weaker than the areas of late wood with thick-walled small cells. Properties are more homogeneous in the diffuse system.

The *specific weight* of a dry wood is a good criterion for its hardness and resistance, although to assess the latter, it is necessary to consider the type of force exerted on the wood (continuous effort, torsion, shocks, etc.).

Moreover, the structure and the chemical composition of the different types of wood is of considerable interest in the evaluation of their response to corrosive agents (humidity, temperature, microorganisms) and mainly condition the fixation and penetration of the protective substances. In this respect, resinous woods of Gymnosperms are less liable to putrefy together with the woods richest in tannins *(Quercus)*. Finally, in general, the woods are more easily worked out when they have a more *homogeneous* structure, but other properties, well known by craftsmen, interfere with this structural feature.

7.6 The Xylem and the Systematics of Vascular Plants

The xylem is the only tissue for which the taxonomic and phylogenic importance can be compared to that of the reproductive apparatus.

Earlier in this chapter, we have mentioned the significance of the metaxylem which, alone, serves to characterize the great branchings and the groups of vascular plants: Pteridophytes (scalariform tracheids), Gymnosperms (areolate tracheids, no true vessels) and Angiosperms (tracheids and often, true reticulated or pitted vessels).

The secondary xylem existing presently in Gymnosperms and Dicotyledons easily allows one to distinguish these two groups since only Dicotyledons have true vessels. But the distinction is subtle in the most primitive forms of the latter, the so-called Dicotyledons "without vessels" (e.g. Winteraceae). However, the xylem of such groups is well individualized. Thus, the wood of the Ranales "without vessels" generally includes multiseriate rays, whereas uniseriate rays only occur in the wood of Gymnosperms.

Among the Chlamydosperms, in Gnetales, cells with areolate pits of Gymnospermian type and with helical or ray thickenings are found, but these cells become perforated and produce true vessels, which does not occur in Gymnosperms *sensu stricto* (foraminate perforations; Fig. 7.27).

Next to the Chlamydosperms (with the Piperaceae), the Juglandaceae have a very primitive wood.

In each of these great groups, the histological features of the xylem serve to define the orders, families and sometimes the genera themselves. In *Araucaria*, tracheids bear several longitudinal rows of pits, one against the other, on the radial faces. In *Pinus,* they usually have a single row of areolae, more spaced on each radial face.

436

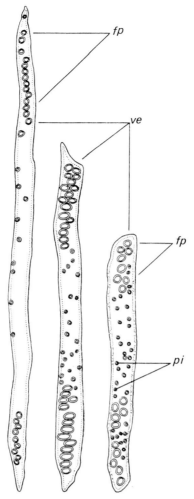

Fig. 7.27. *Ephedra californica,* secondary xylem, more or less fusiform vessel elements *ve; fp* terminal foraminate perforations; *pi* bordered pits along lateral cell walls; (after Esau 1965)

In the secondary xylem of relatively primitive ligneous Dicotyledons, there are intermediate structures between the tracheids of Gymnosperms and the most evolved vessels. According to species, these vascular elements become less and less elongated, their end walls become less and less oblique, with perforations wider and less numerous. The genera *Magnolia, Liriodendron, Betula, Populus, Quercus, Acer,* etc. belong to this evolutionary order (Fig. 7.28).

In more developed forms, the vascular elements are shorter with transversal walls, which are completely resorbed or reduced to a single narrow marginal ring (single perforation; *Quercus, Aesculus,* etc. Fig. 7.26). As mentioned earlier in this chapter, a short terminal appendage, "the tail", is the vestige of apical growth (Fig. 7.21 d).

In herbaceous species, the vascular elements are even shorter; their end walls are transversal and disappear almost completely, leaving only narrow rings against

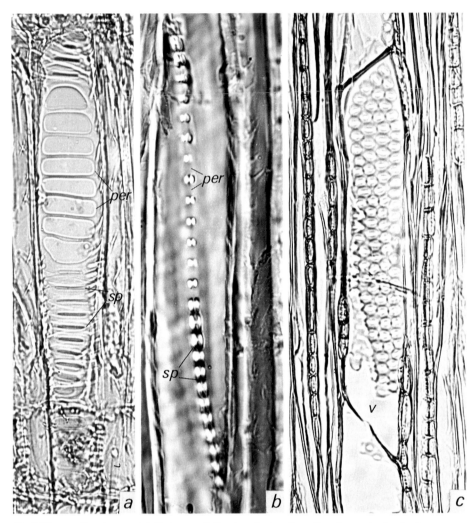

Fig. 7.28 a–c. Developmental features of the terminal cell walls of vessels in Angiosperms.
a, b *Magnolia grandiflora,* considered as a little developed species; *per* strongly oblique scalariform perforations of terminal cell walls; *sp* non-perforated scalariform pits on terminal cell walls.
a Front view (radial section); **b** profile (tangential section); **c** *Populus alba,* moderately developed species, vessel element with simple perforation of terminal cell walls, which are less oblique than those of *Magnolia;* relatively small length; alternate arrangement of pits (tangential section)

the vessel wall; in some cases this structure can be found in striped vessel elements (Fig. 7.29).

Particularly in Dicotyledons, the xylem structure gives ground to the idea that herbaceous species are, statistically, more evolved than arborescent ones.

The evolution going on in the *primary xylem* of the herbaceous species can be compared to that which occurs in the secondary xylem of the arborescent species.

The observation particularly shows the formation of vessels with scalariform perforations, then foraminate ones, earlier or later in the metaxylem, depending

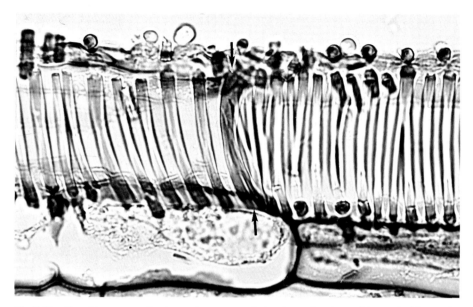

Fig. 7.29. *Cucurbita pepo;* striped vessel; transversal simple perforation, with no significant rim *(arrows),* hardly distinct from striped ornamentations

on the degree of species evolution (CHEADLE, BAILEY). The "precocity" of the formation of the first true vessels from the secondary xylem to the earlier metaxylem and sometimes to the helical elements of the protoxylem is also a general feature of evolution. With regards to this feature, Graminaceae are among the most advanced (CHEADLE 1943 a, b, 1944, 1956). In Monocotyledons, true vessels first occur in the primary xylem of roots, then in stems (CHEADLE 1944).

Finally, it should be noted that the evolutionary tendencies are the same in all the great groups of vascular plants which, in this respect, differ mainly in the significance of the most evolved structures:

1. In Phanerogams, Gymnosperms *sensu stricto* do not produce true vessels, even in the secondary xylem, but only tracheids. Conversely, the Gnetales, which have long been considered as particularly developed Gymnosperms, and which have presently been separated from them and classified in Chlamydosperms, generate vascular elements with areolate pits of Gymnospermian type, but in which perforations of such areolae occur in the secondary xylem of relatively poorly evolved genera *(Ephedra)* and already in the metaxylem of more advanced forms. In *Gnetum,* simple perforations occur in the vessels of the secondary xylem.

2. In Dicotyledons, true vessels appear in the metaxylem, earlier or later, depending on the degree of species evolution, but they are mainly abundant in the secondary xylem.

3. In Monocotyledons, which do not produce any wood, true vessels are numerous in the metaxylem. Finally, in the present species of Pteridophytes, devoid of secondary xylem, the histological development does not induce the formation of true vessels except in rare, interesting cases. First, among the anatomically most

advanced Filicales, in two species, *Pteridium aquilinum* and *Athyrium filix-foemina,* the scalariform cells become perforated on their end walls: true scalariform vessels are thus formed in these ferns. Among the Lycopodinae, true scalariform vessels occur also in the genus *Selaginella,* in some "homophyll" species only (e. g. *Selaginella rupestris;* DUERDEN 1934, 1940).

Therefore, the differentiation of the vascular elements is hardly completed in Pteridophytes, not completed in Gymnosperms *sensu sticto* and more and more completed from Chlamydosperms to the anatomically most developed Angiosperms: herbaceous Dicotyledons and Monocotyledons.

7.7. The Xylem and the Phylogenic Trends of Vascular Plants

The previous section shows the interest of the xylem histology for the systematics of vascular plants and strongly suggests that the xylem structure reflects the coherence of phylogenic evolution. Beyond doubt, the first vessels appeared in the ancient secondary xylem and along with the evolution processes, the differentiation of true vessels first reached the latest elements of the metaxylem (scalariform and reticulate tracheids), then the helical tracheids with a less and less tight pitch, in the species most developed in this respect, including various Monocotyledons (CHEADLE 1943 a).

Depending on species, the earliness of the perforations in the tracheids of the primary or secondary xylem is concomitant with the shortening of the cells derived from the procambium or cambium, and the other cells of the xylem seem to follow the same evolutionary trends (BAILEY 1944 b).

In the comparison of the great divisions of the vascular plants, those evolutionary features are certainly of value with respect to phylogeny. However, it is also obvious that in each group, *parallel* evolutionary processes occur; this is one reason for being careful when establishing the filiations; many of which were formerly suggested and can no longer be retained.

The histological and paleobotanical studies lead to the concept that the evolution of the vascular cells cannot return to its former processes. The most specialized forms should not give rise again to more ancient structures. Thus, vessels cannot change into tracheids. For example, Dicotyledons *without vessels* (arborescent Ranales considered as primitive: Winteraceae, *Trochodendron, Tetracentron,* etc.; cf. BAILEY 1944 a, b, 1953) and more generally Angiosperms cannot be derived from Chlamydosperms such as Gnetales which have true vessels. In the same way, Monocotyledons in which the xylem shows an original development, cannot generate from herbaceous Dicotyledons in which the xylem is highly specialized and which are probably derived from bushy Dicotyledons (BAILEY 1944 b, 1953; CHEADLE 1953). It seems that the latter might have evolved either into herbaceous species, or into trees.

Finally, the use of the xylem histology for phylogeny may be valuable provided that it is associated to the evolutionary data obtained from the other features of vascular plants. In particular, it is certain that the morphology and biology of flowers are far from being always harmonized with vascular evolution (BAILEY 1944b). In fact, phenomena of evolutionary convergence occur in both. The comparison between the evolution of the various features of vascular plants yields many lessons for the phylogenists: a certain degree of independence between these evolution processes reflects the multiplicity of the physiological adaptations to the continental medium which ensured the development of the vascular plants as well as a kind of "selection" that, in various phylums, was more concerned with some organs and tissues and neglectful of others (see BAILEY 1953; CHEADLE 1955).

References

Alibert G, Ranjeva R, Boudet AM (1977) Organisation subcellulaire des voies de synthèse des composés phénoliques. Physiol vég 15: 279–301

Avanzi S, Maggini F, Innocenti AM (1973) Amplification of ribosomal cistrons during the maturation of metaxylem in the root of *Allium cepa*. Protoplasma 76: 197–210

Bailey IW (1913) The preservative treatment of wood. II. The structure of the pit membranes in the tracheids of conifers and their relation to the penetration of gases, liquids and finely divided solids into green and seasoned wood. For Quart 11: 12–20

Bailey IW (1920) The cambium and its derivative tissues. II. Size variations of cambial initials in gymnosperms and angiosperms. Am J Bot 7: 355–367

Bailey IW (1936) The problem of differentiating and classifying tracheids, fiber tracheids, and libriform wood fibers. Trop Woods 45: 18–23

Bailey IW (1944a) The comparative morphology of the Winteraceae. III. Wood. J Arnold Arbor Harv Univ 25: 97–103

Bailey IW (1944b) The development of vessels in angiosperms and its significance in morphological research. Am J Bot 31: 421–428

Bailey IW (1953) Evolution of the tracheary tissue of land plants. Am J Bot 40: 4–8

Bailey IW, Faull AF (1934) The cambium and its derivative tissues. – IX. Structural variability in the redwood, *Sequoia sempervirens,* and its significance in the identification of fossil woods. J Arnold Arbor Harv Univ 15: 233–254

Bailey IW, Kerr T (1937) The structural variability of the secondary wall as revealed by lignin residues. J Arnold Arbor Harv Univ 18: 261–272

Bannan MW (1951) The annual cycle of size changes in the fusiform cambial cells of *Chamaecyparis* and *Thuya*. Can J Bot 29: 421–437

Bannan MW (1953) Further observations on the reduction of fusiform cambial cells in *Thuya occidentalis* L. Can J Bot 31: 63–74

Bannan MW (1960) Ontogenic trends in conifer cambium with respect to frequency of anticlinal division and cell length. Can J Bot 38: 795–802

Bannan MW (1963) Cambial behaviour with reference to cell length and ring width in *Picea*. Can J Bot 41: 811–822

Bannan MW (1968) Anticlinical divisions and the organization of conifer cambium. Bot Gaz 129: 107–113

Barghoorn ES (1940) The ontogenetic development and phylogenetic specialization of rays in the xylem of dicotyledons. I. The primitive ray structure. Am J Bot 27: 918–928

Barghoorn ES (1941a) The ontogenetic development and phylogenetic specialization of rays in the xylem of dicotyledons. II. Modifications of the multiseriate and uniseriate rays. Am J Bot 28: 273–282

Barghoorn ES (1941b) The ontogenetic development and phylogenetic specialization of rays in the xylem of dicotyledons. III. The elimination of rays. Bull Torrey Bot Club 68: 317–325

Barlow PW (1985) The nuclear endoreduplication cycle in metaxylem cells of primary roots of *Zea mays* L. Ann Bot 55: 445–457

Bauch J, Liese W, Schultze R (1972) The morphological variability of the bordered pit membranes in gymnosperms. Wood Science Technol 6: 165-184

Beckman CH (1971) The plasticizing of plant cell walls and tylose formation, a model. Physiol Plant Pathol 1: 1-10

Benayoun J, Catesson AM, Czaninski Y (1981) A cytochemical study of differentiation and breakdown of vessel end walls. Ann Bot 47: 687-698

Botosso PC, Gomes AV (1982) Radial vessels and series of perforated ray cells in Annonaceae IAWA Bull 3: 39-44

Brower DL, Hepler PK (1976) Microtubules and secondary wall deposition in xylem; the effects of isopropyl N phenylcarbamate. Protoplasma 87: 91-111

Buvat R (1956) Variations saisonnières du chondriome dans le cambium du *Robinia pseudo-acacia*. CR Acad Sci Paris 243: 1908-1911

Buvat R (1964a) Infrastructures protoplasmiques des vaisseaux du métaxylème de *Cucurbita pepo* au cours de leur différenciation. CR Acad Sci Paris 258: 5243-5246

Buvat R (1964b) Comportement des membranes plasmiques lors de la différenciation des parois latérales des vaisseaux (métaxylène de *Cucurbita pepo*). CR Acad Sci Paris 258: 5511-5514

Buvat R (1964c) Observations infrastructurales sur les parois transversales des éléments de vaisseaux (métaxylème de *Cucurbita pepo*) avant leur perforation. CR Acad Sci Paris 258: 6210-6212

Catesson AM (1983) A cytochemical investigation of the lateral walls of *Dianthus* vessels. Differentiation and pit-membrane formation. IAWA Bull 4: 89-101

Catesson AM, Czaninski Y (1967) Mise en évidence d'une activité phosphatasique acide dans le reticulum endoplasmique des tissus conducteurs de Robinier et de Sycomore. J Microsc 6: 509-514

Catesson AM, Czaninski Y (1979) Dynamical cytochemistry of wall development during vessel differentiation. IAWA Bull 2, 3: 36

Catesson AM, Moreau M (1985) Secretory activities in vessel contact cells. Isr J Bot 34: 157-165

Cheadle VI (1943a) The origin and certain trends of specialization of the vessel in the Monocotyledonae. Am J Bot 30: 11-17

Cheadle VI (1943b) Vessel specialization in the late metaxylem of the various organs in the Monocotyledonae. Am J Bot 30: 484-490

Cheadle VI (1944) Specialization of vessels within the xylem of each organ in the Monocotyledonae. Am J Bot 31: 81-92

Cheadle VI (1953) Independent origin of vessels in the monocotyledons and dicotyledons. Phytomorphology 3: 23-44

Cheadle VI (1955) The taxonomic use of specialization of vessels in the metaxylem of Graminaceae, Cyperaceae, Juncaceae and Restionaceae. J Arnold Arbor Harv Univ 36: 141-157

Cheadle VI (1956) Research on xylem and phloem, progress in fifty years. Am J Bot 43: 719-731

Coppick S, Fowler WF Jr (1939) The location of potential reducing substances in woody tissues. Pap Trade J 109: 81-86

Cronshaw J (1965a) Cytoplasmic fine structure and cell wall development in differentiating xylem elements. In: Cellular ultrastructure of woody plants. WA Côté (ed) Syracuse University Press, Syracuse NY, pp 99-124

Cronshaw J (1965b) The formation of the wart structure in tracheids of *Pinus radiata*. Protoplasma 60: 233-242

Cronshaw J, Bouck GB (1965) The fine structure of differentiating xylem elements. J Cell Biol 24: 415-431

Cronshaw J, Wardrop AB (1964) The organization of cytoplasm in differentiating xylem. Aust J Bot 12: 15-23

Czaninski Y (1964a) Variations saisonnières du chondriome dans les cellules du parenchyme ligneux vertical du *Robinia pseudoacacia*. CR Acad Sci Paris 258: 679-682

Czaninski Y (1964b) Variations saisonnières du chondriome et de l'amidon dans les fibres libriformes du xylème du *Robinia pseudoacacia*. CR Acad Sci Paris 258: 5945-5948

Czaninski Y (1965) Différenciation saisonnière du xylème de *Robinia pseudo-acacia*. Cellules des parenchymes vertical et horizontal. CR Acad Sci Paris 260: 639-642

Czaninski Y (1966) Aspects infrastructuraux de cellules contiguës aux vaisseaux dans le xylème de *Robinia pseudo-acacia*. CR Acad Sci Paris 262: 2336-2339

442

Czaninski Y (1967) Observations infrastructurales sur les fibres libriformes du xylème du *Robinia pseudoacacia*. CR Acad Sci Paris 264 sér D: 2754-2756

Czaninski Y (1968a) Etude du parenchyme ligneux du Robinier (parenchyme à réserves et cellules associées aux vaisseaux) au cours du cycle annuel. J Microsc 7: 145-164

Czaninski Y (1968b) Etude cytologique de la différenciation cellulaire du bois de Robinier. I. Différenciation des vaisseaux. J Microsc 7: 1051-1068

Czaninski Y (1968c) Cellules oxalifères du xylème de Robinier *(Robinia pseudoacacia)*. CR Acad Sci Paris 267 D: 2319-2321

Czaninski Y (1970) Etude cytologique de la différenciation du bois de Robinier. II. Différenciation des cellules du parenchyme (cellules à réserves et cellules associées aux vaisseaux). J Microsc 9: 389-406

Czaninski Y (1972a) Observations ultrastructurales sur l'hydrolyse des parois primaires des vaisseaux chez le *Robinia pseudoacacia* L. et *l'Acer pseudoplatanus* L. CR Acad Sci Paris 275 D: 361-363

Czaninski Y (1972b) Mise en évidence de cellules associées aux vaisseaux dans le xylème du Sycomore. J Microsc 13: 137-140

Czaninski Y (1973) Observations sur une nouvelle couche pariétale dans les cellules associées aux vaisseaux du Robinier et du Sycomore. Protoplasma 77: 211-219

Czaninski Y (1974) Formation des thylles dans le xylème de *Daucus carota* L. - Etude ultrastructurale. CR Acad Sci Paris 278 D: 253-256

Czaninski Y (1977a) Vessel associated cells. IAWA Bull 3: 51-55

Czaninski Y (1977b) Etude cytochimique ultrastructurale de la paroi des cellules de transfert dans la tige du Blé. Biol Cell 29: 221-224

Czaninski Y (1978) Localisation ultrastructurale d'activités peroxydasiques dans les parois du xylème du Blé pendant leur différenciation. CR Acad Sci Paris 286 D: 957-959

Czaninski Y (1979) Cytochimie ultrastucturale des parois du xylème secondaire. Biol Cell 35: 79-102

Czaninski Y, Catesson AM (1969) Localisation ultrastructurale d'activités peroxydasiques dans les tissus conducteurs végétaux au cours du cycle annuel. J Microsc 8: 875-888

Czaninski Y, Monties B (1982) Etude cytochimique ultrastructurale des parois du bois de Peuplier après extraction ménagée. CR Acad Sci Paris 295 sér III: 551-556

Czaninski Y, Catesson AM, Moreau M, Peresse M (1973) Elaboration de matériel pariétal par des cellules associées aux vaisseaux en réponse à l'infection de l'oeillet par le *Phialophora cinerescens* (Wr) van Beyma. CR Acad Sci Paris 277 D: 405-407

Czaninski Y, Monties B, Roland JC, Catesson AM (1982) Localisation de polysaccharides dans les parois lignifiées après extractions ménagées. Actes Colloque Sciences et Industries du bois Grenoble, 11 p

Duerden H (1934) On the occurrence of vessels in *Selaginella*. Ann Bot 48: 459-465

Duerden H (1940) On the xylem elements of certain ferns. Ann Bot 54: 523-531

Esau K (1943) Origin and development of primary vascular tissues in seed plants. Bot Rev 9: 125-206

Esau K (1965) Plant anatomy. Wiley, New-York

Esau K, Charvat I (1978) On vessel member differentiation in the bean (*Phaseolus vulgaris*, L) Ann Bot 42: 665-677

Esau K, Hewitt WMB (1940) Structure of end walls in differentiating vessels. Hilgardia 13: 229-244

Esau K, Cheadle VI, Gill RH (1966a) Cytology of differentiating tracheary elements. - I. Organelles and membrane systems. Am J Bot 53: 756-764

Esau K, Cheadle VI, Gill RH (1966b) Cytology of differentiating tracheary elements. II. Structures associated with cell surfaces. Am J Bot 53: 765-771

Fahn A, Arnon N (1962) The living wood fibres of *Tamarix aphylla* and the changes occurring in them in transition from sapwood to heartwood. New Phytol 62: 99-104

Fahn A, Leshem B (1962) Wood fibres with living protoplasts. New Phytol 62: 91-98

Falconer MM, Seagull RW (1985) Xylogenesis in tissue culture: taxol effect on microtubule reorientation and lateral association in differentiating cells. Protoplasma 128: 157-166

Falconer MM, Seagull RW (1986) Xylogenesis in tissue culture II. Microtubules, cell shape and secondary wall patterns. Protoplasma 133: 140-148

Frenzel P (1929) Über die Porengröße einiger pflanzlicher Zellmembranen. Planta (Berl) 8: 642–665

Fukuda H, Komamine A (1983) Changes in the synthesis of RNA and protein during tracheary element differentiation in single cells isolated from the mesophyll of *Zinnia elegans*. Plant Cell Physiol 24 (4): 603–614

Goodwin RH (1942) On the development of xylary elements in the first internode of *Avena* in dark and light. Am J Bot 29: 818–828

Gunning BES, Pate JS, Green LW (1970) Transfer cells in the vascular systems of stems: taxonomy, association with nodes and structure. Protoplasma 71: 147–171

Harris JM (1954) Heartwood formation in *Pinus radiata* D. Don. New Phytol 53: 517–524

Hepler PK, Newcomb EH (1964) Microtubules and fibrils in the cytoplasm of *Coleus cells* undergoing secondary wall deposition. J Cell Biol 20: 529–533

Hepler PK, Rice RM, Terranova WA (1972) Cytochemical localization of peroxidase activity in wound vessel members of *Coleus*. Can J Bot 50: 977–983

Huber B (1935) Die physiologische Bedeutung der Ring- und Zerstreutporigkeit. Ber Dtsch Bot Ges 53: 711–719

Jacquiot C (1955) Atlas d'anatomie des bois des Conifères. Centre technique du bois, Paris

Koernike M (1905) Über die Wirkung von Röntgen- und Radiumstrahlen auf die Pflanzen. Ber Dtsch Bot Ges 23: 404–415

Larson PR (1982) The concept of cambium in new perspectives in wood anatomy. Bass P (ed) Martinus Nijhoff W Junk, pp 85–121

Liberman-Maxe M (1982) La différenciation des trachéides du Polypode. Etude cytologique et cytochimique. Ann Sc Nat Bot Paris 13e sér 4: 91–111

Liese W (1963) Tertiary wall and warty layer in wood cells. J Polym Sci Part C 2: 213–229

Liese W (1965) The fine structure of bordered pits in softwoods. In: Côté WA Jr (ed) Cellular ultrastructure of woody plants. Syracuse Univ Press, Syracuse NY pp 271–290

Liese W, Bauch J (1967) On the closure of bordered pits in conifers. Wood Science Technol 1: 1–13

Liese W, Hartmann-Fahnenbrock M (1953) Elektronenmikroskopische Untersuchungen über die Hoftüpfel der Nadelhölzer. Biochim Biophys Acta 11: 190–198

Liese W, Ledbetter MC (1963) Occurrence of a warty layer in vascular cells of plants. Nature 197 (4863): 201–202

Money LL, Bailey IW, Swamy BGL (1950) The morphology and relationships of the Monimiaceae. J Arnold Arbor Harv Univ 31: 372–404

Naegeli CW (1858) Das Wachstum des Stammes und der Wurzel bei den Gefäßpflanzen und die Anordung der Gefäßstränge im Stengel. Beitr Z Wiss Bot Heft 1: 1–156

Newcomb EH (1963) Cytoplasm-cell wall relationships. Annu Rev Plant Physiol 14: 43–64

O'Brien TP (1970) Further observations on hydrolysis of the cell wall in the xylem. Protoplasma 69: 1–14

O'Brien TP, Thimann KV (1967) Observations on the fine structure of the oat coleptile. III. Correlated light and electron microscopy of the vascular tissues. Protoplasma 63: 443–478

Parameswaran N, Liese W (1981) Torus-like structures in interfibre pits of *Prunus* and *Pyrus*. IAWW Bull 2: 89–93

Pate JS, gunning BES (1972) Transfer cells. Annu Rev Plant Physiol 23: 173–196

Perrin A (1971) Présence de "cellules de transfert" au sein de l'épithème de quelques hydathodes. Z Pflanzenphysiol 65: 39–51

Picket-Heaps JD (1966) Incorporation of radioactivity into wheat xylem walls. Planta (Berl) 71: 1–14

Schmid R (1965) The fine structure of pits in hardwoods. In: Côté WA (ed) Cellular ultrastructure of woody plants. University Press, Syracuse NY, pp 291–304

Sinnot EW (1952) Reaction wood and the regulation of tree form. Am J Bot 39: 69–78

Srivastava LH, O'Brien TP (1966) On the ultrastructure of cambium and its vascular derivatives. I Cambium of *Pinus strobus*. Protoplasma 61: 257–276

Stewart CM (1968) Excretion and heartwood formation in living trees. Science 153: 1068–1074

Thiéry JP (1967) Mise en évidence des polysaccharides sur coupes fines en microscopie électronique. J Microsc 6: 987–1018

Vestal PA, Vestal MR (1940) The formation of septa in the fiber tracheids of *Hypericum androsemum* L. Harv Univ Bot Mus Leaflet 8: 169–188

444

Vian B, Roland JC (1972) Différenciation des cytomembranes et renouvellement du plasma-lemme dans les phénomènes de sécrétions végétales. J Microsc 13: 119–136

Wardrop AB (1964) Cellular differentiation in xylem. In: Côté WA (ed) Cellular ultrastructure of woody plants. Syracuse University Press, Syracuse NY, pp 61–97

Wardrop AB, Davies GW (1962) Wart structure of gymnosperm-tracheids. Nature 194: 497–498

Wooding FBP, Northcote DH (1964) The development of the secondary wall of the xylem in *Acer pseudoplatanus*. J Cell Biol 23: 327–337

Zimmermann MH (1983) Xylem structure and the ascent of sap. Springer, Berlin Heidelberg New York

CHAPTER 8

Supporting Tissues

A. Collenchyma

Collenchyma is a living tissue of the aerical organs of vascular plants, mainly characterized by thick, *pecto cellulosic primary cell walls,* typically non-lignified. Collenchyma characteristically occurs in a peripheral position in stems and leaves, particularly in petioles.

In view of several features, collenchyma resembles parenchymas and in the same organ some transitional cells are often found between the collenchyma and the parenchyma which surrounds it and lies under it. Moreover, collenchyma cells can dedifferentiate as easily as parenchyma cells. In this case, reversible changes occur in the wall thickness.

8.1 Cytological Structures of the Differentiated Collenchyma Cells

Collenchyma is a simple tissue composed of only one type of cell, the "collocytes". They are generally more slender and longer than the neighbouring parenchyma cells. In some cases, they are very long and fusiform, as in fibres, even in herbaceous plants (*Ballota,* ROLAND 1967; Fig. 8.1) but some of them are moderately elongated and the end walls are then transverse or slightly oblique (Fig. 8.3 E). Some are divided again by a non-thickened transverse septum (Fig. 8.1).

8.1.1 Structure of the Cell Wall

The walls of collenchyma cells are sometimes very characteristic in some plants. The wall thickening is usually uneven and it is sometimes more marked either at angles or selectively on some walls (MÜLLER 1890; Figs. 8.2 and 8.3).

The wall thickening at the angles *(angular collenchyma)* is frequently found. In transections, the walls of the cells assume then an elegant, more or less star-shaped pattern (Figs. 8.2b, 8.3 A, D); e.g. *Cucurbita, Rumex, Ampelopsis, Begonia,* etc. If the wall thickening is very active, the wall faces also are thickened, in addition to the angles, and in transections the cell lumen assumes a circular outline *(annular collenchyma)* (Figs. 8.2a, 8.3c). This modification occurs frequently in Umbelliferae. In such cases, there are many transitional modifications between the angular collenchyma and the annular collenchyma. Still in *Compositae, Labiae, Malvaceae,* another form of collenchyma is characterized by lacunae which occur in the

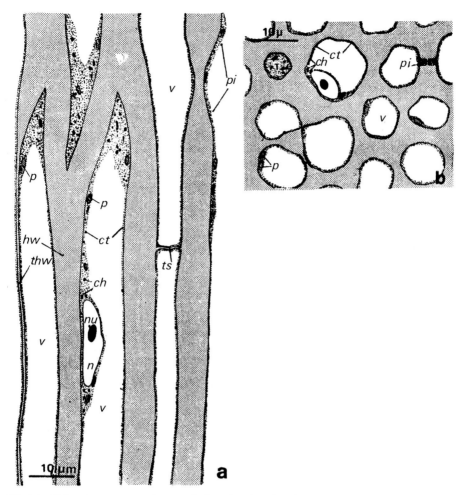

Fig. 8.1. Light microscope cytology of *Ballota foetida* collenchyma cells; **a** longitudinal section; **b** transversal section. *ch* Chondriosomes, most of them in the short shape of mitochondria; *ct* cytoplasm reduced to a thin pellicle against the cell walls, except at the extremities of the "collocytes", where it is more abundant; *n* nucleus, *nu* nucleoli, *p* plastids; *hw* hypertrophied cell wall; *thw* cell wall remaining thin; *pi* pits; *ts* transversal septum; *v* vacuoles. Fixation: REGAUD – staining: hematoxylin; (after ROLAND 1967)

bundles of cells in the process of collenchymatous differentiation, lacunae which are more important that those commonly found in parenchyma, and the developmental thickening is then more pronounced on the walls facing the intercellular spaces (Fig. 8.2 c) *(lacunar collenchyma).*

Finally, in other cases, the wall thickening occurs chiefly on the walls parallel to the surface of the organ (tangential faces). The collenchyma assumes a more or less striated or "storied" aspect, e.g. *Sambucus, Eupatorium (lamellar collenchyma)* (Figs. 8.2 d, 8.3 B).

447

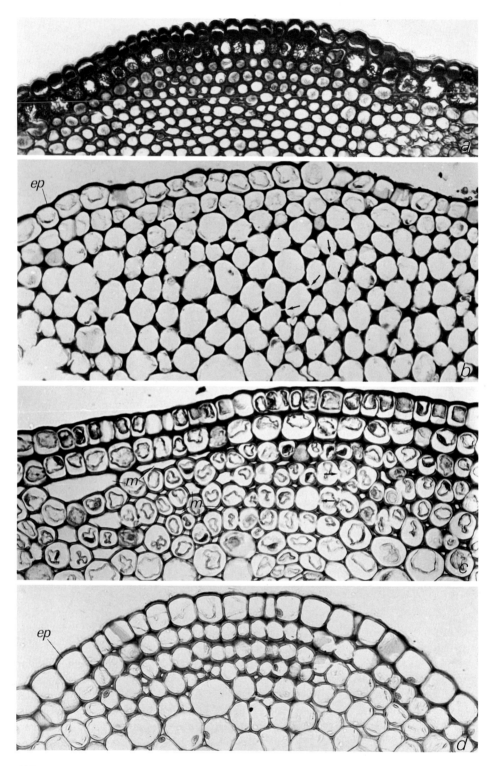

Fig. 8.2 a–d. Some examples of collenchyma. **a** *Onobrychis sativa,* stem, round or annular collenchyma; **b** *Cucurbita pepo,* petiole, angular collenchyma; **c** *Helianthus tuberosus,* stem, lacunar collenchyma (with meatus); **d** *Sambucus nigra,* young petiole, tangential collenchyma. *ep* Epidermis; *m* meatus; *arrows* indicate pits

These variations in wall thickening are less characteristic in longitudinal sections. The lateral walls are thinner or thicker depending on the locus of sections. Transverse or slightly olique end walls are generally thin. In fusiform cells, the thickening may involve the pointed ends (Fig. 8.1 a).

Primary pit fields occur in collenchyma cells in both the thin and the thick parts of walls (Figs. 8.2 b, c, arrows; 8.3 C, E). Plasmodesmata occur in each pit area, sometimes grouped in bundles or branched, which seem to become connected into a kind of strongly stained nodule in the middle lamellae (Plate 8.1 a, b).

The mature walls of collocytes consist mainly of pecto cellulosic material, they contain much water, about 60 to 70% of their content, with respect to the weight of fresh wall. ROLAND's works described the organization of such walls. The three main polysaccharide components consist of pectic and cellulosic material together with a great quantity of hemicelluloses. In fact, the amounts of these components vary during the differentiation process and as the collenchyma grows older. Thus, in the collenchyma strands of celery petioles, the pectic material ranges from

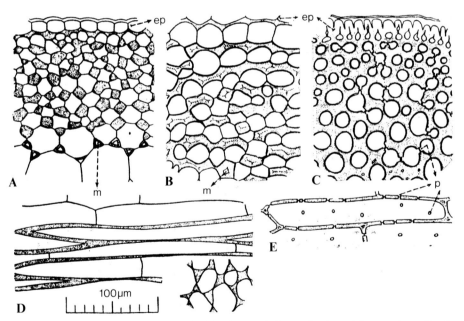

Fig. 8.3 A–E. Main histological types of collenchyma. **A** *Beta vulgaris,* petiole, angular collenchyma; **B** *Sambucus nigra,* stem, tangential collenchyma; **C** *Nerium oleander,* leaf vein, annular collenchyma; **D** *Cucurbita pepo,* stem, transversal and longitudinal sections of angular collenchyma; **E** *Ficus elastica,* petiole, longitudinal view of a short collocyte, the end cell walls of which are transversal. *ep* Epidermis; *m* meatus; *p* pits; (after DUCHAIGNE 1955)

449

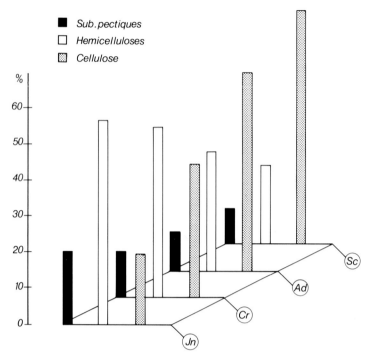

Fig. 8.4. Quantitative variations of the main constituents of celery collenchyma cell wall. *Jn* Beginning of growth; *Cr* end of growth; *Ad* adult state; *Sc* senescence; (after J. C. ROLAND 1966)

about 20 to 10% of the weight of dry wall between the onset of growth and the mature state; meanwhile the rate of hemicelluloses varies from about 60 to 30%, whereas the cellulose rate increases from about 20 to 60% (ROLAND 1966; Fig. 8.4).

In light microscopy, using convenient preparations (the effect of sodium hydroxide, iodide sulphuric acid, selective dissolution of hemicelluloses) one can see that the wall which seems to be evenly arranged in natural light, is in fact stratified and lamellar (Plate 8.2 b–d).

After the extraction of the components different from cellulosic material and chromium shadowing, the use of electron microscopy reveals that cellulose layers themselves are composed of elementary layers of cellulose fibrils. These layers are well individualized within the lateral layers near the cellular cavity but become

---

Plate 8.1 a–c. Primary pits and plasmodesmata in terminal (= transversal) cell wall of collocytes.

a *Sambucus nigra;* grouped plasmodesmata *Pl,* anastomosed at the middle lamella level, in a primary pit field; some are branched; × 9500.

b *Daucus carota:* grouped plasmodesmata *Pl,* as in **a,** at the bottom of a pit field *Pi; P* cell wall; × 9500. Fixation: KMnO₄ for **a** and **b**; (after J. C. ROLAND 1967)

c *Sambucus nigra;* transversal (= terminal) cell wall of a collocyte in front view; chromium shadowing after elimination of constituents others than cellulose fibrils. The density of the cellulose fibril is lower in the areas containing plasmodesmata *(pl);* × 30 000 (courtesy of J. C. ROLAND 1966)

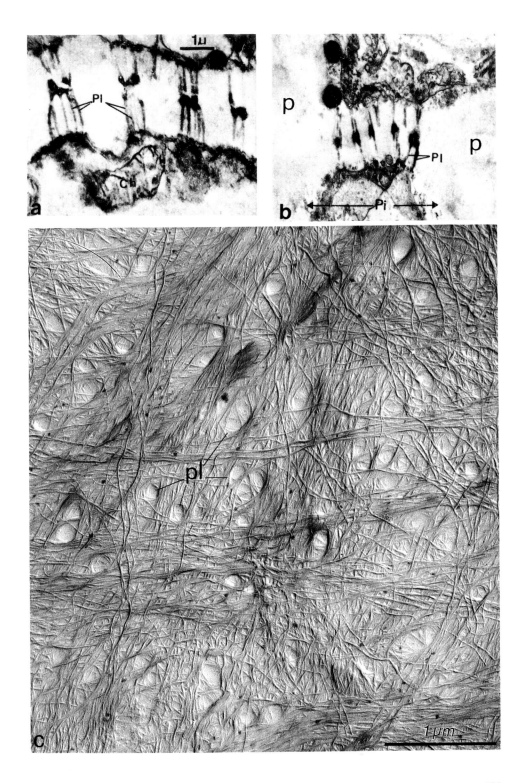

451

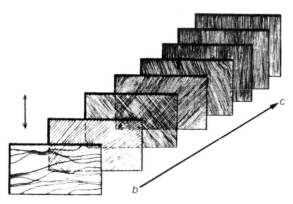

Fig. 8.5. Re-orientation of cellulose fibrils in the successive strata of the collocyte-thickened cell wall, from the transversal or oblique deposit, preceding the growth in length (in *b*), up to the longitudinal and parallel orientation of fibrils which have undergone that growth (in *c*). The *double arrow* indicates the direction of elongation; (after J.C. ROLAND 1966)

progressively disorganized towards the middle lamella (Plate 8.2 e). Cellulose fibrils are longitudinal (following the elongation of the organ) in the middle part of the wall thickening. Conversely, the most internal fibrils beside the plasmalemma are transverse or oblique (Plates 8.2 e, f; 8.3). Thus, during the elongation of the organ, the transverse cellulose fibrils of collocytes become progressively straightened following the elongation to become longitudinal (ROLAND 1966; Fig. 8.5). In transverse septums, in which the surface remains practically unchanged, cellulose fibrils form a network, clearer at the sites of plasmodesmata (Plate 8.1 c).

As for the hemicelluloses, ROLAND distinguished interlamellar hemicelluloses *hi*, separating the cellulose layers, and hemicelluloses *hl* which encompass the elementary cellulose layers and become associated in the formation of heterostruc-

---

Plate 8.2 a–f. *Sambucus nigra*, collenchyma, cellulose arrangement and texture.

a–d Light microscopy, transversal sections showing the different areas of the collenchymatous cell wall. **a** In natural light; **b** beginning of effect of cold 4% sodium hydroxide solution: displaying the middle lamella *lm* and the *internal bordering zone li; p* thickened and stratified cell wall. **c** After specific dissolution of interlamellar hemicellulose; individualization of concentric cellulose lamellae *sc,* and persistence of the middle lamella *lm.* **d** Action of iodate sulphuric acid, decreasing staining from the **D** to the **A** regions, suggesting a centripetal destruction of the cellulosic layers; × 1500.

e Electron microscopy, chromium shadowing; after dissolution of non-cellulosic constituents with hydrogen peroxide (H_2O_2) at 70 °C; view of a parietal angle. Visualization of heterostructured strata *sc* in which elementary strata *se* become visible at some sites, showing a layer of cellulose fibrils, marked by *arrows.* *f* Disorganized cellulose fibrils on the side of the middle lamella; *fl* longitudinal fibrils present in the major part of the cell wall thickness; *ft* transversal fibrils on the side of the cell cavity *L.*

f Detail of the inner side (**D** region) near the cell cavity; the more recent fibrils *ft* are oblique or transversal, and superimposed to the longitudinal and parallel fibrils *fl;*

original lettering of the author; × 18000; (courtesy of J.C. ROLAND 1966)

452

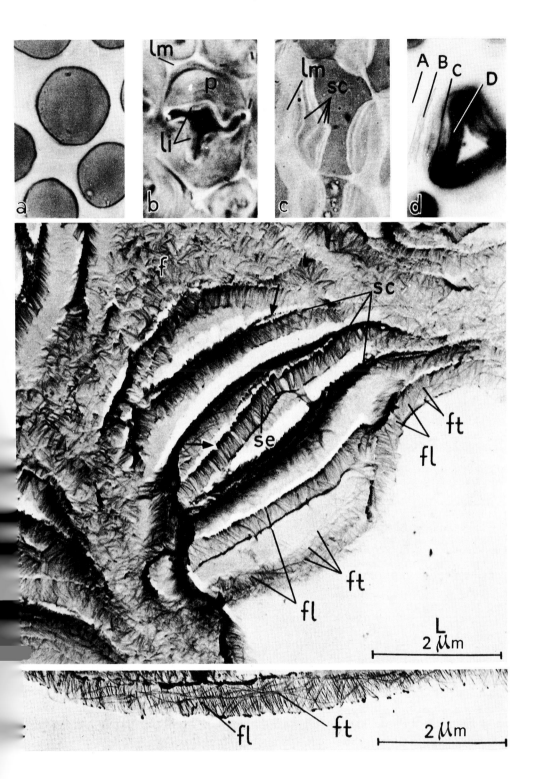

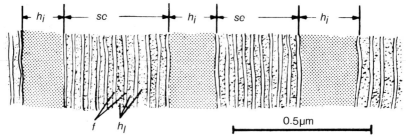

Fig. 8.6. "Heterostructured lamellae" *sc* of the collocyte cell wall. *f* Cellulose fibrils, wrapped in intrallamellar hemicelluloses h_l, probably distinct from the interlamellar hemicelluloses h_i; (after J.C. ROLAND 1966)

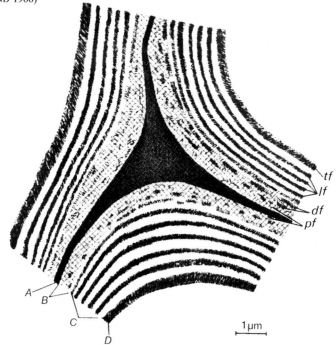

Fig. 8.7. Scheme of the different parietal layers of collocytes. *A* Pectic middle lamella; **B** *deep zone, with more or less disorganized microfibrils, wrapped in complexes of hemicelluloses and esterified pectic substances; C* main parietal zone, composed of an alternation of heterostructured lamellae (see Fig. 8.6) and hemicelluloses; *D* internal bordering zone. *df* Microfibrils proceeding from the disorganization of the ancient heterostructured lamellae; *lf* longitudinal microfibrils; *pf* ancient microfibrils, occurring at the beginning of the primary cell wall deposit; *tf* recent microfibrils, transversely or obliquely oriented, with regard to the elongation direction of the collocyte; (after J.C. ROLAND 1966)

▷

Plate 8.3. *Petasites niveus* (Baumg); arrangement of cellulose fibrils in collocyte cell wall thickness (transversal section).

B, C and *D* correspond to the regions similarly lettered in Plate 8.2 **d.** In the *B* area, neighbouring the middle lamella, the arrangement of cellulose fibrils *fd* is disorganized; in the *D* area, near the plasmalemma, fibrils *ft* are more or less transversal; ×25000. Chromium shadowing after elimination of non-cellulosic parietal constituents; original lettering of the author; (courtesy of J.C. ROLAND 1966)

454

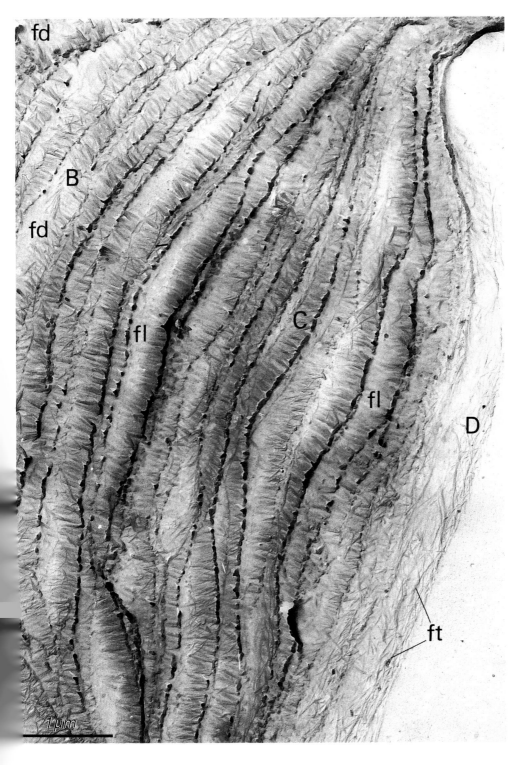

fd

B

fd

fl

C

fl

D

ft

1 μm

455

tured layers recognizable in electron microscopy (Fig. 8.6). The scheme of the collocyte wall is shown in Fig. 8.7. This peculiar structure allows collenchyma cells to adjust the elongation of organs and to be capable of differentiating *simultaneously*. Besides, this feature gives ground to consider the walls of this tissue as *primary* walls regardless of the thickness variations.

8.1.2 The Protoplasma of Collocytes

It seems that MAGIN (1956) was the first to perform the cytological study of the protoplasma of collocytes in *Lamium album* with light microscopy. Each cell encloses a single vacuole, and this fact seems to be general, so that the nucleus is shifted away to the wall into the cytoplasmic film applied against it. This lenticular

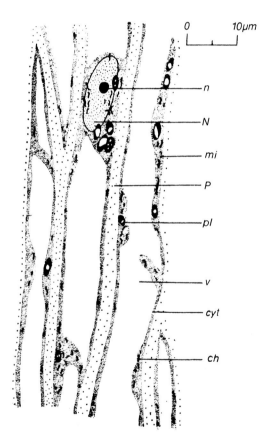

Fig. 8.8. End of differentiation of *Lamium album* collocytes. Considerable lengthening of cells, the extremities of which are stretched out and fusiform; thin pellicle of cytoplasm, applied against the cell walls, surrounding a unique vacuole and the elongated or lentil-shaped nucleus. Chloroplasts retain starch and are only moderately differentiated. During this evolution, nucleoli maintain a relatively large and approximately constant volume (diameter 1.7 to 2 µm). *ch* chondriocontes; *cyt* cytoplasm; *mi* mitochondria; *N* nucleus; *n* nucleolus; *P* thickened primary cell wall; *pl* plastids; *v* large vacuole. Fixation: REGAUD – hematoxylin; (after TH. MAGIN 1956)

456

nucleus is larger and contains a more voluminous nucleolus than in the neighbouring cells of the cortical parenchyma (Fig. 8.8). Relatively few chloroplasts occur in the cytoplasm, less developed than those which occur in parenchyma and slightly amyliferous. Punctiform mitochondria and short chondriocontes are scattered within the cytoplasm. A similar chondriome was recognized in the collenchyma of *Ballota foetida* (ROLAND 1967) with light microscopy and electron microscopy (Plate 8.4 a, b). Electron images also display numerous, apparently active dictyosomes and a well-developed endoplasmic reticulum with cisternae parallel to the plasmalemma (Plate 8.4 a, b). Chloroplasts, of common granar type, occur in mature cells but the swelling of thylacoids in the older collenchyma indicates a degeneration process.

8.2 Origin and Setting in Place of the Collenchyma

The origin of initial cells of the collenchyma tissue, or colloblasts, must be found in the apical meristem of stems. The setting of collenchyma strands are related to the birth and the development of leaf elements (leaves and leaf segments) from the initiating ring (cf. p. 109).

In most species, collenchyma strands originate in the marginal meristem of leaf buttresses and appear to be continuous along "leaf horns", from the lateroventral edges of petioles, then all along the foliar segments (Labiateae, in particular; Figs. 8.9 and 8.10 A); in other species, collenchyma strands appear on the median ribs of the petiole (*Clematis vitalba, Sambucus nigra, Daucus carota;* Fig. 8.10 B–D). Still more collenchyma strands may appear between the previously mentioned strands, particularly near secondary veins *(Daucus)* and finally, in some

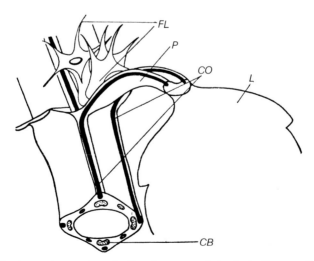

Fig. 8.9. Localization and extension of collenchyma strands in the leafy stem of *Lamium album;* continuity between cauline and petiolar strands; *CO* collenchyma strands; *L* leaf; *CB* conducting bundles; *FL* flowers; *P* petiole; (after TH. MAGIN 1956)

457

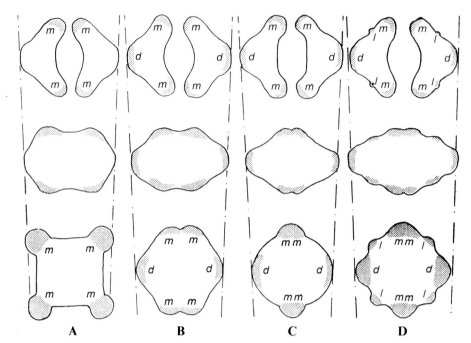

Fig. 8.10 A–D. Origin and setting of the collenchyma in four species with opposite leaves. **A** *Glechoma hederacea;* two pairs of marginal petiolar bundles *m* occur successively in the foliar segment, without merging or anastomosing, resulting in the four stem bundles. **B** *Clematis vitalba;* two marginal bundles *m* and one dorsal bundle *d* of the petiole extend into each foliar segment, as in **A**. So, the stem contains six non-equivalent collenchyma bundles. **C** *Mercurialis annua;* as in **B** for the petiole, but the marginal bundles *m* of the opposite petioles merge when extending into the stem, which then possesses four bundles, two of which *mm* result from the merging of marginal petiolar bundles. **D** *Sambucus nigra;* petiolar bundles *m* and *d* extend as in **B** and **C**, but in addition there are two laterodorsal bundles *l* which extend separately and in continuity into the internode; (after J.C.ROLAND 1967)

Plate 8.4a, b. Fine structure of collocyte cytoplasm.

a *Ballota foetida;* thin pellicle of cytoplasm applied against the cell wall *p*, surrounding a large and unique vacuole *V; ch* abundant, but short chondriosomes (mitochondria); *G* Golgi apparatus, numerous dictyosomes surrounded by Golgi vesicles; *re* long parallel saccules of endoplasmic reticulum with scattered dilatations in the shape of terminal bulbs *at*, and parallel to the cell wall; *pe* plasmalemma; *t* tonoplast; *cd* dense bodies, unidentified. **b** *Sambucus nigra;* structures similar to those of **a**, but with particularly significant Golgi vesicles; original lettering of the author. Plastids are not represented in these cytoplasmic areas; × 22000.

Fixations: KMnO₄; (courtesy of J.C.ROLAND 1967)

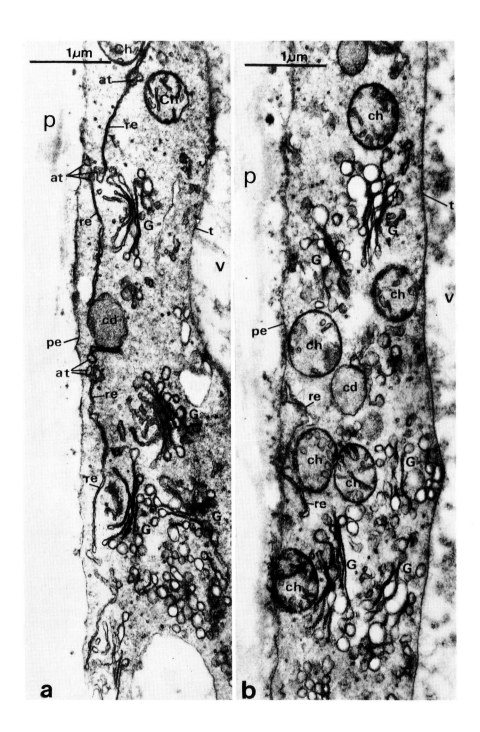

459

cases, the lateral strands of opposite leaves merge into a single bundle in the internode (*Mercurialis annua;* Fig. 8.10 C, D).

Furthermore, from the origin, collenchyma cells occur in a peripheral position and are separated by cortical parenchyma from the conducting tissues, when the latter become differentiated. However, in very young leaf elements, still completely meristematic, the initials of collenchyma are not always clearly separated from the procambium or from the initials of the dorsal parenchyma of the buttresses and the cortical parenchyma of the foliar bases. This fact explains that several histologists concluded that the collenchyma tissue was derived from the procambium (AMBRONN 1881; MAJUMDAR 1941). However, when the ontogenesis of the collenchyma occurs late enough, parenchyma cells separate future collocytes from conducting tissues, and, e.g. in celery, by secretory ducts which differentiate outside the conducting bundles (ESAU 1936). In *Lamium album,* as in general, the collenchyma originates from a *pericline proliferation of the subepidermal layer.* Its origin is therefore more external that that of the procambium and independent of the latter.

In fact, collenchyma is a cortical and peripheral tissue; its differentiation occurs later than that of parenchymas (MAGIN 1956). The study of cellular organelles confirms the separation of the initials of collenchyma (the colloblasts) from those of the procambium. With a comparable cytological analysis, ROLAND (1967) reported the same results with regards to *Glechoma hederacea, Daucus carota* and *Sambus nigra* in which the beginning of parenchymatization, at least, separates the initials of collenchyma in all cases.

8.3 Differentiation of the Collocytes

In fact, after arising, the colloblasts either immediately enter the initiating phase or postpone their initiating activity (ROLAND 1967).

In the first case (e.g. *Glechoma)* they maintain their typical primary meristematic condition: basophilic cytoplasm, voluminous nucleus and nucleolus, reduced vacuoles, undifferentiated proplastids (Fig. 8.11 A).

In the second case, the colloblasts undergo a more or less stressed parenchymatization (e.g. *Daucus carota, Sambus nigra,* mediodorsal collenchyma; Fig. 8.11 B, C). The following phase is an initiating phase characterized by the proliferation, at first pericline, of the colloblasts, followed by variably oriented longitudinal mitoses, then resulting in a homogeneous strand of *procollocytes.* Further on, whereas the proliferation may go on, the procollocytes undergo cytoplasmic and wall differentiation which change them into *collocytes.*

Because of longitudinal divisions, moreover accompanied by a few transversal cleavages, the transection of cells tends to decrease, whereas the elongation is highly increased. Through this activity, the procollenchyma assumes an aspect similar to that of the procambium, causing some confusion (Fig. 8.12 a, b). Once started, the differentiation continues while the organs are growing.

The cytological aspects of the ontogenesis of collocytes were studied by MAGIN (1956) in *Lamium album* with light microscopy and by ROLAND (1967) in *Gle-*

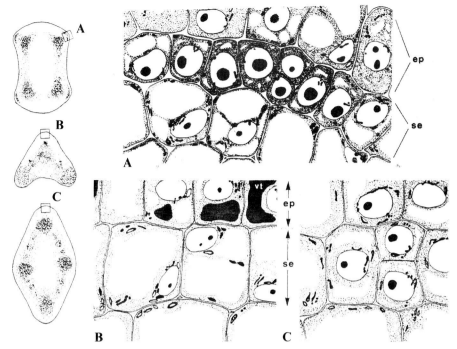

Fig. 8.11 A-C. Onset of differentiation of colloblasts, according to three specific modalities. **A** *Glechoma hederacea;* highly meristematic characters of original cells are maintained and allow one to distinguish these cells from more internal cortical cells, up to the ontogenic initiation of the collenchyma. **B** *Daucus carota;* the ontogenesis of collenchyma is delayed; parenchymatization of subepidermal cells at the site of a future collenchyma bundle. **C** *Sambucus nigra;* as in **B,** but the differentiation of subepidermal initials is less pronounced. *ep* epidermis; *se* subepidermal layer; *vt* tanniniferous vacuole. Fixation: REGAUD – staining: hematoxylin; × 1600; (after J.C. RO-LAND 1967)

choma hederacea and *Sambucus nigra,* as well as with electron microscopy in the latter species. Leaf horns maintain a primary meristematic state during the proliferation of the subepidermal layer *(Lamium, Glechoma)* or this state is restored through a dedifferentiation process if the colloblasts have undergone the beginning of differentiation (*Daucus,* mediodorsal collenchyma of petiole). In the same way, in young internodes of *Lamium,* the future procollenchymatous cells undergo the beginning of vacuolization before being activated. Then, they also resume the typical primary meristematic condition (MAGIN 1956).

The differentiation does not present many characteristics with regards to the cytoplasm. As it is generally noted, meristematic vacuoles increase and unite into a single large vacuole (Fig. 8.12 d). Mitochondria remain globular or form short chondriocontes; proplastids are moderately and later differentiated into small chloroplasts, sometimes after a transient period of amylogenesis (*Lamium;* Fig. 8.12 c). Conversely, the cytoplasm has a high density of ribosomes and thus remains more basophilic than the cells of the neighbouring parenchyma (ROLAND 1967).

461

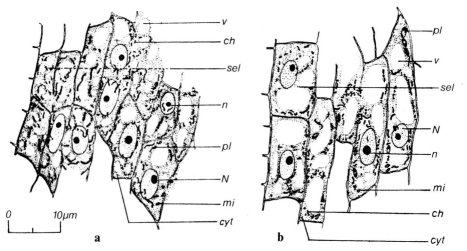

Fig. 8.12 a–d. Light microscope cytology of *Lamium album* collenchyma differentiation (beginning). **a** Phase of cell divisions; transversal sizes of the future collocytes tend to decrease; vacuolar apparatus still divided into small vacuoles; undifferentiated plastids; nuclei at the centre of the cells. **b** Onset of procollocyte elongation; enlargement of vacuoles; plastids begin to grow and to elaborate starch. **c** Continuation of vacuole development and plastid amylogenesis, mainly around nuclei which become more elongated. **d** Phase of confluence of vacuoles, pushing away the nuclei against the cell wall, where these nuclei become lentil-shaped; the cell walls remain thin. *am* Amyloplast; *sel* subepidermal layer; *ch* chondrioconte; *cyt* cytoplasm; *mi* mitochondria; *N* nucleus; *n* nucleolus; *pl* young plastid; *v* vacuole. Fixation: REGAUD – hematoxylin; (after TH. MAGIN 1956)

There are two essential characteristics in the differentiation of the collocytes: on the one hand, the nuclei as well as the nucleoli, which are more voluminous than in the neighbouring parenchyma cells and, on the other hand, the lateral wall formations. The diameter of the nucleoli of collenchyma in *Lamium* is about 1.6 to 1.8 μm, whereas it is 1.2 μm in the parenchyma (MAGIN 1956).

The considerable growth in length of the cells, hence of their surface, and the considerable thickening of the walls, implies an outstanding activity concerned with the elaboration of wall components. Thus, collenchyma, as the cells of the root cap, constitutes a valuable material for studying the mechanisms through which plant walls are generated (VIAN and ROLAND 1972, cf. p. 208).

Although the collocyte walls may become sclerified later (WENT 1924; DUCHAIGNE 1953, 1954), collenchyma is typically a living tissue in which the walls become thickened but which are composed of pectic and cellulosic material. These features ensure the *plasticity* of collenchyma, plasticity which generally disappears in the cells producing secondary walls. Thus, in the perennating stems of lignified Dicotyledons, collenchyma is capable of adjusting to the transversal growth of the organ for a time: the collenchymatous cells themselves enlarge and their walls become thinner (*Tilia, Acer, Aesculus;* DE BARY 1884).

Moreover, and this is a third feature of primary nature of cell walls, the wall thickenings of collenchymatous cells may regress and the cells themselves resume the meristematic state. This dedifferentiation process occurs naturally, when a

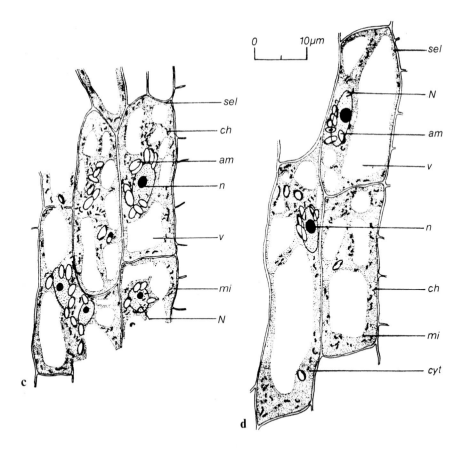

phellogen arises within collenchyma, or when collenchyma cells respond to injuries with the production of wound-healing periderm.

Experimental observations along with the observation of natural cases (VENNING 1949) showed that the medium conditions affect the developmental thickening of walls: the wall thickness is increased if, during the development, the plants are exposed to wind.

8.4 Anatomical Localization

Collenchyma is the first supporting tissue which differentiates during the development of aerial organs (stems and leaves) in the plants which are capable of generating such a tissue. Collenchyma occurs exceptionally in roots and in other underground organs, for which this supporting tissue is not necessary.

Collenchyma is typically a *peripheral tissue*. Most often, it is present beneath the epidermis of stems or leaves. In rare cases, it may be separated from the epidermis by some layers of parenchyma, and more frequently the walls of the internal faces of the epidermal cells become thickened at the contact of collenchyma

463

strands. Sometimes, the entire epidermal cell becomes thickened. In stems, collenchyma occurs either in the form of a continuous cylinder, often uneven in thickness, or in strands, particularly in the protruding ribs of grooved stems (Figs. 8.9, 8.10; Labiatae, Umbelliferae, Equisetaceae, etc.). In leaves it may be present on both poles of the vascular bundles of the veins, or on one side only, where veins protrude on the leaf lamina.

Supporting tissues similar to collenchyma occur also, in some stems, at the distal parts of phloem and xylem bundles or around the conducting strands. In fact, they correspond to differentiations of the parenchyma, which may or may not become lignified, and which differ, in view of histogeny, more from collenchyma than from sclerenchyma.

The peripheral position of collenchyma is therefore remarkable. Mechanically, it it the *most valuable* anatomical localization for the supporting role of this tissue.

8.5 Mechanical Properties

Collenchyma is capable of adjusting to the growth and *of strenghtening* young organs. The observations of MAJUMDAR (1941) and MAJUMDAR and PRESTON (1941) in *Heracleum* showed that thick-walled collenchyma cells were already found in young internodes that were intended to extend several times in length during their growth. The mechanical properties of collenchyma strands were studied on dissected specimens (AMBRONN 1881; ESAU 1936). They showed the plasticity and resistance of tension of this tissue. Collenchyma has been found capable of supporting 10 to 12 kg mm^{-2}, whereas fibre strands support 15 to 20 kg mm^{-2}. But after they have been subjected to 15 to 20 kg mm^{-2} tension, the fibres regain their original length, they are *elastic,* whereas collenchyma strands remain permanently extended after they have been made to support 1.5 to 2 kg mm^{-2}. They are not as elastic as the fibres, but their *plasticity,* demonstrated by the importance of the elongation before they rupture, is largely superior (ROLAND 1967; Fig. 8.13).

Due to these features of resistance to rupture, i.e. plasticity and therefore *flexibility,* collenchyma is particularly well adapted for supporting the more or less fragile, growing aerial organs.

8.6 Collenchyma in the Series of Vascular Plants

Collenchyma, a typical tissue in young or herbaceous stems, is in fact absent or rare in Monocotyledons, plants which are most typically herbaceous, in which ground parenchyma differentiates early into sclerenchyma.

Monocotyledons are predominantly sclerenchyma plants, although sclerenchyma can be found in all groups.

Collenchyma occurs mainly in young stems, in the adult stems of herbaceous species and in the leaves of Dicotyledons.

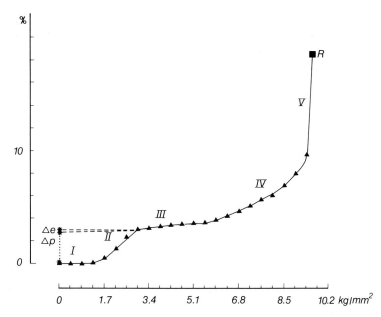

Fig. 8.13. *Ballota foetida;* stretching of an isolated collenchyma bundle submitted to increasing loads expressed in kg mm^{-2} cell wall. Δe Elastic deformation (reversible); Δp plastic deformation (irreversible); R rupture (breaking).

 I load lower than 1.5 kg mm^{-2}, no obvious extension;

 II load between 1.5 and 3 kg mm^{-2}, extension proportional to the load;

 III load between 3 and 6 kg mm^{-2}, weak extension;

 IV load between 6 to 10 kg mm^{-2}, more and more important extension;

 V load about 10 kg mm^{-2}, rapid extension and rupture; (after ROLAND 1967)

  In Gymnosperms, lignified plants, the intense sclerification reduces the importance of collenchyma. It may also be discrete in *Filicineae* which have underground or arborescent stems (sclerified); but it largely appears in the upright stems of *Equisetum* which are thin and rigid, although they are lacunar, in which it constitutes the main part of grooved tissues. It should be noted that in *Equisetum*, plants with collenchyma, phloem and xylem formations have been found similar to those of Monocotyledons, plants with sclerenchyma and without collenchyma.

B. Sclerenchyma

8.7 General Characteristics and Various Types

Sclerenchyma together with collenchyma are the supporting tissues of plants. The main distinction with parenchyma consists in the fact that the walls of sclerenchymatous cells *become* usually *lignified* sooner or later.

  Another distinction is the anatomical position in the plant, usually more internal than the position of collenchyma, and above all more diversified.

465

In fact, sclerenchymatous cells differ according to their origin, their mode of differentiation, their morphological features and their position, and many classifications have been suggested (FOSTER 1944; TOBLER 1957; ESAU 1965).

For this study, it has been more convenient to exclude from sclerenchyma all the supporting elements related to conducting tissues (phloem and xylem fibres) already discussed in the previous chapters.

Then, one can separate the sclerenchymatous tissues, in which the differentiation is more or less direct and stressed, that occur in the form of well-specialized anatomical elements, from the formations which differentiate later and arise through the secondary development of tissues previously differentiated into parenchyma, collenchyma or epidermis. In particular, the ground sclerified tissue that is so commonly found in Monocotyledons belongs to the last type.

Sclerenchyma that differentiates directly may produce compact tissues or isolated cells scattered within the ground parenchyma tissues, i.e. *sclereids*. Both cases will be referred to as *primary sclerenchyma*.

However, most English speaking anatomists (if not all) use the term "sclereids" to refer to all sclerified cells which are not elongated in the form of fibres (cf. ESAU 1965). In the following, most of these cells arise in late sclerenchyma resulting from the evolution of parenchyma or of already differentiated cells; for example, this is the case of the mature pericarp of dried fruits, which results from the sclerification of parenchymatous cells, and of the epidermis of the ovarian wall.

8.8 Sclerenchyma of "First Formation"

8.8.1 Compact Sclerenchymatous Tissues

Usually the cellular development reveals, prior to the changes in the cell wall structure, the future evolution of the cells of young organs which will further differentiate into sclerenchyma. In transections, these cells, as those which will produce the collenchyma, are narrower than the neighbouring parenchymatous cells (Fig. 8.14) or become narrower through longitudinal mitoses.

Conversely, they demonstrate a tendency to elongate without dividing transversally. The cells may assume a fusiform feature like that of the fibres previously discussed in conducting tissues (Fig. 8.21 b). But the elongation may be limited to general growth without intrusive formation and the end walls remain transverse or slightly oblique.

The cells of this sclerenchyma of "first formation" are arranged into strands often associated with the conducting strands, in particular along the leaf veins, or they appear as a continuous cylinder outside the conducting tissues (*Cucurbita;* Fig. 8.15). The diameter and the arrangement of strands often assume shapes and distributions favourable to a better rigidity.

Moreover, these tissues are found to be more internal than the collenchyma strands which can sometimes occur in the same organs (Fig. 8.15).

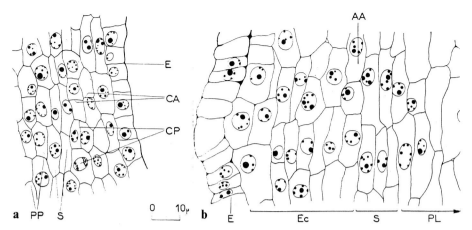

Fig. 8.14 a, b. *Cucurbita pepo,* young foliar bases, longitudinal radial sections. **a** Second foliar base; *S* sclerenchyma initials, relatively internal in the cortex, narrower than cortical parenchyma cells; *E* epidermis; *PP* so-called pericyclic parenchyma (in fact, in the protophloem); *CA* anticlinal cleavages; *CP* periclinal cleavages. **b** Sixth foliar base; more elongated sclerenchyma initials *S,* remaining narrower than cortical cells *Ec* and cells of parenchymatized protophloem *PL; AA* amyliferous layer; *E* epidermis; original lettering of the author (after J. MOURRÉ 1958)

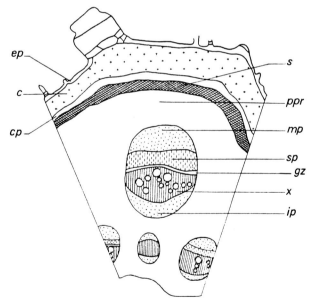

Fig. 8.15. *Cucurbita pepo,* stem, transversal section, interpreted according to J. MOURRÉ (1958); *s* unbroken ring of sclerenchyma, more internal than the collenchyma *c,* and bordering the parenchymatized protophloem *ppr; ep* epidermis; *mp* metaphloem; *cp* cortical parenchyma; *ip* internal phloem; *sp* secondary phloem; *x* xylem; *gz* cambium (generating zone); (after J. MOURRÉ 1958)

467

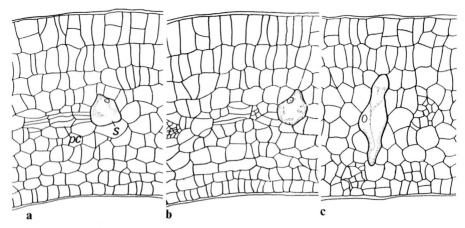

Fig. 8.16 a–c. *Mouriria huberi* (Cogn.), very young leaves, transversal sections. **a, b** Initial cells of sclereids, *s*, situated at the extremity of vascular bundles, still in the state of procambium *pc*. **c** Onset of sclereid growth, which extends towards both the ventral and dorsal epidermis (adaxial and abaxial); ×340; (after FOSTER 1947)

8.8.2 Sclereids

In addition to sclerenchyma, which forms compact tissues in stems, leaves and often in roots, supporting cells more or less isolated occur in numerous parenchyma. They seem to grow independently from the neighbouring ones as if they escape to the coordination processes of growth which harmonize the development. In fact, these cells often become very large and their shapes seem most arbitrary. These cells, called *sclereids,* can be distinguished early from the ground cells of paren-

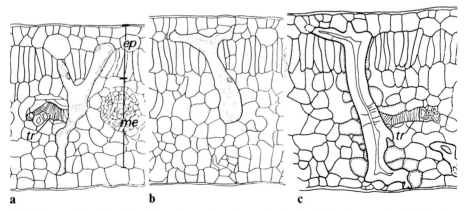

Fig. 8.17 a–c. *Mouriria huberi* (Cogn.); continuation of leaf sclereid development. **a, b** Intrusive extensions due to a kind of dichotomy between multiseriate adaxial epidermal cells (partial sections); *tr* terminal tracheids at the end of minor leaf veins, at the contact of the sclereid; *ep* adaxial multiseriate epidermis; *me* mesophyll at the beginning of lacunae formation; ×226. **c** Completed sclereid in a mature leaf remaining in contact with terminal tracheids *tr* of minor leaf veins; the sclereid extensions thread their way between cells of the two epidermis; ×175; (after A. FOSTER 1947)

chyma in which they will differentiate. They can be singled out long before the end of growth and *a fortiori* of the differentiation. Thus, FOSTER (1947) demonstrated that the sclereids of the leaves of *Mouriria,* localized at the extremities of the conducting strands which end up in the mesophyll, are quite recognizable, whereas the small veins are still entirely procambial and the intercellular spaces have not yet been differentiated in the mesophyll (Fig. 8.16).

This fact shows that such cells differentiate directly. If this term did not refer to the particular circumstances of the differentiation of the conducting tissues, this type of sclerenchyma could be designated as primary sclerenchyma, in contrast to the tissue which arises through the late sclerification of cells which have already differentiated into another tpye of tissue (parenchyma, epidermis, etc.).

Sclereids occur mainly in the tissues which are not very rigid such as the leaf mesophyll, the aeriferous parenchymas of aquatic plants and the pulpy tissues of fruit. They are characterized by their size and variety of shapes. They can increase, in a manner similar to fibres, through processes of intercalary growth and above all through intrusive growth. Excessive outgrowths occur then, which invade the intercellular spaces or the middle lamellae of the parenchymatous cells. In some cases they may penetrate the epidermal cells in parting the anticlinal walls (Figs. 8.17 and 8.18; *Trochodendron, Mouriria;* FOSTER 1945, 1947). Sometimes shapeless and relatively large, the sclereids can be very numerous and may occupy a large volume in the organ.

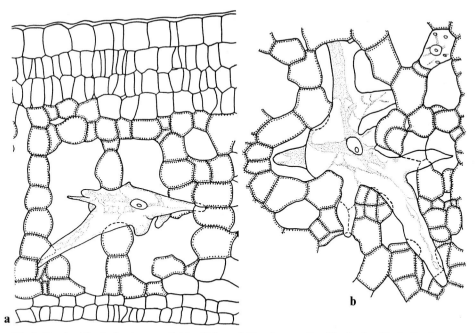

Fig. 8.18 a, b. *Trochodendron aralioides* (Silb and Zucc); growth of leaf sclereids in the lacunae and between the spongy parenchyma cells; **a** transversal section; **b** paradermal section; ×360; (after A. FOSTER 1945)

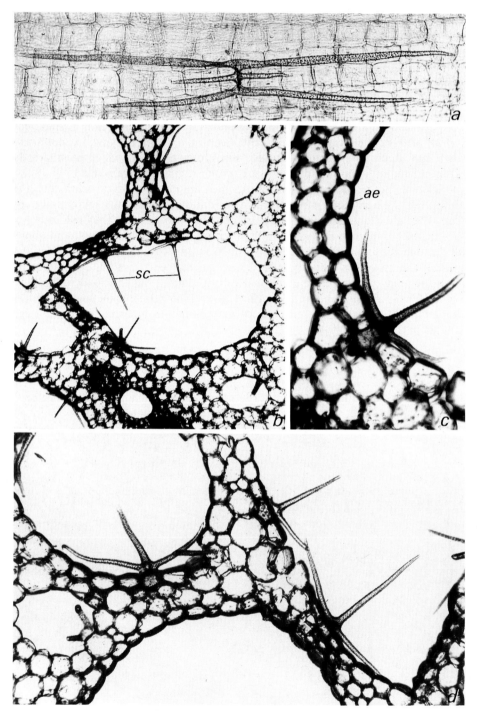

Fig. 8.19 a–d. Sclereids of *Nuphar* sp. petiolar aerenchyma; **a** *in toto* aspect in longitudinal view; **b–d** thick transversal sections of the petiole, showing lateral branches of sclereids *sc* resting on aerenchyma septa *ae*

The direction of growth and the shape of sclereids is not always uncoordinated and individual. Thus, in aerenchymas (aeriferous parenchyma) of aquatic plants, the sclereids which occur where several lamina of parenchyma cells meet (Fig. 8.19) assume the shape of a star, the branches of which, parallel two by two, encompass the lamina of cells, thus ensuring them an efficient support.

8.9 Origin and Differentiation of the Cells of First Formation Sclerenchyma

8.9.1 Differentiation of Compact Sclerenchyma

In the same organs, compact sclerenchyma is always found to be more internal than collenchyma. Whereas the initials of collenchyma can be recognized beneath the epidermis and from the apical meristem, it is sometimes difficult to visualize the future sclerenchymatous cells in the vicinity of the vegetative apices. However, the first recognizable cells of sclerenchyma are usually closely associated with *procambium* and are in relationship or in contact with the phloem tissue. In numerous Dicotyledons, sclerenchyma covers up the phloem of the conducting strands or surrounds them. However, these relations can be obscured by the development of the protophloem parenchyma, as in the stem of *Cucurbita* (MOURRÉ 1958). Such a parenchyma, usually acknowledged as cortical, contains sieve tubes of protophloem (Fig. 8.15).

Therefore, one can consider that the compact sclerenchyma is entirely or at least partially derived from the procambium. It is difficult to assess this origin because the specificity of the future cells of sclerenchyma is only demonstrated with delay. In fact, one of the characteristics of sclerenchyma is the mode of differentiation: it begins with a considerable and continued process of growth which occurs before the formation of the secondary cell layers and their lignification. Conversely, during the differentiation of collenchyma, the (primary) wall thickenings begin before the end of the growth process.

8.9.1.1 Differentiation of the Protoplasma

The two successive phases of differentiation, growth then cell wall differentiations, are obviously the result of the activity of the living content of the cells which requires further cytological studies of the protoplasm.

In light microscopy, the differentiation of the stem and petiole sclerenchyma of *Cucurbita pepo,* observed by J. MOURRÉ (1958), can be used as an example. The future cells of sclerenchyma are recognizable from the fifth internode. They have then an average length of 25 µm and and average width of 5 µm. The growth phase results in fusiform cells which can be 5000 µm in length and about 30 µm in width. Most of this volume increment is induced by the development of the vacuolar system which goes from the state where it is divided, in young cells (Fig. 8.20) to the state of a single vacuole pushing back the protoplasm against the cell wall

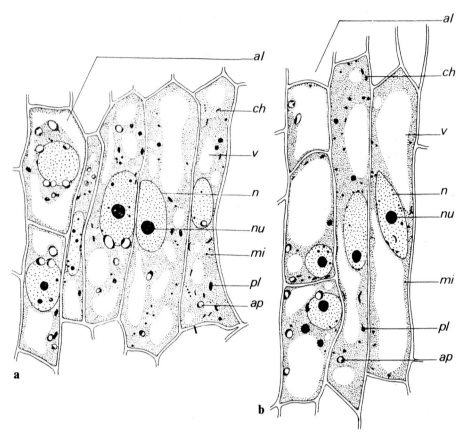

Fig. 8.20 a, b. *Cucurbita pepo,* differentiation of the sclerenchyma (beginning). **a** In the fifth internode; future sclerenchymatous cells, which have kept some meristematic characters; *v* numerous small vacuoles; *n* nucleus in the cell centre; *ch* short chondriosomes; *mi* mitochondria; *pl* small plastids, often amyliferous *(ap)*; *al* amyliferous layer, more internal than the sclerenchyma cells; *nu* nucleolus. **b** Seventh internode; future sclerenchyma cells; confluence of the numerous small vacuoles *v* into two large ones on each nucleus side; *ap* few amyloplasts; the cells are more elongated than in **a**; other indications as in **a**; (after J. MOURRÉ 1958)

(Figs. 8.20 and 8.21). But the extent of growth also implies an increased amount of protoplasm. The increase is obvious with regards to the nucleus which, in some cells reaches unusual sizes of about 40 μm (Fig. 8.22) and acquires a lobate and swollen aspect, suggesting an endopolyploid state (MOURRÉ 1958). This state has been acknowledged in the tomato collenchyma (BUVAT 1944) and it is known that phloem fibres may become multinucleated (HABERLANDT 1909, 1914; ESAU 1943b, 1965). Along with the nucleus growth, nucleoli grow, increasing from 1.2 μm in the first internode to 5 and even 7 μm (4 μm average), just before the deposit of secondary cell wall layers, whereas in the cells of the neighbouring parenchyma, the nucleolar diameter hardly reaches 1 μm. Further on, during the differentiation phase of the cell walls, the nucleolar volume decreases (MOURRÉ 1958). There are no significant variations in the chondriome. It is composed of short elements and

472

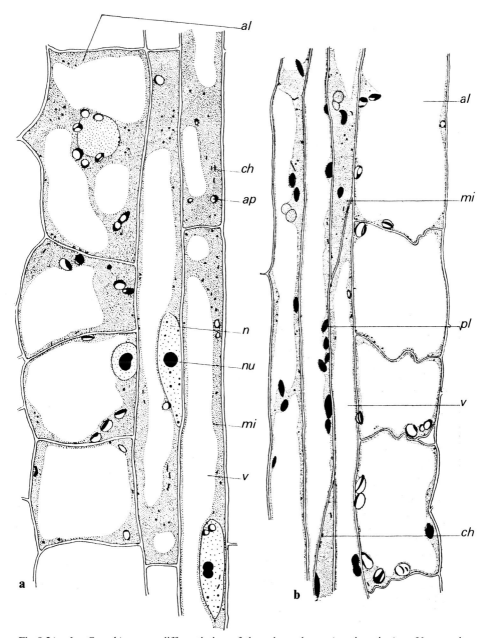

Fig. 8.21 a, b. *Cucurbita pepo,* differentiation of the sclerenchyma (continuation). **a** Young sclerenchyma of the tenth internode; the future sclerenchyma cells (length about 70 µm) as well as their nucleus (about 12 µm) have considerably extended; confluence of vacuoles into a large unique one; chondriosomes more divided in mitochondria; plastids almost unchanged, contrarily to those of layer *al,* which are differentiated in amyliferous chloroplasts. **b** Fourteenth internode; sclerenchyma cells are considerably elongated, and their extremities become oblique (fusiform cells); most chondriosomes are globular mitochondria; plastids have increased and are differentiated in chloroplasts. In some of these cells the nucleus becomes hypertrophied and contains several nucleoli of uncommon diameter (up to 7 µm, see Fig. 8.22); lettering as in Fig. 8.20; (after J. MOURRÉ 1958)

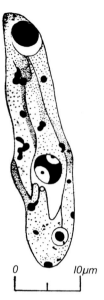

Fig. 8.22. Giant polyploid nucleus in a future sclerenchyma cell, developed just before the differentiation of secondary cell walls; (after J. MOURRÉ 1958)

mitochondria and seems to divide into even more slender granules at the end of the growth phase (Fig. 8.21).

Usually plastids manifest a period of amylogenesis, then assume a structure of lentil-shaped chloroplasts, including very little or no starch (Figs. 8.20b; 8.21) at the end of the growth. During the thickening and lignification of the cell walls, they decrease in number and size.

8.9.1.2 Differentiation of Cell Walls

The significant growth which precedes the deposit of secondary cell wall layers implies a strong secretory activity of the pectocellulosic precursors of the middle lamella and the primary wall. Both layers, rich in pectic compounds are, in fact, hardly distinguishable and form a "complex middle lamella-primary wall" (MOURRÉ 1958). At the same time as the elongation becomes harmonized with that of the internodes, the cells, at first prismatic, become fusiform probably due to an intrusive growth, similar to that of phloem and xylem fibres (Fig. 8.21 b).

At the end of the elongation process, in some places of the sclerenchymatous ring of the *Cucurbita* stem, lignification processes begin in the middle lamellae and in the wall which is still exclusively primary. Then, deposits of secondary layers occur and become lignified almost simultaneously.

In the areas where the primary complex is not yet lignified, the very quick secondary deposit is first pectocellulosic and the lignification is less immediate but it always begins with the primary wall.

474

The thickening goes on during the whole summer in the matured internodes, and ceases when the plant dies. The wall thickness, about 1.3 μm in July, reaches 2.8 μm in September. The cytochemical tests (phloroglucine-HCl, ruthenium red) indicate that the pectose compounds can no longer be found in the primary cell wall complex, but that they remain visible in the secondary layers (MOURRÉ 1958).

The secondary thickenings form concentric layers usually very visible, two or three in number, separated by a dark line.

The differentiated cell of the sclerenchyma ring of *Cucurbita* is thus a very long fusiform cell, in which the cytoplasm is reduced to a thin layer applied against the lignified wall; it includes a usually flattened nucleus which is reduced in size from the beginning of the secondary wall formations (about 20 μm in length) but proportionally much less than nucleoli (diameter: 1.3 to 1.4 μm).

The cytoplasm, denser in the pointed extremities, includes a few mitochondria and more or less degenerated plastids, depending on the size of the screen formed by the walls to the penetration of light.

8.9.2 Differentiation of Sclereids

The cytology of the sclereid differentiation has not been thoroughly studied; FOS-TER's works (1945, 1947) dealt with their ontogenesis in the leave of *Trochodendron aralioides* and *Mouriria huberi*. In the lacunar parenchyma of the petiole or the leaf lamina of *Trochodendron,* the sclereids are derived from polyhedral cells which apparently differ from the mesophyll cells only by the larger size of their nucleus (Fig. 8.16a, b). The branches of these sclereids usually start growing across the aeriferous lacunae of the tissue, then between the walls of the mesophyll cells. During the whole growth period, sclereids keep a single but huge nucleus (Fig. 8.18). Small plastids occur in these original "idioblasts"; they tend to degenerate in developed sclereids. When growth is completed, these cells form thickened secondary walls.

The sclereids of *Mouriria* (FOSTER 1947) have the peculiarity that they are in contact with the tracheids, the most external of the extremity of leaf veins. They develop from cells which are at first not very different from the mesophyll cells and then they lengthen, branching roughly perpendicular to the surface of the lamina until reaching the two epidermis (Fig. 8.17). Then they elaborate a thick secondary wall. During their intrusive growth between the young cells of the mesophyll, still without lacunae, sclereids have a single nucleus, like in *Trochodendron*.

However, this single-nucleated state is not general; according to STERLING (1947), the sclereids of the cortex and the medulla of *Pseudotsuga taxifolia* are multinucleated.

8.10 Late Sclerenchyma or Sclerenchyma of "Second Formation"

These tissues are composed of cells which have first functioned as differentiated cells i. e. part of other tissues, mainly parenchyma and epidermal tissues.

In most cases, these cells maintain approximately the same size and shape that they acquired during the first phase of differentiation, but, in some cases, they undergo a *new growing phase* which changes them more or less.

The cells of ground parenchyma, particularly medullary parenchyma, belong to the first case. They become sclerified later, a particularly frequent process in the stems of Monocotyledons (see Fig. 6.22).

When they are of the so-called sclerophytic type (e. g. leaves of *Psamma;* Fig. 8.23), most of the various parenchymas of the plants adapted to dry medium conditions, also become sclerified without changing much in shape.

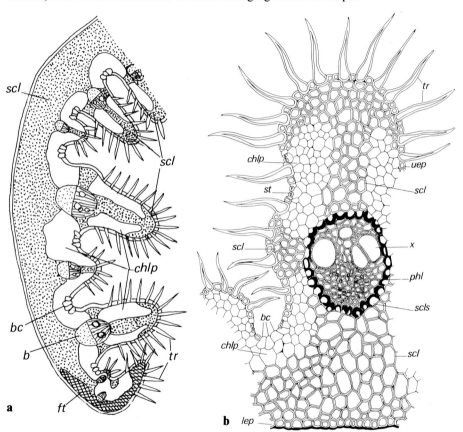

Fig. 8.23 a, b. *Psamma arenaria,* transversal section of the leaf. **a** Scheme of the tissue distribution; *bc* bulliform cells (see p. 270); *b* conducting bundle, phloem is *dotted,* xylem is *stripped; chlp* chlorophyllous parenchyma; *scl* sclerenchyma proceeding from the sclerification of an important part of the mesophyll *(dotted); ft* fibrous tissue bordering the leaf blade; *tr* trichomes. **b** Anatomical detail of a crest and a groove; *lep* lower (= dorsal) epidermis; *uep* upper (= ventral) epidermis; *scls* sclerified perifascicular sheath; *phl* phloem; *x* xylem; *st* stoma; other indications as in **a**

In the second case, the late sclerification occurs at the same time as the general development of the organ; it is more pronounced than in the previous case. Several examples can be found, in particular, in the modification processes which change the ovary into a fruit and the ovule into a seed. As for dehiscent dried fruits, all the layers of the ovarian wall become sclerified after *they have enlarged,* and the cellular shapes vary during this new growing phase. Sometimes the mesocarp cells (more or less similar to the mesophyll cells) lengthen and become oriented in such directions as to allow the dehiscence after the sclerification (e.g. cloves of certain Papilionaceae: *Lathyrus*). While the ovule develops into a seed, in the same way the *tegument cells,* which are derived from the epidermis, increase and are sclerified while changing shapes (Fig. 8.24).

Finally, the sclerification may be preceded by a recurrent phase of proliferation of the cells which are part of a matured tissue. This is particularly the case when the internal epidermis of the carpel becomes a lignified endocarp along with the development of the drupes. The epidermal layer produces many layers which are strongly sclerified later (Fig. 8.25). Thus, this last sclerenchyma of lignified endocarps, derived from cells which proliferate again after a period of dormancy, results from histogenetic processes comparable to those which take place in the sec-

Fig. 8.24a, b. Sclerified tegument of *Malus communis* seed (apple pip). **a** Transversal section of sclerenchyma cells covered by an epidermis, *ep,* the cells of which are not lignified, but have thickened cell walls and a thick cuticle. **b** Longitudinal section showing the elongated shape of sclerenchyma and epidermal cells

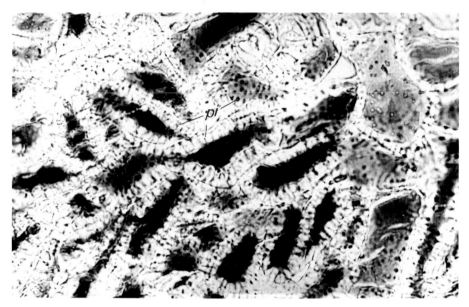

Fig. 8.25. *Persica vulgaris,* lignified endocarp, the cells of which have strongly thickened cell walls, but have preserved numerous pits *pi* between them

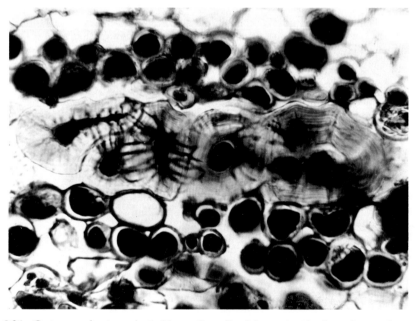

Fig. 8.26. *Castanea vulgaris,* ancient phloem; the cells of the ancient horizontal parenchyma are changed into sclereids, the cell walls of which are considerably thickened and show numerous secondary layers, hence their lamellar aspect. Numerous canalicular pits are clearly visible; nearby parenchymatous cells are strongly tanniniferous

478

ondary conducting tissues; thus the term secondary sclerenchyma would not be misused.

The secondary growth of the cells which develop into late sclerenchyma may occur in the same two ways as for the other sclerenchyma: general growth and intrusive growth. The secondary parietal formations may be very poorly developed (sclerified medullary parenchymas) or, on the contrary, highly developed (lignified endocarp). All the intermediary forms can be found. The secondary walls have numerous pit fields. Despite this fact, some cells issued from the late sclerification may degenerate and die rather quickly, even if the walls are not very thick (sclerified medulla). However, the circumstances which determine the late sclerification are extremely varied and the consequences related to cell survival are not well established. Thus, the sclerification of the medullary parenchyma seems to be a sign, a consequence rather than a cause of the cell degeneration. This is not always the case, mainly in fruits and in certain late sclereids of the phloem tissue. Indeed, it is striking to note that certain tissues may favour the sclerification of the neighbouring tissues or some of their own cells. An example is shown with the phloem fibres of hemp and ramie. In some cases, bicollateral conducting strands of plants, including internal phloem, are also known to differentiate a so-called perimedullary sclerenchyma associated with this internal phloem.

In the old phloem of various arborescent species *(Robinia, Castanea)* outside the phloem fibres, the parenchyma cells which are in contact with these strands may strongly sclerify in some cases, particularly the cells of horizontal parenchyma (Fig. 8.26). There are many examples (see Fig. 6.28) of such sclereids resulting from the late sclerification of phloem parenchyma cells. Despite the pronounced thickening of walls on which the striation is clearly visible, these sclereids contain some obviously normal protoplasm with nucleus, plastids and mitochondria; they stay alive. Secondary canalicular pits, branched in some cases, maintain the exchanges between cells (see Fig. 6.28 a, b).

In the organs and tissues which remain functional the sclerenchyma cells are thus usually alive; they can even accumulate storage material, like the fibres of conducting tissues.

However, when phellogen layers arising from the phloem isolate the sclerified layers of old phloem, these tissues are integrated into the rhytidome. The latter is thus composed of the association of the peridermal formation derived from a former generating layer and sclerified formations of old phloem, the cells of which die and will be exfoliated in most species (ESAU 1964; cf. Chap. 5).

In fact, the late sclerenchyma can be considered as a tissue which is not autonomous, but which is only the result of the development of other tissues, either parenchymatous or epidermal. However, when deep modifications (proliferation, growth) occur after the primary differentiation, it is rather a new tissue which is formed. Moreover, in many circumstances, sclerenchyma development depends closely on external *ecological* conditions rather than related to the intangible development of organ morphogenesis.

The significance of this tissue is more *ecological* than *systematic* (e.g. sclerophytes). Physiologically, the development of this tissue seems to be stimulated by the abundance of glucids, a fact suggested by the relationship between sclerenchyma and phloem tissues, and by the lack of water.

8.11 Mechanical Properties

We have already compared (cf. p.464) the mechanical properties of collenchyma with those of sclerified fibres. The resistance to rupture is higher than that of collenchyma, but always inferior to twice this resistance. The main difference concerns the *elasticity* which is much more important in sclerified cells. Conversely, sclerenchyma has much less plasticity than collenchyma: the lignification of thickened walls would prevent it from being harmonized to organ growth if it has differentiated too much before the end of the elongation phase. The reduced plasticity due to lignification may be roughly shown if we compare the materials made with non-lignified fibres of linen and cotton with the materials made with lignified fibres of hamp and jute.

The lignin is the main impregnating substance for both sclerenchyma and wood. Both tissues have similar mechanical properties but those of sclerenchyma, a simple tissue, are not as sophisticated as those of wood, a complex and more heterogeneous tissue. In some cases, it may be not very resistant, i.e. softer than wood (medullary sclerenchyma of *Agave,* medulla of *Sambucus*) or, on the contrary, extremely hard (bamboo stem, endocarps of peach or apricot).

References

Ambronn H (1881) Über die Entwicklungsgeschichte und die Mechanischen Eigenschaften des Collenchyms. Pringsh Jahrb Wiss Bot 12: 473–541
Bary A De (1884) Comparative anatomy of the vegetative organs of the phanerogams and ferns. Clarendon, Oxford
Buvat R (1944) Recherches sur la dédifférenciation des cellules végétales. Ann Sci Nat Bot 11ᵉ sér 5, see pp 61 and 81–82
Duchaigne A (1953) Sur la transformation du collenchyme en sclérenchyme chez certaines Ombellifères. CR Acad Sci Paris 236: 839–841
Duchaigne A (1954) Nouvelles observations sur la sclérification du collenchyme chez les Ombellifères. CR Acad Sci Paris 238: 375–377
Duchaigne A (1955) Les divers types de collenchymes chez les dicotylédones; leur ontogénie et leur lignification. Ann Sci Nat Bot Paris 11 ème sér 16: 455–479
Esau K (1936) Ontogeny and structure of collenchyma and of vascular tissue in celery petioles. Hilgardia 11: 411–476
Esau K (1943) Vascular differentiation in the vegetative shoot of *Linum.* III. The origin of the bast fibers. Am J Bot 30: 579–586
Esau K (1964) Structure and development of the bark in dicotyledons. In: Formation of wood in forest trees. Academic Press, London, pp 37–50
Esau K (1965) Plant anatomy. Wiley New York
Foster AS (1944) Structure and development of sclereids in the petiole of *Camellia japonica* L. Torrey Bot Club Bul 71: 302–326
Foster AS (1945) Origin and development of sclereids in the foliage leaf of *Trochodendron aralioides* Sieb and Zucc. Am J Bot 32: 456–468
Foster AS (1947) Structure and ontogeny of the terminal sclereids in the leaf of *Mouriria huberi* Cogn. Am J Bot 34: 501–514
Haberlandt G (1909) Physiologische Pflanzenanatomie. Engelmann, Leipzig, 650 p – Engl ed: Physiological Plant anatomy. London, Macmillan and Cᵒ, 1914
Magin T (1956) L'ontogénie du collenchyme chez *Lamium album* L. Rev Cytol Biol Vég 17: 219–258

Majumdar GP (1941) The collenchyma of *Heracleum sphondylium* L. Leeds Phil Lit Soc 4: 25–41

Majumdar GP, Preston RD (1941) The fine structure of collenchyma cells in *Heracleum sphondylium* L. Proc R Soc Lond B Biol Sci 130: 201–217

Mourré J (1958) L'ontogénie du sclérenchyme chez *Cucurbita pepo* L. Rev Cytol Biol Vég 19: 99–149

Müller C (1890) Ein Beitrag zur Kenntnis der Formen des Collenchyms. Ber Dtsch Bot Ges 8: 150–166

Roland JC (1966) Organisation de la membrane paraplasmique du collenchyme. J Microsc 5: 323–348

Roland JC (1967) Recherches en microscopie photonique et en microscopie électronique sur l'origine et la différenciation des cellules du collenchyme. Ann Sci Nat Bot Paris 12 ème sér 8: 141–214

Sterling C (1947) Sclereid formation in the shoot of *Pseudotsuga taxifolia*. Am J Bot 34: 45–52

Tobler F (1957) Die mechanischen Elemente und das mechanische System. In K Linsbauer, Handbuch der Pflanzenanatomie 2nd ed Bd 4, t. 6

Venning FD (1949) Stimulation by wind motion of collenchyma formation in celery petioles. Bot Gaz 110: 511–514

Vian B, Roland JC (1972) Différenciation des cytomembranes et renouvellement du plasmalemme dans les phénomènes de sécrétions végétales. J Microscopie 13: 119–136

Went FA (1924) Sur la transformation du collenchyme en sclérenchyme chez les Podostémonacées. Recl Trav Bot Neerl 21: 513–526

CHAPTER 9

Secretory Cells and Secretory Tissues

9.1 Concept and Various Types

Secretory cells and tissues are concerned with the accumulation of metabolism by products which are not used as reserve substances. Most secretory cells are specialized cells derived from elements belonging to other tissues, mainly epidermis or parenchymatous tissues. In such cases, they do not constitute true independent tissues and it is justified to speak of secretory parenchymas or secretory epidermis.

The term *secretion* per se is rather inaccurate; it is frequently used to include the production and the release of molecules which can be useful to the organism where it occurs, but which becomes separated from the cell. An example of such a process is the secretion of hormones. But cells can also *release* subsidiary or secondary metabolism by-products which are non-utilizable or harmful for the plant body; the separation of products eliminated from metabolism is *excretion*. Only the physiological role and significance of the substances released by cells could allow one to distinguish the two concepts, but most often the role of many by-products is only hypothetical. It is therefore more convenient to use the term secretion for all the releasing processes of substances out of the cells producing them.

This concept should even be more extended. In fact, one of the general properties of the living cell is concerned with the release of products ingested through endocytosis or elaborated in its protoplasm, i.e. in its vacuoles. Among other functions, vacuoles are the seats of accumulation of storage material and food or of excreted substances. They are a kind of appendage of the extracellular medium.

Intravacuolar storage, either useful or non-utilizable, may be considered as *intracellular secretions*. The hydrolases, which permit the intravacuolar digestions and are secreted through the Golgi apparatus, are an example of internal secretions useful for cells.

Conversely, tanniniferous cells, in groups or isolated among cells without tannins, retain the substances they have secreted in their vacuoles throughout their life; the use of these substances is not clearly established, particularly if we consider their significant amount. It is the same with the calcium oxalate crystals. Finally, the various categories of laticifers retain their secretion in their vacuoles or in their cytoplasm.

From a cytological standpoint, two categories of cells can be distinguished: cells that release their excretion to the surface of the plant or into the plant. In each group, the histological organization and the nature of secreted substances allow the determination of several histocytological types as shown in Table 9.1.

Some secretory structures mentioned in Table 9.1 correspond more or less to late specializations of cells which belong to definite tissues (e.g. epidermis) and

Table 9.1 Schematic classification of the main categories of secretory cells or secretory tissues

Cells that release the secretion out of their protoplasm

Histological type	Prevalent secretion
1. Secretory hairs	Essential oils, balsams, resins, mucilages, salts
2. Secretory epidermis	Essential oils, balsams, resins, mucilages, salts, polysaccharides, lipophilic substances
3. Secretory cavities	Essential oils, balsams, mucilages
4. Secretory canals	Essential oils, balsams, mucilages
5. Hydathodes	Water
6. Nectaries	Glucids

Cells that retain their secretion in an intercellular container

Histological type	Prevalent secretion
1. Isolated essential oil cells	Essential oils
2. Tanniniferous cells and tissues	Polyphenolic substances (tannins)
3. Articulated laticifers	Polyterpenes, lipophilic substances (latex, alkaloids)
4. Non-articulated laticifers	Lipids, polyterpenes (latex, alkaloids)

which keep their cellular morphology. But when these structures are organized into thoroughly differentiated systems (secretory hairs or canals) and particularly if their differentiation occurs as early as that of the neighbouring tissues, they are much more independent and deserve the term *secretory tissues*. This is mainly the case of non-articulated laticifers which will be described in Section 9.12.

9.2 Secretory Epidermis

Secretory epidermis has already been described in Chapter 5. Generally all the ground cells of epidermis become secretory. Their excretion occurs first as hydrophobic enclaves within the cytoplasm itself. Then, usually these products either accumulate between the external membrane and the cuticle which become more or less distended or they diffuse through the cuticle (Fig. 9.1).

The substances produced by secretory epidermis are, in most cases, essential oils, i.e. compounds of terpene carbides of low molecular weight (particularly $C_{10} H_{16}$ and $C_{15} H_{24}$) and aromatic substances produced from their oxidation or hydroxylation; this is the case of the epidermis of the petals of various flowers to which they owe their fragrance: e.g. the upper epidermis of petals of *Rosa, Syringa, Jasminum, Viola odorata*, etc., the lower epidermis of petals of *Hyacinthus* or *Convallaria*.

Secretory epidermis can also be found on stems or leaves, e.g. *Lavandula*.

Then the sticky substance which covers certain tree buds *(Salix, Populus, Aesculus)* is an oleo-resin secreted by the epidermal cells of the scales. Similar substances produced by epidermal tissues make the stems of *Silene nutans* and other Caryophyllaceae gummy.

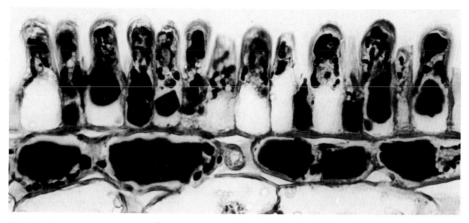

Fig. 9.1. Petal of *Rosa,* secretory epidermis, after osmium tetroxide fixation. The vacuolar apparatus, partly tanniniferous, is strongly contrasted

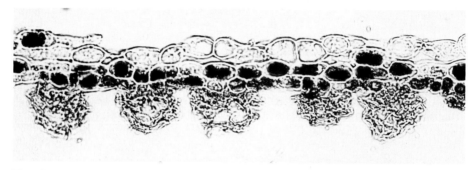

Fig. 9.2. *Aesculus hippocastanum,* bud scale; mucilage secretory glands, originating from the epidermis. Secretory cells degenerate precociously, contrarily to banal epidermal cells

Secretory epidermis may be localized differentiations of common epidermis, like on the petioles and stipules of *Salix* or supported by emergences of subepidermal tissues in the shape of papillae or some kinds of mamillated and peltate secretory glands (*Nerium* petioles; cf. Fig. 5.13) or scales of winter buds in *Betula* or *Aesculus* (Fig. 9.2).

The stigmatic secretions of flowers are also produced by epidermal tissues; the secretion product is often complex and variable, depending on the species. They are mainly compounds of lipophilic substances (glycerides, unsaturated fatty acids) with polysaccharides and polyphenols (tannins, flavonoids), e.g. *Forsythia intermedia* (DUMAS 1973a, b, c; 1975, 1978).

In some cases, secretory epidermal cells are more highly differentiated in the glandular apparatus with a precisely defined structure, including underlying cells physiologically associated with them. Examples of such structures are *salt glands* which remove sodium chloride in halophilic plants such as the *Tamarix aphylla* (SHIMONY and FAHN 1968, Fig. 9.3a).

484

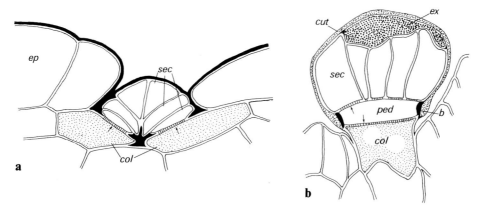

Fig. 9.3 a. *Tamarix aphylla* L., salt gland composed of eight epidermal cells, six of which are secretory cells *sec,* and two forming the lower tier are called *collecting cells, col; ep* epidermis. *Arrows* indicate cell walls, crossed by numerous plasmodesmata, between collecting cells and lower secretory cells; (after SHIMONY and FAHN 1968) **b** *Avicennia marina,* salt gland constituted by a *secretory trichome,* composed of two to four basal cells, embedded into the epidermis, the collecting cells *col* (only one visible on the figure), of a single pedicel cell *ped,* surrounded by a parietal differentiation *b,* resembling the endodermal Casparian bands and of a head, most often formed of eight secretory cells *sec* (five are visible on the section). *cut* Cuticle; *ex* exudate accumulated between the distal cell walls of the secretory cells and the cuticle which is loosened from the cell walls at these sites; (interpreted after a micrograph of SHIMONY et al. 1973)

Histologically, salt glands are intermediate structures between epidermis and secretory trichomes, the latter being typically pedicelled as shown in the comparison of Fig. 9.3 a, b. In some cases secretory epidermis includes several layers of secretory cells derived from periclinal divisions of initial epidermal cells; e.g. the secretory surfaces of various flower stigmates (*Forsythia intermedia, Narcissus jonquilla, Salvia pratensis,* etc.; DUMAS 1975). In contrast to common epidermal cells, secretory cells usually retain cytological features of the meristematic type. They are highly basophilic due to their large content of ribosomes and their vacuome is divided into small vacuoles, including osmiophilic precipitates; plastids are poorly lamellated; the chondriome is divided into short mitochondria (Plate 9.1 A).

9.3 Secretory Trichomes

The genesis of epidermal trichomes and their various shapes has been described previously (cf. p. 265). Secretory trichomes develop in the same way. When they become secretory, the distal cells become specialized and elaborate the secretion, mainly composed of essential oils in most species. Secretory trichomes show various degrees of complexity; the simpler trichomes are composed of a single file of cells, the last cell being secretory (*Cistus, Primula sinensis, Pelargonium zonale;* Figs. 9.4 b, c; 9.5 a–d).

Other trichomes may have an uniseriate stalk with a multicellular head at the end (Solanaceae, including *Hyocyamus*). They may also be more massive, with a

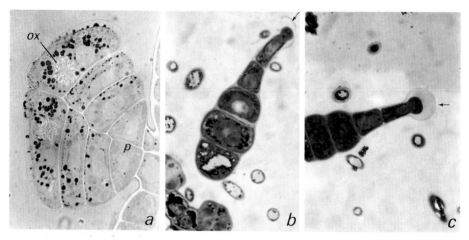

Fig.9.4a–c. Two examples of secretory trichomes. **a** *Artemisia campestris* L.; trichome with a globular head composed of three tiers with two cells each and borne by a short pedicel *p; ox* twinned calcium oxalate in apical cells. Semi-thin section, after glutaraldehyde – OsO_4 fixation and Sudan Black B staining; × 1250; (courtesy of L. Ascensão 1985). **b,c** *Cistus monspeliensis,* secretory trichomes composed of a file of cells, the vacuoles of which contain an oleo-resin. The cuticle of the end cell, detached from the cell wall, is torn in **b,** still intact in **c** *(arrows);* × 675; same technique as in **a**; (courtesy of L. Ascensão and M. S. Pais 1981)

thick pedicel (*Cannabis;* numerous Labiae and Compositae). Typically the secretory hairs of Labiae and various Compositae have a head composed of four to eight cells radially arranged around the axis (Figs. 9.4a; 9.5e).

The secretion accumulates between the cuticle and the external wall of secretory cells. It may be excreted through the cuticle or released after rupture of the cuticle which extends considerably (Fig. 9.4). It may become regenerated and the secretion accumulates again, but the rupture may induce the degeneration of the trichome after a single act of excretion (Stahl 1953).

The trichomes of various plant species contain particular substances, different from essential oils. For example, in carnivorous plants *(Drosera, Pinguicula)* trichomes secrete proteolytic enzymes combined with mucilages and nectar. Trichomes secreting sodium chloride occur in halophilic plants such as *Avicennia marina* (Shimony et al. 1973; Fig. 9.3b) and *Atriplex halimus* (Smaoui 1975; Fig. 5.12). In this species the salt is removed through the exfoliation of trichomes which regenerate throughout leaf growth; in this case the secretion is of a holocrine type.

Fig. 9.5a–e. Scanning microscopy of secretory trichomes. **a,b** *Petunia violacea,* trichomes with monocellular head, secreting mucilages on the calix epidermis; × 250 (**a**) and 420 (**b**). **c,d** Calix of *Thymus vulgaris,* globular secretory trichomes, secreting oil substances, intergrade with protecting trichomes; in **d,** cuticular folds are visible; × 330 (**c**) and 600 (**d**). **e** *Mentha-x-piperita,* leaf trichome with multicellular head arranged in a crown; × 1600; (courtesy of A. Perrin)

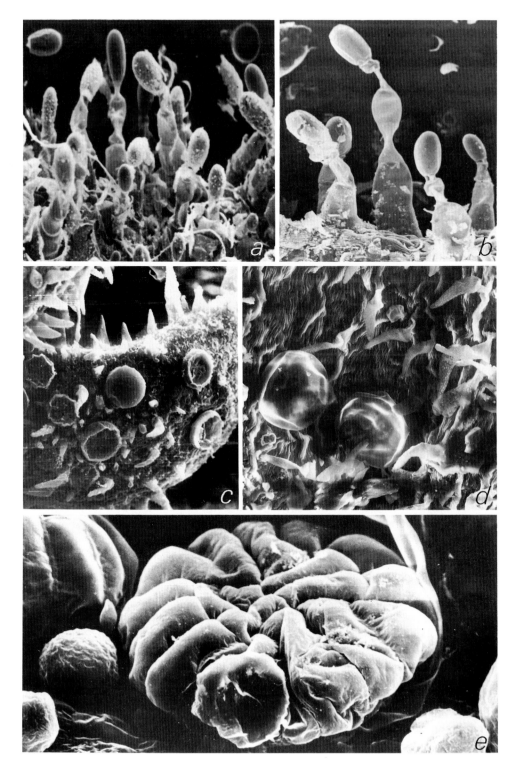

487

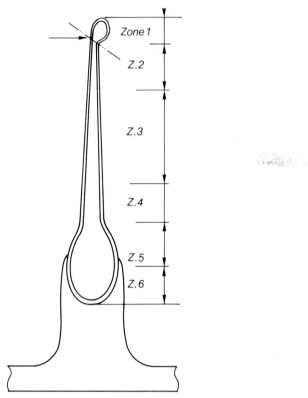

Fig. 9.6. Scheme of *Urtica dioica* stinging trichome. Parietal distribution of calcium and silicon along the trichome, according to zones *1* to *6* (THURSTON 1974); a basifugal gradient characterizes the silicon amount (almost no silicon below zones *1* and *2;* on the contrary, calcium is only present in the cell wall at the trichome base, and practically disappears in zones *2* and *1* (SAMAT 1984). *Arrow* indicates the breaking line

The stinging trichomes of the nettle *(Urtica urens)* differentiate in a particular way. The swollen base, vesicular, is embedded in epidermal cells which rise above the surface. Above the base, a narrow tube ends with a kind of small bud (Figs. 9.6 and 9.7). This bud breaks off along a predetermined line when the hair comes in contact with an object (or with the skin) leaving a sharp edge capable of penetrating human skin (Fig. 9.8). The pressure exerted on the base releases the liquid content. It is a complex secretion including histamine, acetylcholine (FELDBERG 1950) and serotonine (COLLIER and CHESCHER 1956). The wall of stinging trichomes is calcified at its lower end and silicified at its upper end, with inverse gradients (SAMAT 1984). Thus, the apical rupture occurs in the silicified part. With electron microscopy, ultrastructural study (MARTY 1968a) reveals the presence of all the usual organelles in trichomes. The huge distal vacuole which contains the stinging secretion results from the fusion of numerous fusiform vacuoles which develop at the lower end, in the bulbous area where the endoplasmic reticulum, arranged in parallel tiers or in nodular complexes, produces some series of characteristic dilatations (Plate 9.1 B). The plasmalemma of the proximal base, in contact with the sur-

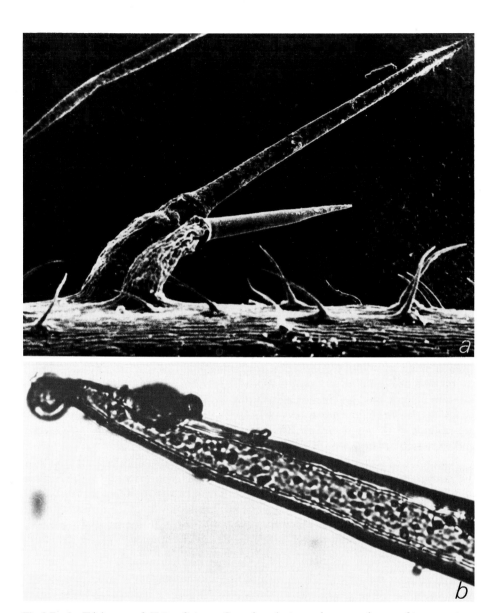

Fig. 9.7 a, b. Trichomes of *Urtica dioica*. **a** Scanning electron microscope image of two secretory trichomes, the extremity of which is devoid of the apical bulbs. Several smaller protective trichomes are visible. **b** Extremity of a secretory trichome, bearing its terminal bulb (light microscopy) (courtesy of M. SAMAT 1984)

Fig.9.8. Schematic representation of the breaking of a stinging trichome extremity, releasing a sharp point, and allowing the stinging liquid to pour out; (after SAMAT 1984)

rounding epidermal cells, is rough and rich in plasmodesmata, suggesting that intercellular exchanges are intense at these sites. The action of calcium chloride allows the localization of acetylcholine in the vacuoles, particularly those of the base, and in cavities of the endoplasmic reticulum (MARTY 1968A; Plate 9.1 C).

Like in the secretory epidermis, most often the secretory cells of trichomes maintain a somewhat meristematic infrastructure with an abundance of ribosomes, numerous small vacuoles, poorly differentiated plastids and short rod mitochondria or in the shape of granules (Plate 9.1 A). However, there are numerous exceptions in which a huge vacuole occurs in the end cells, shifting the cytoplasm against the wall, where it is applied as a thin film. This aspect is shown in trichomes secreting sodium chloride in *Atriplex halimus* (SMAOUI 1975; Plate 9.2).

Plastids are more or less differentiated, depending on the species and on the location of cells in the gland. For example, the cells of the base of the secretory heads of *Artemisia campestris* trichomes differentiate into chloroplasts, whereas the plastids of the upper end cells remain in the state of leucoplasts (ASCENSÃO and PAIS 1985; Plate 9.3 A).

The walls of secretory cells are frequently uneven, with protuberances towards the cytoplasm, similar to those found in transfer cells, which icnreases the surface of contact between the plasmalemma and the wall (e.g. *Cistus monspeliensis*, ASCENSÃO and PAIS 1981; Plate 9.4 A).

We shall deal again with secretory cells with respect to the sites of syntheses and the secretion mechanisms (cf. p.509).

———————————————————————————————————▷

Plate 9.1 A–C. *Urtica urens*, stinging trichome ultrastructures.

A Basal part of the young stinging trichome apical cell *st*, ribosome rich; basal cytoplasm, surmounted by a large vacuole *V*, extended up to the cell tip; other small vacuoles *v* are scattered in the cytoplasm; *m* mitochondria; *n* nucleus, more or less lobed; *p* plastid; *er* parallel sheets of endoplasmic reticulum; ×9125. Fixation: glutaraldehyde – OsO_4; contrast according to REYNOLDS (1963). **B** Enlargement of endoplasmic reticulum profiles with associated ribosomes; *der* dilated endoplasmic reticulum; *V* central vacuole; ×49000. Fixation: OsO_4; contrast: $KMnO_4$. **C** Opaque crystalline precipitates *pr* in the cavity of the endoplasmic reticulum, obtained by addition of calcium chloride to the fixative, and owing to the presence of acetylcholine, in comparison with other techniques; ×27000.

Fixation: glutaraldehyde – OsO_4 – $CaCl_2$; contrast according to REYNOLDS; (courtesy of F.MARTY 1968a)

490

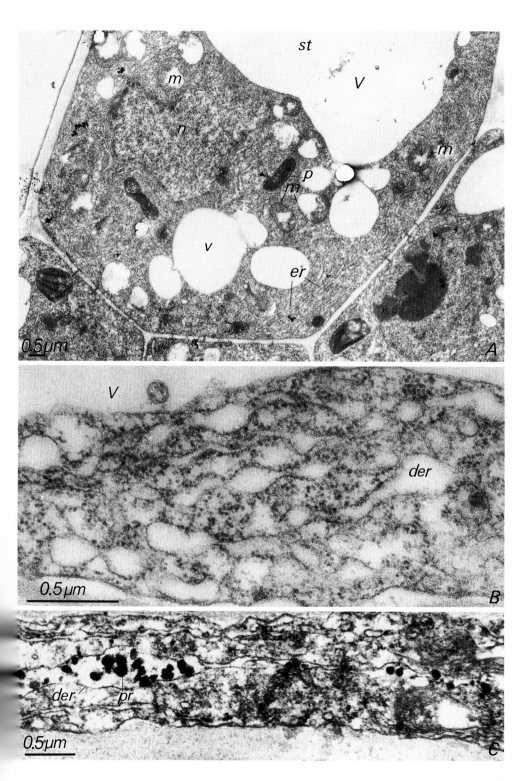

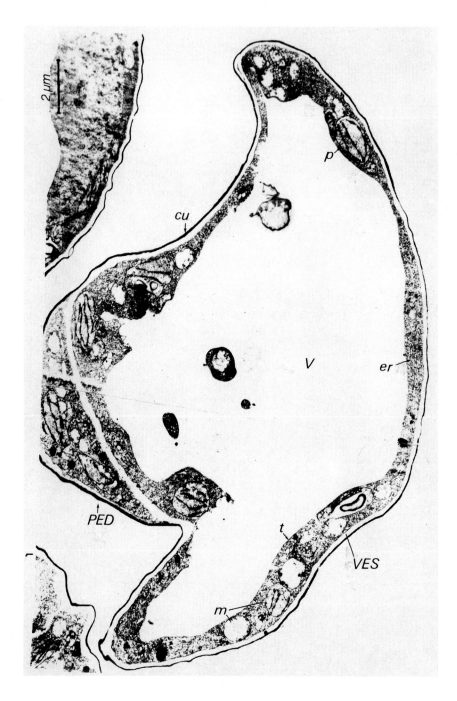

Plate 9.2. *Atriplex halimus* L., trichome vesiculous cell *VES,* with a large vacuole *V* in which the sodium chloride concentrates.

PED pedicel cell; *cu* cuticle; *m* mitochondria; *p* poorly lamellated plastid; *er* endoplasmic reticulum; *t* tonoplast; ×7000. Electron micrograph of SMAOUI (1975)

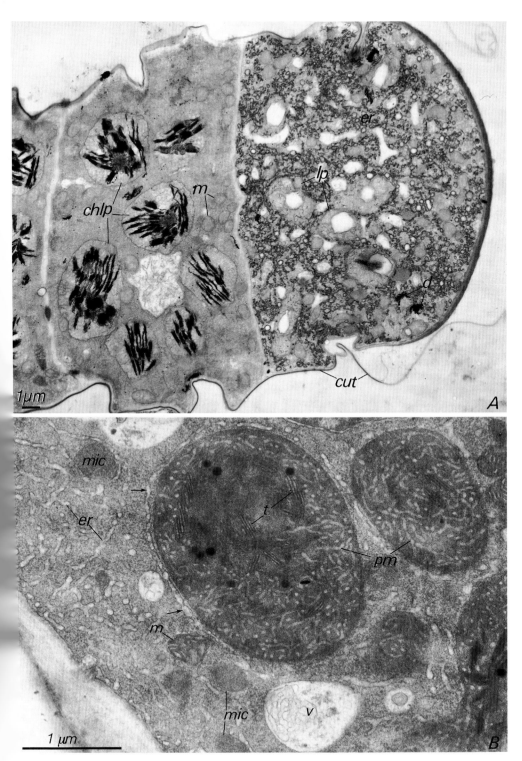

Plate 9.3. Legend see p. 494

Plate 9.3 A, B. *Artemisia campestris* L., ssp. *maritima;* cytology of the secretory trichome.

A Top of a young trichome in the secretion phase; *lp* leucoplasts *in the apical cell* (without chlorophyll); *chlp* chloroplasts in the underlying cell; *cut* cuticle, loosened from the external cell wall by the secretion bulge; *er* endoplasmic reticulum, particularly developed, in the shape of a network of anastomosed tubules in the apical cell; *d* dictyosomes; ×4600. Zinc iodine – O_5O_4 impregnation technique, according to NIEUBAUER et al. (1969). **B** Trichome in secreting phase, cytoplasmic area of the subapical cell; very abundant endoplasmic reticulum, profiles of which tightly encircle the chloroplasts *(arrows);* these chloroplasts possess a particularly developed plastidal meshwork *pm; m* mitochondria; *mic* microbodies (= peroxisomes); *t* thylacoids; *v* vacuole; ×27000.

Fixation: glutaraldehyde – O_5O_4; contrast: uranyle acetate-lead citrate; (courtesy of L. ASCENSÃO 1985)

◁——

9.4 Nectaries

Nectaries are secretory structures which excrete sugars, called the nectar, and play an important part in the relationship between the so-called zoidophil plants and the animals which ensure their pollination. In most cases, insects (Hymenoptera, Lepidoptera, Diptera) are attracted both by the nectar of the so-called entomophil plants and by the fragrant secretion whether pleasant or not for human smell (unpleasant smell of Aristoloche flowers which attract Diptera). Nectaries occur mainly on flowers, but also on the vegetative organs of some species, for example on the foliar stipules of *Vicia faba* or on the petiole of the *Passiflora* leaves. Floral nectaries occupy various positions on the flower, and have different shapes: a ringlike nectary at the base of stamens, below the androecium (Caryophyllaceae, Polygonales, Chenopodiales) a ring or a disc at the base of the ovary (Ericales, Solanales, Labiateae), between the androecium and the gynoecium, i.e. on the floral receptacle, where they may be regularly convex pads, sometimes lobed or ornamented with secretory tubers like in many *Dialypetales,* grouped under the name of *Disciflores* due to their nectaries. Moreover, nectaries may be modified stamens, the staminodes, as in *Parnassia palustris* and, among others, in various Ranales.

Nectary tissues occur also on different parts of the perianth: sepals of broom, of lime tree; bulged base of Crucifer sepals; spur of the *Tropaeolum* calyx; spurs of the petals of *Delphinium, Aquilegia, Viola,* various *Orchis,* etc. (Fig. 9.9).

Secretory cells of nectaries may be exclusively epidermal or combined with underlying cell layers. In this case, the cells of these layers form a compact tissue similarly to that of a meristem due to the small size of the cells, the basophilia of the cytoplasm and the high nucleocytoplasmic ratio (EYMÉ 1963; Fig. 9.10).

The histological complexity of nectaries has been compared with the evolutionary features of close species (FAHN 1953).

Food is supplied to nectaries either through the phloemian and vascular tissues which are merely those of the organ bearing them, or through their own conducting bundles (FREI 1955).

Nectaries secrete a higher concentrated sucrose solution when the conducting tissue consists of phloem elements only or when it is dominant in conducting bun-

494

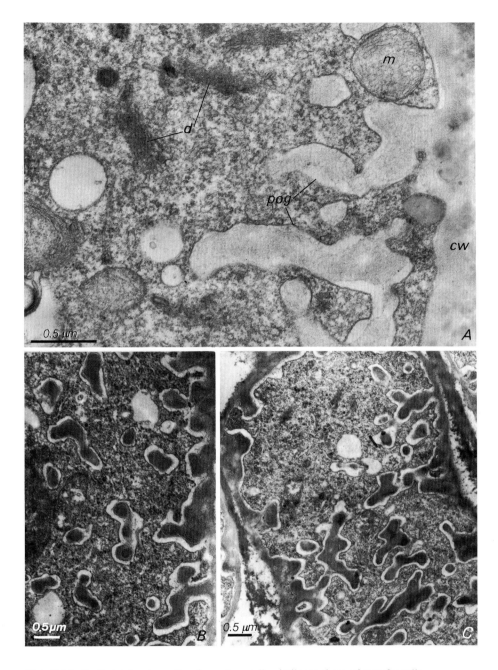

Plate 9.4 A–C. Parietal outgrowths of secretory cells, similar to those of transfer cells.

A *Cistus monspeliensis,* secretory trichomes; *d* dictyosomes; *m* mitochondrion; *cw* cell wall; *pog* parietal outgrowths; × 34 500; (courtesy of L. ASCENSÃO and S. PAIS 1981).
B, C *Cichorium intybus,* hydathode cells, parietal excrescences similar to transfer cell walls; × 14 000 (**B**) and 13 000 (**C**); (courtesy of A. PERRIN 1972)

Fixations: glutaraldehyde – O$_S$O$_4$; contrast according to REYNOLDS (1963)

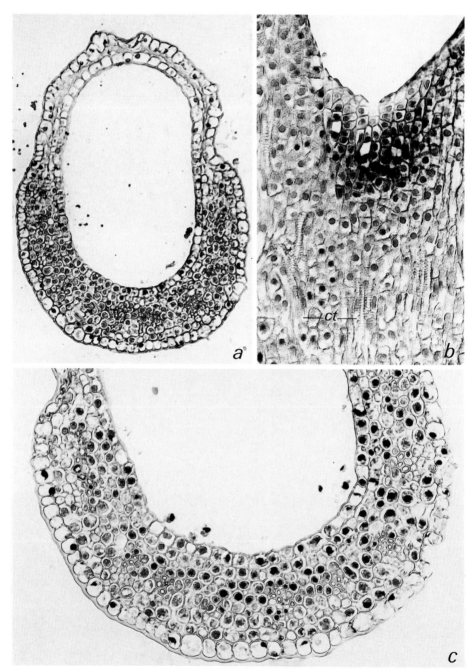

Fig. 9.9 a–c. *Delphinium consolida,* nectariferous spur. **a** Transversal section in the spur distal zone; the nectary secretory part is thickened and below the epidermis includes a chromophil tissue, composed of cells smaller than those which occur in the non-secretory part. **b** Oblique section in the spur distal zone; the secretory part is made of very chromophil isodiametric small cells; *ct* trails of conducting tissues below the nectary. **c** Enlargement of the nectariferous spur secretory area; isodiametric small cells resembling a false meristematic structure

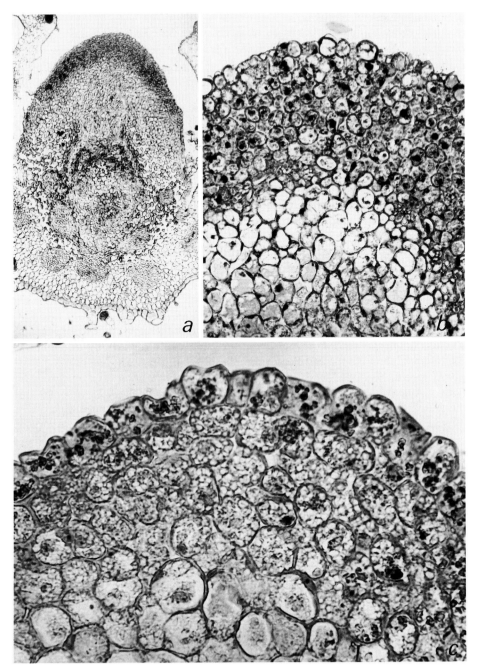

Fig. 9.10 a–c. *Salvia sclarea,* nectariferous tissue situated at the lower part of the ovary. **a** Oblique section passing through the base of the ovary (*lower part* of the figure) and in a nectariferous pad (*upper part* of the figure, strongly chromophil). **b** More enlarged part of the top of the nectariferous tissue, composed of small isodiametric cells. **c** Cytological aspect of nectariferous cells; cytoplasm containing numerous small vacuoles

497

dles, than when xylem tissue is predominant (FREY-WYSSLING 1955). When the bundles which supply the nectaries consist mainly of xylem, these nectaries are physiologically intermediate forms between true nectaries and hydathodes which excrete water.

The nectar is excreted through diffusion across the external cell wall followed by ruptures of the cuticle, through apoplasmatic progression in anticline walls, or, in the less highly specialized nectaries, through stomata (FAHN 1953). In some nectaries, these stormata are modified and become uncapable of closing.

The differentiation of cells secreting nectar involves all the cellular organelles as well as the wall. For example, in the nectaries of the corolla tube of *Lonicera japonica,* the wall bears numerous internal protuberances, while the cuticle is detached from the external wall (FAHN and RACHMILEVITZ 1970; Plate 9.5 A). The vacuome of these cells increases at the beginning of nectary ontogenesis, then decreases along with the secretion, while the cytoplasm increases. The granular endoplasmic reticulum develops into parallel layers producing numerous vesicles on their fringes, which reach the wall protuberances and seem to merge with the plasmalemma (Plate 9.5 A). Mitochondria become more numerous at the beginning of the secretion phase and starch is transiently elaborated by plastids. The Golgi apparatus is particularly active at the beginning of differentiations, when the protuberances develop on the wall; the authors conclude that it is mainly concerned with the elaboration of polysaccharide material, precursors of parietal substances, whereas the nectar secretion would be essentially due to the activity of the endoplasmic reticulum (FAHN and RACHMILEVITZ 1970). In previous data on the nectaries of several Dicotyledons *(Helleborus niger, H. foetidus, Diplotaxis erucoides, Ficaria ranunculoides)* EYMÉ (1966, 1967) showed the abundance of dictyosomes and the accumulation, striking in some cases, of Golgi vesicles at the beginning of the differentiation and the secretory phase, like in *Ficaria* (Plate 9.6 A). These vesicles certainly play a part in the growth of walls, but also in the storage of part of the glucid material, a component of the nectar secretion.

The endoplasmic reticulum deserves special attention. On the one hand, when its granular type is well characterized, it develops in parallel tiers edged with bulgings which lose their ribosomes and separate in vesicles similar to those fusing with the plasmalemma in *Lonicera japonica.* Here also, one can suppose that the nectar is produced by the Golgi apparatus and the endoplasmic reticulum.

On the other hand, the exuberant development of the endoplasmic reticulum gives rise to particular inclusions which, on thin sections, have a "coat of mail" aspect composed of wavy cytomembranes which are differentiations of the common reticulum, with which they are in continuity (Plate 9.6 B; EYMÉ 1966, 1967).

Finally, in some places the plasmalemma of nectary cells forms pinocytose pockets including membranous elements; these pockets are isolated in the cytoplasm and captured by vacuoles (Plate 9.5 B, C). These endocytosis processes occur along with the excreting activity of nectaries (EYMÉ 1966, 1967).

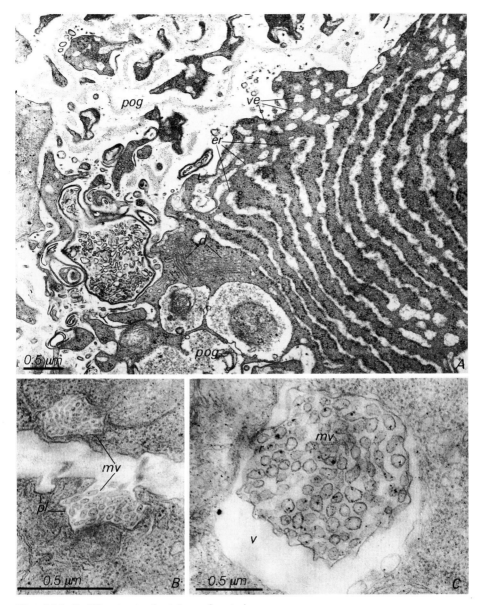

Plate 9.5 A–C. Ultrastructural cytology of nectaries.

A *Lonicera japonica,* secretory cell of the floral nectary; cytoplasmic area crossed out by numerous parallel sheets of rough endoplasmic reticulum *er,* the margins of which are dilated and produce vesicles *ve* which tend to associate with the plasmalemma; *d* dictyosomes; *pog* parietal outgrowths, overdeveloped, as in various other secretory cells; × 22300. Fixation: glutaraldehyde – O$_S$O$_4$; contrast according to REYNOLDS (1963) (courtesy of A. FAHN and T. RACHMILEVITZ 1970).

B *Diplotaxis erucoides,* nectariferous cells; formation of pinocytosis sacs containing membranous trabeculae and spherules, and becoming isolated in the cytoplasm, in the shape of membranar vesicles *mv; pl* plasmalemma; × 51000.

C Id membranar vesicles *mv* captured by a vacuole *v* in which the content will be lysed, sometimes with myelinic features; × 35000. Fixations: glutaraldehyde – O$_S$O$_4$; (courtesy of J. EYMÉ 1967)

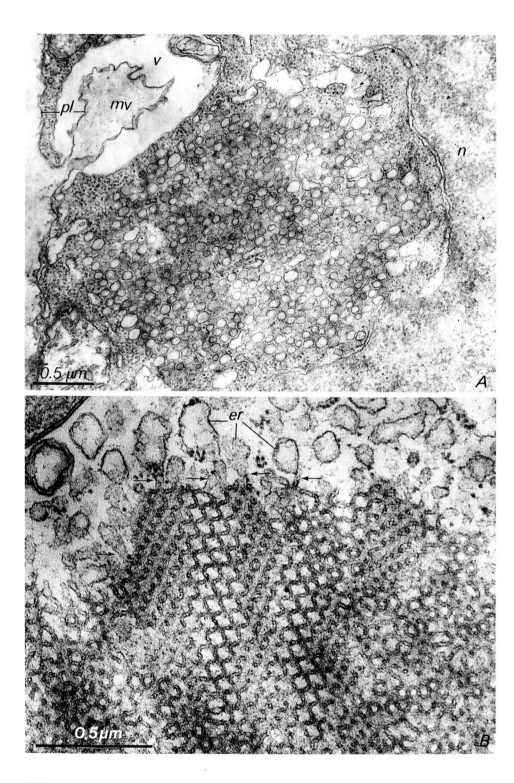

500

9.5 Hydathodes

Hydathodes are structures specialized in the secretion of water, a secretion in droplets on the surface of leaf laminae, particularly on the edges of leaves or at the upper end of the lamina (Graminaceae). This process, called "guttation" has been observed since the 17th century (see DUCHARTRE 1859, for older works) and more precisely studied by DE BARY (1877) and HABERLANDT (1894, 1909); the term hydathode was provided by the latter author (1894).

Under this heading, HABERLANDT referred to various histological structures, such as the epidermal cells or the trichomes concerned with guttation, mainly of water, or comparable to the "salt glands" mentioned above *(active epidermal hydathodes)*. In agreement with A. PERRIN (1972) the use of this term has been limited to hydathodes in relationship with vascular tissues, the secretion of which is pushed forward by the radicular pressure *(passive hydathodes* of HABERLANDT).

Sensu sticto, these hydathodes are functional combinations of three categories of elements: one or several aquiferous stomata, the end of one or several vascular bundles composed of a few tracheids, and a body of characteristic cells forming the epithem (DE BARY 1877) or *hydather tissue* (PERRIN 1972). The latter appears as a mesophyll differentiation in contact with the terminal tracheids and the aquiferous stomata, establishing a relation between them (Fig. 9.11).

Stomata have two active chlorophyll cells including a large nucleus. They are different in shape from the aeriferous stomata of the same leaf, particularly in Graminaceae in which they occur at the lamina end. Moreover, there are transitional forms of stomata in these plants. The ostiole is distorted in some cases, and its more or less ample movements do not allow complete closure (PERRIN 1972).

In most species, the epithem is composed of cells with complicated shapes, leaving wide intercellular spaces between them, enlarged in *aquiferous chambers* below the epidermis and the aquiferous stomata (Fig. 9.11 and Plate 9.7). The hydathode cells react positively to acid phosphatase, mainly at the level of the plasmalemma. Although it is still difficult to interpret these data, they suggest that the epithem cells are perhaps not completely passive with respect to the water which moves along through this tissue (PERRIN 1972).

Some arguments in favour of this hypothesis can be inferred from the variable amount of the products dissolved in the excreted liquid compared to the tracheid sap.

◁────────────────────────────────────

Plate 9.6 A, B. Cytological characteristics of nectaries.

A *Ficaria ranunculoides,* nectarigen cell of a very young floral bud; significant accumulation of Golgi vesicles between the cell wall and the nuclear envelope; *n* nucleus; *pl* plasmalemma, forming a membranar vesicle *mv,* already encircled by a vacuole *v;* ×33000.

B *Ficaria ranunculoides,* nectarigen cell; local differentiations of the endoplasmic reticulum: close associations of waving tubules, resulting in the formation of the structure termed "cotte de mailles" (coat of mails); *er* profiles of endoplasmic reticulum showing continuity with the coat of mails structure *(arrows);* ×63000.

Fixations: O_sO_4 (Courtesy of J. EYME 1967).

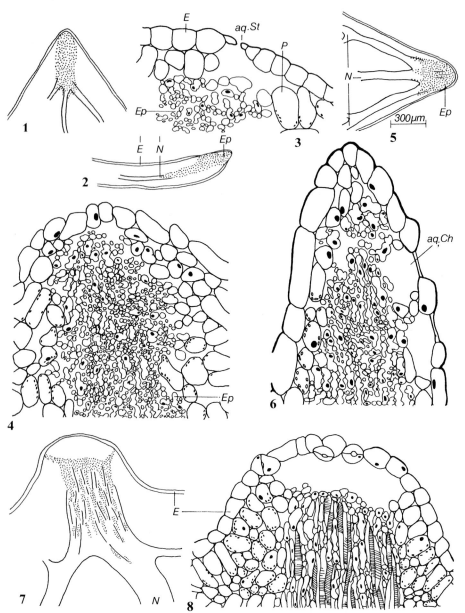

Fig. 9.11. Examples of hydathodes. **1-4** *Stellaria media;* **1** morphological aspect on a section parallel to the plane of the leaf lamina; **2** id, in a plane perpendicular to the lamina; *3* aquiferous stoma, *aqst,* and substomatal chamber; *4* section parallel to the plane of veins, and near the upper epidermis; abundance of intercellular spaces and sinuous shapes of the epithem cells *Ep.* *E* epidermis; *P* parenchyma. **5,6** *Cerastium aquaticum;* **5** topographic relations between the epithem *Ep* and the vein endings *N;* *6* organization of the apical leaf extremity, in longitudinal section, perpendicular to the leaf surface; *aqCh* aquiferous chamber, relatively large. **7,8** *Geranium robertianum,* hydathode, longitudinal sections parallel to the leaf lamina; example of a little developed epithem, crossed by numerous isolated tracheids, prolonging the veins *N;* very large aquiferous chamber; *E* epidermis; (courtesy of A. PERRIN 1972)

502

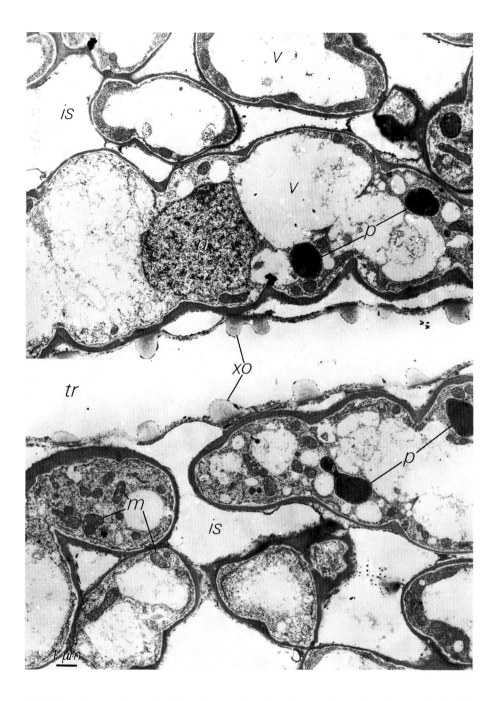

Plate 9.7. *Papaver rhoeas*, hydather tissue cells, situated around a xylem element (tracheid) *tr*; sinuous elongated shapes, parallel to the hydathode axis, leaving large intercellular spaces *is* between them; *m* mitochondria; *n* nucleus; *xo* xylem ornamentations (secondary cell wall); *p* plastids; *tr* tracheid (or vessel element); *v* vacuoles; × 5500. Fixation: glutaraldehyde – O$_S$O$_4$; contrast according to REYNOLDS (1963); (courtesy of A. PERRIN 1972)

In fact, in the guttation water, not only mineral salts can be found, but also carbonates, chlorides, nitrates, etc. in which potassium is the more abundant cation followed by calcium, magnesium, sodium, iron, etc. and furthermore numerous organic molecules, i.e. glucids, such as glucose, galactose, fructose, saccharose, xylose, arabinose; and numerous amino acids and vitamins as well as various hydrolases (RNAase, DNAase, proteases).

But most of these chemicals are more diluted in the guttation water than in the sap of tracheids and vessels. Therefore, hydathode cells may participate in the retention and absorption of one part of these aqueous solutions, thus manifesting an original physiological activity which, however, is not yet established.

Using electron microscopy, extensive studies on the infrastructure of the epithem of numerous species (PERRIN 1972) showed, in fact, that the plasmalemma of these cells is usually very sinuous and may play a part in the exchanges between the cytoplasm and the extracellular medium, in the shape of invaginations similar to plasmalemmasomes (Plate 9.8 A–C).

Besides the nucleus, which is voluminous, the ultrastructural peculiarities of these cells concern the very abundant chondriome, the elements of which are either short or elongated and rich in cristae, and the plastids. These have variable aspects; they may produce starch inclusions but are very poor in chlorophyll or have none at all, although they sometimes develop lamellae similar to chloroplast thylakoids. The cytoplasm also includes peroxisomes, amorphous or with a crystalline nucleoid (Plate 9.8 D). There are many microtubules in these cells (PERRIN 1972). Some hydathodes (Compositae, Papaveraceae) have transfer cells; their walls form cristae or finger like protrusions in the cellular lumen, thus increasing the surface of the plasmalemma (Plate 9.4 B, C).

Guttation processes probably partially condition the external medium in the neighbourhood of plant leaves, a medium called "phyllosphere". With this, they play an ecological part, particularly phytopathological, in stimulating infections, bacterial or fungal proliferations. Moreover, they can increase the toxicity of phytosanitary pesticides.

9.6 Secretory Cavities

The secretory cavities result from the development of intercellular spaces between the cells which become secretory (Figs. 9.12, 9.13). These cells are derived from an initial and their development results in two categories of secretory cavities.

9.6.1 Schizogenous Cavities

The parenchymatous initial cell undergoes several divisions in a radial orientation and the middle lamella opens in a central lacuna which enlarges, while mitoses are in process, radially oriented with respect to this central space. A cavity is thus constituted with a single layer of cells which become secretory (Fig. 9.12). In some cases, but rarely, pericline cleavages result in schizogenous cavities with several layers.

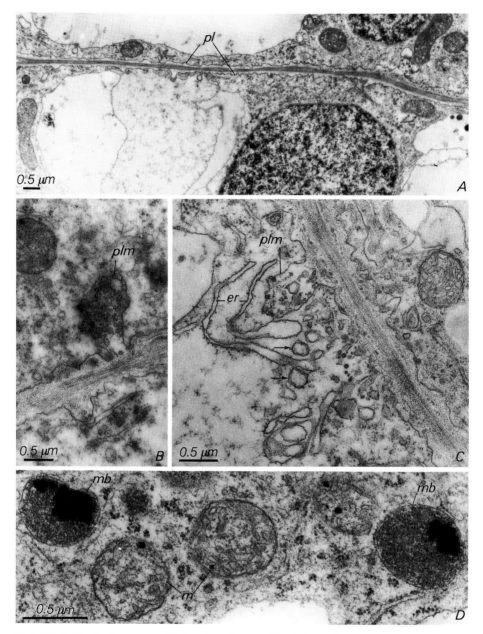

Plate 9.8. A–C. *Saxifraga oppositifolia,* plasmalemma of hydather cells.

A Waving plasmalemma *pl,* forming here and there invaginations towards the cytoplasm; **B** Formation of plasmalemmasomes *plm,* enclosing light vesicles and numerous smaller ones, the content of which is dense; **C** Plasmalemmasome, as in **B**, surrounded by endoplasmic reticulum profiles *er* which tend to produce light vesicles *(arrows);* ×8500 (**A**), 15 900 (**B**), 20 500 (**C**). **D** *Saxifraga cuneifolia;* "microbodies" *mb* (peroxisomes) in the hydather cells; *m* mitochondria; ×34 500.

Fixations: glutaraldehyde – OsO_4; contrast according to REYNOLDS (1963); (courtesy of A. PERRIN 1972)

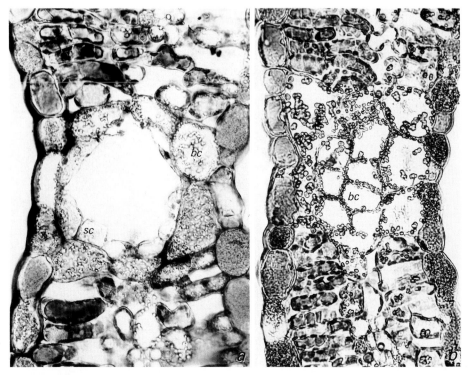

Fig. 9.12 a, b. *Hypericum perforatum,* transversal sections of leaf lamina. **a** Schizogenous secretory cavity, comprising one layer of secretory cells *sc,* one layer of bordering cells *bc* non-chlorophyllous, surrounding the preceding one; the refringent globules are tanniniferous. **b** Subtangential section, showing the bordering cells, tanniniferous but non-chlorophyllous, extending from one epidermis to the other. Fixation: REGAUD - staining: hematoxylin

First, secretory cells accumulate, within their own cytoplasm, the essential oils that they have elaborated and then these products pass through the wall into the cavity.

Schizogenous apparatus occur in Myrtaceae (*Eucalyptus,* cloves) in Hypericaceae in which the secretory cavities of the leaves form, in transparence, translucid points (hence, the names of "Millepertuis" or *Hypericum perforatum* for St. John's work). Other less classical families are characterized by schizogenous cavities occurring also in some plants (isolated with respect to these cavities) which belong to families in which such cavities cannot usually be found, e.g. the leaves of certain Gymnosperms, the Compositae *Tagetes* (French marigold).

9.6.2 Schizolysigenous Cavities

They develop at the beginning as the previous ones but then, pericline cleavages multiply the layers of bordering cells, thus producing several layers of cells which become secretory and accumulate essential oils within their cytoplasm; then the

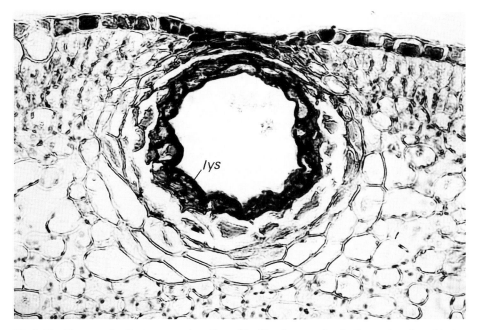

Fig. 9.13. *Citrus medica* L., transversal section of leaf lamina, passing in the centre of a schizoly-sigenous secretory cavity; repeated periclinal divisions of peripheral cells give rise to abundant cellular material, and the more internal cells are progressively lysed; *lys* lysed cells. Fixation: RE-GAUD – staining: hematoxylin

more internal degenerate and are lysed, the walls are partially diluted and the fragments of these cells fall into the cavity. The lysis progresses to the periphery, and a big drop of essential oils combined with cellular fragments fills the cavity (Fig. 9.13).

Such schizolysigenous cavities containing essential oils are characteristic of Rutaceae.

Apparatus resulting from the gelation of membranous substances of isolated or grouped cells (cavities containing mucilages or balsams in Malvaceae and Tiliaceae) can be compared to secretory cavities. In fact, one cannot speak of secretions, since the main part of these cavity contents result from the lysis of membrane substances.

9.7 Secretory Canals

Their mode of formation is similar to that of the schizogenous secretory cavities from which they differ only by their elongated shape. The lumen of the secretory canals results from the lysis of the pectose compounds included in the middle lamellae between cells which become secretory. In the roots of germinations of *Pinus halepensis,* dictyosomes produce dense vesicles which are transferred into the wall through exocytosis, in the shape of invaginations into the plasmalemma; these vesicles could induce the dissolution of the pectose substances of the middle lamella and the formation of the canal lumen (FAHN and BENAYOUN 1976). In many cases, two layers of cells occur on each side of the canals (conifers) only the internal layer being the secretory layer (Fig. 9.14).

The secretion product varies, depending on the species. It is often basically composed of essential oils (Umbelliferae) or resins (conifers, Terebinthaceae) both types of products being often combined in various Terebinthaceae and conifers; in other cases, the canals contain combinations of products including gums (gum resins in certain Umbelliferae). These substances are prevalent in "gumming canals" (Sterculiaceae). In their wood, cortex and medulla, many species of tropical Cesalpineae differentiate anastomosed secretory tubes which produce a resin used in the making of varnish, the *copal* (MOENS 1955).

Independently of Gymnosperms, secretory canals can be found in various families of Dicotyledons, mainly in Araliaceae, Umbelliferae, Papilionaceae, radiated Compositae (Plate 9.9 A, B) Terebinthaceae and a few others. Among Monocotyledons, only Aroideae seem to have such apparatus.

Secretory canals occur in almost all parenchyma tissues (cortical, phloem, lignified, medullary, etc.). They are, however, exceptional in the mesophyll of leaves (*Pinus silvestris,* Fig. 9.14), but frequent in the petioles and veins. Their position is

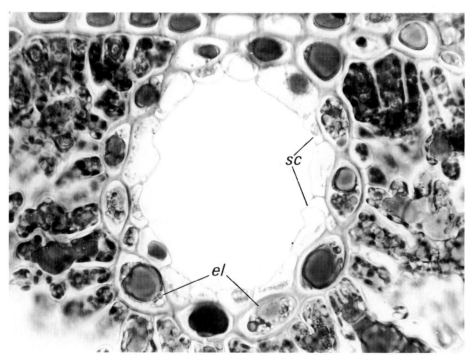

Fig. 9.14. *Pinus silvestris* needle; secretory canal, composed of two cell layers, the internal layer only being secretory. *el* External layer; *sc* secretory cells of the internal layer (transversal section). Fixation: Navaschin – staining: hematoxylin

508

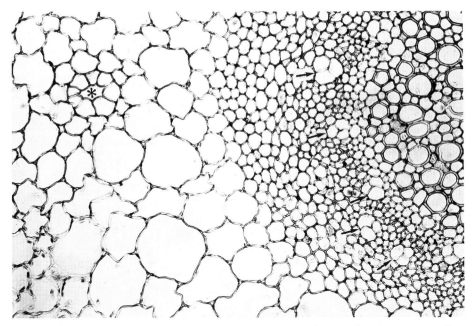

Fig. 9.15. *Petroselinum sativum* (parsley), petiole; transversal section showing the two types of secretory canals, one in the cortical parenchyma facing the conducting bundles *(asterisk)*, the others in the phloem *(arrows)*. Fixation: REGAUD – staining: hematoxylin

often characteristic of the family. For example, in Umbelliferae, two systems of secretory canals can be found, particularly in petioles: canals of relatively large size, located below the collenchyma and into the cortical parenchyma, and smaller ones in the phloem (Fig. 9.15).

9.8 Sites of Synthesis, Transport and Cytological Mechanisms of Excretion

Sites of synthesis, where products are elaborated by secretory cells, are variable depending on the secretates. The secretates are often mixtures or combinations of several categories of molecules, their precursors arise in different sites, which supposes an intracellular and/or intercellular biochemical evolution resulting from a transport process.

When the secretion product is of polysaccharide nature, the endoplasmic reticulum and the dictyosomes are involved and their role is comparable to the role of both anabolism and transport played by these membranes in the precursor deposits of parietal components (VIAN and ROLAND 1972).

As for the lipophilic secretion products, fatty acids, terpenes, sterols, flavonoids, polyphenols, the plurality of infrastructures in which they can be found, e.g. endoplasmic reticulum, plastids, mitochondria and vacuoles, demonstrates

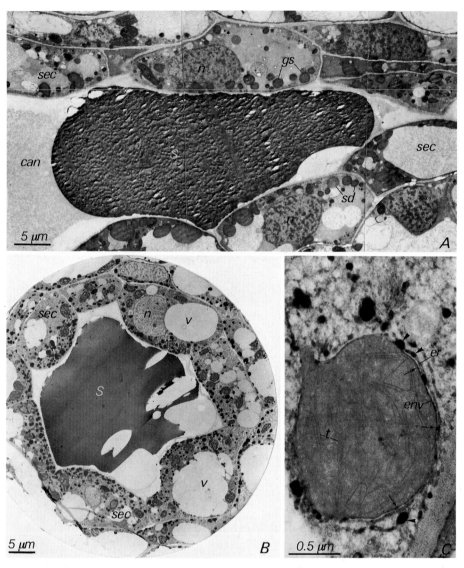

Plate 9.9 A–C. *Artemisia campestris,* ssp. *maritima;* cytology of secretory canals in the secretion phase.

A Longitudinal section of a secretory canal part; *sec* secretory epithelial cells bordering the canal cavity *can; sd* secretion droplets, concentrated against the canal bordering cell wall; *n* nuclei of epithelial cells; *S* secretate bulk in the canal cavity; × 1980.

B Transversal section of a secretory canal; accumulation of secretion droplets towards the canal cavity; *v* vacuoles; other indications a in **A;** × 1380.

C Secretory cell plastid closely surrounded by endoplasmic reticulum profiles; these profiles contain secretion droplets which locally dilate them *(arrowheads);* in addition, some secretion products are found in the space between the two membranes of the plastid envelope *env (arrows); t* thylacoids little developed; × 27 500.

Fixations: glutaraldehyde – O$_S$O$_4$; contrast according to Reynolds (1963); (courtesy of L. As-censão 1985)

510

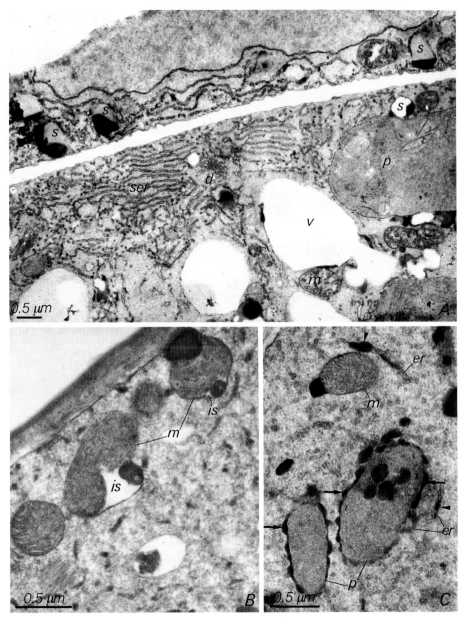

Plate 9.10 A–C. *Artemisia campestris* L., ssp. *maritima;* cytology of the secretory apparatus.

A Development of the smooth endoplasmic reticulum *ser* of the secretory canal epithelial cells; *d* dictyosomes; *m* mitochondria; *p* plastids; *s* secretion droplets; *v* vacuole; × 13 500.
B Trichome apical cell in the secretion phase; heap of secretion material in the dilated intermembranar space of the mitochondria *m;* × 30 000.
C Trichome apical cell in the secretion phase; secretion material is present in dilated parts of the endoplasmic reticulum *(arrowheads)* and in the intermembranar space of the plastid envelope *(arrows);* *m* mitochondrion containing a secretion droplet; × 26 000.

Fixations: gluraraldehyde – O$_S$O$_4$; contrast according to Reynolds (1963); (courtesy of L. Ascensão 1985)

511

probably both a biochemical variety and a developmental maturation of the secretate between the sites of synthesis of the precursors and the sites of transport to the plasmalemma against which these products accumulate.

Such evolutions are suggested, for example, in the secretory trichomes of Compositae such as *Inula viscosa* (WERKER and FAHN 1981), *Chrysanthemum morifolium* (VERMEER and PETERSON 1979) and *Artemisia compestris* (ASCENSÃO and PAIS 1985). These organs are composed of a two-celled pedicel over two basal cells and a secretory head including two apical cells and two layers of two underlying cells (Fig. 9.4a). The four last cells differentiate chloroplasts, whereas the two apical cells contain leucoplasts. Some authors (ASCENSÃO and PAIS) consider that chorophyll plastids synthesize the fatty acids which are a component of the secretate, as well als the precursors of monoterpenes and sterols, the biosynthesis of which could be completed in the non-chlorophyllous apical cells. The two types of cells differentiate a *"smooth endoplasmic reticulum"* which is considerably developed (Plate 9.3 A, B) and may play a role in the transport of the secretate elements. The significant network of the smooth reticulum of the apical cells is comparable to the structures existing in the animal cells which produce steroid hormones (CHRISTENSEN and FAWCETT 1960). This smooth reticulum development can also be found in secretory canals (Plate 9.10 A).

Moreover, secretate granules occur in mitochondria which carry them to the plasmalemma concurrently with the reticulum (ASCENSÃO and PAIS 1985; Plate 9.10 B).

The cytological modalities of transport between cells or cellular infrastructures are difficult to observe in electron microscopy. Accumulations of lipophilic substances have been observed in the plastids of numerous lipophilic glandular tissues (oil, terpene or flavonoid glands). They are usually accompanied by close associations between plastids and endoplasmic reticulum (DUMAS 1974b; ASCENSÃO 1985; ASCENSÃO and PAIS 1985; Plates 9.9 C and 9.10 C). These figures suggest possible passages between the plastidal stroma, the envelope of plastids and the associated reticulum; in fact, with various techniques of preparation, inclusions with the same characteristic as the secretate are visible in the three structures (Plates 9.9 and 9.10 C). The unusual development of the peripheral plastidal reticulum gives further ground for this assumption (ASCENSÃO and PAIS 1985; Plate 9.3 B).

▷

Plate 9.11 A–C. *Atriplex halimus* L., salt secretory trichome.

A Tangential section of the cell wall separating the pedicel cell from the vesicular apical cell; particular abundance of plasmodesmata *Pd;* × 54200. Fixation: glutaraldehyde – paraformaldehyde – O$_S$O$_4$.

B Localization of Cl$^-$ ion (as Cl Ag); Cl$^-$ ion moves notably across plasmodesmata *Pd* of the separating cell wall between the pedicell cell *P* and the vesicular apical cell *V; Pl* plasmalemma; × 31000. Fixation: O$_S$O$_4$; reactive: NO$_3$ Ag.

C Pedicel cell, *P,* distal zone, marked by the silver chloride precipitate; × 8200; same technique as for **B**

(Courtesy of A. SMAOUI 1975)

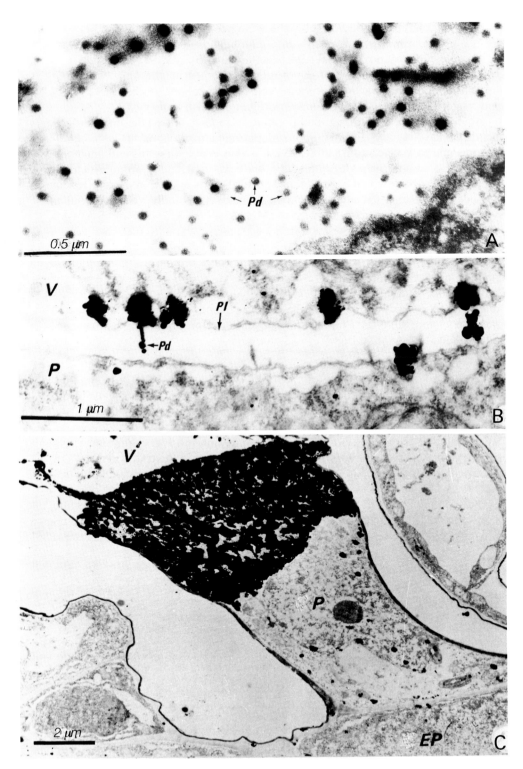

Other infrastructural figures show the existence of narrow relations between layers of endoplasmic reticulum with an osmiophilic content and vacuoles including lipophilic inclusion which have the same feature (DUMAS 1974a).

Unfortunately, the passage of the osmiophilic content of the reticulum to the tonoplast and into the vacuole, or the reverse process cannot be observed in electron microscopy, suggesting molecular processes of active transport.

Among the data concerning the lipophilic secretions, one of the most striking facts is the development of the smooth endoplasmic reticulum, particularly significant in some cases. According to DUMAS, it is unprobable that this endomembranous system results from degranulations of the typical endoplasmic reticulum (DUMAS 1973a, b). It would rather be an organization distinct from the true reticulum, comparable to Golgi and provacuolar formations, i.e. to the "tubulo-reticular system" from which vacuoles are derived (MARTY 1973a). GIFFORD and STEWARD (1967) have even identified the membranes of smooth reticulum with the vacuole tonoplasts.

In the salt secretory trichomes of *Atriplex halimus,* the progression and the distribution of ions was observed by SMAOUI (1975) by precipitating them under an insoluble form opaque to electrons, through treatment with silver nitrate for Cl^- ions and with potassium pyroantimoniate for Na^+ (and Ca^{2+}) ions. Moreover, the transport of inorganic salts was acknowledged, after absorption of lanthane nitrate. Results are as follows:

1. Cl^- ions and Na^+ ions are concentrated in the apical vesicular cell (Plate 9.12 A);
2. Lathane salts, Na^+ ions, progress, partially at least, in the cellular walls of the mesophyll, the pedicel and the vesicular end cells; therefore, it is an apoplasmatic mode of transport;
3. On the other hand, Cl^- and Na^+ ions move along in the cytoplasm, going through plasmodesmata which occur in great number in the transverse walls of the pedicel (Plate 9.11 A–C and Fig. 9.16); this is a symplasmatic mode of transport;
4. Before accumulating in the vacuole of the apical cell, the ions which penetrate into the wall of this cell by the apoplasmatic route pass through the plasmalemma, which is often very rough. The author has shown that the penetration is

---

Plate 9.12 A–D. *Atriplex halimus* L., salt secretory trichome.

A Na^+ ion detection by potassium pyroantimoniate treatment; the sodium antimoniate precipitate is concentrated in the vacuole *V* of the differentiated vesicular apical cell; × 17000. Fixation: glutaraldehyde - potassium pyroantimoniate - O_SO_4.

B–D Lanthanum nitrate circulation in the pedicel and the vesicular cells. B Strong lanthanum labelling of the cell wall between these two cells, and penetration into the cytoplasm by means of pinocytosis vesicles *pv;* × 45500. C Plasmalemma *(pl)* invaginations of the vesicular cell, in which parietal lanthanum precipitates penetrate; *er* endoplasmic reticulum profiles, coming in contact with plasmalemma invaginations; × 48000. D Young pedicel cell; thin lanthanum precipitates are visible in Golgi vesicles *gv* and in small bodies resembling lysosomes *l;* × 35000. Fixations: glutaraldehyde - paraformaldehyde - O_SO_4, after in vitro incorporation of lanthanum nitrate

(Courtesy of A. SAMOUI 1975)

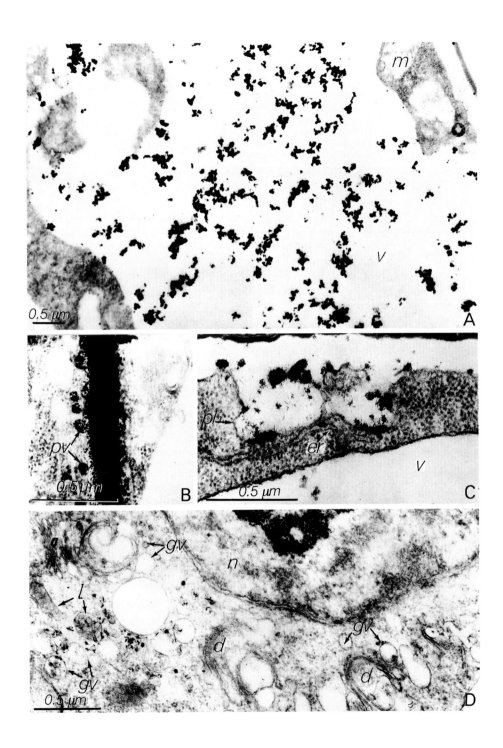

0.5 μm

m

V

A

pv

0.5 μm

B

pl

er

V

0.5 μm

C

gv

l

n

d

gv

d

gv

0.5 μm

D

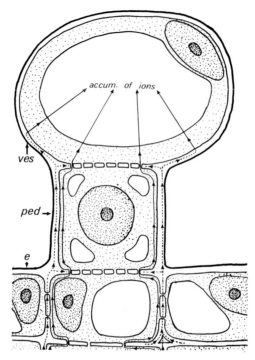

Fig. 9.16. Routes of ionic transport in cells of the vesicular trichome of *Atriplex halimus* L. *e* Epidermis; *ped* pedicel; *ves* vesicular cell; *dotted lines* apoplasmatic route; *unbroken lines* symplasmatic route; (after SMAOUI 1975)

partially achieved by pinocytosis (Plate 9.12 B, C); pinocytosis vesicles are then captured by the vacuoles. This process does not exclude the existence of penetrations induced by mechanisms of active transport; thus, ions could be carried away to the vacuole through the endomembranous system including the Golgi apparatus (Plate 9.12 D and Fig. 9.17).

Furthermore, in cells which excrete their products in a living state, the secretate granules are conveyed to the plasmalemma, against which they accumulate (Plate 9.9 A, B) whether they are free in the hyaloplasm or included in a cytomembrane derived from the endoplasmic reticulum, usually smooth, or from the dictyosomes. But the mechanisms permitting the passage of these substances out of the plasmalemma are not as yet clearly established. They are certainly very variable and SCHNEPF (1969) distinguished several modes of secretions:

1. *The granulocrine mode:* secretion products are kept in vesicles produced by the emergences of the endoplasmic reticulum or of the Golgi apparatus where they may be synthesized. These vesicles are in contact with the plasmalemma and their content passes into the periplasmic space through an exocytosis similar to that occurring in wall elaborations (according to VIAN and ROLAND 1972, see p. 208). The stigmatic secretion of *Verbascum phlomoides*, and to a lesser degree,

516

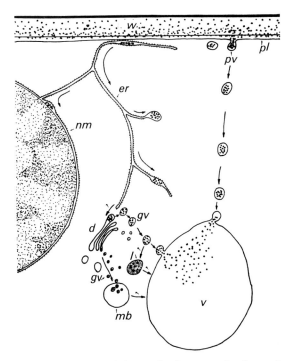

Fig. 9.17. Progression of lanthanum precipitates after incorporation into cells of an *Atriplex halimus* vesicular trichome. Lanthanum particles *(⬚)* pass through the plasmalemma *pl,* either in the form of pinocytosis vesicles *pv* or in passing into the endoplasmic reticulum *er,* according to a non-elucidated mechanism. They are carried away either towards dictyosomes *d,* or directly towards the vacuole *v.* In the first case, Golgi saccules discharge these particles by means of Golgi vesicles *gv,* later incorporated into lysosomes *l* or directly into the vacuole. On the other hand, Golgi vesicles loaded with hydrolases merge into multivesicular bodies *mb* which bring these enzymes to the vacuoles. Finally, the vacuole concentrates the precipitates (see Plate 9.12 A). *nm* Nuclear membrane, sometimes containing lanthanum precipitates; (after SMAOUI 1975)

    that of *Forsythia intermedia* could be of a granulocrine mode insofar as the secretion is partly polysaccharide (DUMAS 1975).

2. *The eccrine mode:* in most cases, the passage of the exudate through the plasmalemma, as well as the above mentioned transfers between cellular components, cannot be observed with the electron microscope. In that case, one can think that this passage occurs in the state of scattered molecules, either passively if there is a concentration gradient or by means of transport agents, i.e. active transport. According to DUMAS (1974c), this type of excretion is prevalent in the stigmatic secretory cells of *Forsythia intermedia.* In fact, in species in which holocrine and eccrine modes of excretion are combined, the exudates secreted through one mode or the other may be of a different nature.

3. *The holocrine mode:* the secretate is eliminated by the removal of cell fragments or through the thorough degeneration of entire cells. This is the case of the sodium chloride secretory trichomes of the *Atriplex halimus* leaf (SMAOUI 1971, 1975).

9.9 Isolated Secretory Cells

In the above mentioned structures, the secretion products of the secretory cells are released from the protoplasm. In other cases, non-utilized substances are kept inside the cells as long as they live. Thus, isolated cells become secretory but there are no intercellular spaces around them to retain their products.

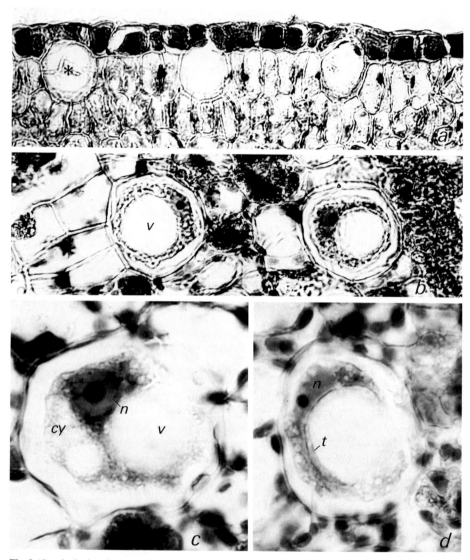

Fig. 9.18 a–d. Isolated "essential oil" secretory cells. **a** *Citrus* leaf (lemon tree), subepidermal cells (one of them marked with an *asterisk*) interrupting the palisade parenchyma. **b** *Laurus nobilis* L.; leaf; essence secretory cells in the spongy mesophyll. **c,d** Cytological aspects of essence secretory cells of *Laurus nobilis* leaf; *cy* spongy cytoplasm; *n* nucleus; *t* tonoplast; *v* large central vacuole. Fixation: REGAUD, staining: hematoxylin

518

In some cells, the secretion products are stored within the vacuome where emulsions occur if they are hydrophobic, whereas others retain them in the cytoplasm.

These latter cells are mainly *oil cells*. In most cases these cells are morphologically differentiated along with their chemical differentiation. They are scattered in various parenchyma tissues (mesophyll, cortical, parenchymas, fruits, seeds, etc.). They differ by a larger size than the neighbouring cells, by a thickened cell wall in some cases and a round outline. In addition, they include very refringent globules of essential oils which are combinations of terpene carbides of low molecular weight, mainly $C^{10} H^{16}$ and $C^{15} H^{24}$ and by-products (Fig. 9.18).

They occur in Aroideae, Piperaceae, Radiae and Compositae and they characterize quite well a group of Ranales, including particularly the Magnoliaceae, Calycanthaceae, Myristicaceae and Lauraceae. Among Lauraceae, in the *Laurus camphora* species, cells elaborate a terpene cetone, the camphor.

9.10 Tanniniferous Cells

Tanniniferous cells are so frequent that one can hardly refer to them as secretory cells. However, in parenchymatous cells, isolated or in small chains, the vacuole may be filled with tannins contrarily to the vacuole of neighbouring cells (medulla

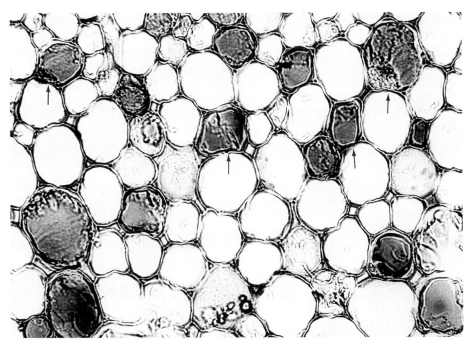

Fig. 9.19. Isolated tanniniferous cells (idioblasts) in the leaf parenchyma of *Ficus elastica (arrows)*. Fixation: REGAUD, staining: hematoxylin

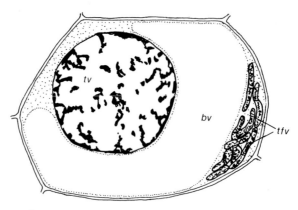

Fig.9.20. *Mimosa pudica,* motor swelling cell, containing one banal vacuole, without tannin, *bv,* one globulous tanniniferous vacuole, *tv,* in which the tannin is precipitated, and several filamentous tanniniferous vacuoles *tfv,* in the peripheral cytoplasm; (after G.MANGENOT 1927)

of *Rosa* stem, *Ficus* mesophyll, Fig.9.19). Moreover, these products seem to stay unused as the cell reaches the end of its development and large amounts frequently accumulate in dead tissues (wood core, cork). They occur frequently in cells with a decreasing proliferating activity as well as in tissues with a high physiological activity (leaf mesophyll). Most living cells may store tannins, which are put aside in vacuoles in the shape of colloidal pseudosolutions. Tanniniferous cells are easily stained cherry red by neutral red. Some cells contain both vacuoles which are tanniniferous and vacuoles deprived of tannins (cells of the motor organs of *Mimosa pudica;* Fig.9.20). Most cytological fixatives precipitate tannins as large precipitates either spherical or in the shape of the vacuole or yet as aggregates of small granules, in particular with chromed fixatives which stain them brown-yellow (Figs.9.19 and 9.20).

Differentiation of Tanniniferous Cells. In the cells of root extremities of *Euphorbia characias* (MARTY 1974) the polyphenolic secretions occur in provacuolar vesicles (see MARTY 1973a) derived from dictyosomes. These provacuoles incor-

──▷

Plate 9.13A, B. *Euphorbia characias,* differentiation of tanniniferous root cells.

A Meristematic cell at the onset of formation of the provacuolar tubulo-vesicular system; small dense spiny vesicles *(arrows)* penetrate into light larger ones, which constitute *"multivesicular bodies",* mb, in the vicinity of Golgi zones *gz.* These vesicles merge, giving birth to provacuoles *pv,* some of which soon show tannin reactions. During the course of differentiation, the tanniniferous provacuoles *tpv* are converted into vacuoles; *m* mitochondria; *n* nucleus; *p* plastids; × 10500.

B Arrangement of tanniniferous provacuoles *tpv* into a sequestration sheet; around this provacuole group, several tanniniferous enclaves, *ten,* may be interpreted either as resulting from the fusion of provacuoles or as young vacuoles; *d* dictyosomes producing light vesicles and spiny vesicles with a dense outline *(arrows); m* mitochondria; *pr* proplastid; × 35500.

Fixations: paraformaldehyde - glutaraldehyde - O_sO_4; contrast; $KMnO_4$; (courtesy of F.MARTY 1974)

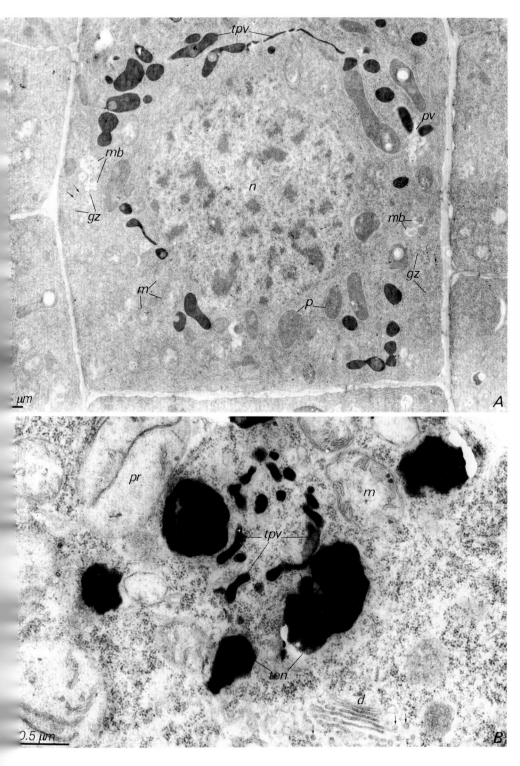

porate alveolate (or spiny) Golgi vesicles (BUVAT 1981); their content becomes denser and denser. They quickly acquire the osmiophilic features of tannins (Plate 9.13 A).

Tanniniferous provacuoles develop as other provacuoles; they become arranged in sequestration sheets forming autophagic vacuoles. They become tanniniferous vacuoles through the usual processes (Plate 9.13 B).

Tanniniferous vacuoles of lytic origin are usually significantly enlarged by tanniniferous provacuoles which do not undergo the autophagic phase and merge with them directly (MARTY 1974; Plate 9.13 B).

9.11 Pseudolaticifers

9.11.1 Concept

The secretory systems called pseudolaticifers are closer to the former secretory cells, of which they have many features, than to the real laticifers, which we shall describe further.

The pseudolaticifer cells are indeed cells of a parenchyma type, specialized in the making of *hydrophobic* products, usually retained in the cytoplasm, which looks like an emulsion.

9.11.2 Anatomical Distribution and Arrangements

These cells are either scattered in the ground parenchyma (Fumariaceae) or they constitute lines and networks when they are in contact by their extremities or laterally. Finally, the intermediate cell walls of the secretory cells sometimes resorb, and the cellular contents anastomose. These possibilities permit one to distinguish several morphological types of pseudolaticifers. For instance, the *Allium* pseudolaticifers constitute lines of longitudinal cells that do not anastomose; these are called *"non-anastomosed or non-articulated pseudolaticifers"*. In contrast, the pseudolaticifers of the Chicoraceae parenchyma (*Cichorium, Taraxacum, Sonchus,* etc.) form a network due to the anastomosis of the cells, these are called *"anastomosed in network or articulated pseudolaticifers"*.

Between these two forms of pseudolaticifers, we find the pseudolaticifers with cells, forming lines, connected only by their terminal cell walls, forming parallel tubes without lateral anastomoses *("anastomosed in file pseudolaticifers")*.

The intercellular communications are often made by perforations that resemble the phloem sieve areas (Fig. 9.21; *Chelidonium*), otherwise the intermediate walls disappear completely (Chicoraceae, in general).

The cell walls of the pseudolaticifers are generally as thin as those of the adjacent parenchyma; they can, however, be slightly thicker, but remain cellulosic. When the cells anastomose, the vacuoles sometimes become connected in a central tube, but, in most cases, the vacuolar system remains divided into multiple vacuoles (*Lactuca sativa;* GIORDANI 1981b).

522

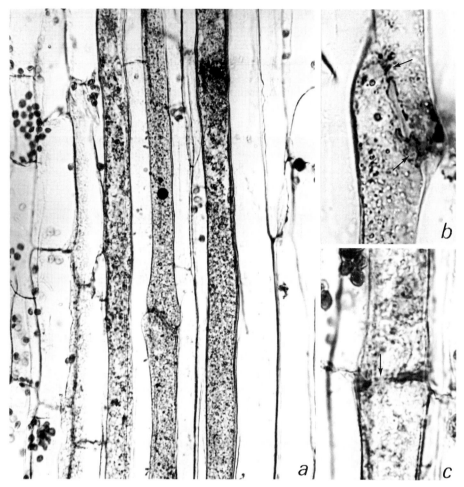

Fig. 9.21 a–c. *Chelidonium majus,* pseudolaticifers; secretory cells anastomosed in files, in the petiole periphloemian parenchyma. **a** Topographic aspect; **b,c** perforated terminal cell walls *(arrows)* between two laticifer cells. Fixation: REGAUD, staining: hematoxlin

The latex composition of these systems varies, depending on the species, but is relatively close to the latex of the true laticifers; this will be examined after a discussion of the latter.

9.11.3 Origin and Mode of Growth

The development of pseudolaticifers has been observed among several Chicoraceae, *Lactuca* and *Taraxacum,* in *Hevea brasiliensis, Papaver somniferum* and *Manihot* (bibliography in GIORDANI 1981).

The initials of pseudolaticifers seem to be lacking among non-germinated seeds of several species: *Taraxacum koksaghiz* (SAVCHENKO 1940), *Papaver somni-*

ferum (CLAUTRIAU 1889; MIKA 1955; SARKANY et al. 1970). They are recognized in the hypocotyl and the cotyledons of other Chicoraceae (SCOTT 1882; BARANOVA 1935) as longitudinal rows of cells, but their end walls are still intact; and in the cotyledons of the seed of *Hevea brasiliensis* (BOBILIOFF 1923).

The laticifer cells become distinct from the first stage of germination; the end walls are lysed and the pseudolaticifers are extended towards new tissues, through differentiation of new cells generated by the apical meristems.

A system of *primary laticifers* is then developed preferably in the surrounding tissues and outside of the primary phloem. Other groups of pseudolaticifers are produced in the secondary phloem; they result from the evolution of long cambial initials similar to those of the sieve cells, and this suggests that they could come from sieve cells transformed into secretory cells, a hypothesis that has not been confirmed.

Possibly, these *secondary pseudolaticifers* grow through resorption of the end walls or through lateral perforations followed by mergings of protoplasm (*Tragopogon*, SCOTT 1882, *Taraxacum*, ARTSCHWAGER and McGUIRE 1943). There seems to be no anatomical relationship between the primary and secondary pseudolaticifers. The latter are usually extended in the phloem of all the plant organs.

9.11.4 Systematic Repartition

There is not much difference between the *"non-anastomosed"* pseudolaticifers and the *anastomosed in network or file pseudolaticifers*. All the intermediate features are encountered and also certain families appear to develop several types of pseudolaticifers. The Papaveraceae are a good example:

1. Isolated cells, therefore without anatomosis: *Glaucium*.
2. Anastomosed cells in files (perforations): *Chelidonium*.
3. Anastomosed cells in network: *Papaver*.

A comparison of the pseudolaticifers in the representatives of the same family suggests possible series of increasing specialization and reveals a phylogenetic significance.

In 1884, DE BARY noticed that in the Aroideae, certain species appear to lack laticifers, others have longitudinal rows of cylindrical cells, with no perforations, and still others contain pseudolaticifers with perforations. Similarly, the Cruciferae also have idioblastic cells resembling laticifers (SPERLICH 1939), but some authors deny this assimilation as laticifers because their content cannot be properly called latex. The Fumariaceae, closely allied plants, have similar cells that appear in certain species to intergrade progressively with the laticifers of the Papaveraceae.

So, the phylogenetic relationship between these three families is also manifested in the more and more prominent specialization of their secretory system. These considerations also show that it is difficult to determine the level differentiation from which an isolated secretory cell can be classified as pseudolaticifer.

Apart from these families, we find pseudolaticifers mainly in the following:

1. *Pseudolaticifers in lines* (intact or anastomosed cells), Convolvulaceae *(Convolvulus, Ipomea)*, Sapotaceae *(Achras sapota, Palaquium* or *Gutta* tree), Liliaceae *(Allium)*, Musaceae *(Musa)*, etc.
2. *Articulated anastomosed pseudolaticifers:* Campanulaceae, Compositae, Cichoraceae, Caricaceae *(Carica papaya)*, Euphorbiaceae *(Hevea*, see Fig. 9.29, *Manihot)*.

9.11.5 Infrastructures of Pseudolaticifers

The infrastructures of the pseudolaticifers have been observed on the *Hevea*, various species of the Chicoraceae: *Taraxacum, Cichorium intybus, Lactuca sativa* and on *Papaver somniferum* (references in GIORDANI 1981b).

9.11.5.1 Vacuolar Apparatus

The vacuome of the species mentioned above remains continuously in a state of multiple enclaves which do not unite in a single vacuole as in the true laticifers (Plate 9.14 A). These vacuoles would be the sites of alkaloid synthesis such as morphine of *Papaver somniferum* (DICKENSON and FAIRBAIRN 1975), the sanguinarine and helerytine of *Chelidonium majus*, which stains the latex of this species yellow (MATILE et al. 1970).

Some data suggest that Golgi vesicles migrate to these vacuoles (THURESON-KLEIN 1970) which is consistent with the data of the cytochemical study of their content. In *Chelidonium*, MATILE et al. (1970) detected a series of hydrolase activities which permit the assimilation of these vacuoles with *lysosomes*, like common vacuoles, to which these enzymes are brought through the Golgi apparatus (COULOMB and COULOMB 1973). Moreover, at the onset of laticifer differentiation, autophagic phenomena occur in the vacuoles and take part in this differentiation.

The more extensive studies concerning pseudolaticifer vacuoles deal with the *lutoids* of the latex of *Hevea*. These spherical, yellow particles of about 2 to 10 μm in diameter are positive to vacuolar vital staining (RUINEN 1950). They maintain an acid pH in accumulating organic acids such as citric acid (RIBAILLER et al. 1971). Numerous acid hydrolases characteristic of lysosomes have been detected along with the phenomenon of enzymatic latency of lysosomes. They do not seem to normally assume autophagic processes and have therefore been considered as primary lysosomes (PUJARNISCLE 1971). Their colour would be due to carotenoids. Electron microscope preparations of young laticifers show helical microfibrils in lutoids. These microhelices are grouped in bundles; they are composed of an acid protein and seem to be lacking in the lutoids of old laticifers (ARCHER et al. 1963), their significance is not established.

In very young laticifer elements of *Lactuca sativa*, vacuoles result from sequestration processes of cytoplasm fragments by provacuolar vesicles of Golgi origin, followed by autophagic processes of the isolated content. At this moment, dictyosomes are significantly active. Later, cytoplasm protrusions may be ingested and lysed by the vacuoles (GIORDANI 1981 b).

9.11.5.2 Other Organelles

All the other usual organelles can be found in pseudolaticifers. There are more mitochondria in young elements. Plastids are poorly structured, without thylacoids but contain an osmiophilic inclusion in *Papaver somniferum* (NESSLER and MAHLBERG 1979) or several in *Lactuca sativa* (GIORDANI 1981 b; Plate 9.14 B–E). The latter are lipid materials surrounded by a simple membrane (possibly a crista). The *Hevea* latex includes particles which have been variously interpreted, the "particles of Frey Wyssling" which are composed of lipids and carotenoids and are limited by a double membrane, features which are usually found in chromoplasts (MARTY 1974). The cytoplasm of *Lactuca* pseudolaticifers also includes peroxisomes (GIORDANI 1981 b).

Nuclei assume a lobed shape in some cases; they do not seem to divide but may pass from one cell to the other through lateral anastomoses and there may be several nuclei in one cell. Phenomena of nuclear degeneration along with diffusion of the nucleolar material may involve part of the nuclei of the laticifer system (BRUNI et al. 1974).

9.11.5.3 Secretion Grains and Latex

Secretion grains of pseudolaticifers are apparently synthesized within the cytoplasm where they are stored. Already in 1926, using double staining with neutral red and indophenol blue, POPOVICI showed that the secretion grains of the *Chelidonium majus* laticifers remain in the cytoplasm and are well distinguished from the numerous vacuoles (Fig. 9.22). This localization was confirmed on an ultrastructural scale by GIORDANI (1979, 1981 b; Plate 9.15 A, B). *Lactuca sativa* grains of secretion are mainly lipidic (GIORDANI 1981 b) but the composition of the latex of pseudolaticifers is very variable, with a combination of lipids and hydrophobic

---

▷

Plate 9.14 A–E. *Lactuca sativa,* ultrastructures of articulate pseudolaticifers (**I**).

A Vacuolar apparatus, remaining at the state of numerous non-confluent vacuoles *v;* cytoplasm dispersed between the vacuoles and scarce; × 11 500.

B Young plastid *p,* little structured, enclosing a voluminous osmiophilic inlusion; *m* mitochondrion; × 37 000.

C Lipase enzymatic disgestion of the osmiophilic plastidal inclusion *(asterisk)* showing its lipid nature; × 37 000.

D Scattering of the osmiophilic substance into multiple granules and tubules, surrounded by a membrane, as if they where inside the plastidal cristae; × 37 000.

E Stretched plastid, frequent in pseudolaticifers, where they are often distorted; × 37 000.

Fixations: glutaraldehyde – paraformaldehyde – OsO_4; contrast according to REYNOLDS (1963); (courtesy of R. GIORDANI 1981 b)

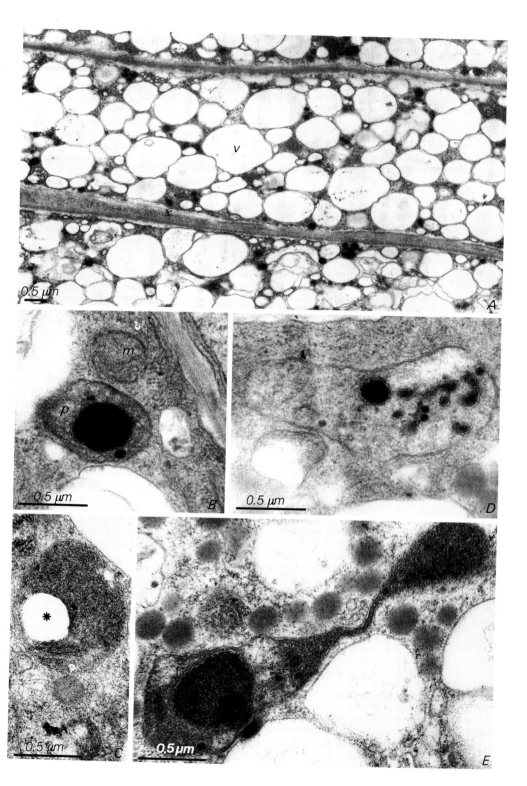

527

Fig. 9.22. *Chelidonium majus,* pseudolaticifer, double vital staining with neutral red-indophenol blue. Vacuolar apparatus divided into numerous small vacuoles *v,* stained with neutral red, and essence droplets *e,* within the cytoplasm, stained with rising indophenol blue; (redrawn after H. Popovici 1926)

terpene compounds: rubber, resins, essential oils, carotenoids. Moreover, numerous substances in solution can be found, mainly in vacuoles: hydrolases, peroxidases, catalase, alkaloids, proteases such as the papain of *Carica papaya.*

In fact, the latex released by pseudolaticifers when they are cut open is composed of the whole protoplasm together with its inclusions and is not of a pure vacuolar content, as in true laticifers.

9.11.5.4 Mechanisms of Cell Wall Perforation

The so-called "articulated" pseudolaticifers consist of cells which anastomose after partial resorption of the walls. Perforations occur on end walls through partial or complete resorption in anastomosed "in file" pseudolaticifers; they occur on both end and lateral walls in "anastomosed in network" laticifers.

Perforations either on end or lateral walls begin by a thinning of the wall (Plate 9.15 C–E), sometimes from a plasmodesma (KARLING 1929; SASSEN 1965), however, the presence of a plasmodesma is not necessary or usual (VERTREES and MAHLBERG 1978). GIORDANI's works (1977, 1979, 1980, 1981a) showed that the parietal perforations imply a significant activity of the plasmalemma and numerous

---

Plate 9.15 A–E. *Lactuca sativa,* ultrastructures of articulate pseudolaticifers (II).

A Localization of secretion grains *sg,* loose in the cytoplasm; *v* multiple subspherical vacuoles, forming a scattered vacuolar apparatus; × 20000.

B Accumulation of osmiophilic secretion grains in the pseudolaticifer central cytoplasm; × 11000.

C Separating terminal cell wall between two secretory cells, before its perforation; the future lysed zone becomes thinner, but is still very reactive to the acid polysaccharide test, according to THIÉRY (1967); the plasmalemma *pl* is already puffed up; × 12800.

D Accentuated thinning of the perforation zone, and accentuation of plasmalemma blisters, which seem to fall down into the cytoplasm in the shape of vesicles *(arrows); m* mitochondria; *n* nucleus; *v* vacuole; × 20000.

E Annular perforation of the terminal cell wall *tw;* the detached fragment is still visible in the cytoplasm *(asterisks),* but is only weakly reactive to the polysaccharide test; × 20000.

Fixations: glutaraldehyde – paraformaldehyde – OsO_4; contrasts: $KMnO_4$ (**A, B**); silver proteinate, according to THIÉRY's technique (1967; **C, D** and **E**); (courtesy of R. GIORDANI 1981b)

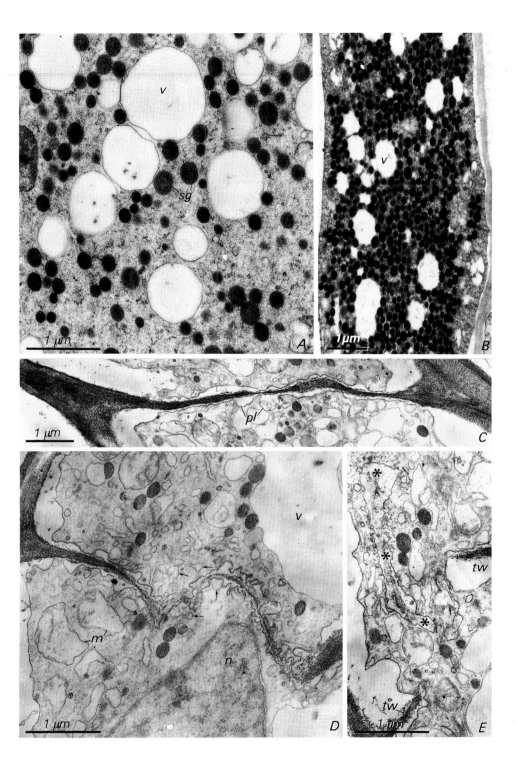

529

enzymes. The pectic substances of the middle lamella are the first to be eliminated, probably by pectinolytic phosphatase (Plate 9.16 A–C); acid and neutral ATPases, thiolacetic esterase, etc. are clearly stimulated as well as the peroxidase activity (Plate 9.16 D–E). It is still impossible to determine the role of these enzymes in the degradation of the wall, but these activities are similar to those acknowledged by several researchers in the processes of leaf abscission, perforation of the phloem sieve areas and parietal destruction during the differentiation of the vessel elements. All these activities could have a common explanation (GIORDANI 1981b).

The plasmalemma remains in contact with the cell wall throughout the destruction; it forms several sacs which encompass parietal remnants and become isolated in the cytoplasm in the shape of vesicles (Plate 9.16 G, H). The latter progress in the hyaloplasm and are finally sequestrated by vacuoles (Plate 9.17 A, B). In the vesicles and vacuoles in which they are enclosed, a cellulolytic activity has been detected, whereas such an activity has not been noted at the level of the wall in situ (GIORDANI 1980, 1981a).

The parietal enzymatic equipment may be provided by Golgi vesicles which become capable of anastomosing with the plasmalemma (VIAN and ROLAND 1972). Generally, they take part in the regeneration of this wall and restore the parts destroyed by the formation of the endocytosis vesicles. Here, they participate in the differentiation of the pseudolaticifers and thus, this differentiation is an example of the relationship between endocytosis and exocytosis processes, as well as of a general characteristic of the vacuoles, i.e. the capture and digestion of the products yielded through endocytosis (see DE DUVE 1974).

Among enzymes which intervene in parietal lysis of pseudolaticifer walls, a strict β-fucosidase has been recently isolated and characterized. [GIORDANI R and NOAT G (1988) Isolation, molecular properties of a strict β-fucosidase from *Lactuca sativa* latex. Its possible role in the cell wall degradation of articulated laticifers. Eur J Biochem, in press]

Plate 9.16 A–H. *Lactuca sativa,* pseudolaticifers; cytochemistry of the terminal cell wall perforation (I).

A Acid phosphatase activity, particularly intense on the ring of a future perforation; × 7600 (GOMORI's technique, 1952).
B,C Acid phosphatase activity at the level of an annular perforation (B) and resorption of the parietal disc, isolated in the cytoplasm (C); × 5000 (B) and 5600 (C) (GOMORI's technique, 1952).
D Peroxidase activity at pH 7.6 in the terminal cell wall, to be noted on the edge, the reactivity of the plasmalemma; × 16000.
E Intensification of the peroxidase activity in the thinned cell wall; to be noted on the edge, the reactivity of the plasmalemma; × 16000.
F Perforation limited by a parietal ring, strongly positive to the peroxidase test, according to GRAHAM and KARNOWSKY (1966); × 10400.
G Thinned terminal cell wall just before its perforation; numerous polysaccharide "parietal vesicles" *pv* in the neighbourhood; × 32800; THIÉRY's technique (1967) with silver proteinate.
H After the perforation; concentration of "parietal vesicles" *pv* near terminal cell wall remains; *m* mitochondria; *vtw* vestigial terminal cell wall; × 32800; same technique as for G.

(Courtesy of R. GIORDANI 1980, 1981a, b)

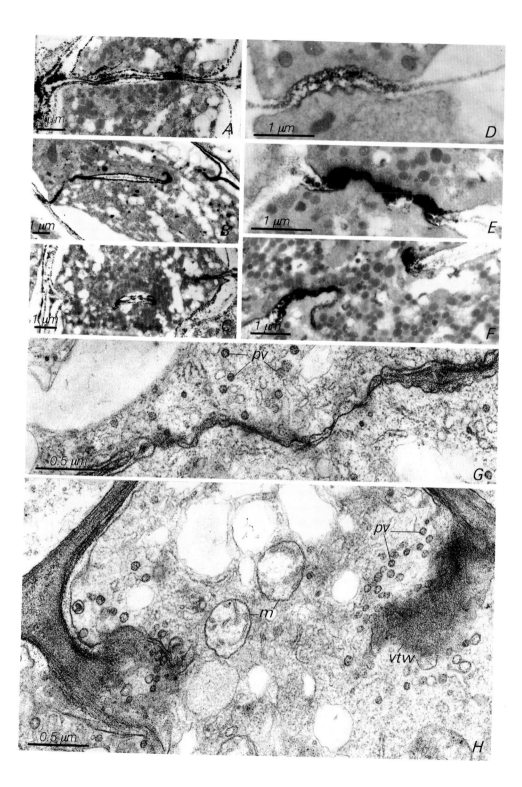

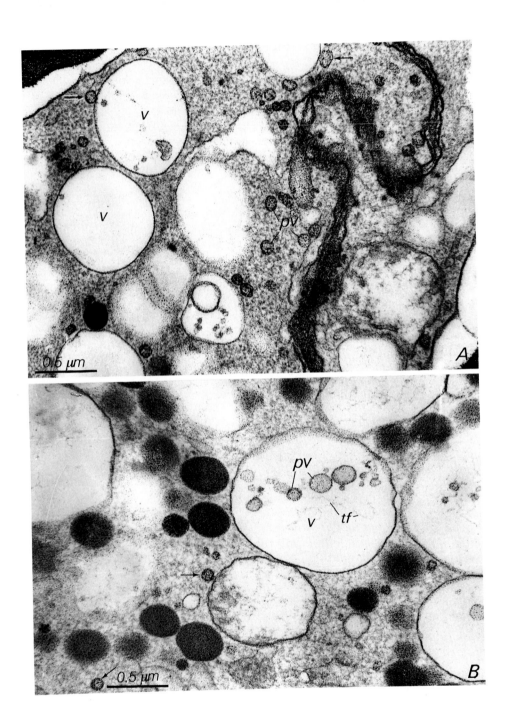

A

0.5 µm

v

v

pv

B

pv

v   tf⁻

0.5 µm

9.12 True Laticifers

9.12.1 Origin and Development

True laticifers are best characterized by their origin and development. In fact, cells are already recognizable in the embryo before seed maturity. These primordia are arranged in circles in the nodal plane of cotyledons; each species has a determined number of cells, e.g. *Euphorbia engelmanni,* 4; *E. peplus* and *E. exigua,* 8 (MAHLBERG and SABHARWAL 1968); *Euphorbia marginata,* 12 (CAMERON 1936); *Asclepias curassavica,* 16 (CHAUVEAUD 1891); *Nerium oleander,* 28 (MAHLBERG 1961).

When the seed is mature, the embryo already includes a significant system of laticifers permeating the whole embryo (Fig. 9.23).

Each laticifer of this system results from the development of one primordial cell which has grown while the nucleus mulitplies without dividing. Each one consists of a tubular system, branched in all organs, which may intergrade with the branchings generated by another primordial cell, but which *remains independent,* i.e. no anastomosis occurs between the productions of each primordial cell. Later, during the development of the plant derived from this embryo, each laticifer keeps on growing and branches penetrate the tissues in all organs. By repeated karyokineses, not followed by cytodiereses (TREUB 1880; MAHLBERG 1959) multinucleated systems occur with a tube structure capable of reaching several kilometres in length. At the onset of development, these huge multinucleated cells are typically independent and *do not anastomose.*

Thus, the development of laticifers implies a significant intrusive growth. The tips extend to the vicinity of the meristem apices, growing in harmony with the elongation zone of the organs.

The mechanisms of laticifer intrusive growth is probably enzymatic rather than mechanical and implies lytic processes in the middle lamella of the cells permeated by laticifers (MAHLBERG 1963). The presence of an acid pectinase (optimum: pH 5.2) in the latex of *Asclepias syriaca* gives ground for this hypothesis (WILSON et al. 1976).

A precocious origin in the course of embryogeny together with a mode of development which preserves the independence of each laticifer are considered as characteristic of true laticifers. However, these features do not appear to be quite generalized. Thus, in the *Vinca* or *Cannabis* genera, in the young stem, primordia

◁————————————————————————————————

Plate 9.17 A, B. *Lactuca sativa,* pseudolaticifers; cytochemistry of the terminal cell wall perforation (**II**).

A Scattering of parietal vesicles *pv* containing polysaccharides in the cytoplasm where they progress towards the vacuoles *v (arrows).*

B Penetration of parietal vesicles *pv* into vacuoles, where they are lysed; *tf* thin filamentous polysaccharide threads in the vacuole; other polysaccharide vesicles are still to be found in the cytoplasm *(arrows);* × 32 500.

Fixations: glutaraldehyde – paraformaldehyde – O_5O_4; acid polysaccharide test according to THIÉRY (1967); (courtesy of R. GIORDANI 1981 a)

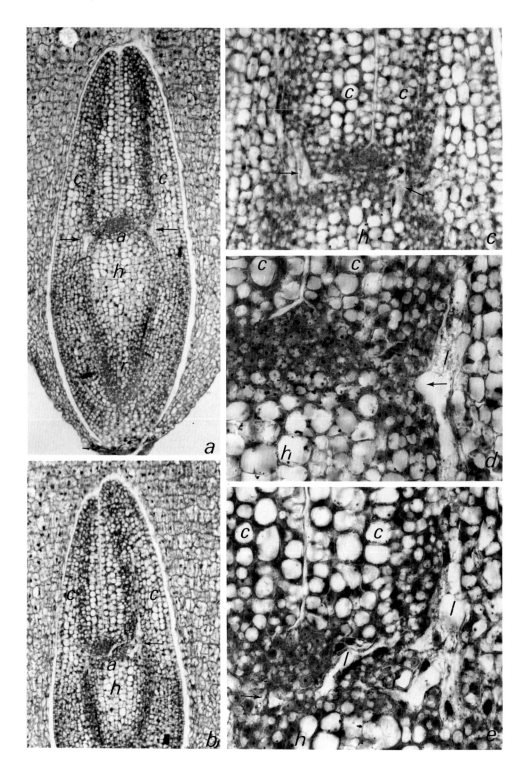

534

Fig. 9.23 a–e. *Euphorbia serrata,* embryo; origin and onset of development of laticifers from the plantlet cotyledonary node. **a, b** Longitudinal sections of the embryo showing the start of laticifers *(arrows)* beside the apical meristem *a.* **c, d** Development of laticifers, spreading out of the cotyledonary node *(arrows).* **e** Enlargement of the apical area of **b.** *c* Cotyledons; *h* hypocotyl; *l* extension of laticifers towards the cotyledons and the hypocotyl; *r* radicle. Fixation: REGAUD, staining: hematoxylin

◁——

have been singled out only after germination. They only produce simple, *non-branched* laticifers, which become multicellular in a few species only. Moreover, beneath apical meristems and in developing leaves, *new independent* laticifer cells occur constantly. The distinction between true laticifers and pseudolaticifers, which seems obvious when comparing the systems of *Euphorbia* with those of Chicoraceae, would not be so clear if the above mentioned structures were really intermediate. In fact, the cytological development and the relationship of the secretory products with respect to cytoplasmic structures would be the only data permitting the determination of doubtful cases in the classification of laticifers.

9.12.2 Differentiation and Cytological Structures

9.12.2.1 Ancient Works in Light Microscopy

Ancient works (see MOLISCH 1901) have not provided very accurate data on the relationship between the latex and the cytoplasm and have not clearly distinguished both types of laticifers.

In *Ficus carica* laticifers, POPOVICI (1926) showed that the cytoplasm is preserved as a thin pellicle against the wall, including numerous nuclei and mitochondria. It encompasses a large canalicular vacuole containing latex. Secretion droplets of oleo-resin occur in the cytoplasm and are released in the vacuole in which they constitute the scattered phase of the emulsion (Fig. 9.24). These structures were confirmed later (see FREY-WYSSLING 1935; MOYER 1937).

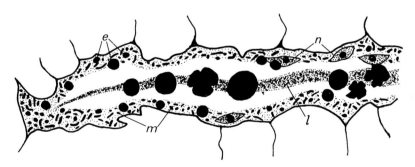

Fig. 9.24. *Ficus carica,* laticifer, longitudinal section. Peripheral cytoplasm, enclosing numerous mitochondria *m,* several nuclei *n* and oleo-resin droplets *e,* which are excreted into the central vacuole, where they constitute a part of the latex *l.* Fixation according to MEVES; staining: acid fuchsin; (drawn after H. POPOVICI 1926)

535

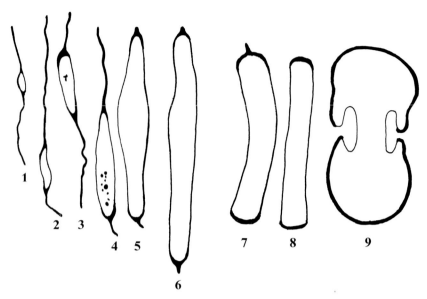

Fig. 9.25. Starch elaboration in *Euphorbia* laticifer plastids. **1-6** *Euphorbia heterophylla;* **1-3** concretion of a starch grain in a filamentous proplastid; **4,5** distension of the elaborating substance; **6** localization of this plastidal substance at the two extremities of the grain. **7,8** *Euphorbia peplus:* starch grains bearing a plastidal cap on the two extremities, where the starch grains grow over a longer time, hence their shapes. **9** *Euphorbia resinifera:* dumbbel-shaped starch grains; the elaborating plastidal substance is exclusively localized on the two swellings; (redrawn after G. MANGENOT 1925)

The laticifers of various *Euphorbia,* mainly of perennating species, differentiate amyliferous plastids in which the starch grain may assume unusual shapes (Figs. 9.25 and 9.26). These grains result from a prolonged elaboration, lasting several years in some cases. The distended plastidal stroma is localized on one part of the grains surface or is divided into several discontinued and polarized small masses under which the grain continues developing (MANGENOT 1925). At the end of their development, these amyloplasts may be released in the vacuole, like the secretion droplets.

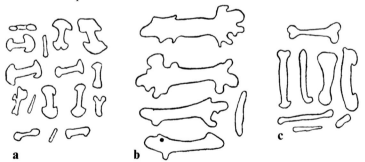

Fig. 9.26a-c. Starch grains elaborated in *Euphorbia* laticifer plastids, then released into the vacuole with the latex. Each group comprises young growing grains in the cytoplasm, and completed grains with characteristic shapes such as they are found in the latex. **a** *Euphorbia resinifera;* **b** *E. grantii;* **c** *E. mammilaris;* (in GUILLIERMOND et al. 1933; after MANGENOT 1930)

9.12.2.2 Cell Wall Differentiation

The walls of true laticifers of *Euphorbia* may be thicker than the walls of the adjacent parenchyma cells (Fig. 9.27). They contain cellulose and pectoses and maintain the features of a primary wall, capable of growing in unison with the organs. During growth, the cellulose microfibrils which, at first, have been arranged in helices with a tight pitch, therefore almost transversally oriented, are reoriented according to the schema of the "multinet type". The walls of the laticifers are highly hydrated and contain a high proportion of pectic substances and hemicelluloses (MOOR 1959).

The differentiation of the laticifers has been studied with electron microscopy in many species: *Nerium oleander* (MAHLBERG 1963); *Euphorbia marginata* (MAHLBERG and SABHARWAL 1968); *E. characias* (MARTY 1970 a, b, 1971 a, b, 1974); *Asclepias syriaca* (WILSON and MAHLBERG 1978); *Asclepias curassavica* (GIORDANI 1975, 1978, 1981 b); among others. The data obtained in the *Euphorbia*

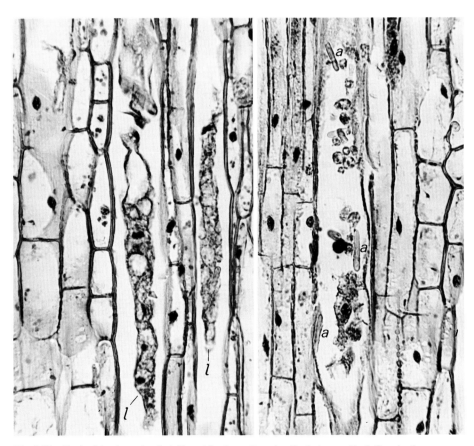

Fig. 9.27. *Euphorbia characias,* laticifers, histology. Longitudinal views of laticifers in the stem periphloemian parenchyma; *a* rodlike starch grains; *l* coagulated latex. Fixation: Navaschin, staining: hematoxylin

characias will be the main example, as F. MARTY carried out particularly extensive cytological and cytophysiological research in *Euphorbia characias;* in fact, these results should be considered as the basis of generalizations in other specimens.

9.12.2.3 Origin of the Vacuolar Apparatus

Most publications acknowledge the absence of a large vacuole in the laticifers of embryos, young germinations and the growing typs which, in adult laticifers, penetrate into the newly formed tissues of the leaf axis and buttresses. In this case, the cytoplasm is dense, of meristematic aspect, rich in ribosomes and encloses many small vesicles tentatively designated as vacuoles by some authors (WILSON and MAHLBERG 1978).

Publications also acknowledge the differentiated areas in which a huge, axial, undelimited vacuole shifts the protoplasm away towards the periphery. However, there is still discussion concerning the mode of passage from the meristematic state to the differentiated state. WILSON and MAHLBERG described the development of a system containing large elongated vacuoles which would derive from dilatations of the smooth or "degranulated" endoplasmic reticulum in *Asclepias syriaca.*

In *Euphorbia characias,* MARTY (1970a, 1974) considered the elongated enclaves, possibly similar to those of *Asclepias,* as systems of flattened cavities, or double membranes of sequestration, which isolate large protoplasmic areas. Firstly, independent vesicles occur, arranged in provacuoles of sequestration, then they become united and become autophagic vacuoles (Plate 9.18 A). They encompass part of the protoplasm, which will be lysed (Plate 9.18 and 9.19). After the dislocation of their internal membrane, the digestive enclaves become true vacuoles, the further fusion of which results in the large axial vacuole of the laticifer (MARTY 1970a, 1974). Thus, such a process would be a remarkable example of differentiation along with a significant autophagy, whereas, as mentioned above, more limited autophagic processes occur in the ontogenesis of pseudolaticifers (ESAU and KOSAKI 1975; GIORDANI 1979, 1981b). The autophagic origin of the meristematic vacuoles has now been acknowledged in the primary meristems, where these phenomena are only of smaller sizes (MARTY 1973a, 1978; BUVAT and ROBERT 1979, etc.).

───▷

Plate 9.18 A, B. *Euphorbia characias,* laticifers.

A Laticifer part in the course of differentiation; the axial cytoplasm is divided into islets, separated by double sequestration membranes *sm* (see Plate 9.20 **A**). In the islet of this figure, other double membranes *(arrows)* redivided the cytoplasm into smaller fragments, which will be lysed; × 38 000. Fixation: OsO_4; contrast: uranyl acetate – lead citrate.
B Longitudinal section of an area in the course of differentiation; all the axial cytoplasm *ac* is isolated by a large vacuole which will become an autophagic vacuole *av,* and it separates the axial cytoplasm from the peripheral cytoplasm *pc;* only the latter will persist after the lysis of the sequestrated part; × 10 500. Fixation: paraformaldehyde – glutaraldehyde; contrast: $KMnO_4$. (courtesy of F. MARTY 1970a)

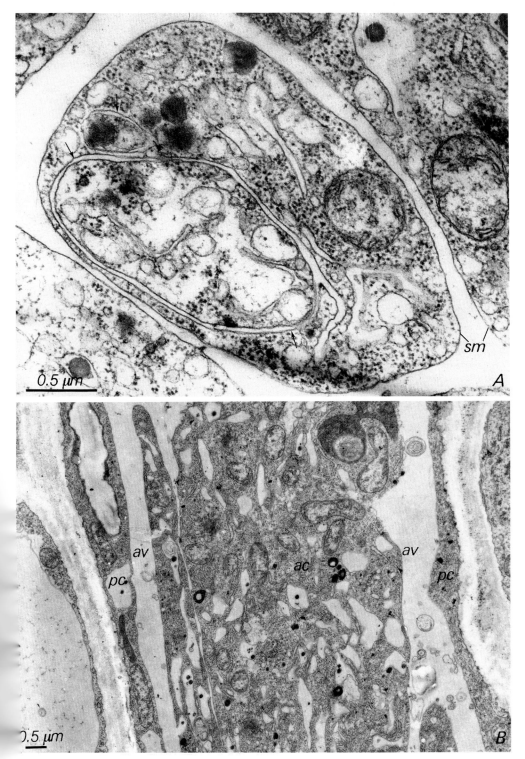

0.5 μm

sm

A

av

pc

ac

av

pc

0.5 μm

B

539

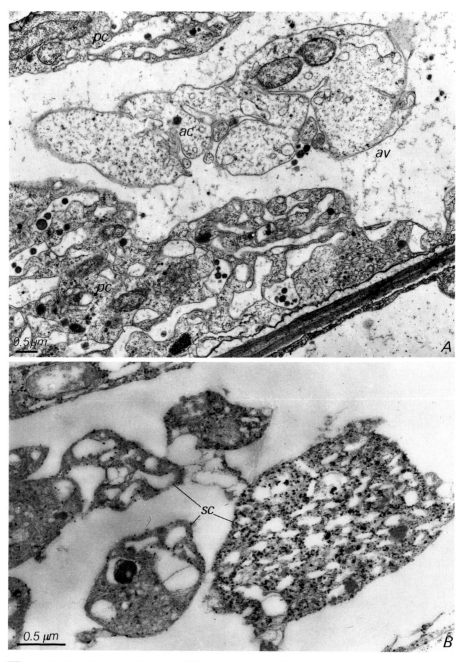

Plate 9.19 A, B. *Euphorbia characias,* differentiation of laticifers.

A Islet of axial cytoplasm, *ac,* sequestrated into the autophagic vacuole *av; pc* peripheral cytoplasm; ×12000. Fixation: O$_S$O$_4$; contrast: uranyle acetate – lead citrate.

B Test of acid phosphatase activity, showing opaque lead phosphate precipitates, which detect the presence of this hydrolase in the sequestrated and degenerating fragments of cytoplasm, *sc;* ×26000. Gomori's technique, 1952. (Courtesy of F. Marty 1970a, 1974)

Moreover, several overlapping systems of double membranes of sequestration develop in the sequestrated area of the laticifers of *Euphorbia* (MARTY 1970a, 1974; Plate 9.20 A). The repeated branching of the double membranes of sequestration is still more significant in *Asclepias curassavica* (GIORDANI 1981b) in which the internal membrane of the initial saccule generates more or less concentric, flattened layers which thoroughly divide the cytoplasm destined to disappear (Plate 9.20 B). The selective impregnation with osmium iodide-zinc-tetroxide (MAILLET 1963; MARTY 1973b) gives evidence of the sequestration feature of the concentric layers of double membranes (GIORDANI 1981b; Plate 9.21 A).

The autophagy of these systems is demonstrated by the acid phosphatase test which is positive in the sequestration cavities (GIORDANI 1981b; Plates 9.19, 9.21 B). Moreover, the vacuolar and lysosomal nature of the latex of *Asclepias curassavica* was confirmed by means of biochemical analysis which allowed the detection of several typical enzymes of lysosomes: acid phosphatase, RNAase, acid protease (GIORDANI et al. 1982).

These results might not be inconsistent with those obtained by WILSON and MAHLBERG (1978). The formation of vacuoles from the endoplasmic reticulum has been acknowledged in some cases (BUVAT 1971), sometimes denied (MORRE and MOLLENHAUER 1974) or limited to the formation of provacuoles and vacuoles of small size (WHALEY et al. 1964). But with regards to the *so-called* smooth reticulum, it is difficult to separate the typical endoplasmic reticulum from the *smooth* double membranes which belong to the GERL system (NOVIKOFF 1964; NOVIKOFF and SHIN 1964) and which take part in the productions of the Golgi apparatus, the whole set inducing the genesis of vacuoles with a lysosomal content.

9.12.2.4 Cytology of the Latex Secretions

The peripheral cytoplasm of the differentiated laticifers of *Euphorbia characias* contains many canaliculi with a more or less circular section, wholly parallel to the elongation of the organ and the laticifer and limited by a simple membrane. These canaliculi include polyterpenic granules. Other similar granules occur within the hyalophasm, mainly in the neighbourhood of dictyosomes (MARTY 1968b, Plates 9.21 C, D and 9.22 A). The tubules with a polyterpenic content are not in continuity with the endoplasmic reticulum (Plate 9.22 B); their membrane is different from that of the reticulum, but similar to that of clear vesicles without inclusions, sent out by the dictyosomes. Near the Golgi saccules, the existence of vesicles, presenting an aspect of transition between the large Golgi vesicles and the latex-loaded canaliculi, in which polyterpenic particles are captured, suggests that these canaliculi are homologous to the *provacuoles*.

Contrarily to their topographic independence regarding the endoplasmic reticulum, the canaliculi anastomose, through their membrane with the tonoplast of the large axial vacuole, in which they might release their content (Plate 9.22 A, *arrows*). This fact was checked by means of autoradiographic techniques using tritiated acetate, a precursor of polyterpenic molecules and L-leucine ^{3}H, a precursor of isopentenyl-pyrophosphate which is an intermediate of the terpene synthesis (MARTY 1974). After a period of exposition to the radioactive metabolite, the spec-

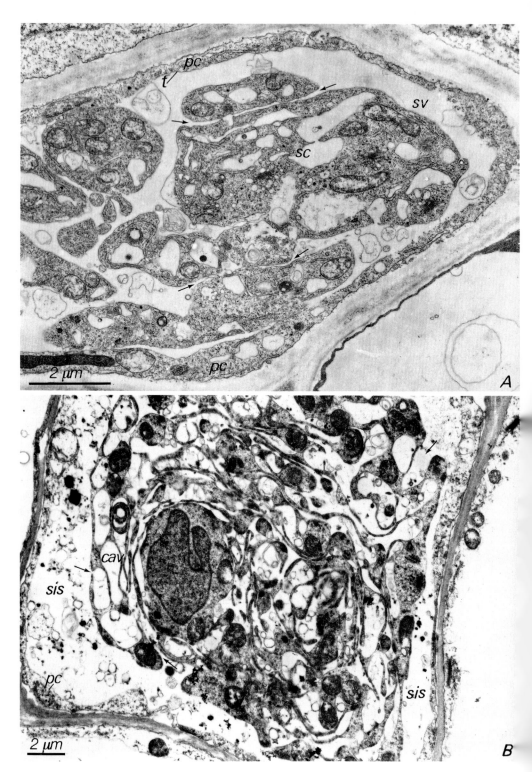

imens (root tips) are sunk into solutions of "cold" metabolites (non-radioactive). In summary, the author observed that the polyterpenic particles of the hyaloplasm, followed by those of the canaliculi and later by those of the axial vacuole are successively and progressively labelled, the labelling intensity increasing with the duration. One can conclude that latex granules are synthesized within the hyaloplasm, then incorporated in the canalicular system through a mechanism which involves the clear Golgi vesicles; latex granules are carried along by the canalicular system to the large vacuole where they are released.

9.12.2.5 Autophagic Phenomena in Differentiated Laticifers

Autophagic vesicles issued from dictyosomes occur in some areas of the peripheral cytoplasm of differentiated laticifers. Through a kind of internal pinocytosis, these provacuoles ingest fragments of cytoplasm including polyterpene particles and destroy them together with the hyaloplasm remnants and the captured ribosomes. Then such vesicles, with a more or less damaged content, are encountered in the axial vacuoles (MARTY 1971b; Plate 9.22 C).

Moreover, regularly, the tonoplast of this vacuole produces invaginations in which the cytoplasm penetrates, including various bodies, mainly mitochondria and leucoplasts. These cytoplasm protrusions are constricted at the base, isolated and fall into the vacuole where they are digested (Plate 9.23 E), resulting in the partial destruction of the perivacuolar cytoplasm and the encroachment of the tonoplast surface. This encroachment is restored by the membrane deposits resulting from the anastomoses of the tonoplast with the membranes of canaliculi when they release their content in the vacuole (MARTY 1974). These mechanisms represent a renewal of the tonoplast ensured through the provacuolar system, a process which has remained unknown up to now, but similar to the plasmalemma renewal through the membranes of polysaccharide Golgi vesicles (VIAN and ROLAND 1972).

Due to all the provacuolar and lytic deposits in the laticifer vacuole, in addition to polyterpene particles, the latex contains several substances producing an emulsion, the dispersion phase of which is very complex.

Plate 9.20 A, B. Sequestration vacuoles in the course of laticifer differentiation.

A *Euphorbia characias,* breaking up of the sequestrated axial cytoplasm *sc,* due to the confluence of internal double membranes with the membrane of the initial large sequestration vacuole *sv* *(arrows); pc* peripheral cytoplasm; *t* tonoplast; × 10900. Fixation: paraformaldehyde – glutaraldehyde; contrast: $KMnO_4$; (courtesy of F. MARTY 1970a).

B *Asclepias curassavica* L.; extreme dilaceration of the central cytoplasm of a laticifer, first isolated from the peripheral cytoplasm by a sequestration initial saccule *sis.* Here, the cytoplasm is crossed out by numerous anastomosed double membranes, which are connected with the initial saccule *(arrows); cav* cavities formed by swelling of the double sequestration membranes; × 5000. Fixation: glutaraldehyde – paraformaldehyde – O_sO_4; contrast: $KMnO_4$; (courtesy of R. GIORDANI 1978)

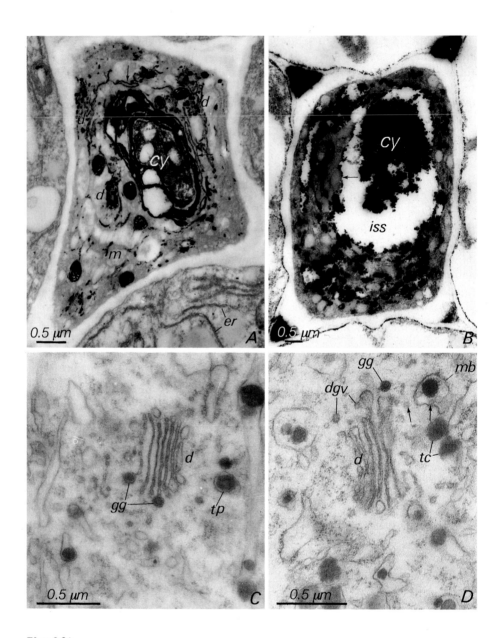

Plate 9.21.

A, B *Asclepias curassavica* L., repeated sequestrations followed by lysis of the sequestrated areas, giving rise to the central vacuole. **A** Zinc iodide-osmium tetroxide impregnation, contrasting the systems of concentric sequestration saccules; *cy* sequestrated central cytoplasm; *d* dictyosomes, strongly impregnated, as well as Golgi vesicles; *m* mitochondrion, non-reactive, as well as the endoplasmic reticulum *er* of neighbouring cells; *arrows* indicate marked saccules probably pertaining to the multisaccule sequestration system; × 15500. Fixation: glutaraldehyde – paraformaldehyde; technique of NIEUBAUER et al. (1969) and MARTY (1973b); Micrograph of R. GIORDANI (1981b). **B** Test of acid phosphatase activity; central cytoplasm *cy* strongly reactive, as well as numerous enclaves in the peripheral cytoplasm; the *arrow* shows the location of lead phosphate pre-

544

9.12.2.6 Behaviour of Other Organelles

In the apical areas, sequestration membranes isolate both cytoplasm and *nuclei* which degenerate *slowly* during the hydrolysis of the isolated area. The chromatin becomes clearer and disappears, while the nucleoli become hypertrophied and persist in the floating nuclear envelope for a long time. Conversely, the nuclei of the peripheral protoplasm, which have not been encircled, multiply through karyokinesis not followed by cytodieresis, and persist until the death of the laticifer, concomitant with that of the organ.

The cytoplasmic sheath which surrounds the axial vacuole of the laticifer contains all the usual organelles (Plate 9.22 A, B). The endoplasmic reticulum occurs in multiple sheets parallel to the wall; ribosomes are abundant, many of them being associated in polysomes. Chondriosomes also are very numerous, some of them in the shape of ovoid mitochondria, others elongated in filiform and sinuous chondriocontes. Dictyosomes are well represented in the meristematic areas and in the differentiated areas where their distal face produces relatively large vesicles. Plastids and peroxisomes (microbodies) are more remarkable.

Laticifer plastids of *Euphorbia characias* differentiate in leucoplasts from proplastids which have about the same size as mitochondria but are deprived of cristae or stromatic lamellae. These leucoplasts constitute two ontogenically different sets (MARTY 1971 a). Some of them, non-amyliferous, acquire a large size, a few and short cristae, with filamentous and sometimes ramified shapes, similar to those of the neighbouring chondriocontes (Plate 9.23 A) but well distinct due to their ultrastructure. Other leucoplasts produce a starch grain early. In the beginning, the grain is located at one of the extremities, but it grows in the shape of a rod completely surrounded by the substance which elaborates it. This substance is denser at the extremities of the grain surface where the amylogenesis goes on (Plate 9.23 B, C). Both categories of leucoplasts contain scattered phytoferritin particles.

In the differentiated areas of the laticifers, leucoplasts, amyliferous or not, enclosed in a thin pellicle of cytoplasm, protrude into the vacuole, shifting the tonoplast. The protrusion is constricted at the base, becomes separated from the vacuolar membrane and falls in the latex. In the vacuole, the stroma of these plastids

⊲───

cipitates against the membrane of the initial sequestration saccule *iss;* × 9400. Fixation: glutaraldehyde - paraformaldehyde; GOMORI's technique (1952); (courtesy of R. GIORDANI 1981 b).
C, D *Euphorbia characias* L., suggesting configurations of the participation of the cytoplasm in the synthesis of laticifer terpenic particles, and the participation of dictyosomes in the synthesis and transport of these particles. **C** Golgi granules *gg,* similar to terpenic particles *tp,* enclosed in a vesicle; × 34 200. **D** Multivesicular body *mb,* containing a dense granule, probably terpenic, and several small vesicles, such as the vesicle marked by an *arrow,* similar to small Golgi vesicles (other *arrow*); *dgv* dense Golgi vesicles (alveolate) equally similar to those which are found near the multivesicular body *mb,* which suggests that they may be incorporated into it later; *tc* terpenic particle, free in the cytoplasm; × 37 200. Fixation: glutaraldehyde - OsO_4; contrast: $KMnO_4$; (courtesy of F. MARTY 1968 b)

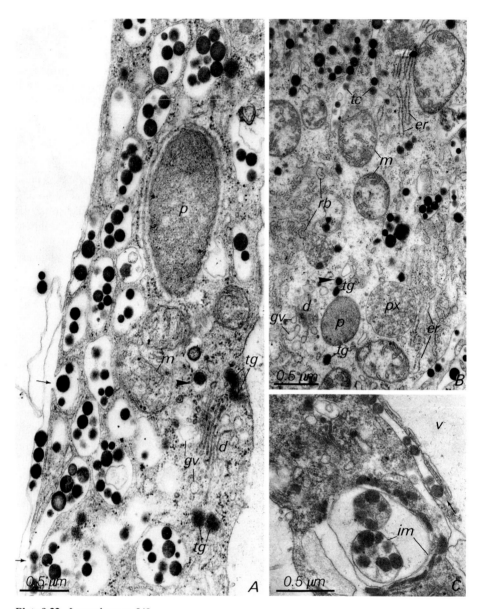

Plate 9.22. Legend see p. 548

Plate 9.23. Legend see p. 548

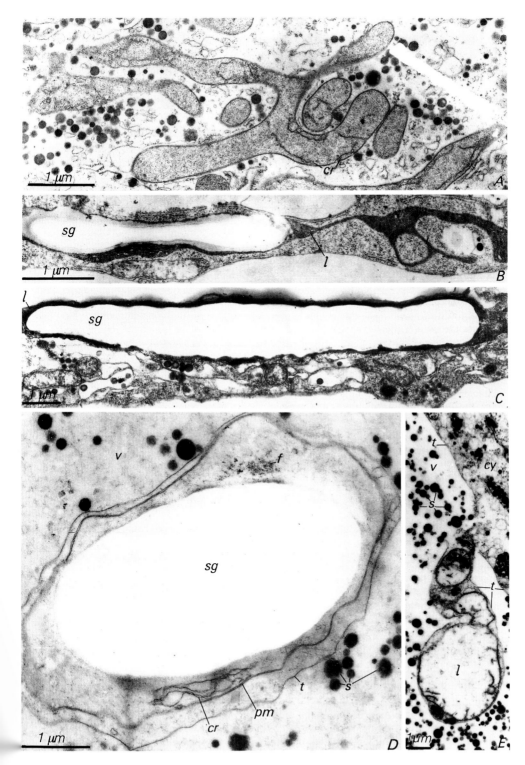

become clearer, a sign of degeneration, but they are still surrounded by the membrane derived from the tonoplast (Plate 9.23 D). This was demonstrated with electron microscopy, whereas the presence of starch grains with specific shapes in the latex of *Euphorbia* was well observed with light microscopy (MANGENOT 1925).

The differentiated laticifers of *Euphorbia characias* are also characterized by the significant amount of peroxisomes, as numerous as mitochondria (Plate 9.24 A; MARTY 1970b). These ovoid "microbodies", in ultrathin sections about 0.5 to 0.8 μm in diameter, are limited by a simple membrane; their content is finely granulous, sometimes including a small osmiophilic nucleoid (Plate 9.24 B). Peroxisomes are selectively located in the neighbourhood of endoplasmic reticulum saccules either near the profile extremities or against them (Plate 9.24 B). They are rare in the meristematic areas, more and more numerous and voluminous in the differentiated areas. According to the test of GRAHAM and KARNOWSKY (1966),

◁——

Plate 9.22 A–C. *Euphorbia characias;* cytology of the latex secretion.

A Area of peripheral cytoplasm, crossed by numerous canaliculi enclosing terpenic granules; some of these canaliculi run into the central vacuole, where they discharge their contents *(arrows);* other terpenic granules *tg* are free in the hyaloplasm, notably near the dictyosome *d,* which emits transparent vesicles *gv;* a terpenic granule is found in a vesicle similar to *gv (arrowhead);* × 26 200.
B Independence of terpenic canaliculi *tc* and endoplasmic reticulum *er; d* dictyosome, surrounded by Golgi vesicles *gv* and by terpenic granules *tg,* free in the hyaloplasm, with the exception of one of these, situated in a vesicle, perhaps Golgi vesicle *(arrowhead); m* mitochondria; *px* peroxisome; *p* plastid; *rb* ribosomes, associated in polysomes; × 23 200. Fixations: paraformaldehyde – glutaraldehyde – O_SO_4; contrast according to REYNOLDS (1963).
C Autophagic vacuole in the peripheral cytoplasm of a differentiated laticifer; *im* isolation membrane; the *arrow* indicates the opening of a terpenic canaliculus in the central vacuole *v;* × 30 800. Fixation: glutaraldehyde O_SO_4; contrast: KMnO$_4$.

(courtesy of F. MARTY 1971b)

Plate 9.23 A–E. Differentiation of plastids in *Euphorbia characias* laticifers.

A Leucoplasts, branched and sinuous, devoid of starch and containing only short cristae *cr;* × 17 800.
B Young leucoplast *l,* containing a starch grain *sg* in the course of development; × 19 000. Fixations: glutaraldehyde – O_SO_4; contrast: KMnO$_4$.
C Leucoplast *l,* the starch grain *sg* of which has reached its characteristic rodlike shape; × 10 500. Fixation: O_SO_4; contrast: uranyle acetate – lead citrate.
D Leucoplast transferred in to the vacuole, in slightly oblique section, enclosed in part of the tonoplast *t* which surrounded it at the moment of its protrusion into the vacuole; *cr* short plastid cristae; *f* accumulation of phytoferritin; *pm* double plastid membrane; *s* droplets of secretion; *v* vacuole; × 18 300. Fixation and contrast as in **A, B.**
E Leucoplast *l,* protruding into the vacuole *v,* but still enveloped in the tonoplast *t* and some cytoplasmic vestiges; × 6400. Fixation: O_SO_4; contrast: KMnO$_4$.

(Courtesy of F. MARTY 1971a)

——▷

Plate 9.24 A, B. *Euphorbia characias,* peroxisomes of the peripheral cytoplasm.

A Cytoplasmic area showing the whole set of organelles of the laticifer; the peroxisomes *(asterisks)* are apparently as numerous as mitochondria *m; ct* canaliculi containing terpenic granules; *d* dictyosomes, surrounded by Golgi vesicles *gv,* one of them enclosing a terpenic granule; *tg* terpenic granules, free in the hyaloplasm; *l* leucoplast; *er* rough endoplasmic reticulum; × 14 500.

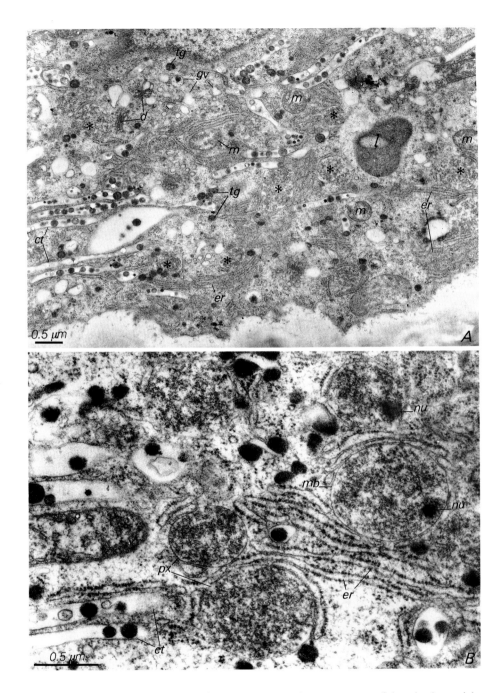

B Detail of a cytoplasmic area containing numerous peroxisomes *px*, most of them in close neighbourhood with rough endoplasmic reticulum sheets *er;* several peroxisomes enclose an osmiophilic "nucleus" *nu* in a slender granular stroma; *mb* simple cytomembrane delimting the peroxisome stroma; *ct* canaliculi loaded with terpenic granules; ×34000.

Fixations: glutaraldehyde – O_5O_4; contrast: $KMnO_4$; (courtesy of F. MARTY 1970b)

549

they are positive to the detection test of peroxidases, using 3-3' diaminobenzidine associated with hydrogen peroxide, at pH 7.6, which detects a peroxidase activity due to the catalase. The reaction is selective and the precipitate covers the granulous areas and the nucleoids.

The physiological significance of the abundance of peroxisomes in the cytoplasm of these laticifers remains hypothetical. MARTY (1974) suggested their participation in the change of acetate into an intermediate of the polyterpene synthesis in the hyaloplasm, for example the mevalonic acid.

9.12.3 Anatomical Localizations

In the *Euphorbia* in which CAMERON (1936) accurately studied the distribution of laticifers, the author showed that canals, located at the periphery of the vascular cylinder of the hypocotyl and the root, and others, thinner, at the periphery of the cortex (Fig. 9.28) arise from primordial cells of the cotyledonary node. Another set of tubes is prolonged into the cotyledons, whereas others spread into the gemmule, towards and near the apical meristem. The laticifers overgrow in this way most of the primary tissues of the young plant, by growing simultaneously with it. One can speculate as to what happens when the thickening of the organs occurs, due to the cambium activity.

BLAZER (1945) and ARTSCHWAGER (1946) demonstrated in their research that in *Cryptostegia grandiflora* the tubes that extend from the cortex to the pith pass through regions that will become interfascicular spaces, and are not ruptured by the occurrence and activity of the cambium; on the contrary, they follow the thickening of the cambium by resuming an *intercalary growth*. Moreover, there is not a single new laticifer cell in the secondary tissues that are developed, but the cells contained in the cortex and the primary phloem penetrate the secondary phloem by ramified extensions and overgrow it by *intrusive growth*.

These processes suggest that even in some tissues that have completed their differentiation (and therefore no longer grow), the concerned parts of laticifers retain their possibility of resuming growth. This characteristic has appeared under several experimental circumstances (SCHAFFSTEIN 1932). For example, when ger-

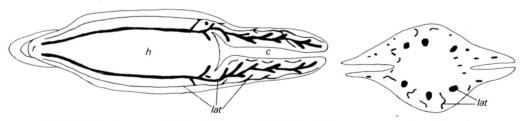

Fig. 9.28. Development of the laticifer system in the *Euphorbia marginata* embryo. From the embryonic initials of the cotyledonary node, three series of ramified extensions (lat) invade: **1** the periphery of the conducting tissues of the hypocotyl *h* and the radicle *r*; **2** the periphery of the cortex (thinner canals); and **3** the cotyledons *c*. In addition, some extensions reach the gemmule apex; (after CAMERON 1936)

minations of *Euphorbia* are decapitated and therefore involve the growing of adventitious shoots on the hypocotyl, the laticifers of this organ, which have terminated their growth, penetrate the new shoots. One observes a similar growth on the adventitious roots obtained by the cutting of shoots, or on the callus formed in grafting (SCHAFFSTEIN 1932).

In electron microscopy, MOOR (1959) indicated that differentiated regions of laticifers, especially their wall, can give rise to new lateral branches in actively growing tissues, whereas in the absence of such tissues, the branches of the laticifers cease to grow and die with the surrounding tissues.

These facts suggest that the growth and activity of the laticifers are stimulated by those of the surrounding tissues, and that correlations harmonize their relationship with these tissues.

These considerations show that laticifers are spread throughout the organs of the vegetative system and in most tissues. That is the case in the Euphorbiaceae but the most important and numerous branches are mainly formed in the phloem. In some Apocynaceae, Asclepiadaceae and Moraceae, the spreading is even more significant and reaches the xylem in particular.

In some Euphorbiaceae, studied by SPERLICH (1939), laticifers send out branches in the epidermis. Some of these branches penetrate between the epidermal cells and extend even between their external membrane and the cuticle. In

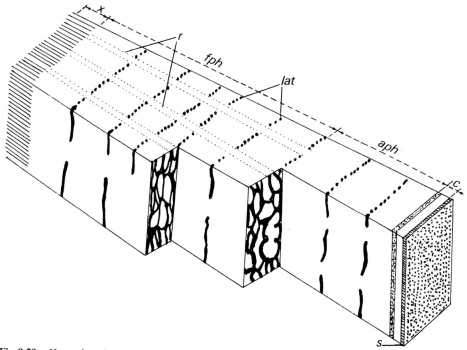

Fig. 9.29. *Hevea brasilensis,* topography of the pseudolaticifer meshwork; *c* cortex; *lat* anostomosed pseudolaticifers, forming concentric meshworks, particularly in the functional secondary phloem *fph; aph* ancient (non-functional) phloem, containing numerous sclereids (not represented); *r* phloem rays; *s* cork (suber); *x* xylem; (after VISCHER 1923)

contrast, the unbranched laticifers of *Vinca* and *Cannabis* are only found in the *primary tissues* (ZANDER 1928; SCHAFFSTEIN 1932). The anastomosed laticifers of the *Hevea* are mainly localized, as are the pseudolaticifers of the Chicoraceae, in the secondary phloem, where they are arranged in tangential strips (i.e. in concentric cylinders, Fig. 9.29).

In leaves, laticifers follow the bundles of the veins and send out branches or unbranched prolongations into the mesophyll.

9.12.4 Composition and Physical State of the Latex

The physico-chemical characteristics of all types of latex are roughly common to pseudolaticifers and laticifers.

The previous cytological study showed that the latex of laticifers is nothing else but the vacuolar content of such laticifers. The ground substance which constitutes the dispersion phase is an aqueous solution similar to any vacuolar liquid. Therefore, it contains various crystalloids in solution, e.g. glucids, organic acids, mineral salts, etc. and various organic substances in colloidal suspension, e.g. tannins, sterols, phospholipids, mucilages, alkaloids, etc.

In this complex phase, the products which give the liquid the aspect of an emulsion are dispersed in more voluminous particles. They are sometimes about the same as the previous ones, mainly lipids, but they also include such substances as essential oils, resins, camphor, carotenoids (often mixed with lipids) and finally the most interesting product, rubber.

The main dispersed constituents belong to the carbide family of *terpenes*, the basic components of which are the acetylenic carbides with the formula $C_5 H_8$ such as *isoprene;* in addition, numerous substances derived from the hydrocarbons through oxidation, e.g. alcohols, aldehydes, terpenic cetones. Finally, resins are, at least partially, more complex products derived from essential oils through oxidation. These essential oils are compounds mainly including terpenic carbides of low molecular weight: $(C_5 H_8)$ 2 or $C_{10} H_{16}$ (dimer carbides) and $(C_5 H_8)$ 3 (trimer carbides).

Polyterpenes $(C_5 H_8)$ n constitute the gutta and the rubber. "n" is probably low in the components of the latex "in situ", but these components become polymerized when the latex, released from the plant, coagulates.

The various substances occur in various amounts, depending on the species. In *Hevea,* rubber may constitute 40 to 50% of the latex; other latexes contain no polyterpenes.

The latex of some plants contains more specific substances which do not occur commonly in laticifers, e.g. proteins, *Ficus callosa;* sugar, Compositae; tannins, Aroidae; alkaloids, Papaveraceae among which *Papaver somniferum* including morphine (opium is dry latex of *Papaver somniferum*); various enzymes such as papain, a proteolytic enzyme of *Carica papaya.*

Crystals of organic acid salts (oxalates and malates) may be abundant in latex. Certain plants contain starch grains in the laticifers such as *Euphorbia,* sometimes together with amylase.

552

The dispersed particles have variable sizes from a fraction of μm to a few μm. The milkiness of latex results from the difference between the refractive indices of the particles and the dispersion medium. The latex of various plants, wrongly named in this case, may be clear (*Morus, Nerium oleander*).

9.12.5 Possible Functions of Laticifers

This brief review does not allow an extensive explanation of the laticifer functioning. More recent research has confirmed that the substances accumulated in these systems are not concerned with the plant metabolism. Therefore, the content of laticifers can only be considered as a secretory product.

References

Archer B, Barnard D, Cockbain E, Dickenson PB, McMullen A (1963) - Structure, composition and biochemistry of *Hevea* latex. In: Bateman (ed) The chemistry and physics of rubber-like substances. Mac Laren, London

Artschwager E (1946) Contribution to the morphology and anatomy of *Cryptostegia (Cryptostegia grandiflora)*. US Dept Agric Tech Bull 915

Artschwager E, Mc Guire RC (1943) Contribution to the morphology and anatomy of the Russian dandelion *(Taraxacum koksaghyz)*. US Dept Agric Tech Bull 843

Ascensão L (1985) Estruturas secretoras em *Artemisia campestris* L Differenciação e processo secretor. Thesis Dr Biol, Univ Lisboa

Ascensão L, Pais MSS (1981) Ultrastructural aspects of secretory trichomes in *Cistus monspeliensis*. In: Components of productivity of mediterranean climate regions. Junk, The Hague, pp 27-38

Ascensão L, Pais MSS (1985) Différenciation et processus sécréteur des trichomes d'*Artemisia campestris*, ssp. *maritima* (Compositae). Ann Sci Nat Bot 13e sér 7: 149-171

Baranova EA (1935) Ontogeny of the laticiferous system of *tau-saghyz. (Scorzonera tau - saghyz* Lipsch et Bosse) Bot Zh SSSR 20: 600-616

Bary A De (1877) Vergleichende Anatomie der Vegetationsorgane der Phanerogamen und Farne. Engelmann, Leipzig, 663 p. Engl ed: Comparative anatomy of the vegetative organs of the phanerogams and ferns. Clarendon Oxford (1884)

Blazer HW (1945) Anatomy of *Cryptostegia grandiflora* with special reference to the latex. Am J Bot 32: 135-141

Bobilioff W (1923) Anatomy and physiology of *Hevea brasiliensis*. Part I. Anatomy of *Hevea brasiliensis*. Art Institut Orell Füssli, Zürich

Bruni A, Dall'olio G, Fasulo M (1974) Morphological aspect of the nuclei in mature articulated laticifers of *Calystegia soldanella*. Experientia (Basel) 30: 1390-1391

Buvat R (1971) Origin and continuity of cell vacuoles. In: Reinert I and Ursprung H (ed) Origin and continuity of cell organelles. Springer, Berlin Heidelberg New York, pp 127-157

Buvat R (1981) Vésicules alvéolées et vésicules épineuses dans les racines de l'Orge *(Hordeum sativum)*. CR Acad Sci Paris sér III 292: 825-832

Buvat R, Robert G (1979) Vacuole formation in the actively growing root meristem of barley *(Hordeum sativum)*. Am J Bot 66: 1219-1237

Cameron D (1936) An investigation of the latex systems in *Euphorbia marginata*, with particular attention to the distribution of the latex in the embryo. Bot Soc Edinb Trans Proc 32: 187-194

Chauveaud MG (1891) Recherches embryogéniques sur l'appareil laticifère des Euphorbiacées, Urticacées, Apocynées et Asclépiadacées. Ann Sci Nat Bot sér 7 14: 1-161

Christensen AK, Fawcett DW (1960) The fine structure of testicular interstitial cells in the opossum. Anat Rec 136: 333

Clautriau G (1889) La localisation des alcaloides du Pavot. J Pharm Chim 20: 161 (Cited by Giordani, 1981b)

Collier HOJ, Chescher GB (1956) Identification of 5 - hydroxytryptamine in the sting of the nettle *Urtica dioica*. Brit J Pharmacol 11: 186–189

Coulomb C, Coulomb P (1973) Participation des structures golgiennes à la formation des vacuoles autolytiques et à leur approvisionnement enzymatique dans les cellules du méristème radiculaire de la Courge. CR Acad Sci Paris sér D 277: 2685–2688

Dickenson PB, Fairbairn JW (1975) The ultrastructure of the alkaloidal vesicles of *Papaver somniferum* latex. Ann Bot 39: 707–712

Duchartre P (1859) Recherches physiologiques, anatomiques et organogéniques sur la Colocase des Anciens, *Colocasia antiquorum* Schott. Ann Sci Nat Bot 12: 272–279

Dumas C (1973a) Contribution à l'étude cytophysiologique du stigmate. III. Evolution et rôle du réticulum endoplasmique au cours de la sécrétion chez *Forsythia intermedia* Z; étude cytochimique. Z Pflanzenphysiol 70: 119–130

Dumas C (1973b) Contribution à l'étude cytophysiologique du stigmate. IV. Réticulum endoplasmique et sécrétion granulocrine chez *Forsythia intermedia* Z. J Microsc 17: 46a

Dumas C (1973c) Contribution à l'étude cyto-physiologique du stigmate: essai de caractérisation cytochimique du contenu du réticulum endoplasmique et des grains de sécrétion chez *Forsythia intermedia* Z. CR Acad Sci Paris sér D 277: 1479–1482

Dumas C (1974a) Contribution à l'étude cytophysiologique du stigmate. VII. Les vacuoles lipidiques et les associations réticulum endoplasmique - vacuole chez *Forsythia intermedia* Z. Botaniste 56: 59–80

Dumas C (1974b) Contribution à l'étude cytophysiologique du stigmate. VIII. Les associations réticulum endoplasmique-plastes et la sécrétion stigmatique. Botaniste 56: 81–102

Dumas C (1974c) Mise en évidence histochimique de la nature essentiellement lipidique de l'exsudat de *Forsythia intermedia* Z, en microscopies photonique et électronique. Acta Histochem 48: 115–123

Dumas C (1975) Le stigmate et la sécrétion stigmatique. Thèse Doct Sci Nat, Université Claude Bernard, Lyon

Dumas C (1977) Lipochemistry of the progamic stage of a self-incompatible species: neutral lipids and fatty acids of the secretory stigma during its glandular activity, and of the solid style, the ovary and the anther in *Forsythia intermedia* Zab (heterostylic species). Planta 137: 177–184

Dumas C (1978) Stigmates sécréteurs et lipides neutres sécrétés. Soc Bot Fr Actual Bot 1, 2: 61–68

Duve C De (1974) Les lysosomes. Recherche (Paris) 49: 815–826

Esau K, Kosaki H (1975) Laticifers in *Nelumbo nucifera* Gaertn: Distribution and structure. Ann Bot 39: 713–719

Eymé J (1963) Observations sur les nectaires de trois Renonculacées (*Helleborus foetidus* L., H. niger L et *Nigella Damascena* L.). Botaniste 41: 137–178

Eymé J (1966) Infrastructure des cellules nectarigènes de *Diplotaxis erucoides* D. C., *Helleborus niger* L. et H. foetidus L. CR Acad Sci Paris sér 162 D: 1629–1632

Eymé J (1967) Nouvelles observations sur l'infrastructure des tissus nectarigènes floraux. Botaniste 50: 169–183

Fahn A (1953) The topography of the nectary in the flower and its phylogenetic trend. Phytomorphology 3: 424–426

Fahn A, Benayoun J (1976) Ultrastructure of resin ducts in *Pinus halepensis*. Development, possible sites of resin synthesis, and mode of its elimination from the protoplast. Ann Bot 40: 857–863

Fahn A, Rachmilevitz T (1970) Ultrastructure and nectar secretion in *Lonicera japonica*. In: New research in plant anatomy, pp 51–56. (Supp to Botanical J of the Linnean Soc 63)

Feldberg W (1950) The mechanism of the sting of common nettle. Brit Sci News 3: 75–77

Frei E (1955) Die Innervierung der floralen Nektarien dicotyler Pflanzenfamilien. Ber Schweiz Bot Ges 65: 60–114

Frey-Wyssling A (1935) Die Stoffausscheidung der höheren Pflanzen. Monographien aus dem Gesamtgebiet der Physiologie der Pflanzen und der Tiere, Band 32. Springer, Berlin

Frey-Wyssling A (1955) The phloem supply to the nectaries. Acta Bot Neerl 4: 358–369

Giaquinta RT (1983) Phloem loading of sucrose. Annu Rev Plant Physiol 34: 347–387

Gifford EM, Stewart KD (1967) Ultrastructure of the shoot apex of *Chenopodium album* and certain other seed plants. J Cell Biol 33: 131–142

Giordani R (1975) Différenciation des cellules sécrétrices du latex chez l'*Asclepsias curassavica* L. et l'*A. Syriaca* L. Thèse Doct Spéc Biol Cell, Aix Marseille II

Giordani R (1977) Dégradation des parois terminales durant la différenciation des laticifères articulés anastomosés de *Lactuca sativa* L. CR Acad Sci Paris 284 D: 569-572

Giordani R (1978) Autophagie cellulaire et différenciation des laticifères non articulés chez une asclépiade. Biol Cell 33: 253-260

Giordani R (1979) Ultrastructure des laticifères articulés de la Laitue. CR Acad Sci Paris 288 D: 615-618

Giordani R (1980) Dislocation du plasmalemme et libération de vésicules pariétales lors de la dégradation des parois terminales durant la différenciation des laticifères articulés. Biol Cell 38: 231-236

Giordani R (1981a) Activités hydrolasiques impliquées dans le processus de dégradation pariétale, durant la différenciation des laticifères articulés. Biol Cell 40: 217-224

Giordani R (1981b) Etude comparée de deux types de laticifères: *Asclepias curassavica* (laticifères non articulés); *Lactuca sativa* (laticifères articulés). Thèse Doct Sci Univ Aix - Marseille II

Giordani R, Blasco F, Bertrand JC (1982) Confirmation biochimique de la nature vacuolaire et lysosomale du latex des laticifères non articulés d'*Asclepias curassavica*. CR Acad Sci Paris sér III 295: 641-646

Gomori G (1952) Microscopic histochemistry; principles and practices. University of Chicago Press, Chicago

Graham RC, Karnowsky MJ (1966) The early stages of injected horseradish peroxidase in the proximal tubules of mouse kidney: ultrastructural cytochemistry by a new technique. J Histochem Cytochem 14: 291-302

Guilliermond A, Mangenot G, Plantefol L (1933) Traité de Cytologie végétale. Le François (ed) Paris

Haberlandt G (1894) Über Bau und Function der Hydathoden. Ber Dtsch Bot Ges 12: 367-378

Haberlandt G (1909) Physiologische Pflanzenanatomie. Engelmann, Leipzig, 650 p. Engl ed: Physiological plant anatomy. Macmillan, London (1914)

Karling JS (1929) The laticiferous system of *Achras sapota* L. I. A preliminary account of the origin, structure and distribution of the latex vessels in the apical meristem. Am J Bot 16: 803-824

Mahlberg P (1959) Karyokinesis in the non-articulated laticifers of *Nerium oleander* L. Phytomorphology 9: 110-118

Mahlberg P (1961) Embryogeny and histogenesis in *Nerium oleander* L. II. Origin and development of the non-articulated laticifers. Am J Bot 48: 90-99

Mahlberg P (1963) Development of non-articulated laticifers in seedling axis of *Nerium oleander*. Bot Gaz 124: 224-231

Mahlberg P, Sabharwal P (1968) Origin and early development of non-articulated laticifers in embryos of *Euphorbia marginata*. Am J Bot 55: 375-381

Maillet M (1963) Le réactif au tetroxyde d'osmium - iodure de zinc. Z Mikrosk Anat Forsch (Leipz) 70: 397-425

Mangenot G (1925) Sur le mode de formation des grains d'amidon dans les laticifères des Euphorbiacées. CR Acad Sci Paris 180: 157-160

Mangenot G (1927) Sur la présence de vacuoles spécialisées dans les cellules de certains végétaux. CR Soc Biol Paris 97: 342-345

Mangenot G (1930) Données morphologiques sur la matière vivante. Guillon éd. Paris

Marty F (1968a) Infrastructures des organes sécréteurs de la feuille d'*Urtica urens*. CR Acad Sci Paris 266 D: 1712-1714

Marty F (1968b) Infrastructure des laticifères différenciés d'*Euphorbia characias*. CR Acad Sci Paris 267 D: 299-302

Marty F (1970a) Rôle du système membranaire vacuolaire dans la différenciation des laticifères d'*Euphorbia characias* L. CR Acad Sci Paris 271 D: 2301-2304

Marty F (1970b) Les peroxysomes (microbodies) des laticifères d'*Euphorbia characias* L. Une étude morphologique et cytochimique. J Microsc 9: 923-948

Marty F (1971a) Différenciation des plastes dans les laticifères d'*Euphorbia characias* L. CR Acad Sci Paris 272 D: 223-226

Marty F (1971b) Vésicules autophagiques des laticifères différenciés d'*Euphorbia characias* L. CR Acad Sci Paris 272 D: 399-402

Marty F (1973a) Mise en évidence d'un apparail provacuolaire et de son rôle dans l'autophagie cellulaire et l'origine des vacuoles. CR Acad Sci Paris 276 D: 1549-1552

Marty F (1973b) Sites réactifs à l'iodure de zinc - tétroxyde d'osmium dans les cellules de la racine d'*Euphorbia characias* L. CR Acad Sci Paris 277 D: 1317-1320

Marty F (1974) Vacuome et sécrétion intracellulaire chez *Euphorbia characias* L. Thèse Doct Sci, Université d'Aix-Marseille II

Marty F (1978) Cytological studies on GERL, provacuoles and vacuoles in root meristematic cells of *Euphorbia*. Proc Natl Acad Sci USA 75: 852-856

Matile P, Jans B, Rickenbacher R (1970) Vacuoles of *Chelidonium* latex: lysosomal property and accumulation of alkaloids. Biochem Physiol Pflanz (BPP) 161: 447-458

Mika ES (1955) Studies on the growth and development and morphine content of opium poppy. Bot Gaz 116: 323-339

Moens P (1955) Les formations sécrétrices des Copaliers congolais. Etude anatomique, histologique et histogénétique. Cellule 57: 33-64

Molisch H (1901) Studien über den Milchsaft und Schleimsaft der Pflanzen. Fischer, Jena

Moor H (1959) Platin-Kohle-Abdruck - Technik angewandt auf Feinbau der Milchröhren. J Ultrastruct Res 2: 293-422

Morré DJ, Mollenhauer HH (1974) The endomembrane concept: a functional integration of endoplasmic reticulum and Golgi apparatus. In: Robards A (ed) Dynamic aspects of plant ultrastructure. Mc Graw Hill, London

Moyer LS (1937) Recent advances in the physiology of latex. Bot Rev 3: 522-544

Nessler C, Mahlberg P (1979) Plastids in laticifers of *Papaver*. I. Development and cytochemistry of laticifer plastids in *P. somniferum* L. (Papaveraceae). Am J Bot 66: 266-273

Nieubauer G, Krawczyk WS, Kidd RL, Wilgram GF (1969) Osmium zinc-iodide reactive sites in the epidermal Langerhans cell. J Cell Biol 43: 80-89

Novikoff AB (1964) GERL, its form and functions in neurons of rat spinal ganglia. Biol Bull 127: 358 A

Novikoff AB, Shin WY (1964) The endoplasmic reticulum in the Golgi zone and its relations to microbodies, Golgi apparatus and autophagic vacuoles in rat liver cells. J Microsc 3: 187-206

Perrin A (1972) Contribution à l'étude de l'organisation et du fonctionnement des hydathodes: recherches anatomiques, ultrastructurales et physiologiques. Thèse Doct Sci, Université Cl Bernard, Lyon

Popovici H (1926) Contribution à l'étude cytologique des laticifères. CR Acad Sci Paris 183: 143-145

Pujarniscle S (1971) Etude biochimique des lutoïdes du latex d'*Hevea brasiliensis* Müll. Arg. Différences et analogies avec les lysosomes. Mémoire ORSTOM n° 48, Paris

Reynolds ES (1963) The use of lead citrate at high pH as an electron opaque stain in electron microscopy. J Cell Biol 17: 208-212

Ribailler D, Jacob J, d'Auzac J (1971) Sur certains caractères vacuolaires des lutoïdes du latex d'*Hevea brasiliensis* Müll. Arg. Physiol Vég 9: 423-437

Ruinen J (1950) Microscopy of the lutoids in *Hevea* latex. Ann Bogor 1: 27-45

Samat M (1984) Contribution à l'édude du genre *Urtica*. Thèse Fac Pharmacie, Marseille

Sarkany S, Michels-Nyomarkay K, Venzar-Petri G (1970) Über die histologischen und Feinstrukturellen Beziehungen und in die Frage der Alkaloidbildung im Samen und in die Keimpflanzen von *Papaver somniferum* L. Pharmazie 10: 625-629

Sassen M (1965) Breakdown of the plant cell wall during the cell fusion process. Acta Bot Neerl 14: 165-196

Savchenko NL (1940) Entwicklung und Anordung des Milchsaftgefäßsystems bei *Taraxacum kok-saghyz*. Dokl Akad Nauk SSSR 27: 1052-1055

Schaffstein G (1932) Untersuchungen an ungegliederten Milchföhren. Bot Centralbl Beih 49: 197-220

Schnepf E (1969) Sekretion und Exkretion bei Pflanzen. Protoplasmatologia 8

Scott D (1882) The development of articulated laticiferous vessels. QJ Microsc Sci 22: 136-153

Shimony C, Fahn A (1968) Light and electron-microscopical studies on the structure of salt glands of *Tamarix aphylla* L. J Linn Soc Lond Bot 60 (383): 283-288

Shimony C, Fahn A, Reinhold L (1973) Ultrastructure and ion-gradients in the salt glands of *Avicenna marina* (Forsok) Vierh. New Phytol 72: 27-36

Smaoui A (1971) Différenciation des trichomes chez *Atriplex halimus* L. CR Acad Sci Paris 273 D: 1268-1271

Smaoui A (1975) Les trichomes vésiculeux d'*Atriplex halimus* L. - Modalités de sécrétion saline d'une plante halophile. Thèse spécialité, Aix-Marseille II

Sperlich A (1939) Das trophische Parenchym. B. Exkretionsgewebe. In: Linsbauer, Handbuch der Pflanzenanatomie, Band 4, Lief 38 Borntraeger Berl

Stahl E (1953) Untersuchungen an den Drüsenhaaren der Schafgarbe (*Achillea millefolium* L.) Z Bot 41: 123-146

Thiéry JP (1967) Mise en évidence des polysaccharides sur coupes fines en microscopie électronique. J Microsc 6: 987-1018

Thureson-Klein A (1970) Observations of the development and fine structure of the articulated laticifers of *Papaver somniferum*. Ann Bot 34: 751-759

Thurston EL (1974) Morphology, fine structure and ontogeny of the stinging emergence of *Urtica dioica*. Am J Bot 61: 809-817

Treub M (1880) Sur des cellules végétales à plusieurs noyaux. Arch Neerl Sci 15: 39-69

Vermeer J, Peterson RL (1979) Glandular trichomes on the inflorescence of *Chrysanthemum morifolium* c.v. Dramatic. (Compositae). II. - Ultrastructure and histochemistry. Can J Bot 57: 714-729

Vertrees G, Mahlberg P (1978) Structure and ontogeny of laticifers in *Cichorium intybus* (Compositae). Am J Bot 65: 764-771

Vian B, Roland JC (1972) Différenciation des cytomembranes et renouvellement du plasmalemme dans les phénomènes de sécrétions végétales. J Microsc 13: 119-136

Vischer W (1923) Über die Konstanz anatomischer und physiologischer Eigenschaften von *Hevea brasiliensis* Müll. Arg. (Euphorbiaceae). Verh Naturforsch Ges Basel 35: 174-185

Werker E, Fahn A (1981) Secretory hairs of *Inula viscosa* (L.) Ait. Development, ultrastructure and secretion. Bot Gaz 142: 461-476

Whaley WG, Kephart JE, Mollenhauer HH (1964) The dynamics of cytoplasmic membranes during development. In: M. Locke (ed) Cellular membranes in development. Academic Press, London, pp 135-173

Wilson K, Nessler C, Mahlberg P (1976) Pectinase in *Asclepias* latex and its possible role in laticifer growth and development. Am J Bot 63: 1140-1144

Wilson KJ, Mahlberg PG (1978) Ultrastructure of non-articulated laticifers in mature embryos and seedlings of *Asclepias Syriaca* L. (Asclepiadaceae). Am J Bot 65: 98-109

Zander A (1928) Über Verlauf und Entstehung der Milchröhren des Hanfes *(Cannabis sativa)*. Flora 23: 191-218

PART III

Conclusions

1 The Differentiation of Plant Cells

This book shows that the organism of vascular plants may have a great number of cell types and that some of the tissues composed of these cells are very complex. However, the scientists dealing with animal or human histology would assess this complexity and diversity as being much less pronounced than those existing in their scope. They would probably consider that plant cells are much less differentiated than some animal cells, for example, nervous cells or striated muscle cells. So, they can conceive that plant cells sufficiently rich in living material can dedifferentiate and produce young cells again, whereas a real dedifferentiation does not commonly exist in animals.

In fact, such deductions are consistent with the morphological aspects of cells and tissues, but if we consider the physiological and biochemical features of plant cells, we cannot help thinking that, in some cases, these cells are not less evolved than animal cells. The chlorophyllous cells of autotrophic plants, endowed with a faculty of synthesis which cannot be found in animal cells, are necessarily supplied with particularly complex photochemical and enzymatic equipment.

The differentiation characteristic of green plants is manifested through a cytological differentiation which does not occur in animal cells, that of *plastids*. But the faculty to achieve difficult syntheses, absent in animals, does not always result in cytological structures visible with a microscope, even an electron microscope: the apparent cytological simpleness of bacteria does not prevent some of them from using molecular nitrogen to produce their own living substance.

Therefore, the morphology of cells is only an inconstant indication of their basic complexity.

2 Dedifferentiation: Common Process of Vascular Plant Ontogeny

The development of plants contrasts with that of animals because, generally, in contrast to animals, plants can continuously generate, through dedifferentiation, meristematic cells capable of giving rise to new organs: leafy stems, roots or reproductive organs.

The specificity of plant cells, including those with a high metabolic activity, is not irreversible and moreover, there are no distinct and independent germinal lineage and somatic lineage. Germinal cells can arise from derivatives of parenchyma cells or epidermal cells of a vegetative organ.

The regeneration of meristematic cells necessary for the continued growth of plants is thus linked to the dedifferentiation, a characteristic of plant cells. One can consider that, on the one hand, indefinite growth and dedifferentiation and, on the other hand, completed growth and irreversible differentiation, refer to a notion of generalized opposition between animal and plant kingdoms.

3 Fundamental Tissues and Accessory Specializations

The development of tissues composed of numerous dead cells or senescent cells is another characteristic of vascular plants also related to indefinite growth. Because of this phenomenon, such tissues contrast with the tissues considered as fundamental due to their biochemical activity, i.e. with meristems and parenchymas. Their main role, a role of adaptation (to aerial or underground conditions of life) seems to consist mainly in providing parenchymatous or meristematic cells with the substances they need and in maintaining conditions of an "internal medium" for these cells. For example, this is the case with food-conducting and protective tissues. With such a role, they appear rather as subordinate tissues, although they have often been considered as fundamental by anatomists on grounds of systematics or phylogeny. However, one should keep in mind that if the more active cells play the most important role in the functioning of beings, the organism appears as a whole, in which each type of tissue is necessary to the others or has been so in a previous time.

4 The Pectocellulosic Walls, a Specific Feature of the Plant World

The most prominent feature of plant histology is certainly the existence of the skeletal wall which is typically pectocellulosic and delimits each cell. In the past, the extracellular skeleton has been more often studied than the living content of the cells themselves. Moreover, it was the subject of morphological analysis in which scientists were more concerned with geometry than with the understanding of the functioning and growth of the organs.

In dealing selectively with the physiological relationship between the cell walls and the cytoplasm, for instance in trying to specify the role of each structure in the transport of metabolites through the symplasmatic and apoplasmatic routes (see GIAQUINTA 1983) as well as in determining the nature of the reciprocal exchanges through endocytosis and exocytosis between the cytoplasm and the parietal space (see VIAN and ROLAND 1972), modern works have shown that the cell walls of plants are not only plain supporting structures of cells, but that, mainly in continental plants, they represent the equivalent of the internal medium of animals as well as closely contributing to the physiology of cells, tissues and organisms.

Generally, plant histology and physiology must now remain closely related, in particular at the scale of organisms. The cytological and cytochemical study of the tissues, which is far from being completed, but for which the present ultrastructural techniques are very helpful, must accompany the modern physiological research.

Author Index

Subject Index

Page numbers in *italics* refer to illustrations

567

578